Organic Carcinogens in Drinking Water

Organic Carcinogens in Drinking Water

Detection, Treatment, and Risk Assessment

EDITED BY

NEIL M. RAM
Manager, Analytical Laboratory Department
GCA Corporation, Technology Division, Inc.
Bedford, Massachusetts

EDWARD J. CALABRESE
Professor of Public Health
University of Massachusetts
Amherst, Massachusetts

RUSSELL F. CHRISTMAN
Professor and Chairman of
the Department of Environmental Sciences and Engineering
University of North Carolina at Chapel Hill
Chapel Hill, North Carolina

A Wiley-Interscience Publication
JOHN WILEY & SONS
New York · Chichester · Brisbane · Toronto · Singapore

Library of Congress Cataloging in Publication Data:

Organic carcinogens in drinking water.

  "A Wiley-Interscience publication."
  Includes bibliographies and index.
    1. Drinking water—Contamination.   2. Carcinogens.
3. Organic water pollutants.   4. Health risk
assessment.   I. Ram, Neil M., 1952–
II. Calabrese, Edward J., 1946–   III. Christman,
R. F. (Russell F.), 1936–
RA591.5.074   1986   363.6′1   86-5558
ISBN 0-471-80959-4

Printed in the United States of America

10 9 8 7 6 5 4 3 2 1

Don't assume the cloak of omniscience. Society must make choices despite uncertainty, and frequently this has to be done at the highest levels of government on scanty scientific evidence. Standards are not immutable.

—Abel Wolman, 1980

CONTRIBUTORS

Shirley A. A. Beresford
Department of Epidemiology
School of Public Health
University of North Carolina at Chapel Hill
Chapel Hill, North Carolina

Herbert J. Brass
Office of Drinking Water
Technical Support Division
Environmental Protection Agency
Cincinnati, Ohio

R. J. Bull
College of Pharmacy
Washington State University
Pullman, Washington

Edward J. Calabrese
Division of Public Health
University of Massachusetts
Amherst, Massachusetts

Andrew T. Canada
Division of Public Health
University of Massachusetts
Amherst, Massachusetts

Kenneth P. Cantor
Environmental Epidemiology Branch
National Cancer Institute
National Institutes of Health
Bethesda, Maryland

Joseph A. Cotruvo
Criteria and Standards Division
Office of Drinking Water
Environmental Protection Agency
Washington, D.C.

Francis A. DiGiano
Department of Environmental Science and Engineering
School of Public Health
University of North Carolina at Chapel Hill
Chapel Hill, North Carolina

James K. Edzwald
Department of Civil Engineering
University of Massachusetts
Amherst, Massachusetts

Charles E. Gilbert
Division of Public Health
University of Massachusetts
Amherst, Massachusetts

William H. Glaze
Environmental Science and Engineering Program
University of California
Los Angeles, California

Susan Goldhaber
Criteria and Standards Division
Office of Drinking Water
Environmental Protection Agency
Washington, D.C.

Jerome J. Healey
Water Management Division
Environmental Protection Agency
Boston, Massachusetts

Charles D. Hertz
Environmental Studies Institute
Drexel University
Philadelphia, Pennsylvania

David W. Hosmer
Division of Public Health
University of Massachusetts
Amherst, Massachusetts

Marty S. Kanarek
Department of Preventive Medicine
Medical School
University of Wisconsin
Madison, Wisconsin

Barbara A. Kingsley
SRI International
Menlo Park, California

D. S. Millington
Division of Genetics and Metabolism
Department of Pediatrics
Duke University Medical Center
Durham, North Carolina

J. C. Morris
Alkamadelaan 1074
2597 BK
den Haag, The Netherlands

D. L. Norwood
Department of Environmental Sciences and Engineering
School of Public Health
University of North Carolina at Chapel Hill
Chapel Hill, North Carolina

Neil M. Ram
Manager, Analytical Laboratory Department
GCA Corporation, Technology Division, Inc.
Bedford, Massachusetts

Alan A. Stevens
Drinking Water Research Division
Municipal Environmental Research Laboratory
Environmental Protection Agency
Cincinnati, Ohio

I. H. (Mel) Suffet
Environmental Studies Institute
Drexel University
Philadelphia, Pennsylvania

James M. Symons
Environmental Engineering Program
University of Houston
Houston, Texas

Robert G. Tardiff
Environ Corporation
Washington, D.C.

Barbara B. Taylor
Cambridge Analytical Associates
Environmental Studies
Boston, Massachusetts

E. M. Thurman
U.S. Geological Survey
Arvada, Colorado

Craig Vogt
Criteria and Standards Division
Office of Drinking Water
Environmental Protection Agency
Washington, D.C.

Theresa B. Young
Center for Human Systems
Institute for Environmental Studies
Madison, Wisconsin

Susan H. Youngren
Environ Corporation
Washington, D.C.

PREFACE

Progress has been made since the passage of the Safe Drinking Water Act (SDWA) in 1974 toward identifying the major organic contaminants in drinking water and in assessing their hygienic significance. Engineering accomplishments and advances during the past decade, coupled with increased federal monitoring and regulation of microorganic substances, have resulted in an increased growth in treatment methods to remove or control such contaminants and in epidemiologic, toxicologic, and chemical studies related to the significance and formation of these materials in water. Yet basic questions remain with respect to the possible ocurrence of water-borne organic contaminants not identified to date, their associated health risks, the fundamental processes and conditions related to their formation and persistence, and their removal by treatment technology.

These complex issues require expertise in a variety of disciplines, including chemistry, engineering, epidemiology, toxicology, and environmental law. In this book, we have made an effort to bring these disciplines to bear upon the chemical phenomena affecting the formation and identification of these contaminants, the health considerations arising from the presence of such substances in our drinking water, and the treatment and regulatory approaches in use or under consideration. In our opinion, factual evidence and experience in a number of these issues presently constitute a significantly critical mass to justify this effort. This book has been compiled for those concerned with drinking water quality who might benefit from an integrated presentation of the state of the art.

The goal of the SDWA is unchanged since enactment by Congress in 1974: to regulate those contaminants that may have an adverse effect on human health. Despite continued uncertainty about the chemical quality of drinking water, our society is inclined to establish criteria and standards and to recommend technologies for distributing safe and palatable water to the public even if such decisions are based upon scanty evidence. The magnitude of the potential adverse effects requires our continued concern and attention to this problem, despite ambiguity regarding the effects of chronic exposure to low dosages of organic substances.

The book has been divided into five sections appropriate to current drinking water problem areas: (1) overview chapters, (2) identification methods, (3) water treatment processes that prevent or remove trihalomethanes (THMs) and other organic contaminants in drinking water supplies, (4) procedures used in assessing the risks associated with organic contaminants in drinking water, and (5) legislative and regulatory aspects of organic contaminant control.

PART 1

Chapter 1 is an overview of qualitative and quantitative generalities regarding volatile and nonvolatile organic compounds that have been found in drinking water supplies. It reviews the natural and human sources of organic compounds, the level and composition of these materials in water, surrogate parameters, analytical schemes, and major federal actions in controlling organic substances in public water supply sources and drinking water. Chapter 2 is an overview of the chemistry of aqueous chlorine and its use in the disinfection of drinking water supplies, with particular attention to chloroorganic product formation. It discusses the chemistry of aqueous chlorine, the treatment of water supplies with aqueous chlorine, reactions of aqueous chlorine with aromatic substances, the haloform reaction, reactions of aqueous chlorine with amino acids and heterocyclic compounds, and the chlorination of natural waters. Chapter 3 is a review of what is known regarding dissolved organic carbon levels in natural water and their composition. The role of various aquatic natural product organic substances as precursors of chloroorganics is also discussed.

PART 2

Chapter 4 is a review of the methods for isolation and concentration of aquatic organic solutes including resin sorption, extraction, lyophilization, and membrane procedures. It describes major problems facing analytical chemists: analyte recovery, maintenance of original composition, and artifact formation. Chapter 5 discusses gas chromatography/mass spectrometry (GC/MS) procedures in use today for analyzing complex environmental samples with emphasis on drinking water. It compares these procedures with ideal criteria for certainty of structural identification using electron ionization, chemical ionization, accurate mass measurement, and for certainty in quantitative analysis (e.g., matrix effects and proper use of internal standards). Chapter 6 is a review, with examples, of a variety of useful analytical techniques for identification of trace organic compounds that are complementary to GC/MS: high-performance liquid chromatography, total organic halide (TOX) analysis, liquid chromatography/mass spectrometry, and closed loop stripping analysis. It discusses such important problem areas as the fate of nonvolatile TOX and transient species. Chapter 7 presents the role and importance of quality assurance programs in surveys of organic water quality. The uses of second-column GC/MS confirmation, duplicate, split, and blind samples, and quality control are discussed as well as the levels of quality assurance in existing data bases from national surveys of surface and ground water.

PART 3

Chapter 8 presents current research and engineering practice on the removal of THMs and their precursors by coagulation and filtration. The chapter describes the principal methods for the removal of humic substances by alum coagulation, the effects of alum dosage and water characteristics on this removal, and the results of a 3-year evaluation of two full-scale conventional-type water treatment plants. Particular attention is focused on direct filtration. Chapter 9 reviews current research and engineering practice on the activated carbon removal of THMs and their precursors with particular attention to humic and fulvic acids. Topics emphasized include applications of the adsorption process, factors controlling adsorbability of THM precursors, measurement of adsorption effectiveness for THM precursor removal, competitive adsorption, and process costs. Chapter 10 presents the use of alternative disinfection processes in light of concerns regarding secondary reaction products, disinfection efficiency, and engineering practicability. The chapter presents general considerations of disinfectant application, as well as information on the formation of THMs by alternatives to free chlorine, the formation of other by-products, and relevant engineering considerations.

PART 4

Chapters 11 and 12 present an integrated synthesis of how organic contaminants are evaluated for their potential impact on humans. Recognized experts in toxicology and epidemiology have critically evaluated the role of their discipline in risk assessment of organics in drinking water, including their inherent strengths and limitations as well as complementary aspects provided by these approaches. Chapter 13 is a general orientation toward the principles of quantitative risk assessment (QRA) and the regulatory implications of QRA. It is not an esoteric treatment of QRA designed exclusively for statisticians. The chapter establishes the historical and theoretical foundations of QRA in predicting cancer risks, including an overview of the general types of models (i.e., stochastic and tolerance distribution). Chapter 14 is an assessment of the published literature of both an epidemiologic and toxicologic nature concerning the carcinogenic hazards associated with consuming chlorinated drinking water. The chapter is written in an integrated fashion and is intended to give a clear indication of how the principles established in Chapters 11 and 12 may be applied to a current issue of societal importance. Chapter 15 is an epidemiologic assessment of health risks associated with the presence of organic contaminants in drinking water. The chapter has a broad focus since it includes both volatile and non-volatile organic contaminants, and it presents historical perspectives of hypothesis-generating studies, hypothesis-testing studies, as well as case-control studies. Evidence of a cancer risk associated with certain contaminated drinking waters is presented. Chapter 16 discusses the public health significance of organic substances in drinking water and presents the scientific process by which risk to—and, by contrast, safe exposure conditions for—humans can be described and estimated using a variety of information. The chapter presents various approaches to health risk assessment hazard identification, dose-response, and exposure assessment and risk characterization. Chapter 17 discusses some presently unresolved issues associated with quantitative risk assessment, such as genotoxic versus epigenetic carcinogens. The problem is complicated by the fact that humans are actually exposed to complex chemical mixtures about which toxicologic data are very limited and that some members of the population are more sensitive than others.

PART 5

Chapter 18 examines basic issues and regulatory decisions used in assigning "tolerable" risks associated with exposure to environmental hazards. It presents information on the THM MCL in light of exposure to other environmental risks. Topics include the determination of the risks associated with ingesting organic contaminants and the problems associated with relative risk determinations. Chapter 19 discusses the roles of state and federal regulatory agencies in monitoring and controlling organic contamination in public drinking water supplies. A case study is used to highlight the interactive political, economic, and engineering concerns with regulatory compliance. Finally, Chapter 20 is an explanation and evaluation of the criteria that EPA uses to decide whether to evaluate drinking water contaminants. This is followed by a critical assessment of how organics in drinking water fit into this regulatory scheme. This chapter demonstrates how monitoring programs, occurrence data, and risk assessment procedures may be meaningfully synthesized to provide the basis for regulation of organic contaminants in drinking water.

NEIL M. RAM
EDWARD J. CALABRESE
RUSSELL F. CHRISTMAN

CONTENTS

Part 1   Introductory Papers   **1**

1. Environmental Significance of Trace Organic Contaminants in
   Drinking Water Supplies / 3
   Neil M. Ram

2. Aqueous Chlorine in the Treatment of Water Supplies / 33
   J. C. Morris

3. Dissolved Organic Compounds in Natural Waters / 55
   E. M. Thurman

Part 2   Identification Methods   **93**

4. Detection of Organic Carcinogens in Drinking Water:
   A Review of Concentration/Isolation Methods / 95
   Charles D. Hertz and I. H. (Mel) Suffet

5. Application of Combined Gas Chromatography/Mass
   Spectrometry to the Identification and Quantitative
   Analysis of Trace Organic Contaminants / 131
   D. S. Millington and D. L. Norwood

6. Alternatives to Gas Chromatography/Mass Spectrometry for the
   Analysis of Organics in Drinking Water / 153
   William H. Glaze

7. Quality Assurance Programs in the Analysis of Trace Organic
   Contaminants / 173
   Herbert J. Brass and Barbara A. Kingsley

Part 3   Water Treatment Processes That Prevent or Remove Trihalomethanes and Other Organic Contaminants in Drinking Water Supplies   197

8. Conventional Water Treatment and Direct Filtration: Treatment and Removal of Total Organic Carbon and Trihalomethane Precursors / 199
   James K. Edzwald

9. Removal of Organic Contaminants in Drinking Water by Adsorption / 237
   Francis A. DiGiano

10. Alternative Disinfection Processes / 265
    Alan A. Stevens and James M. Symons

Part 4   Assessing the Risks Associated with Organic Contaminants in Drinking Water   291

11. Toxicologic Assessment of Organic Carcinogens in Drinking Water / 293
    Edward J. Calabrese and Andrew T. Canada

12. Epidemiologic Approaches to the Assessment of Carcinogens in Drinking Water / 317
    Kenneth P. Cantor, Marty S. Kanarek, and Theresa B. Young

13. Principles of Quantitative Risk Assessment and Their Application to Drinking Water / 339
    David W. Hosmer

14. Carcinogenic Hazards Associated with the Chlorination of Drinking Water / 353
    R. J. Bull

15. Epidemiologic Assessment of Health Risks Associated with Organic Micropollutants in Drinking Water / 373
    Shirley A. A. Beresford

16. Public Health Significance of Organic Substances in Drinking Water / 405
    Robert G. Tardiff and Susan H. Youngren

17. Unresolved Issues in Risk Assessment / 437
    Edward J. Calabrese and Charles E. Gilbert

**Part 5   Legislative and Regulatory Aspects of Organic
Contaminant Control**   461

18. The Relative Risks Associated with Ingesting Organic
Contaminants / 463
Barbara B. Taylor

19. Technical and Enforcement Aspects of Organic
Contamination in Drinking Water / 493
Jerome J. Healey

20. Regulatory Significance of Organic Contamination in the
Decade of the 1980s / 511
Joseph A. Cotruvo, Susan Goldhaber, and Craig Vogt

Index / 531

PART ONE

Introductory Papers

CHAPTER 1

Environmental Significance of Trace Organic Contaminants in Drinking Water Supplies

Neil M. Ram

Manager, Analytical Laboratory Department
GCA Corporation, Technology Division
Bedford, Massachusetts

1.1 INTRODUCTION

During the past decade, a growing awareness of the possible toxicological hazards and carcinogenic risks associated with long-term consumption of drinking water containing trace quantities of organic compounds has resulted in a nationwide effort to identify and quantify the full spectrum of organic compounds reaching the water consumer. Initially, attention focused on the discovery by Rook (1) and later by Bellar et al. (2) that chloroform and other halogenated methanes are formed during the chlorination of water for disinfection. Dihaloacetonitriles (DHANs) were later identified as yet another chlorinated disinfection by-product of human health concern (3,4). Synthetic chemicals were additionally shown to be contaminants in our nation's water resources (5,6). In addition to compounds known to be deleterious to human health, other compounds such as tri- and di-chloroacetic acid, having unknown health effects, may be just as ubiquitous and found at even higher concentrations than chloroform (7,8).

The impact of these discoveries was a rapid growth in the development of analytical procedures to isolate and concentrate individual microorganic drinking water contaminants and in improved instrumentation to identify and quantify these substances at the parts per trillion level. Paralleling this growth was an acceleration of both epidemiological and toxicological studies to assess the health risks associated with ingesting microorganic chemicals over a life time of exposure, and of research to elucidate the conditions of chloroorganic formation as well as to develop treatment methods for their

3

control or removal. The detection and identification of the myriad organic chemicals in the nation's water supplies was largely due to the application of analytical methodology, primarily gas chromatography/mass spectroscopy (GC/MS), that did not exist 10 years earlier. As a consequence of such "discoveries," the precise meaning of the frequently used term "contamination" was thrown open to question. The findings themselves have further catalyzed development of even more sophisticated analytical methods, to the extent that reports of organic chemicals in drinking water have outpaced our abilities to place much of the data into a meaningful conceptual framework.

Close to 1000 organic contaminants, present at nanogram to microgram concentrations, have now been identified in drinking water for all communities investigated. Conditions and mechanisms of trihalomethane (THM) formation have been studied and treatment methods to remove them or their precursors have been developed. Yet the rate of this data accumulation has defied our ability to assimilate these data. This is due partially to the uncertainty of the data, but largely to our inability to integrate the vast array of information into an intelligible whole. It is clear that drinking water contains a greater variety of organic substances than had been previously realized. The human origin of many of these compounds is also clear. But what is still uncertain is their contribution to the body burden relative to other environmental pollutant exposures. Much remains unknown and the actual threat to health remains questionable.

Despite ambiguity regarding the effects of chronic exposure to low dosages of specific organic micropollutants, the magnitude of the potential adverse effect has stimulated continued research on this topic and has resulted in the enactment of legislation to control microorganic contaminants in drinking water. It is hardly surprising that the Environmental Protection Agency's (EPA) scientific advisory board (9) has concluded that an unquantifiable human health risk exists from consumption of organically polluted drinking water. The risks have been summarized as follows (10):

1. Chemicals which have been shown to cause cancers in animal studies are commonly found in drinking water in small amounts.
2. Some known human carcinogens have been found in drinking water.
3. Exposure to even very small amounts of carcinogenic chemicals poses some risk and repeated exposures amplify the risk.
4. Cancers induced by exposure to small amounts of chemicals may not be manifested for 20 or more years and thus are difficult to relate to a single specific cause.
5. Some portion of the population that is exposed is at greater risk because of other contributed factors such as prior disease states, exposure to other chemicals, or genetic susceptibility.

Such conclusions support the need to define further the extent and seriousness of this threat so that corrective measures can be taken. Doing this requires knowledge of the following: (1) the sources, levels, and composition of the contamination; (2) the chemical characteristics and fate of these contaminants; (3) the treatment technology needed to effectively remove these pollutants or their precursors; and (4) the human response to long-term, low-level exposure to these organic substances. The funds available to investigate these issues, however, are not without limits. The costs can only be justified by improvement in water quality and resulting decrease in human cancer risk.

1.2 NATURAL AND HUMAN SOURCES OF ORGANIC COMPOUNDS

Organic substances occur in the environment either as the result of natural processes or their introduction by human activity. Natural sources contribute the majority of organic material in natural waters via decay of vegetation and animal tissues, animal excretion, photosynthetic by-products, and extracellular release of organic matter by plankton and aquatic macrophytes (11). Humic substances are by far the most frequently occurring natural material in drinking water supplies. Organic contaminants originating in human activity include: organic residuals in sewage and industrial waste waters and organic compounds in dispersed sources (e.g., pesticides). These compounds enter our water resources via domestic, agricultural and industrial wastes, accidental spillages, rainfall, seepage, non-point-source runoff, and improper land disposal of wastes. In addition there are the special groups composed of (1) synthetic chemicals added or formed during water treatment processes (e.g., THMs, chlorinated acids, and DHANs), and (2) substances added during distribution arising from contact with constructional materials and protective coatings (e.g., tetrachloroethylene).

The presence of naturally or human-derived organic compounds in drinking water has been under examination since the turn of the century when the legitimacy of chlorination as a method of water treatment was established. Early public concern over disagreeable tastes arising from water chlorination prompted some attention to this problem. The occurrence of tastes and odors in chlorinated water was attributed in part to "substitution compounds" of chlorine with naturally occurring organic substances (12). At that time, however, there was little specific knowledge about the nature of organic matter and little means of determining its composition analytically. Early studies were directed toward phenols and cresols and their chlorinated products because of their medicinal taste in water supplies (12,13). Phenols are derived from natural products such as "color macromolecules" (14) and have been isolated in urine from domestic animals (15). They are also by-products of coal distillation processes (12). Although a recommended limit for phenol of 0.001 mg/L in public water supplies was proposed by the National Academies of Sciences and Engineering (16), no maximum contaminant level (MCL) has been established by Federal actions to date. An MCL for pentachlorophenol is under current review (17).

There are many human-derived organic contaminants in drinking water owing to the proliferation of synthetic compounds used in industrial and agricultural processes. Several of these deserve special attention because of their toxicological properties or their widespread distribution. Polychlorinated biphenyls (PCBs) constitute a class of pollutants identified in drinking water (18,19) that are characterized by their somewhat environmental persistence (20) and their toxicity to humans and animals (16,21). These compounds are complex mixtures of polychlorinated derivatives of biphenyls having specified chlorine content and are usually referred to by the Arochlor number under which they were commercially marketed. Polynuclear aromatic hydrocarbons (PAHs), which enter the aquatic environment as the result of the leaching of coal-tar products used in tank coatings and pipe linings, are also of concern because of their carcinogenic properties (22). Additional substances under current examination for possible regulation by the EPA include: phthalates, adipates (esters of phthalic acid and adipic acid, respectively), acrylamides (a polyelectrolyte used in the water treatment process), and several synthetic volatile organic chemicals such as trichloroethylene, trichloroethane, and tetrachloroethylene (17).

The presence of chloroorganic compounds in drinking water has been under increasingly close scrutiny because of the possible carcinogeneity of some chemicals within this group. Recently, two classes of chloroorganic compounds have received particular attention: THMs and DHANs. THMs are members of the family of organohalogen compounds that are derivatives of methane in which three of the four hydrogen atoms have been replaced by three atoms of chlorine, bromine, or iodine. Ten distinct compounds are possible by various combinations of the three halide atoms. Chloroform ($CHCl_3$) is the most common THM encountered in drinking water and it is usually found in a higher concentration than the other haloforms (23). More recently, DHANs that are mutagenic in the Ames test (24) have been identified in drinking water. Their carcinogenic activity also has been confirmed in mouse skin initiation promotion studies (25). These compounds originate either in the reaction between chlorine and certain amino acids or polypeptide containing the corresponding amino acid residues on the amino terminus, or in the reaction between chlorine and humic and fulvic substances having amino acid moieties appended to the ring system. Another class of chlorinated organic compounds of environmental significance are di- and trichloroacetic acid and other chlorinated acids. These substances recently have been reported to comprise a greater percentage of the organic halide formed upon the chlorination of extracted fulvic acid than chloroform does (7,8,26,27). The human health effects of these compounds, however, currently are not known. Other chlorinated products formed during water disinfection include: chlorinated phenols, benzoic acids (28), methybutane derivatives (29), and chlorinated polynuclear aromatic compounds (30). Current research does not offer a firm basis for the belief that the most worrisome chlorination by-products have been identified because total measured concentrations represent only small percentages of the total organic carbon content and often are not even the majority of the total organic halogen concentration.

The majority of synthetic chloroorganic compounds formed in water treatment facilities are as yet unidentified and many will probably remain so because of structural, stability, and solubility factors. Furthermore, the chemical distribution and total quantity of these chlorinated by-products is highly dependent upon the precursor concentration as well as process factors such as pH, temperature, Cl_2-to-compound molar ratio, and reaction time.

The following description of the overall formation of synthetic organic compounds during the disinfection of water supplies depicts current approaches to understanding the environmental significance of this phenomenon:

$$\text{Precursor} + \underset{(2)}{\begin{array}{c}\text{Disinfectant} \\ (Cl_2, O_3, \\ ClO_2, UV, \\ \text{etc.})\end{array}} \xrightarrow[\text{conditions}]{\text{reaction}} \underset{(3)}{} \underset{(4)}{\begin{array}{c}\text{Product} \\ \text{composition} \\ \text{and yield}\end{array}} \begin{array}{c} \text{Toxicological} \\ \nearrow \text{assessment} \\ \\ \searrow \text{Epidemiological} \\ \text{assessment} \\ (5) \end{array}$$

(1)

Several key issues are highlighted:

1. Much of the mutagenic activity associated with organic by-products of disinfection results from the reaction of the disinfectant with precursor organic matter in the source water, which is of unknown chemical structure.
2. All disinfectants produce organic by-products. The structures of some by-products are known.
3. The extent and rate of reaction is dependent upon water characteristics.
4. There may be a wide variety of possible reaction pathways and products. Identification of specific by-products is difficult and costly.
5. The total environmental impact of the reaction by-products is not yet established and is under active investigation.

This reaction further illustrates current treatment methods to control organic by-product formation. These include: (1) precursor removal, (2) use of disinfectants other than chlorine that do not generate THM by-products, (3) establishment of reaction conditions that minimize by-product formation, and (4) direct removal of the by-products formed.

Current estimates on the percentage of chlorine used in the chlorination of natural waters are: halogen formation, 0.5–5%; oxidation of organic compounds to CO_2, 50–80%; haloacetonitrile formation, 0–5%; formation of non-haloform organic halogens, 1–6%; and halogenated phenol formation, 0.1% (31). More recent studies indicate that chloroacetic acids are also produced at levels significantly greater than the concentration of chloroform (7,8,26). Once organic compounds enter a water supply, some attenuation through biodegradation, sorption, volatilization, and photodecomposition does occur, but only to a minimal extent (32). The fate of such compounds is unclear owing to the numerous chemical alterations and reactions that may occur. Water treatment is then the only dependable approach to prevent such compounds from reaching the water consumer.

1.3   LEVEL AND COMPOSITION OF ORGANIC CONTAMINANTS

The identification of organic compounds in drinking water is a formidable task. Organic assemblages in environmental samples are very diverse, having complicated and heterogeneous structures, and individual compounds occur but only in trace quantities. Such chemical complexity requires a variety of analytical schemes using state-of-the-art materials and instrumentation to identify and quantify correctly a compound at the ppb level. Nonetheless investigators in search of micropollutants in drinking water have rarely been disappointed. The number of compounds detected in a sample is related to the detection level. As this level decreases by an order of magnitude, there is a corresponding increase in the number of compounds detected.

The identification of microorganics in water involves a series of analytical steps, each of which are relatively sophisticated and in an active state of development. These include:

1. *Pretreatment or isolation* techniques to remove extraneous or possible interfering compounds from solution.
2. *Concentration* of compounds to detectable levels.
3. *Separation* of the complex mixture of organic compounds into individual constituents.
4. *Identification* of resolved compounds.
5. *Quantification* of identified materials.

The success of these analytical steps is largely attributable to the exponential development of accurate and dependable methods coupled with sensitive and sophisticated instrumentation and detectors, some of which will be discussed in Part Two of this book. Such procedures invariably result in lists of compounds and associated detection levels. The utility of such data is, however, plagued with uncertainty about precision and accuracy, false positives and negatives, quality control, artifacts of analytical procedures, and questions of percent recoveries. One possible exception is data obtained from isotope dilution GC/MS.

1.3.1 National Surveys

The development of a comprehensive list of organic contaminants in water supplies was mandated by the 1974 Safe Drinking Water Act (33) which directed the EPA to regulate contaminants in drinking water on the basis of their adverse human health effects. The Act directed EPA to, "conduct a comprehensive study of public water supplies and drinking water sources, and to determine the nature, extent, sources, and means of control of contamination by chemicals or other substances suspected of being carcinogens."

In response, a series of studies were initiated to determine the frequency and levels of organic contaminants occurrence and to provide a data base for the establishment of MCLs and implementation of treatment technologies. These studies included the National Organics Reconaissance Survey (5), National Organics Monitoring Survey (6), Volatile Organic Carbon (VOC) Groundwater Survey (34), National Screening Program (35), and Community Water Supply Survey and State data (36). These efforts produced a significant increase in the number of identified compounds, a result due more to improved analytical methods of detection than an increase in actual contamination. The "discovery" of organic contaminants in drinking water, then, represents the uncovering of an old water quality problem rather than the sudden emergence of a new one. Much of these data have now been compiled in the detailed data base, "WaterDROP" (37).

1.3.2 Surrogate Parameters

The analytical difficulty, as well as the high cost and labor requirements needed to identify specific chemical constituents in the vast array of substances found in surface, ground, or drinking waters, has led to the use of surrogate parameters to evaluate the general organic composition in these sources. Although the identification of individual compounds in aqueous organic assemblages may be the most thorough method to assess the environmental significance of such contamination, surrogate parameters provide a more cost-effective and analytically simpler approach to such an evaluation. Furthermore, not all organic substances are amenable to GC/MS or high-performance liquid chromatography (HPLC) methods of identification, and some may be present well below the limit of identification for specific chemical species. An investigator must, therefore, usually chose between surrogate parameters to gain some information about the nature and level of a complex organic contaminant mixture versus expensive and time-consuming sophisticated analytical techniques to obtain data on specific organic constituents comprising only a small fraction of a total organic assemblage. Although surrogate parameters do not provide detailed information on the exact composition of individual compounds, they do offer the advantage of providing some general data on a large percentage of the complex heterogeneous mixture. It really gets down to the question: "Should we gather a lot of data on a small number of individual compounds, or a little data on a large mixture of substances?" This choice has challenged investigators and will probably continue to be an issue of concern for years to come. Table 1.1 shows some of the surrogate parameters that have been suggested as useful alternatives to specific compound identification.

Before 1970, most attempts to evaluate the level of organic contamination largely were based either upon element specific surrogate parameters such as total organic carbon (TOC) or biochemical oxygen demand (BOD),

TABLE 1.1
Surrogate Parameters Used to Evaluate Organic Compositions[a]

1. *Element-specific parameters*
   Total organic carbon (TOC)
   Purgeable organic carbon (POC)
   Nonpurgeable organic carbon (NPOC)
   Total organic halogen (TOX, POX, NPOX)
   Total organic nitrogen (TON)
2. *Procedurally defined parameters*
   Carbon chloroform extract (CCE)
   Carbon alcohol extract (CAE)
   Taste, odor, color
   Oil and greases
   Hexane solubles
   Organic solids
   Chlorine demand
3. *Functional group parameters*
   Total methoxyl
   Total carboxyl
   Alcoholic hydroxyl
   Total sulfhydryl
4. *Chemical class parameters*
   Phenols
   Pesticides
   Polynuclear aromatic hydrocarbons
   Total trihalomethanes
   Detergents
5. *Physical/chemical property specific parameters*
   UV absorbance
   Fluorescence, phosphorescence
   Metal binding capacity
   Redox potential

[a]Often these categories are not exclusive, and all surrogate methods in popular use are to some degree procedurally defined.

or on procedurally defined parameters such as those listed in Table 1.1. More recently several new surrogate parameters have been added to this list, including fluorescence (38), total organic halide (TOX), purgeable organic halide (POX), non-purgeable organic halide (NPOX), volatile organic carbon compounds (VOC), and total trihalomethane formation potential (TTHMFP) as a measure of THM precursor material. Additional surrogate parameters, as described by Wegman (39), for organic halogenated compounds are shown in Table 1.2. Several recent studies have suggested the use of total organic chlorine (TOCl) as a method for determining the total carbon-bound chlorine (40,41) and as a promising comprehensive and quantitative measurement of the collective chloro-organic compounds present in a sample.

TOX has received particular attention as a direct measure of halogen (Cl, Br, I) covalently bound to organic molecules in a sample. Since most halogenated organics are suspect of being toxic or carcinogenic, the TOX of water may be a useful indicator of toxic contaminants. Currently, the most popular method for TOX analysis in water involves carbon adsorption and oxidative combustion, followed by microcoulometric detection of the formed hydrogen halide (HX). POX is detected by purging a sample aliquot with CO_2. The purged components enter directly into the pyrolysis furnace and are converted to HX. NPOX is then calculated from the difference between TOX and POX values. TOX is directly related to the level of disinfection by-products and may be used in water treatment for unit process design and control. Its use as a surrogate has also been suggested for monitoring individual halogen-containing VOCs in waters intended for drinking (36).

TABLE 1.2
Surrogate Parameters for Organic Halogenated Compounds

Abbreviation[a]	Description
TOX (OX)	Total organic halide
POX	Purgeable organic halide
VOX	Volatile organic halide
NPOX	Non-purgeable organic halide
EOX	Extractable organic halide
HEOX	Hexane extractable organic halide
AOX	Adsorbable organic halide
CAOX	Carbon adsorbable organic halide
PCAOX	Purgeable carbon adsorbable organic halide
NPCAOX	Nonpurgeable carbon adsorbable organic halide
DOX	Dissolved organic halide
TTHMFP	Total trihalomethane formation potential

[a]Substitution of 'Cl' for 'X' in each abbreviation is used to designate organic chlorine compounds.

Stevens et al. (42) reviewed the current uses and future prospects of organic halogen measurements. They reported several studies on TOX as a surrogate for specific VOCs. The data indicate no clear relationship between the concentration of specific organic compounds and their percent recovery as the group organic halogen parameter. The data also exhibit a slight recovery bias in favor of TOX over POX. The ability of these two group parameters to measure halogenated VOCs in a variety of groundwaters and the ruggedness of the methods have not been fully investigated.

Both the National Organics Reconaissance Survey (NORS) and the National Organics Monitoring Survey (NOMS) monitored several surrogate parameters, including non-purgeable total organic carbon (NPTOC), carbon chloroform extract, ultraviolet absorbance, and fluorescence emission (5,6). With the exception of NPTOC these parameters did not correlate with the occurrences of specific organic compounds. Correlation coefficients (Pearson R) between TTHMs and NPTOC, however, ranged from 0.55 to 0.70. Although these data are statistically significant, they are in no way predictive (6,43). Other attempts to correlate and predict the presence of specific chemicals on the basis of TOC measurement have generally not been successful. This is partially attributable to the observation that individual compounds are present at one to two orders of magnitude below TOC values (44). TOC has, however, been suggested as the best current surrogate parameter for the removal of a broad spectrum of synthetic organic chemicals by granular activated carbon (GAC) (23).

Some correlations have been demonstrated between TOX and trihalomethanes in chlorinated drinking water (44). For example, a study of 50 raw and finished drinking water samples from the U.S. found that THMs accounted for an average 50% of the TOCl with ranges between 3 and 100% (45). These values represent current estimates of organic halide composition since future studies may alter them considerably. TOX has additionally been suggested as a group parameter that may relate directly to the toxicity of chlorinated water and wastewater (46).

1.3.3   Individual Compounds Identified

The need to understand the nature of the array of contaminants in our water supplies has resulted in the quantitative determination of some individual molecular species present in the microgram per liter range. The list of compounds actually reported in drinking water is quite extensive and encompasses many of the known classes of organic compounds. The extent and quality of the analytical criteria used in reporting an "identification," however, are extremely variable. Some of the classes of "synthetic" and "natural" organic compounds reported in water are shown in Table 1.3 (34,47–54). There is some overlap in these two categories, however.

Noteworthy are several homogeneous groups of organic chemicals which have been studied in more depth owing to their possible carcinoge-

TABLE 1.3
Classes of Compounds Identified in Water

Naturally occurring compounds

Alcohols	Carbohydrates	Nucleic acids
Aldehydes	Carboxylic acids	Organic acids
Aliphatic acids	Enzymes	Organophosphorus
Alkals	Esters	Peptides
Alkenes	Glucides	Protein
Amides	Heterocycles	Purines
Amino acids	Humic substances	Pyrimidines
Amines	Ketones	Steroids
Aromatic acids	Nitrogenous substances	Vitamins
Aryl alkanes	Nitrosamines	

Synthetic compounds

Aromatic compounds	Mercaptans
Bases and neutral organic compounds	Methane and ethane derivatives
Chemical brighteners	Neutral intractibles
Chlorinated acids	Nitro and Nitroso compounds
Chlorinated aldehydes	Nonvolatile acids
Chlorophenols	Organometallic compounds
Cyanides and azo compounds	Pesticides
Drug metabolites	Pharmaceuticals
Esters	Phenols
Ethers	Pigments
Extractable acids	Plasticizers
Food additives	Polynuclear aromatic compounds
Haloform	Quinones
Halogenated aliphatic compounds	Solvents
Halogenated aromatic compounds	Sulfated products
Halogenated carboxylic acids	Surfactants
Herbicides	Unsaturated hydrocarbons
Household chemicals	Volatile acids
Industrial chemicals	

nicity and consequent public concern: haloforms, di- and trichloroacetic acid, trichloroacetaldehyde (chloral), PAHs, PCBs, and DHANs. Total trihalomethane concentrations ranging from 0.10 to 311 µg/L were observed in the NORS (5,43) of 80 U.S. water facilities and from 0.02 to 550 µg/L in the NOMS (6) of 113 U.S. cities. The three most frequently found synthetic compounds found in the NOMS were pentachlorophenol, dichlorobenzene, and trichloroethylene (23). The finding of the predominance of di- and trichloroacetic acids as well as dichloromaleic and fumaric acids upon reaction of chlorine with aquatic fulvic acid has prompted further study of this occurrence (7,55). Trichloroacetaldehyde (chloral) has been identified at low levels in drinking water and has been postulated to form as an intermediate in the conversion of ethanol to chloroform (2). EPA's recent VOC survey reported the six most frequently encountered ground-water contaminants were: trichlorothylene, 1,1,1-trichloroethane, tetrachloroethylene, *cis/trans*-1,2-dichloroethylene, 1,1-dichloroethane, and carbon tetrachloride (34). Trihalomethane concentrations were generally low in the ground-water survey with median values for chloroform of 1.5 µg/L (maximum = 140 µg/L) and 1.6 µg/L (maximum = 300 µg/L) for supplies serving less than or more than 10,000 persons, respectively. Some ground-water sources, however, contained higher THM values with a maximum reported concentration of 300 µg/L.

In 1983, the American Water Works Association in cooperation with the National Association of Water Companies surveyed 490 utilities from 48 states in an effort to determine the frequency and levels of 138 specific compounds not currently regulated by the U.S. EPA and two surrogate parameters for organic carbon (56). The survey findings were based upon 359 responses. Of the utilities responding, 65, 11, and 24% obtained their raw water from surface, well, or both supplies, respectively. The study found that the vast majority (95%) of these utilities were monitoring for a number of unregulated chemicals. Of the 138 compounds surveyed, 63 were either not detected or were detected by less than 2% of the utilities testing for them. Bromodichloromethane, bromoform, chloroform, dibromochloromethane, endrin, and toxaphene, which are currently regulated by the U.S. EPA, were found by 45, 20, 53, 27, 6, and 7% of the 359 utilities, respectively. Forty-one utilities (11%) detected only one of the TTHMs, usually chloroform (57). The 25 organic compounds projected to be found by 10% or more of the utilities are shown in Table 1.4.

Of the total contaminants in water supplies, on a worldwide basis as of 1981, 2221 organic chemicals were identified of which 765 were found in finished drinking water (58). Forty-three of these are suspected or positive carcinogens and 56 are known to be mutagens. Tardiff (59) reported 128 slightly to very toxic compounds were present among the organic compounds listed as having been found in tap water.

It is important to emphasize that these lists of compounds represent the total number of substances identified in all of the water supplies examined to date. The numbers far exceed those encountered in any single water supply. Thus, while the seemingly ubiquitous distribution of microorganic contaminants is certainly of concern with respect to potential human health risk, the number of contaminants to which individuals using a single drinking water source would be exposed would probably be less than 100. The extent of contamination of a single supply is most likely influenced by the size of the area served by the distribution system and the presence of contaminant sources. EPA's VOC survey (34), for example, found that nonrandom samples taken from water supplies thought to have a higher than normal probability of contamination by VOCs contained two to four times greater contaminant levels than random samples. The study further demonstrated that

TABLE 1.4
Unregulated Organic Compounds Projected to be Detected by
10% or more of 359 Utilities Surveyed (56,57)

(delta)-BHC	1,1-Dichloroethylene
(gamma)-BHC	1,2-Dichloroethylene
Bromodichloromethane	Kepone
Bromomethane	Malathion
Carbon tetrachloride	Methylene chloride
Chlordane metabolites	Phenol
Chlorobenzene	Phthalate esters
Chloroethane	Polychlorinated diphenyl ethers
Chloromethane	Tetrachloroethylene
2,4-D (Me)	Toluene
Diazinon	1,1,1 Trichloroethane
1,1-Dichloroethane	Trichloroethylene
1,2-Dichloroethane	Trichlorofluoromethane

larger water distribution systems contained somewhat higher levels of
VOCs.

1.3.4 Nonvolatile Compounds

Although large strides have been made toward identifying the great many
organic contaminants in water supplies, the identity of only 35% by weight
of the nonhumic material present has been established (60). Table 1.5 shows
the percentages of organic compounds identified in different categories of
materials present in natural waters. About 90% of the volatile organic compounds have been identified as compared with only 30–60% of the
nonhumic, nonvolatile constituents. These materials have only recently begun to receive closer attention. This is somewhat surprising since nonvolatile chlorination products rather than THMs reportedly represent the
majority of bound halogen in chlorinated natural waters (61), and have been
associated with a positive test for mutagenicity by the Ames test (62). THMs
and other specifically identified chlorination products account for less than
2% of the mutagenic activity of chlorinated drinking water (25). Quinn and
Snoeyink (63) found that NPOX comprised about 61% of the TOX of
chlorinated humic or fulvic acid solutions. Kuhn and Sontheimer (64,65)
have shown that nonvolatile, activated carbon, adsorbable organohalogen
compounds exceed volatile organic halogenated substances by a factor of
two to four in typical European surface waters. Glaze (66) and Oliver (67)
used XAD adsorbents to identify nonvolatile compounds. Miller and Uden
(8) found numerous nonvolatile chlorinated organic compounds in chlorinated fulvic acid solutions, including dichloroacetic acid and trichloroacetic
acid. Watts et al. (68) found that the nonvolatile fraction of chlorinated

TABLE 1.5
Composition of Materials Identified in Natural Waters

	Classification		
		Nonvolatile	
	Volatile	Humic	Nonhumic
Approximate percent composition in natural waters	10	75[a]	15
Percent of organic compounds identified in classification	90	The specific structures of the humic substances have not been fully established	5–10 (of total nonvolatile compounds) 33–67 (of nonvolatile nonhumic compounds)
Percent of total compounds identified in natural waters	9	—	4.5–9
Percent of nonhumic compounds identified in natural waters	36	—	18–36

[a] Humic substances comprise the major portion of the organic material in natural waters. Percent composition of total organic material in different waters will vary.

drinking water consisted of many discrete compounds, some of which were almost certainly by-products of disinfection. Studies of the by-products of ozonation of humic and fulvic acid isolates also indicated a similar production of nonvolatile organic substances. Christman and co-workers (69), in comprehensive studies on the chlorination of aquatic humic substances, indicated that saturated and unsaturated chlorinated aliphatic acids, fatty acid methyl esters, and aromatic-derived compounds predominated as products. Glaze et al. (46) found that NPOX represented two to five times the levels of THMs in finished drinking water. Furthermore, Kraybill (70) postulated that the predominant risk associated with the ingestion of drinking water resides in nonvolatile compounds, particularly those arising from the products of disinfection reactions with naturally occurring chemicals. Such studies suggest that less volatile substances are also present in municipal drinking water at relatively high levels. Although these substances have not been extensively identified, their possible toxicological properties warrant concern and further study. NPOC has also been described as being a "reasonable" indicator of THM precursors (5).

1.3.5   Comprehensive Analytical Schemes

The advances in improved detection and identification capabilities have only begun to define the extent and composition of microorganic pollution. Toward this goal, major efforts have been initiated to develop a comprehensive (qualitative and quantitative) scheme that would include analysis of organic compounds of all volatility types and of almost all functional groups and could be applied to almost all types of water.

In 1977 the Office of Drinking Water of the USEPA contracted with SRI International (Menlo Park, CA) to perform the National Screening Program for Organics in Drinking Water. The purpose of this project was to develop a highly detailed protocol capable of identifying selected groups of specific synthetic organic contaminants (35,71) and to provide an occurrence data base from a large supply of U.S. water supplies. Later, in September 1978 EPA initiated the Master Analytical Scheme (MAS) in cooperation with the Research Triangle (Research Triangle Park, NC) and Gulf South Research Institutes (New Orleans, LA). The scheme is capable of identifying all volatility classes that will, or can be derivatized to pass through a gas chromatograph (72,73). The scheme involves the fractionation of a sample into nine groups:

1. Purgeables
2. Extractable acids
3. Extractable bases
4. Extractable neutrals
5. Volatile carboxylic acids
6. Nonvolatile acids
7. Volatile primary and secondary amines
8. Halogenated carboxylic acids
9. Neutral intractables

Other schemes for the comprehensive identification of a broad range of organics have been reported (44,74).

The list of contaminants identified in drinking water will certainly grow with continued scientific research and development and improved analytical techniques. Since the methods for such analysis are sophisticated and are in an active state of development, there may not be a great amount of confidence in the precision and accuracy of some of the data found in the literature. The quality of the data is a function of sampling, analysis (replicates and independence), recovery efficiencies, interferences (selectivity), as well as the limits of detection and determination. False positives as well as false negatives probably occur adding more uncertainty to the data. Nonetheless it is important that the full spectrum of organic contaminants and their levels in drinking water be identified so that their hygienic significance may be better understood.

1.4 LEGISLATION

Drinking water standards in the United States first appeared with the Treasury Standards of 1914, and continued with four revisions, the first in 1925 and the last in 1962. The Public Health Service directed particular attention to synthetic organic chemicals in the early 1960s and attempted to limit their distribution in drinking water through the carbon chloroform extract standard established by the U.S. Public Health Service (PHS) Drinking Water Standards of 1962. Ten years later the National Academy of Engineering set forth "recommended" limits for 15 organic compounds in public water supply sources (16). The widespread concern and public awareness over drinking water contaminants culminated in the passage of the Safe Drinking Water Act (33) in 1974 which required EPA to regulate those contaminants that the Administrator determines "may have" an adverse effect on human health. The wording of this act is very important since conclusive proof of an adverse effect is *not* a prerequisite to regulation. As a consequence, National Interim Primary Drinking Water Regulations (NIPDWR) have been established (75) which currently specify maximum contaminant levels (MCLs) for inorganic and organic chemicals as well as microbiological parameters and certain other classes of water contaminants (turbidity, optional chlorine residual monitoring, and nucleotides, for example). Pesticides were the first class of organic contaminants for which MCLs were established in public water supplies, even though they are generally found in drinking water at levels far below permissible standards (76,77). The NIPDWR (78) currently includes MCLs for four insecticides (endrin, lindane, methoxychlor, and toxaphene), two herbicides [2,4-D and 2,4,5-TP (silvex)], and total trihalomethanes (TTHMs equal to 0.10 mg/L).

Recently, EPA proposed recommended maximum contaminant levels (RMCLs) for nine organic chemicals which "may have an adverse effect upon the health of persons" (79). These include RMCLs of zero for: trichloroethylene, tetrachloroethylene, carbon tetrachloride, 1,2-dichloroethane, vinyl chloride, 1,1-dichloroethylene, and benzene; 0.2 mg/L for 1,1,1-trichloroethane; and 0.75 mg/L for *p*-dichlorobenzene. These compounds were selected on the basis of several criteria:

1. The analytical ability to detect these contaminants in drinking water.
2. The frequency and level of their occurrence and population exposed to these substances.
3. Their potential adverse health effects.

This action was the initial stage in rulemaking for the establishment of primary drinking water regulations for VOCs (80). More recently EPA proposed MCLs and final RMCLs for eight VOCs in drinking water under the Safe Drinking Water Act. In addition, monitoring, reporting, and public notification requirements for these eight VOCs and 51 other VOCs were proposed (81).

TABLE 1.6
Synthetic Organic Chemicals Most Commonly Detected
in Groundwater Proposed for Regulation[a]

Chemical Names	
Trichloroethylene	Benzene
Tetrachloroethylene	Chlorobenzene
Carbon tetrachloride	Dichlorobenzene (s)
1,1,1-Trichloroethane	Trichlorobenzene (s)
1,2-Dichloroethane	1,1-Dichloroethylene
Vinyl chloride	cis-1,2-Dichloroethylene
Methylene chloride	trans-1,2-Dichloroethylene

[a] Presented in U.S. Federal Register, 47 (No. 43), p. 9352 (36), (1982)
and Included for legislative action in the proposed "Safe Drinking Water Amendments of 1984" (81).

The U.S. House of Representatives did pass The Safe Drinking
Water Act Amendments of 1984 (82) by a vote of 366 to 27 (H.R. 5959) which
would have provided the first fundamental changes in the drinking water law
since it was passed in 1962. However, the U.S. Senate counterpart (S. 2649)
did not pass the Senate floor prior to the Senate's 1984 adjournment. The bill
required EPA to set standards for the 14 volatile organic compounds shown
in Table 1.6 within 12 months and 50 additional chemicals within 3 years. It
additionally called for EPA to conduct monitoring programs jointly with the
states on all water systems to look for as-yet-unregulated contaminants. The
bill provided the promulgation of treatment techniques for contaminants that
cannot be accurately measured in drinking water. The bill did not require
that any specified technology, treatment technique, or other means be used
for purposes of meeting maximum contaminant levels. It did, however, call
for required disinfection as a treatment for all public water systems.

The most recent development in drinking water legislation was the passage of an amended Safe Drinking Water Act by both the U.S. House of
Representatives and Senate in May, 1986, directing EPA to establish drinking water standards for 83 contaminants within three years including VOCs,
SOCs, inorganics, microbials, radionuclides, and disinfection by-products.

The monitoring requirements and MCLs in the National Primary Drinking Water Regulations are being reviewed comprehensively for inorganic
and organic compounds, microbiological contaminants, turbidity, and radionuclides. EPA is planning on conducting an assessment of exposure, analytical methods, potential health effects, and the performance and costs of
treatment technologies. EPA is expected to publish the final rulemaking for
Phase I, regulating VOCs, with final MCLs appearing in 1986. Additionally,
EPA has included a proposal for monitoring unregulated VOCs along with
the Phase I rulemaking (81, 83). Phase II rulemaking on synthetic organic
compounds (SOCs) will include proposed RMCLs (1985), final RMCLs and

proposed MCLs (1986), and final MCLs (1987). Phase III rulemaking involves RMCLs for radionuclides. Phase IV rulemaking proposing RMCLs for disinfection by-products is expected to appear in 1986.

A complete chronological listing of major Federal actions in controlling organic substances in drinking water, beginning with the U.S. PHS Drinking Water Standards of 1914 is shown in Table 1.7 (5,6,16,17,21,23,33,36,75–92).

There are three approaches to regulating organic contaminants in drinking water. These include establishing and enforcing:

1. *MCLs for individual compounds.* Water supplies having substances in excess of the MCL are in violation of the law and must take appropriate measures to decrease the contaminant level to acceptable values.
2. *MCLs for group parameters* as a surrogate indicator of individual chemical contamination of human health concern.
3. *Specific treatment standards* where monitoring for individual compounds is not technically or economically feasible.

To date, EPA has used the first two approaches to regulate organic chemicals as illustrated by the MCLs for the herbicides and insecticides, the proposed RMCLs for VOCs, and the MCL of 100 ppb for TTHMs defined as the arithmetic sum of the concentrations of THM compounds (trichloromethane, dibromochloromethane, bromodichloromethane, and tribromomethane) rounded to two significant figures. The TTHM MCL is applicable to all community water systems serving 10,000 or more persons. The law additionally specifies several "Best Treatment Technologies" that a system may be required to use to come into compliance with the TTHM MCL.

Table 1.7 demonstrates a general trend in U.S. legislation in which emphasis was first given to establishing MCLs for broad group parameters such as color, odor, CCE, CAE, and phenol compounds. Later legislation specified MCLs for TTHMs, a homogeneous group parameter, as well as individual substances (six specific pesticides or herbicides). Specific treatment technologies, when required, were also included. This trend from a general approach for organic regulation to specific limits and treatment methods reflects the government's response to data on the occurrence and toxicity of organic contaminants that have been compiled during the last decade, and to research findings indicating the complex interactions and reactions of organic contaminants in our water resources. A simplistic answer can hardly be provided for such a complex problem. Rather definite restrictions and defined methods have been promulgated to cope with the difficult task of ensuring safe and palatable water to the consumer.

The concern about microorganic contaminants in drinking water is shared by other countries. Provisional guideline limits for 21 organic substances were adopted by the World Health Organization (WHO) in 1980 (93). These included 11 pesticides, one polyaromatic hydrocarbon (3,4 benzo[a]pyrene), several chlorinated alkanes, benzene, and several chloro-benzenes. Among the THMs only a guideline for chloroform was recommended, equal to 30 μg/L. More recently the WHO set health-related guideline values for the 17 organic compounds shown in Table 1.8 (94) based upon evidence of toxicity and carcinogenicity. In addition tentative guidelines were set for carbon tetrachloride (3 μg/L), tetrachloroethene (10 μg/L), and trichloroethene (30 μg/L), since the carcinogenicity data did not justify a full guideline value but the compounds were considered to have important health implications when present in drinking water. The recommended guideline value for chloroform was obtained using a linear multistage extrapolation of data obtained from male rats. The WHO cautioned that, although the available toxicological data were only useful in establishing a guideline for chloroform, the concentration of other THMs should also be minimized. Limits ranging from 25 to 250μg/L have been set in several countries for the sum of four specific THMs. The WHO emphasized that absolute safety is an impossible goal in relation to any aspect of life (93). The regulatory approach to microorganic contaminant control, then, requires the careful adoption of certain priorities based upon toxicological and epidemiological data, frequency of occurrence information, cost-benefit analyses, and direct attention to risk assessment.

1.5 PERSPECTIVES

The continued interest in microorganic contamination of water supplies is reflected by active research in this area, general public awareness and concern about the problem, and enactment of new federal regulations to limit and control the levels of trace organic substances reaching the water consumer. There is now general agreement within the scientific community that prolonged exposure to trace organic contaminants in the environment contributes to the incidence of cancer. This results from the convergence of three conditions: (1) significant exposure to the carcinogen measured in terms of concentration and time; (2) the potency of the carcinogen; and (3) the vulnerability of the individual. The concern about contaminated drinking water can be placed into a more realistic perspective by examining each of these three conditions.

First, while individual compounds are generally present at concentrations sufficiently low so as not to be of concern, the aggregate exposure to such compounds during a lifetime is significant with respect to human health. The additional possibility of synergistic interactions among chemicals might enhance the associated risks.

Second, it is largely accepted that some suspected carcinogenic chemicals are formed during the chlorination of drinking water. The potency of such carcinogens is not unequivocally known, as reflected by continuing epidemiological and toxicological studies on the subject. However, since such compounds may have an adverse effect on human health, there is good

TABLE 1.7
Major Federal Actions in Controlling Organic Substances in Public Water Supply Sources and Drinking Water

Year	Agency	Constituent[a]	Maximum Permissible Concentration (mg/L unless otherwise stated)
1914	U.S. PHS Drinking Water Standards	Only bacterial limits prescribed	
1925	U.S. PHS Drinking Water Standards	Color	20 (platinum-cobalt scale)
1962	U.S. PHS Drinking Water Standards	Color	15 (platinum-cobalt scale)
		Odor	3 (threshold odor number)
	Carbon chloroform extract (CCE)	0.20	
1972	National Academy of Sciences and National Academy of Engineering recommended limits for public water supply sources	Alkylbenzene sulfonate (ABS)	0.5
		Odor	Free from objectionable odor
		Oil and grease	Free from oil and grease
		CCE (low flow)	0.3
		Carbon alcohol extract (CAE)	1.5
		Aldrin	0.001
		DDT	0.05
		Dieldrin	0.001
		Chlordane	0.003
		Endrin	0.0005
		Heptachlor	0.0001
		Heptachlor epoxide	0.0001
		Lindane	0.005
		Methoxychlor	1.0
		Toxaphene	0.005
		Carbamate and organophosphorus pesticides	0.1 total
		2,4-D	0.02
		Silvex	0.03
		2,4,5-TP	0.002
		Phenolic compounds	0.001

Date	Action	Details
1974	U.S. Congress Safe Drinking Water Act PL 93-523	(1) Conduct research to identify and measure contaminants in drinking water (2) Determine the nature, extent, sources of, and means of control of contamination by chemicals suspected of being carcinogenic (3) Improved methods of preventing subsurface water contamination (4) Establish MCLs for contaminants having an adverse effect on human health
March 1975	Proposed National Interim Primary Drinking Water Regulations (NIPDWR)	CCE (low flow) 0.7 Endrin 0.0002 Chlordane 0.003 Heptachlor 0.0001 Heptachlor epoxide 0.0001 Lindane 0.004 Methoxychlor 0.1 Toxaphene 0.005 2,4-D 0.1 2,4,5-TP Silvex 0.01
1975	U.S. EPA National Organics Reconnaissance Survey (NORS)	(1) Determine distribution of four trihalomethanes (2) Determine effect of raw water source and treatment practices on THM formation (3) Characterize organic content of finished drinking water in 10 cities (4) Determine correlation between group parameters: nonvolatile TOC, UV absorbance, fluorescence, and CCE
Dec. 1975	U.S. EPA National Interim Primary Drinking Water Regulations	CCE, heptachlor, heptachlor epoxide, chlordane — Eliminated
1976	U.S. EPA Advanced Notice of Proposed Rule Making (ANPRM)	Two regulatory approaches presented: (1) MCLs for specific organic chemicals, (2) designated treatment technology
1976	U.S. EPA, Quality Criteria for Water recommended limits for domestic water supply	2,4-D 0.100 2,4,5-TP 0.010 Silvex 0.010 Endrin 0.0002 Lindane 0.004 Methoxychlor 0.100 Toxaphene 0.005 Phenol 0.001 Aldrin, dieldrin, chlordane, DDT, heptachlor } CAUTION to minimize human exposure Polychlorinated biphenyls }
1977	U.S. EPA, National Secondary Drinking Water Regulations (recommended)	Color 15 (color units) Foaming agents 0.5 Odor 3 (threshold odor number)

TABLE 1.7 (Continued)

Year	Agency	Constituent[a]	Maximum Permissible Concentration (mg/L unless otherwise stated)
1977	U.S. EPA, National Organics Monitoring Survey (NOMS)	(1) Monitor 21 specific organic compounds in 113 community water supplies (2) Monitor group parameters: TOC, CCE, UV absorbance, and emission fluorescence	
Feb. 1978	U.S. EPA IPDWR Control of Chemicals in Drinking Water: Proposed Rule	(1) Total trihalomethanes (TTHMs) (2) GAC filtration to decrease synthetic organic chemicals	0.10^a Proposed
Nov. 1979	U.S. EPA Control of THMs in Drinking Water: Final Rule	TTHMs GAC filtration	0.10 No decision
Aug. 1980	U.S. EPA IPDWR: Final Rule	Special monitoring of organic chemicals No change	May be required
1980, 1982	National Academy of Science, National Academy of Engineering, Institute of Medicine; National Research Council	Organic chemicals	Suggested no adverse response level (SNARL): Acute and chronic exposure levels for organic chemicals presented
1980	U.S. EPA; Quality Criteria for water; recommended limits for public water supply	Organic chemicals	Ambient water criteria for the protection of human health, and incremental increases in lifetime cancer risks with corresponding criteria are presented.
		Noteworthy: chloroform	Incremental cancer risks: 10^{-5}: 0.00190 10^{-6}: 0.00019 10^{-7}: 0.000019
March 1982	U.S. EPA ANPRM: Synthetic Organic Chemicals in Drinking Water	(1) Public comments to enable the U.S. EPA to determine if MCLs should be established for Volatile Organic Compounds (VOCs) and, if so, at what levels (2) VOC groundwater occurrence data presented	
February 1983	NIPDWR Trihalomethanes; Final Rule	(1) TTHMs (2) Best treatment technologies for TTHM control identified: (a) chloramine disinfection, (b) ClO_2 disinfection, (c) improved clarification to reduce THM precursors, (d) moving chlorine application point, (e) PAC for THM precursor or TTHM reduction, (f) five additional methods not determined to be generally available but which may be available to some systems	0.10^b

		RMCLs	Proposed MCLs

October 1983 — U.S. EPA ANPRM: National Revised Primary Drinking Water Regulations

(1) MCLs and RMCLs (recommended maximum contaminants levels) to be specified
(2) Three-tiered approach for contaminant control based upon frequency and health concern
(3) 29 synthetic organic chemicals being considered for inclusion in NPDWR

June 1984 — USEPA NPDWR; Volatile Synthetic Organic Chemicals; Proposed Rulemaking

Recommended maximum contaminants levels of zero for: trichloroethylene, tetrachloroethylene, carbon tetrachloride, 1,2-dichloroethane, vinyl chloride, 1,1-dichloroethylene, and benzene; 0.2 mg/L for 1,1,1-trichloroethane; and 0.75 mg/L for p-dichlorobenzene

Sept. 1984 — Proposed Safe Drinking Water Act ammendments of 1984 (passed by House of Representatives only)

Within 12 months: set standards for 14 VOCs
Within 3 years: set standards on 50 additional chemicals

Nov. 1985 — NPDWR; Volatile Synthetic Organic Chemicals; Final rule and Proposed Rule

(1) Compliance Monitoring:

Compound	RMCLs	Proposed MCLs
trichloroethylene	0	5 µg/L
carbon tetrachloride	0	5 µg/L
1,1,1-trichloroethane	200 µg/L	200 µg/L
vinylchloride	0	1 µg/L
1,2-dichloroethane	0	5 µg/L
benzene	0	5 µg/L
1,1-dichloroethylene	7 µg/L	7 µg/L
p-dichlorobenzene	750 µg/L	750 µg/L

(2) Proposed monitoring for 51 unregulated VOCs
(3) Proposed RMCLs for 26 Synthetic Organic Compounds (SOCs)

May 1986 — Amended Safe Drinking Water Act (passed by both U.S. House of Representatives and Senate)

Directs EPA to establish drinking water standards for 83 contaminants within three years including VOCs, SOCs, inorganics, microbials, radionuclides, and disinfection by-products

[a] For all communities greater than 75,000 people. MCL for communities greater than 10,000 people to be phased in.
[b] For all communities greater than 10,000 people.

TABLE 1.8
Guideline Values for Health-Related Organic
Contaminants (93)

Contaminant	Guideline Value (µg/L)
Aldrin and dieldrin	0.03
Benzene	10
Benzo[a]pyrene[a]	0.01
Chlordane (total isomers)	0.3
Chloroform [a,d]	30
2,4-D	100
DDT (total isomers)	1.0
1,2-Dichloroethane[a]	10
1,1-Dichloroethene[a]	0.3
Heptachlor and heptachlor epoxide	0.1
Hexachlorobenzene[a]	0.01[c]
Gamma-HCH (lindane)	3.0
Methoxychlor	30
Pentachlorophenol	10
2,4,6-Trichlorophenol[a,b]	10

[a] The guideline values for these substances were computed from a conservative, hypothetical, mathematical model that cannot be experimentally verified and therefore should be interpreted differently. Uncertainties involved are considerable and a variation of about two orders of magnitude (i.e., from 0.1 to 10 times the number) could exist.

[b] The threshold taste and odor value for this compound is 0.1 g/L.

[c] Since the FAO/WHO conditional ADI of 0.0006 mg/kg body weight has been withdrawn, this value was derived from the linear multistage extrapolation model for a cancer risk of less than 1 in 100,000 for a lifetime of exposure.

[d] The microbiological quality of drinking water should not be comprised by efforts to control the concentration of chloroform.

reason for their regulation. Abandoning chlorination does not appear to be a viable alternative since it is a proven and time-honored method to control waterborne infections. Additionally there is still no indication that alternative disinfectants are significantly safer than chlorine.

Third, the susceptibility of individuals to cancer is affected not only by their exposure to drinking water contaminants but also by their own genetic makeup as well as by their contact with environmental carcinogens in food and air. Some typical pesticide levels in breast milk, for example, range between 1–6 ppb and 50–200 ppb for dieldrin and DDT (including metabolites), respectively (95), representing about 600 and 4000 times the amount reported in finished drinking waters (19,96–97). Nursing infants could therefore consume in one liter of milk a mass of dieldrin or DDT equivalent to that consumed in drinking water over 2 and 10.8 years, respectively. The potentially significant adverse impact on large human populations that may result from the consumption of contaminated water certainly warrants special attention to this problem.

Basic questions are still unanswered and will probably remain so for years to come: Are other sources of contaminant exposure more important than drinking water? Are we currently unaware of those compounds of greatest hygienic concern? What is the significance of ppb concentrations of chemical contaminants, and what is the quality of this data? A fundamental understanding of processes and conditions related to microorganic contaminants as well as better knowledge of associated health risks and treatment technology are needed. Such knowledge must ultimately indicate the technical feasibility of developing efficient and cost-effective treatment methods and, in addition, provide a sound basis for decisions on proper monitoring and regulation of microorganic contaminants in drinking water.

REFERENCES

1. Rook, J.J., *Water Treat. Exam.* (England), *23*, 234, 1974.
2. Bellar, T.A., Lichtenberg, J.J., and Kroner, R.C., *J. Am. Water Works Assn., 66,* 703, 1974.
3. Trehy, M., and Bieber, T., *Proc. Am. Water Works Assn.,* 1980 Annual Conference, Am. Water Works Assn., Denver, CO, 1980.
4. Trehy, M., and Bieber, T., in Jolley, R.L., et al. (eds.), *Water Chlorination Environmental Impact Health Effects,* Vol. 4, Ann Arbor Sci. Publ. Inc., Ann Arbor, MI, 1983, p. 85.
5. Symons, J., Bellar, T., Carswell, J., DeMarco, J., Kropp, K., Robeck, G., Seeger, D., Sloccum, C., Smith, B., and Stevens, A., *J. Am. Water Works Assn., 67,* 634, 1975; *67,* 708, 1975.
6. U.S. Environmental Protection Agency, *The National Organic Monitoring Survey,* U.S. Environmental Protection Agency, Technical Support Division, Office of Water Supply, Washington, DC, 1977.
7. Christman, R.F., Norwood, D.L., Millington, D.S., and Johnson, J.D., *Environ. Sci. Technol., 17,* 625, 1983.
8. Miller J.W., and Uden, P.C., *Environ. Sci. Technol., 17,* 150, 1983.
9. U.S. Environmental Protection Agency, *Interim Report to Congress: Preliminary Assessment of Suspected Carcinogens in Drinking Water,* U.S. Environmental Protection Agency, Science Advisory Board, Washington, DC, 1975.
10. Upton, A.C., Memorandum to Dr. D. Castle, administrator U.S. Environmental Protection Agency, National Cancer Institute, Dept. Health Education and Welfare, Bethesda, MD (April 10, 1978).
11. Maier, W. J., McConnel, H.L., and Conroy, L.E., Univ. of Minneapolis, Dept. of Civil & Mineral Engineering, *A Survey of Organic Constituents in Natural and Fresh Waters,* National Technical Information Service, Springfield, VA, PB-236 794/4GA, 1974.
12. Donaldson, W., Waterworks Monthly Issue of *Engineering and Contracting,* November, 74, 1922.
13. Streeter, H.W., *Public Health Rep., 44,* 2149, 1929.
14. Christman, R.F., and Ghassemi, M., *J. Am. Water Works Assn., 58,* 723, 1966.
15. Krump, D., *Water Res.* (England), 8, 899, 1974.
16. National Academy of Science, National Academy of Engineering, *Water Quality Criteria 1972,* National Academy of Sciences–National Academy of Engineering, Washington, DC; National Technical Information Service, Springfield, VA, PB-239 199.

17. U.S. Federal Register, *48*, No. 194, pp. 45502–45521 (Oct. 5, 1983).
18. Coleman, W.E., Melton, R.G., Kopfler, F.C., Barone, K.A., Aurand, T.A., and Jellison, M.G., *Environ. Sci. Technol., 14,* 576, 1980.
19. Lin, D.C., Melton, R.G., Kopfler, F.C., and Lucas, S.V., Glass capillary gas chromatographic mass spectrometric analysis of organics concentrated from drinking and advanced waste treatment waters, in Keith L.H., (ed.), *Advances in the Identification and Analysis of Organic Pollutant in Water,* Vol. 2, Ann Arbor Sci. Publ. Inc., Ann Arbor, MI, 1981.
20. Faust, S.D., Aly, O.M., *Chemistry of Water Treatment,* Ann Arbor Sci. Publ. Inc., Ann Arbor, MI, 1983, pp. 77–79.
21. U.S. Federal Register, *46,* No. 156, pp. 79318–79379 (Nov. 28, 1980).
22. Faust, S.D., Aly, O.M., *Chemistry of Natural Waters,* Ann Arbor Sci. Publ. Inc., Ann Arbor, MI, 1981, pp. 40–45.
23. U.S. Federal Register, *43,* No. 28, pp. 5733–5779 (Feb. 9, 1978).
24. Simmon, V.F., Kauhanen, K., and Tardiff, R.G., *Prog. Gen. Toxicol., 2,* 249, 1977.
25. U.S. Environmental Protection Agency, *Research Outlook,* Office of Research and Development, Washington, DC, 1983, p. 52.
26. Uden, P.C., and Miller, J.W., *J. Am. Water Works Assn., 75,* 524, 1983.
27. Johnson, J.D., Christman, R.F., Norwood, D.L., and Millington, D.S., Reaction products of aquatic humic substances with chlorine, in Lucier, G.W., and Hook, G.E.R., *Environmental Health Perspectives,* Vol. 46, U.S. Dept. of Health and Human Services, National Institute of Environmental Health Sciences, Bethesda, MD, 1982, pp. 63–71.
28. Jolley, R.L., Jones, G., Pitt, W.W., and Thompson, J.E., Determination of chlorinated effects on organic constituents in natural and process water using HPLC, in Keith, L.H., *Identification and Analysis of Organic Pollutants in Water,* Ann Arbor Sci. Publ. Inc., Ann Arbor, MI, 1976, pp. 233–246.
29. Christman, R.F., Johnson, J.D., Hass, J.R., Pfaender, F.K., Liao, W.T., Norwood, D.L., and Alexander, H.J., Natural and model aquatic humics: Reactions with chlorine, in Jolley, R.L., et al., (eds.), *Water Chlorination Environ. Impact Health Effects,* Vol. 2, Ann Arbor Sci. Publ. Inc., Ann Arbor, MI, 1978, pp. 15–28.
30. Graft, W., and Nathafft, G., *Arch. Hyg., 147,* 135, 1963.
31. Hileman, B., *Environ. Sci. Techn., 16,* 15A, 1982.
32. Bouwer, E.J., McCarty, P.L., and Lance, J.C., *Water Res.* (England), *15,* 151, 1981.
33. U.S. Public Law 93-523, 93rd Congress, S. 433, *Safe Drinking Water Act,* U.S. Environmental Protection Agency, Washington, DC, 1974.
34. Westrick, J.J., Mello, J.W., and Thomas, R.F., *The Groundwater Survey: Summary of Volatile Organic Contaminant Occurrence Data,* U.S. Environmental Protection Agency, Technical Support Division, Office of Drinking Water, Cincinnati, OH, 1982.
35. Boland, P.A., Kingsley, B.A., and Stivers, D.F., Protocol for the analysis of a broad range of specific organic compounds in drinking water, in Keith, L.H., (ed.), *Advances in the Identification and Analysis of Organic Pollutants in Water,* Vol. 2, Ann Arbor Sci. Publ. Inc., Ann Arbor, MI, 1981, pp. 831–838.
36. U.S. Federal Register, *47,* No. 43, 9349–9358 (March 4, 1982).
37. Carson, B.L., Going, J.E., Lopez-Avila, V., McCann, J.L., Cole, C.J., and Holt, J.L., The distribution register of organic pollutants in water-waterDROP, in Keith, L.H., (ed.), *Advances in the Identification and Analysis of Organic Pollutants in Water,* Vol. 2, Ann Arbor Sci. Publ. Inc., Ann Arbor, MI, 1981, pp. 497–525.
38. Sylvia, A.E., Donlan, R.J., Monitoring trace organics in drinking water by fluorescence, in McGuire, M.J., and Suffet, I.H., (eds.), *Activated Carbon Adsorption of Organics from the Aqueous Phase,* Vol. 2, Ann Arbor Sci. Publ. Inc., Ann Arbor, MI, 1980, pp. 559–565.
39. Wegman, R.C., Determination of organic halogens: A critical review of sum parameters, in *Proceedings of the Second European Symposium: Analysis of Organic Micropollutants in Water,* Commission of the European Communities, D. Reidel Publishing Co., Boston, MA, 1982.
40. Dressman, R., Nafar, B., and Redzikowski, R., The analysis of organohalides (OX) in

water as a group parameter, in *Proceedings 7th Annual Am. Water Works Assn. Water Quality Technology Conference,* Am. Water Works Assn., Denver, CO, 1979.

41. Glaze, W.H., Peyton, G., and Rawley, L., *Environ. Sci. Technol., 11,* 685, 1977.
42. Stevens, A.A., Dressman, R.C., Sorrell, R.K., and Brass, H.J., *J. Am. Water Works. Assn. 77,* 146, 1985.
43. Symons, J.M. et al., *National Organics Reconnaissance Survey for Halogenated Organics in Drinking Water: Pre-Publication Copy,* U.S. Environmental Protection Agency, Office of Research and Development, Cincinnati, OH, 1975.
44. Brass, H.J., *J. Am. Water Works Assn., 74,* 107, 1982.
45. Cotruvo, J.A., *Environ. Sci. Technol., 15,* 268, 1981.
46. Glaze, W.H., Burleson, J.L., Henderson, IV, J.E., Jones, P., Kinstley, W., Peyton, G., Rawley, R., Saleh, F., and Smith, G., *Analysis of Chlorinated Organic Compounds formed During Drinking Water Chlorination,* Draft report for U.S. Environ. Protection Agency Contracts R-803007, R-805822, A.W. Garrison, Project Officer, North Texas State Univ., Denton, TX, 1981.
47. Stander, G.J., Microorganic compounds in the water environment and their impacts on the quality of potable water supplies, *Water South Africa, 6(1),* 1, 1980.
48. Keith, L.H., *Identification and Analysis of Organic Pollutants in Water,* Ann Arbor Sci. Publ. Inc., Ann Arbor, MI, 1976.
49. Keith, L.H., *Environ. Sci. Technol., 15,* 156, 1981.
50. Adams, V., Watts, R., and Pitts, M., *J. Water Pollut. Control Fed., 55,* 577, 1983.
51. DeWalle, F., Norman, D., Sung, J., Chian, E., and Giabbai, M., *J. Water Pollut. Control Fed., 53,* 659, 1981.
52. DeWalle, F., Sund, J., Kalman, D., Chian, E., Giabbai, M., and Denton, M., *J. Water Pollut. Control Fed., 54,* 555, 1982.
53. Chian, E., *J. Water Pollut. Control Fed., 51,* 1134, 1979.
54. Chian, E., DeWalle, F., Meng, H., and Norman, D., *J. Water Pollut. Control Fed., 52,* 1120, 1980.
55. Norwood, D., Johnson, J.D., Christman, R.F., and Millington, D.S., Chlorination products from aquatic humic material, in Jolley, R.L. et al. (eds.), *Water Chlorination Environ. Impact Health Effects,* Vol. 4, Ann Arbor Sci. Publ. Inc., Ann Arbor, MI, 1983, pp. 191–200.
56. Am. Water Works Assn., Congress asked not to legislate SDWA regs., *Mainstream, 27,* No. 10, Am. Water Works Assn., Washington, DC, 1983.
57. Groff, J.B., Letter to D.E. Eckart on Am. Water Works Assn. Survey Findings, Am. Water Works Assn., Washington, DC, Sept. 21, 1983.
58. Kraybell, H.F., *J. Am. Water Works Assn., 73,* 370, 1981.
59. Tardiff, R.G., Health effects caused by exposure to drinking water contaminants, in Mullaney, J.L., and Tardiff, R.G., *Preliminary Assessment of Suspected Carcinogens in Drinking Water: Interim Report to Congress,* U.S. Environmental Protection Agency, Office of Toxic Substances, Washington, DC, 1975.
60. U.S. Federal Register, *42,* No. 132, pp. 35764–35779 (July 11, 1977).
61. Glaze, W., Peyton, G., Salek, F., and Hunang, F., *Int. J. Environ. Anal. Chem., 7,* 147, 1979.
62. Rook, J.J., Removal of chlorinated brominated, and iodinated non-volatile compounds by GAC filtration, in Laird, L.B., chairman, Jolley, R.L., secretary, *Abstracts of Papers,* 181st ACS National Meeting, Atlanta, GA, March 29–April 3, 1981, American Chemical Society, Washington, DC.
63. Quinn, J., Snoeyink, V., *J. Am. Water Works Assn., 72,* 483, 1980.
64. Kuhn, W., Sontheimer, H., *Vom Wasser* (Ger.), *43,* 327, 1974.
65. Kuhn, W., Fuchs, F., and Sontheimer, H., *Z. Wasser Abwasser–Forsch.* (Ger.), *10,* 192, 1977.
66. Glaze, W.H., Henderson, J.E., Smith, G., Analysis of new chlorinated organic compounds formed by chlorination of municipal wastewater, in Jolley, R.L. (ed.), *Water Chlorination*

Environ. Impact Health Effects, Vol. 1, Ann Arbor Sci. Publ. Inc., Ann Arbor, MI, 1978, pp. 139–159.

67. Oliver, B.G., *Canadian Res. 11,* 21, 1978.
68. Watts, C.D., Crathorne, B., Fielding, M., and Killops, S.D., Nonvolatile organic compounds in treated waters, in Lucier, G.W., and Hook, G.E. (eds.), *Environmental Health Perspectives: Drinking Water Disinfectants,* Vol. 46, National Institutes of Health, Research Triangle Park, NC, 1982, pp. 87–99.
69. Christman, R.F., Liao, W.T., Millington, D.S., and Johnson, J.D., Oxidative degradation of aquatic humic material, in Keith, L.H. (ed.), *Advances in the Identification and Analysis of Organic Pollutants in Water,* Vol. 2, Ann Arbor Sci. Publ. Inc., Ann Arbor, MI, 1981, pp. 979–999.
70. Kraybill, H., *Prev. Med. 9,* 212, 1980.
71. Fratoni, S., and Boland, P., Methods used in the national screening program for organics in drinking water, in *Proceedings of the 6th Am. Water Works Assn. Water Quality Technology Conf.,* Louisville, KY, Am. Water Works Assn., Denver, CO, 1978, pp. 2A-2:1–7.
72. Garrison, A., Alford, A., Craig, J., Ellington, J., Haeberer, A., McGuire, J., Pope, J., Shackelford, W., Pellizzari, E., and Gebhart, J., The master analytical scheme: An overview of the interim procedures, in Keith, L.H. (ed.), *Advances in the Identification and Analysis of Organic Pollutants in Water,* Vol. 1, Ann Arbor Sci. Publ. Inc., Ann Arbor, MI, 1981, pp. 17–30.
73. Gebhard, J., Ryan, J., Cox, R., Pellizzari, E., Michael, L., and Sheldon, R., The master analytical scheme: Development of effective techniques for isolation and concentration of organics in water, in Keith, L.H. (ed.), *Advances in the Identification and Analyses of Organic Pollutants in Water,* Vol. 1, Ann Arbor Sci. Publ. Inc., Ann Arbor, MI, 1981, pp. 31–48.
74. Keith, L.H., *Advances in the Identification and Analysis of Organic Pollutants in Water,* Ann Arbor Sci., Ann Arbor, MI, 1981.
75. U.S. Federal Register, *40,* No. 248, pp. 59565–59587 (Dec. 24, 1975).
76. Lichtenberg, J.J., Eichelberger, J.W., Dressman, R.C., and Longbottom, J.E., *Pesticides Monitoring J., 4,* No. 2, 71, 1970.
77. Dudley, D.R. and Karr, J.R., *Pesticides Monitoring J., 13,* 155, 1980.
78. U.S. Federal Register, *45,* No. 168, pp. 57331–57357, (Aug. 27, 1980).
79. U.S. Federal Register, *49,* No. 114, pp. 24330–24335 (June 12, 1984).
80. U.S. Federal Register, *49,*No. 205, pp. 42132–42199 (Oct. 22, 1984).
81. U.S. Federal Register, *50,* No. 219, pp 46902–46933 (Nov. 13, 1985).
82. U.S. Congress, proposed act, "Save Drinking Water Act Amendments of 1984" 2[cd] Session, H.R. 5959, Calendar No. 1220.
83. Cotruvo, J.A., Director, Criteria and Standards Division, Office of Drinking Water, U.S. Environmental Protection Agency, personal communication, May, 1985.
84. U.S. Federal Register, *40,* No. 51, pp. 11990–11998 (March 14, 1975).
85. U.S. Federal Register, *41,* No. 136, pp. 28991–28998 (July 14, 1976).
86. U.S. Environmental Protection Agency, *Quality Criteria for Water,* U.S. Government Printing Office, Washington, DC, 1976.
87. U.S. Federal Register, *42,* No. 62, pp. 17143–17146 (March 31, 1977).
88. U.S. Federal Register, *44,* No. 231, pp. 68623–68642, (Nov. 29, 1979).
89. National Academy of Science, National Academy of Engineering, Institute of Medicine: National Research Council, Drinking Water and Health, *1, 2, 3,* and *4,* National Academy Press, Washington, DC, 1977, 1980, and 1982.
90. U.S. Environmental Control Administration, U.S. Public Health Drinking Water Standards, U.S. Government Printing Office, Washington, DC, 1962.
91. U.S. Federal Register, *48,* No. 40, pp. 8405–8414 (Feb. 28, 1983).
92. U.S. Public Health Service, Drinking Water Standards, U.S. Government Printing Office, Washington, DC, 1925.
93. Waddington, J.I., *Effluent and Water Treatment Journal,* April, 145, 1983.

94. *Guidelines for Drinking Water Quality,* Vol. 1, World Health Organization, Geneva, 1984.
95. Rogan, W.J., Bagniewska, A., and Damstra, T., *N. Engl. J. Med., 302,* 1450, 1980.
96. Junk, G.A., Richard, J.J., Fritz, J.S., and Svec, H.J., Resin sorption methods for monitoring selected contaminants in water, in Keith, L.H. (ed.), *Identification and Analysis of Organic Pollutants in Water,* Ann Arbor Sci. Publ. Inc., Ann Arbor, MI, 1976, pp. 135–153.
97. Keith, L.H., Garrison, A.W., Allen, F.R., Carter, M.H., Floyd, T.L., Pope, J.D., and Thurston, Jr., A.D., Identification of organic compounds in drinking water from 13 U.S. cities, in Keith, L.H. (ed.), *Identification and Analysis of Organic Pollutants in Water,* Ann Arbor Sci. Publ. Inc., Ann Arbor, MI, 1976, pp. 329–373.

CHAPTER 2

Aqueous Chlorine in the Treatment of Water Supplies

J. C. Morris

Emeritus Professor of Sanitary Chemistry
Harvard University
Cambridge Massachusetts

2.1 INTRODUCTION

Aqueous chlorine is the most widely used chemical for the treatment of water supplies. It is employed principally for disinfection because of its potent germicidal activity, but it is also used extensively as an oxidant for inorganic and organic materials. It also serves to suppress undesired biological growths in coagulation chambers, settling basins, filters, and storage tanks.

The popularity of aqueous chlorine for disinfection and other aspects of water treatment is dependent partly on its economy and ease of use in addition to its germicidal and oxidizing potency. However, the property of aqueous chlorine that makes it almost unique among available large-scale water disinfectants is its ability to maintain a sustained, residual disinfecting action during distribution of the water following treatment. Alternate disinfecting agents, such as ozone or ultraviolet light, do not have this ability; indeed it is common for water treatment plants using ozone as the primary disinfectant to add some aqueous chlorine as residual disinfectant at the end of the treatment process. It is primarily because of the residual germicidal action of aqueous chlorine that it is so difficult to find a suitable replacement exhibiting less hazardous side reactions that will not lessen the hygienic security of the distributed water.

Until now continued use of aqueous chlorine with modifications in procedure to minimize the extent of hazardous side reactions has seemed generally to be a more satisfactory solution than substitution of an alternate disinfectant.

2.2 CHEMISTRY OF AQUEOUS CHLORINE

2.2.1 Reactions of Chlorine in Water

Aqueous chlorine is not Cl_2. When elemental chlorine is dissolved in water, it hydrolyzes rapidly and extensively in accordance with the equation

$$Cl_2 + H_2O = HOCl + H^+ + Cl^- \tag{2.1}$$

Shilov and Solodushenkov (1) Eigen and Kustin (2) and others have shown that hydrolytic equilibrium is reached within a few seconds at 0°C and less at higher temperatures. The hydrolysis constant, given by the expression,

$$K_H = (HOCl)(H^+)(Cl^-)/(Cl_2) \tag{2.2}$$

conforms to the equation

$$pK_H = -\log K_H = 3.836 - 0.0216t + 0.000175t^2 \tag{2.3}$$

where t = temperature, in degrees Celsius, between 0 and 45°C (3,4). For example, $pK_H = 3.51$, $K_H = 2.81 \times 10^{-4}$ at 15°C. The values are such that when the concentration of chlorine is not more than 71 mg/L ($10^{-3}M$) hydrolysis at equilibrium is already 99% or more complete in pure water where the H^+ produced decreases the pH to 3. The hydrolysis is considerably greater in the bicarbonate-buffered solutions that constitute most natural waters. Consequently Cl_2 has little part in the reactions of aqueous chlorine at the concentrations commonly used in water treatment.

Hypochlorous acid, the oxidizing hydrolysis product of Cl_2, is a weak acid, ionizing in accordance with the equation

$$HOCl = H^+ + OCl^- \tag{2.4}$$

The ionization equilibrium conforms to the expression

$$K_A = (H^+)(OCl^-)/(HOCl) \tag{2.5}$$

or

$$\log[OCl^-]/[HOCl] = pH - pK'_A \tag{2.6}$$

Values for the ionization constant K_A (5) and its negative logarithm pK_A are given for several temperatures in Table 2.1.

As Eq. 2.6 shows, the distribution between nonionized HOCl and the hypochlorite ion OCl^- is a function of the pH of the aqueous medium. When $pH = pK'_A$, the concentrations of HOCl and OCl^- are equal. At pH values less than pK'_A nonionized HOCl predominates, whereas at pH values greater than pK'_A hypochlorite ion constitutes the major fraction. The distribution

TABLE 2.1
Acid Dissociation Constant for HOCl

Temperature (°C)	$K_A \times 10^8$ (mol/L)	pK_A	pK_A' ($\mu = 0.01$)[a]
0	1.48	7.83	7.79
5	1.78	7.75	7.71
10	2.04	7.69	7.65
15	2.35	7.63	7.59
20	2.63	7.58	7.54
25	2.88	7.54	7.50
30	3.16	7.50	7.46
35	3.39	7.47	7.43

[a] μ = Ionic strength.

shifts from more than 90% HOCl at a pH one unit less than pK_A' to more than 90% OCl^- at pH one unit greater than pK_A'.

This shift in ionization with pH, which is common to all weak acid-base systems, is particularly significant for aqueous chlorine for two reasons. First, the pK_A' values 7.5–7.7 correspond to the middle of the pH range, 6.0–8.5, characteristic of most natural water systems. So, the pH range of natural waters coincides with the pH range of rapidly changing distribution of HOCl and OCl^-. Second, the chemical reactivity and germicidal potency of aqueous chlorine are almost wholly due to nonionized HOCl. Thus, as the HOCl-OCl^- distribution changes with pH, so also does the reactivity of a given concentration of aqueous chlorine.

Table 2.2 presents data for the distribution of the active chlorine species, Cl_2, HOCl and OCl^-, in a number of typical natural waters to which normal doses of chlorine (<10 mg/L) have been added. It is apparent that the fraction of aqueous chlorine present as Cl_2 in chlorinated natural waters is generally less than one in 10^6, except in seawater, and that the fractions as HOCl and OCl^- can vary extensively, depending on the type of water.

In addition to elemental chlorine, bleach solution (5–15% NaOCl solution) and HTH [solid $Ca(OCl)_2$] are used to produce aqueous chlorine in water treatment. These substances when dispersed in water yield directly the hypochlorite ion OCl^-, which then reacts by the reverse of Eqs. 2.4 and 2.1 to give the same equilibrium mixture of active species as that produced by elemental chlorine.

One difference between the sources of aqueous chlorine is that Cl_2, in reacting to form the equilibrium system, produces H^+ which tends to lower the pH, whereas the hypochlorites use up H^+ as they react toward equilibrium and tend to increase the pH. With the usual doses of aqueous chlorine (<10 mg/L) employed in water treatment, the buffering effect of the water alkalinity is generally strong enough to make any pH shift quite small. In any

TABLE 2.2
Distribution of Active Chlorine Species in Chlorination of Typical Natural Waters[a]

Type Water	Upland Stream	Reservoir	Large River	Underground	Sea
Temperature (°C)	10.0	20.0	15.0	10.0	5.0
pH	6.0	7.0	7.3	8.0	7.8
Ionic strength	0.001	0.003	0.012	0.01	0.6
$[Cl^-]$	0.0001	0.001	0.004	0.0001	0.5
Fraction HOCl	0.980	0.782	0.656	0.306	0.263
Fraction OCl^-	0.020	0.218	0.344	0.694	0.737
Fraction $Cl_2 \times 10^6$	0.41	0.22	0.42	0.0012	4.5

[a] Chlorine dosage <10 mg/L.

event, however, the equilibrium distribution achieved corresponds to the final pH of the water. When the pH is the same, any of the sources of aqueous chlorine give solutions with the same reactivity and germicidal potency.

2.2.2   Reactivity of Aqueous Chlorine

The characteristic reactions of aqueous chlorine are redox processes. All of the aqueous chlorine species are good oxidizing agents. Their electrochemical half-reactions and standard potentials are as follows:

$$Cl_2 + 2e = 2Cl^-; \qquad E^\circ_{298} = 1.36 \text{ V}$$
$$HOCl + H^+ + 2e = Cl^- + H_2O; \qquad E^\circ_{298} = 1.49 \text{ V}$$
$$OCl^- + H_2O + 2e = Cl^- + 2OH^-; \qquad E^\circ_{298} = 0.66 \text{ V}$$

Chloride ion is the reduced product for all these redox processes. So, these reactions do not lead to chlorine substitution. Almost all inorganic oxidations follow this pathway for reduction of the chlorine.

The aqueous chlorine species may also act as agents for the substitution or addition of chlorine. The chlorinating agents are then classed as electrophiles, reacting in electrophilic substitution with nucleophilic centers like reduced carbon or nitrogen atoms. The chlorine atom (or one of them for Cl_2) takes on partially the characteristics of Cl^+ and combines with an electron pair in the nucleophilic substrate.

Which of the aqueous chlorine species is the directly reacting substance in a particular situation depends on two factors, the relative specific activity of the species and its fractional abundance in the solution (6). Of the aqueous chlorine species considered, Cl_2 is regarded generally as the most electrophilic or as having the greatest relative specific activity. Nonionized HOCl is next in activity and OCl^- is much the least electrophilic. One other

Table 2.3
Estimated Net Reactivities of Forms of Aqueous Chlorine,
pH 7, 20°C

Species	Estimated Specific Reactivity	Fraction of Total Cl	Net Relative Reactivity
Cl_2	10^3	0.22×10^{-6}	0.0002
$HOCl$	1.0	0.78	0.78
OCl^-	10^{-3}	0.22	0.0002
H_2OCl^+	10^5	10^{-8}	0.001

aqueous chlorine species, H_2OCl^+, which can form from HOCl and H^+ in acid solutions, is even more reactive than Cl_2. Formation of this last species at pH < 5 accounts for the acid catalysis of many aqueous chlorination reactions.

The net reactivity of a particular species is the product of its relative specific activity and its fractional abundance. Unfortunately, relative specific activities vary depending on the substrate with which the reaction is occurring, so that no widely applicable quantitative values can be established. However, when similar types of reaction are involved, relative specific reactivities often maintain themselves to order-of-magnitude precision so that approximate extrapolation to different substrates is feasible.

Table 2.3 presents a sample evaluation of net reactivities for aqueous chlorine species in the reservoir water of Table 2.2

Clearly, HOCl must be regarded as being almost exclusively responsible for the reactivity of dilute ($<10^{-3}$ M) solutions of aqueous chlorine at pH values between 5 and 9. So, the types of reactions in organic solvents, where Cl_2 is predominant and free radical mechanisms are common, are not to be expected in dilute aqueous solution. Instead the reactions and reaction patterns characteristic of the electrophilicity of HOCl are to be anticipated.

2.2.3   Reactions with Some Inorganic Substances

Either the chlorine atom or the oxygen atom may act as the center of electrophilicity for reaction of HOCl with inorganic compounds. For example, Anbar and Taube (7) showed that in the oxidation of nitrite there was direct transfer of oxygen from HOCl in accordance with the equation

$$ClOH + NO_2^- \rightarrow Cl^- + HONO_2 \qquad (2.7)$$

The oxidation of sulfite seemed also to proceed similarly (8) at least in major part, as shown by the equation

$$ClOH + SO_3^= \rightarrow Cl^- + HOSO_3^- \qquad (2.8)$$

Another example of this type of mechanism is the oxidation of bromide, which probably occurs according to the reaction

$$ClOH + Br^- \rightarrow Cl^- + HOBr \qquad (2.9)$$

Since HOBr is nearly as strong an oxidizing agent as HOCl, it responds equally to most of the analytical methods for measuring aqueous chlorine. So, the occurrence of this reaction is not detected in normal water chlorination procedures. Nevertheless it has important consequences.

The oxidation of bromide, reaction 2.9, is quite rapid in neutral and acidic solutions, so rapid that oxidation to HOBr usually takes precedence over reactions with ammonia or organic compounds under such conditions. A result is the occurrence of bromination reactions in place of or in addition to chlorination reactions. Aqueous bromine and chlorine react similarly, but bromination reactions are usually faster than those of chlorination. For simple oxidation processes catalysis is observed in the presence of bromide; for substitution processes bromo derivatives are formed preferentially to chloro derivatives.

The bromide concentrations of most waters are quite low, so that the formation of HOBr is not significant. However, many groundwaters contain appreciable concentrations of bromide; river waters polluted with synthetic chemical wastes have been found to contain several tenths of a milligram per liter of bromide and seawater contains 60 mg/L. Formation of HOBr is often the predominant aqueous chlorine reaction in seawater and saline estuarine waters.

2.2.4   Chloramine Formation: The Breakpoint Reaction

In contrast to the previous examples, reactions of HOCl with ammonia, amino-nitrogen, and nitrogen in many heterocyclic ring systems proceed with the chlorine atom as the electrophilic center. The chlorine takes on partially the characteristics of Cl^+ and combines with an electron pair in the substrate forming a chloro derivative (9). With ammonia the initial reaction (10) is

$$NH_3 + HOCl \rightarrow NH_2Cl + H_2O \qquad (2.10)$$

The product of the reaction, chloramine or monochloramine, retains the oxidizing capacity of the initial HOCl, but the oxidizing power is considerably less. Also, the NH_2Cl has some germicidal activity, but the potency toward most organisms is one to two orders of magnitude less than that of HOCl.

Because chloramine and other N-chlor compounds oxidize KI to I_2 and respond to other analytical reagents for aqueous chlorine, they are lumped together with aqueous chlorine as available chlorine by routine oxidation methods. When differentiating determinations have been made, the aqueous chlorine is called free available chlorine, whereas the chlorine bound to nitrogen is termed combined available chlorine. This distinction is important for any detailed consideration of aqueous chlorination reactions.

Additional electrophilic chlorine substitution on chloramine occurs in the presence of excess chlorine or in acid solutions to give dichloramine, $NHCl_2$, and possibly nitrogen trichloride, NCl_3, in accord with the equations (11)

$$HOCl + NH_2Cl \rightarrow NHCl_2 + H_2O \qquad (2.11)$$

$$HOCl + NHCl_2 \rightarrow NCl_3 + H_2O \qquad (2.12)$$

Nitrogen trichloride is an undesired substance. It imparts obnoxious taste and odor to water, it is a strong eye irritant in bathing pools, and it has toxic neurological properties. Its formation in water treatment is avoided.

Dichloramine is chiefly important because of its instability. In neutral and alkaline solutions it decomposes according to the stoichiometric equation

$$2NHCl_2 + H_2O \rightarrow N_2 + 3H^+ + 3Cl^- + HOCl \qquad (2.13)$$

by which the ammonia-nitrogen is oxidized to elemental nitrogen and equivalent available chlorine is reduced to chloride. This reaction is the governing step in the so-called breakpoint reaction (12).

When varied doses of aqueous chlorine are added to separate portions of an aqueous system with pH 6–8.5 containing a set concentration of ammonia-nitrogen, the mixtures are allowed to stand for a few hours and then the residual available chlorine in each sample is determined, an interesting reaction pattern, shown in Figure 2.1, is found. As long as the added molar concentration of aqueous chlorine is less than that of the ammonia present, the found residual available chlorine is nearly equal to the concentration added, although it is combined chlorine rather than free chlorine. With increased added chlorine greater than a molar ratio of unity to the ammonia, the residual available chlorine decreases sharply until, at a molar ratio equal to about 1.6 all the added chlorine is found to be reduced. At this ratio, called the breakpoint, all the ammonia has disappeared also, having been oxidized predominantly to N_2, with a small fraction to NO_3^- (11).

Clearly the breakpoint is produced by the decomposition of $NHCl_2$, with Eq. 2.13 being the net stoichiometric reaction. As long as the molar ratio of chlorine to ammonia is less than unity, substantially only NH_2Cl is formed and little decrease in available chlorine is observed. However, with molar ratios in excess of unity, the excess forms $NHCl_2$ and loss of available chlorine occurs in proportion as the ratio exceeds unity. Direct reaction between NH_2Cl and $NHCl_2$ is not involved, for the presence of NH_2Cl has no effect on the rate or degree of loss of available chlorine from $NHCl_2$ solutions (13).

At ratios of aqueous chlorine to ammonia greater than that of the breakpoint, the excess chlorine over that required for the breakpoint remains as free available chlorine.

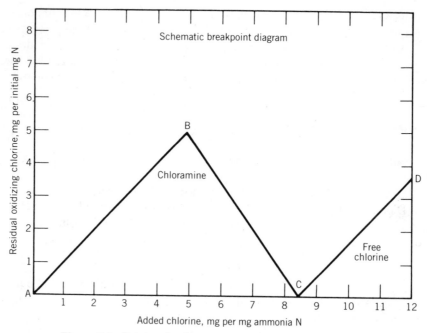

Figure 2.1. Schematic diagram of the breakpoint reaction.

2.2.5 Reactions with Organic Compounds

When HOCl reacts with organic compounds, either the oxygen or the chlorine may serve as the center of electrophilic action. Attack with oxygen results in formation of chloride ion plus an oxidized organic molecule. Attack by chlorine results initially in the formation of a chlorinated organic derivative. Subsequent additional reaction or hydrolysis may give chloride ion in this latter case also. In any event the chlorine is reduced to a conventional -1 oxidation state, for chlorine bound to carbon no longer has oxidizing properties.

This means that the total reaction of aqueous chlorine with organic matter, in the absence of inorganic reducing agents, is given by the decrease in total oxidizing capacity of the added aqueous chlorine. This decrease is termed the chlorine demand, defined by the relation

$$(\text{Cl demand}) = (\text{aqueous Cl added}) - (\text{residual aqueous Cl}) \quad (2.14)$$

Although chlorine demand is not a fixed quantity—it varies with time of contact, temperature, pH, and excess of aqueous chlorine, for example—it is still a useful parameter for control of chlorination in water treatment. Its significance here, however, arises from the further relation:

$$(\text{Cl demand}) = (\text{Cl}^- \text{ increase}) + (\text{organic Cl formed}) \quad (2.15)$$

This relation indicates that the total concentration of chlorine in chlorinated organic compounds formed during chlorination will be some fraction of the chlorine demand. Were it possible to measure the increase in Cl^- (not generally feasible at present), then the total chlorine converted into chlorinated organic compounds could be specified quantitatively.

Direct determinations of total organically bound chlorine could of course give the same result. Some measurements have been made (14,15), but data are relatively scanty and often have not been compared with chlorine demands. What information there is indicates that 80–90% of the reduced chlorine reacts to give chloride ion and that 10–20% forms chlorinated organic compounds of some sort (16,17,18).

The number of organic compounds occurring in natural waters is very large, as pointed out in Chapter 1, and identification of them is far from complete. Therefore it is impossible to describe specifically all of the reactions that may occur. Details of some types of reaction that are expected to be important factors in the overall aqueous chlorination process are given in a later section.

2.2.6 Reactions of Chloramine

Although chloramines retain the ability to oxidize iodide to iodine, they are much weaker oxidants and electrophiles than HOCl. For example, reaction with NO_2^- and Br^- occurs only very slowly. Much less "chlorine demand" is exhibited with chloramines as compared with free aqueous chlorine, and direct electrophilic substitution of chlorine on carbon to form chlorinated organic compounds such as chloroform is not known to occur with chloramine.

2.2.7 Reactions with Nitrogenous Organic Compounds

Nitrogenous organic compounds, including amines, amino acids, pyrimidines, purines, and other heterocyclic compounds, undergo special reactions with aqueous chlorine. The initial step is usually substitution at a nitrogen atom, as with ammonia, typified by the general equation

$$R_1R_2CHNH_2 + HOCl \rightarrow R_1R_2CHNHCl + H_2O \qquad (2.16)$$

Such an initial step is often followed, however, by dehydrochlorination and hydrolysis to yield a carbonyl compound and ammonia, which proceeds through the breakpoint reaction or by additional chlorination, dehydrochlorination, and decarboxylation to give a nitrile or by rearrangement to give chlorine attached to carbon. Whatever the exact processes, reaction with naturally occurring nitrogenous compounds appears to yield both substantial chloride ion and substantial chlorinated organic carbon. Thus, the nitrogenous organic portion of the natural organic matter in water may constitute one of the significant sources of chlorinated organic compounds (19,20).

2.3 TREATMENT OF WATER SUPPLIES

2.3.1 Disinfection by Chlorine

The primary reason for the use of aqueous chlorine for treatment of community water supplies is its potent germicidal activity, which enables it to kill all types of water-borne pathogenic organisms and produce a hygienically safe water at ppm concentrations. For more than 70 years now, the application of aqueous chlorine in sufficient concentration to satisfy the chlorine demand of the water and leave a residual aqueous chlorine of 0.2 mg/L or more has been found effective in preventing the spread of water-borne disease. And, for the past 45 years the use of the breakpoint reaction to accomplish disinfection initially with a relatively great concentration of HOCl and to leave, after extended contact time, an assured residual free aqueous chlorine concentration of a few tenths of a milligram per liter has been recognized as the most effective of chlorination procedures for disinfection (21).

The fundamental equation for the exertion of germicidal activity can be expressed in the form

$$\log_{10}(N_t/N_0) = -kC^n t \qquad (2.17)$$

where N_0 is the initial concentration of microorganisms of a particular type, N_t is the viable concentration remaining after time t, and C is the concentration of germicide. The parameter k then reflects the germicidal potency. The concentration exponent n can often be taken as unity, at least within a limited range (22,23).

The Safe Drinking Water Committee of the National Academy of Sciences has correlated germicidal powers on the basis of the product Ct_{99}, where t is the time in minutes to kill 99% of the organisms at the disinfectant concentration C in milligrams per liter. This product relates to Eq. 2.18 as follows:

$$\log_{10}(N_t/N_0) = -2 = -kCt_{99}$$

$$Ct_{99} = 2/k \qquad (2.18)$$

with n being taken as unity. Because the Ct_{99} products are inversely proportional to k, values are smaller the greater the germicidal potency.

Table 2.4 presents a listing of Ct_{99} values for the action of the components of aqueous chlorine, chloramine, and some alternate disinfectants against most of the type of microorganisms that are causes of water-borne diseases. The listed values are derived primarily from the compilation made by the Safe Drinking Water Committee (24). The effectiveness of HOCl against all types of organisms is shown clearly. Useful comparison with the other major disinfectants O_3 and ClO_2 is also possible.

The Ct_{99} values may be interpreted as the concentrations that will give 99% kill in 1 min. Because most water treatment facilities provide a minimum of 30 min for disinfection, the concentrations thus derived are

TABLE 2.4
**Values of Ct_{99} for Destruction of Classes of Microorganisms
by Several Oxidizing Disinfectants**

Organism	Agent[a]					
	HOCl	OCl$^-$	NH$_2$Cl	O$_3$	ClO$_2$	HOBr
Escherichia coli	0.02	1.0	50	0.005	0.2	0.5
Poliovirus 1	1.0	10	500	0.005	1.5	0.06
Entamoeba histolytica	20	8000	150	2	5	40

[a] pH 7, 20°C.

about a factor of 10 greater than the concentrations theoretically needed for a practical degree of disinfection.

There are, in fact, two types of disinfection involved in the treatment of water supplies. The first is the primary disinfection step where sufficient agent is used actively to kill any pathogenic organisms. Table 2.4 shows that only HOCl, O$_3$, and ClO$_2$ are sufficiently potent against all forms of pathogens to serve generally as primary disinfectants. The species NH$_2$Cl and OCl$^-$ are both too weak, particularly against water-borne viral pathogens, to fill a primary role adequately. The other type is residual disinfection, the maintenance of a small concentration of germicide throughout the distribution system to preserve the hygienicity of the water. Chloramine, as well as HOCl and ClO$_2$, has been found suitable for this purpose.

The fact that chloramine does not react to form chloroform or other chlorinated organic compounds has led to the development of a water chlorination technique in which free aqueous chlorine is applied as primary disinfectant for an hour or so and then ammonia is added to form chloramine as residual disinfectant and inhibit additional formation of haloforms in the distribution system. Such a technique has been found to arrest haloform production and presumably the formation of other chlorinated organic compounds at the point of introduction of the ammonia. Substantial percentage reductions in the concentrations of haloform delivered to consumers have been achieved in this way (25,26).

2.3.2 Other Biological Applications of Chlorination

Aqueous chlorine is active biocidally against other microorganisms than human pathogens. Saprophytic bacteria, slime bacteria and molds, algae, and fungi are all susceptible to its lethal actions. Because growths of many of these types of organisms are prone to occur on the walls of transmission mains and treatment basins, aqueous chlorine has had large-scale use for control of these nuisance organisms.

For example, doses of a few milligrams per liter have been added, in what is called transport chlorination, at the point of abstraction from the source, to maintain cleanliness and high flow in the transmission main to the treatment plant. This type of application also provides maximum contact time for disinfection, of course. More frequently, so-called prechlorination is practiced by adding chlorine or hypochlorite to the water, often in a dosage sufficient to bring about the breakpoint reaction, as the water enters the treatment plant. The residual aqueous chlorine then controls the growth of nuisance organisms in coagulation and flocculation chambers, in settling basins, and in the filtration beds. The prechlorination stage may also serve as the primary disinfectant step, in which case a few hours is usually available for accomplishing disinfection, or a second dose, specifically for disinfection, may be added directly following filtration.

Similarly, the residual, safety chlorine that is maintained in the distribution system for hygienic reasons serves also to repress the growth of nuisance microorganisms there. In the absence of a biocide, odor-releasing bacteria frequently develop on the walls of the water mains as do strands of filamentous organisms that may occasionally dislodge and cause unsightly sediments at consumers' taps.

One of the problems with the use of chloramine for residual disinfection is that it appears to be a less effective suppressant of nuisance growths than is free aqueous chlorine. In some instances operators of plants that have instituted use of chloramine for residual disinfection have found that, after a few years, false positive coliform tests have begun occurring for distributed water at the consumers' taps. Because these positive results are obtained even from water with 1–2 mg/L of residual chlorine, it has been concluded that the coliforms are protected by being imbedded in filamentous clumps that have been able to grow on the walls in the presence of chloramine and then have become detached.

2.3.3 Taste and Odor Control by Aqueous Chlorine

Another function of chlorination in water treatment is the reduction of unpleasant tastes and odors. There are two major sources of natural olfactory substances in water sources: odorous secretions of many species of algae and actinomycetal odors leached from decaying vegetation. Many, though not all, of these substances are susceptible to oxidation to nonodorous or less odorous substances by aqueous chlorine. Ordinarily, the prechlorination step is used to accomplish this taste and odor beneficiation. Breakpoint reaction conditions may be needed during prechlorination for maximum effect. Chloramine is generally ineffective in reducing these tastes and odors.

Another source of tastes and odors is the presence of phenolic substances. Some of these may be derived naturally from decaying forest vegetation, particularly oak and beech leaves, but the major source is industrial pollution. In general, the phenols themselves do not give objectionable tastes and odors. However, when they react with aqueous chlorine, the first products of reaction—the chlorophenols and dichlorophenols—have offensive odors that are perceptible at concentrations as small as 1 μg/L. Excess free aqueous chlorine, such as that in breakpoint chlorination, oxidizes these initial odorous products further to nonodorants. Chloramine is not a strong enough oxidant to eliminate these odors.

2.3.4 Use of Aqueous Chlorine as Oxidant

One other frequent function of the chlorination process is the oxidation of organically bound iron and manganese, along with some oxidation of the colored fulvic and humic acids themselves, to improve the coagulation process. Aqueous iron and manganese are stabilized against simple oxidation by complexation with the fulvates and humates that impart natural color to water. At the same time the complexation tends to stabilize the organic matter against coagulation with filter alum. Oxidative treatment with aqueous chlorine prior to coagulation, that is, in the prechlorination stage, releases the iron and manganese from the organic complexes so that they can be removed during coagulation and conditions the organic matter so that it reacts more readily with the added coagulant. The result is a lessening of the required coagulant dosage and an improvement in the fractional removal of color. Improved clarity of filtered water may also occur.

2.3.5 Summary of Chlorination Practice

As the previous sections have shown, chlorination has numerous functions in water treatment besides disinfection, even though this last is the most important. Aqueous chlorine acts as an oxidant and aid to coagulation, it serves to control tastes and odors in the treated water, it suppresses unwanted biological growths in transmission mains, treatment tank and filters, and it maintains the hygienicity of the treated water in the distribution system.

Chlorination must take place as the raw water enters the treatment plant or soon thereafter to fulfill many of these functions. Enough chlorine or hypochlorite must be added to satisfy the chlorine demand, to carry through the breakpoint reaction, to react with the microorganisms, and to leave a sufficient residual for protection of the water in the distribution system. Moreover, most of the functions require free aqueous chlorine rather than chloramine.

Chlorine demands of waters, which usually consume most of the added chlorine, are roughly related to the color or the total organic content of the water. In addition, the breakpoint reaction consumes about 10 parts of chlorine for each part of nitrogenous content. Natural waters with much color or organic content, or sometimes those with high nitrogenous content, may require application of 10–20 mg/L of chlorine to meet all the conditions of the previous paragraph. Often this total required dosage is split among two or three points of application, for example, a strong prechlorination dose, a booster dose primarily to enhance disinfection immediately following filtration, and a final adjusting dose as the water enters the distribution system. Such multiple application does not affect greatly the total amount of chlorine used, but does permit more precise control of the chlorine concentration at various points.

Since the discovery of the production of haloforms during the chlorination process, variations in this standard procedure have been introduced in an attempt to minimize haloform formation while retaining as many of the benefits of chlorination as possible. The principal modification that has been found helpful is to move the point of strong initial application of chlorine from the entrance of the plant to a point following the coagulation and flocculation stages. Sometimes the chlorine is withheld until after sedimentation has occurred.

In general, the performance of the coagulation-flocculation treatment of water reduces the color and the organic content by 50–60%. The chlorine demand and total trihalomethane formation potential (TTHMFP) are also reduced to approximately the same degree. Accordingly, moving the point of application of chlorine beyond the flocculation stage has usually resulted in halving the amount of haloform produced (27,28). It appears that once the organic matter is tied up in alum flocs, it is no longer readily attacked by aqueous chlorine, so that delay beyond the sedimentation stage is not important.

Postponement of chlorine treatment in this way means giving up some of the benefits of chlorination. Its function of oxidative assistance in coagulation is lost, as is that of inhibiting biological growths in the flocculation basins. So, larger doses of coagulant may have to be used and basins may have to be cleaned more frequently. Moreover, elimination of odorants and completion of the breakpoint may be impaired because of the shortened reaction time available as a result of later application of the chlorine. So, the introduction of modifications needs to be considered in terms of their total impact, not just with regard to their effect on one aspect of the treatment process.

One other modification, already mentioned, to reduce production of haloforms has been the application of ammonia to convert residual free aqueous chlorine to chloramine as the treated water enters the distribution system. Depending on the time of travel in the distribution system and on the magnitude of the previous residual chlorine, reductions of 25–50% in haloforms found at consumers' taps have been experienced with this modification. Difficulties with it have been described earlier (25,26).

A combination of the two modifications has enabled some water utilities to deliver water with about a third of the haloform concentration that was being produced before, bringing the haloform level well within the present U.S. limit of 0.1 mg/L. It is doubtful, however, that such simple modifications would be enough to enable many treated supplies to comply with the recent guideline value of 30 μg/L $CHCl_3$ promulgated by the World Health Organization (29). Also, the effect of these changes on the formation of other chlorinated compounds is not known.

2.4 SOME SPECIFIC CHLORINATION REACTIONS

2.4.1 Reaction at Carbon–Carbon Double Bonds

The carbon–carbon double bond is electron rich and so provides a site for electrophilic attack. A type equation for the overall reaction is

$$R_1 - \overset{\overset{\displaystyle H}{|}}{C} = \overset{\overset{\displaystyle H}{|}}{C} - R_4 + HOCl \rightarrow R_1 - \overset{\overset{\displaystyle H}{|}}{\underset{\underset{\displaystyle OH}{|}}{C}} - \overset{\overset{\displaystyle H}{|}}{\underset{\underset{\displaystyle Cl}{|}}{C}} - R_4 \qquad (2.19)$$

The CHOH group in this type of addition product is then subject to oxidation to a carbonyl group followed by additional chlorination *alpha* to the carbonyl group.

The mechanism of the reaction is not the simple addition reaction shown in Eq. 2.19, however. There is, first, the electrophilic addition of Cl^+ at one end of the double bond. Then the other end of the broken double bond, which has now itself become electrophilic, adds on a nucleophile such as OH^- or Cl^- from the solvent. So, a dichloro derivative can be obtained even though the reactant is HOCl (30).

Reaction at a single double bond probably is not very important in water chlorination, for the double bond is not strongly nucleophilic. Nucleophilicity is increased, however, in a conjugated system of several double bonds.

Oliver and Thurman (31) have suggested that the chromophores in fulvic and humic acids are conjugated double-bond systems with at least four double bonds and that these react with aqueous chlorine as an initial step in the formation of haloforms and other chlorinated organic compounds. They note that color centers in the humic molecules rapidly react with chlorine and fade with a half-life of 1–4 h.

2.4.2 Reactions with the Aromatic Ring

The typical initial reaction of HOCl with the aromatic ring is an electrophilic addition of Cl^+ at a relatively nucleophilic carbon followed by H^+ elimination at the same site. However, benzene and most benzenoid hydrocarbons require a more active electrophile than HOCl, such as H_2OCl^+, for measurable chlorination at ambient temperature. The ring must be "activated" by an appropriate substituted group, such as the OH^- group, in order for facile chlorination to occur (32) with HOCl.

It has been known for a long time that dilute aqueous chlorine reacts with phenolic compounds to form chlorinated derivatives. As early as 1926 Soper and Smith (33) had shown that the reaction was one of electrophilic attack by HOCl on the phenoxide ion, which is much more nucleophilic than the nonionized phenol. Initial substitution is at the positions *ortho* or *para* to the hydroxyl group, for these are the activated sites in the ring. Chlorination continues, after the initial substitution, with approximately the same facility until all open *ortho* and *para* sites have been chlorinated (33).

The accepted detailed mechanism of the reaction is shown by Eq. 2.20.

$$^-O\!\!-\!\!\bigcirc + ClOH \rightarrow O=\!\!\bigcirc\!\!\diagdown{}^H_{Cl} \rightarrow {}^-O\bigcirc Cl + H^+ \quad (2.20)$$

Addition of Cl^+ gives a transitory, quinoid intermediate that decomposes promptly to yield the chlorinated product.

Some rate studies by Lee and Morris (34) have indicated that, at pH 8 and 25°C with 10^{-4} M excess aqueous chlorine, the 90% reaction time for phenol to give monochlorophenol is about 1.2 min.

When the *ortho* and *para* positions of phenolic substances have been fully substituted, further chlorination in dilute aqueous solution becomes difficult. Instead, oxidative rupture of the benzene ring occurs, yielding at first two-carbon residues and eventually CO_2 and H_2O (35,36). A preferred pathway of oxidation for phenol itself appears to be a hydrolytic and oxidative elimination of Cl from the *ortho* position to give an *o*-benzoquinone which then undergoes ring rupture at the unsaturated links. There appear to be no remaining chlorinated organic fragments after exhaustive chlorination of simple phenolic compounds.

Other classes of derivatives, including aromatic aldehydes and carboxylic acids may also be chlorinated in aqueous solutions (37). Aniline, substituted anilines, and anilides comprise another group of compounds that will undergo ring chlorination readily. None of these compounds, however, has been studied enough in detail to know what the ultimate products are in the presence of excess aqueous chlorine.

When there are two hydroxyl groups in *meta* positions on an aromatic ring, then the carbon atom between the hydroxyl substituted carbons is strongly activated and chlorinates very readily. Rook (38) has demonstrated the great reactivity toward aqueous chlorine of resorcinol, *m*-dihydroxybenzene, and a number of related compounds and has shown that their chlorination produces much chloroform as an ultimate product. With resorcinol the yield of chloroform was nearly 1 mol per mol of resorcinol. Boyce et al. (39) showed by ^{13}C study that the chloroform came entirely from the carbon between the *meta* hydroxyls.

Heterocyclic aromatic rings exhibit great variations in reactivity toward chlorine substitution as compared with benzene. Pyridine is greatly deactivated toward electrophilic attack, so that reaction with aqueous chlorine is not expected. The *alpha* hydrogens of pyrrole and that of indole, on the other hand, are quite activated, so that electrophilic chlorine substitution is likely. In accord with this, Baum and Morris (19) found that the amino acids, proline and tryptophane, were sources of chloroform when chlorinated and that chlorophyll, which also contains the pyrrole ring, is another prospective precursor of chloroform. The pyrimidine ring as well was found to undergo chlorination. Reduction of chlorine occurred at pH 6.6 with liberation of chloroform when the pH was raised to 9 or greater.

In general, therefore, aromatic rings, when activated, offer opportunities for the formation of carbon–chlorine bonds. When excess aqueous chlorine is used, the aromatic rings are normally split. Chlorinated aromatic compounds have not been found among the products of such studies.

When simple chlorinated phenols are split, the fragments or ultimate organic products appear to contain no chlorine. However, *meta*-dihydroxybenzene structures, pyrrolic compounds, and pyrimidine derivates are all prospective sources of haloforms.

2.4.3   The Haloform Reaction

The alkaline reaction of aqueous halogens or hypohalites with methyl ketones or compounds oxidizable to methyl ketones to yield haloform is a classic reaction of organic chemistry that has been known since 1822 (40). Although both acid-catalyzed and base-catalyzed forms of the reaction are known, the base-catalyzed reaction pattern is predominant at pH > 5 and so is the one important for water chemistry.

Reactions occur by successive replacement of the hydrogens of the methyl group *alpha* to the carbon with chlorine or other halogen atoms followed by hydrolysis to produce haloform and, generally, a carboxylate. The hydrogens *alpha* to a carbonyl group are activated in much the same way as the *ortho* and *para* hydrogens of phenol (41).

The accepted mechanism for the reaction is shown in Figure 2.2, using acetone as an example. Initially a proton dissociates slowly from the methyl group to yield a carbanion, which is then subject to facile electrophilic attack by HOCl or other halogen. This process is then repeated, more readily, for the second and third hydrogens. Finally a base-catalyzed hydrolysis occurs to yield the haloform.

The reaction of aqueous chlorine with acetone or other simple methyl ketones is too slow at pH 5–9 and 10^{-4} M concentration, however, to account for the formation of chloroform in water chlorination. Some more active structure is required, probably a structure involving a more acidic carbon-hydrogen linkage than those of the CH_3 group in methyl ketones.

Acidic carbon-hydrogen linkages like this are found in methylene groups between two carbonyl groups, such as that in acetylacetone, $CH_3COCH_2COCH_3$. Rook (38) investigated three such compounds: in-

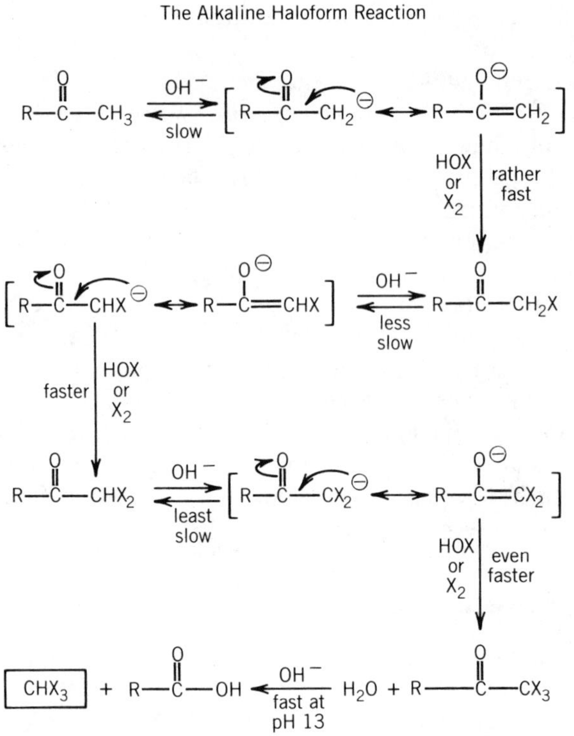

The Alkaline Haloform Reaction

Figure 2.2. Mechanism for the alkaline haloform reaction.

danedione, cyclohexanedione-1,3 and 5,5-dimethylcyclohexanedione-1,3. All of them yielded substantial amounts of chloroform, 0.38–0.52 mol per mol of substrate when allowed to react with 6–8 mM aqueous chlorine at pH 7.8 and 10°C for 4 h. It appears that this diketone structure, either originally present in natural organic matter or formed as a result of oxidative chlorination reactions, is an important source of the haloforms produced in water chlorination. It is difficult to be certain, however, so long as the specific structures of natural humates and fulvates are not known.

Recent studies (42–44) have indicated that trichloroacetic acid and dichloroacetic acid may be as abundant products of the chlorination of fulvate as chloroform is. Whether these acids arise from the same precursor structures as chloroform by parallel or competitive pathways or whether they come from different precursors is not known. There is some indication in the results at different pH values that trichloroacetic acid formation is competitive with chloroform production. This conclusion is also consistent with the results of some preozonation studies. On the other hand the same preozonation studies suggest that dichloroacetic acid is produced from a different precursor structure. All three compounds seem to form similarly as a function of contact time or chlorine concentration.

2.4.4   Reaction with Amino Acids

Amino acids and other nitrogenous compounds may chlorinate either at nitrogen or at carbon. Usually amino-nitrogen is more reactive than carbon so that first reaction occurs at the amino group; amido-nitrogen is not so reactive, however. So it is not always clear where reaction will take place with compounds containing both amido nitrogen and active carbons, such as the pyrimidines and the pyrroles. Both these structures are potential precursors for haloform (19); so either chlorine substitution on carbon occurs directly or there is migration of chlorine from nitrogen to carbon. Incidentally, the results of Morris and Baum (19) with regard to chlorine demand of solutions of uracil, proline, hydroxyproline, and tryptophane indicated that there was eventually complete oxidation of these compounds by aqueous chlorine to carbon dioxide, nitrogen, and water except for 1 mol of chloroform per mol of compound.

There are two principal modes of reaction of simple amino acids with aqueous chlorine. The first involves formation of mono- and dichloramino derivatives, followed by hydrolysis at the C–N bond to give a carbonyl compound plus dichloramine, which then undergoes the breakpoint reaction (45). The type series of reactions is

$$
\begin{array}{c}
\text{H} \\
| \\
\text{R—C—COOH} + 2\text{HOCl} \rightarrow \text{R—C—COOH} + 2\text{H}_2\text{O} \\
| \\
\text{NH}_2
\end{array}
\qquad (2.21)
$$

$$
\begin{array}{c}
\text{H} \\
| \\
\text{R—C—COOH} + \text{HOCl} \rightarrow \text{R—C—COOH} + \text{NHCl}_2 + \text{H}^+ + \text{Cl}^- \\
| \\
\text{NCl}_2
\end{array}
\qquad (2.22)
$$

$$
2\text{NHCl}_2 + 2\text{H}_2\text{O} \rightarrow \text{N}_2 + 3\text{H}^+ + 3\text{Cl}^- + \text{HOCl}
\qquad (2.23)
$$

The fate of the carbonaceous residue depends on the initial amino acid. With alanine the residue is pyruvic acid, a potential precursor of chloroform. With glycine the residue is glyoxalic acid, which oxidizes to carbon dioxide and water.

In the second mode of reaction a decarboxylation follows chloramination rather than a splitting of the C–N bond. The result is ultimately formation of a nitrile. A summarized type reaction sequence is

$$
\begin{array}{c}
\text{H} \\
| \\
\text{R—C—COOH} + 2\text{HOCl} \rightarrow \text{R—C—COOH} + 2\text{H}_2\text{O} \\
| \\
\text{NH}_2
\end{array}
\qquad (2.24)
$$

$$R - \overset{\overset{\displaystyle H}{|}}{\underset{\underset{\displaystyle NCl_2}{|}}{C}} - COOH \rightarrow R - \overset{\overset{\displaystyle }{\|}}{\underset{\underset{\displaystyle NCl}{\|}}{C}} - COOH + H^+ + Cl^- \qquad (2.25)$$

$$R - \overset{\|}{\underset{NCl}{C}} - COOH \rightarrow R - C \equiv N + CO_2 + H^+ + Cl^- \qquad (2.26)$$

Already in 1955, Culver (46) had shown that ClCN was formed by this type of mechanism in the chlorination of glycine in mildly alkaline solutions. More recently Trehy and Bieber (47) have found dihaloacetonitriles as products in the chlorination of natural waters and have concluded that alanine is the precursor, based upon a similar reaction sequence to that shown. It is clear that amino acids and other nitrogenous substances may serve as sources for a variety of chlorinated organic compounds.

2.5 SUMMARY

It is clear that there is a variety of structures that can give rise to chlorinated organic compounds upon reaction with aqueous chlorine. All of these represent partially oxidized organic compounds, because of double bonds, aromaticity, or attached hydroxyl, carbonyl, amino, or similar groups. It is understandable, then, that preoxidative treatments, of whatever sort, may have varied effects on the ability and capacity of organic matter to form chlorinated organic derivatives.

One universal feature of all the reactions to form chlorinated organic compounds is that they consist of electrophilic attack by HOCl on an active carbon or carbanion center. In the search for precursors of chlorinated organic compounds formed in aqueous chlorination, attention must be focussed on structures containing such active centers.

It appears that the so-called haloform reaction in its broadest sense is responsible for a major part of the halogenated organic matter produced in the chlorination of water supplies. It must be remembered that the general definition of the haloform reaction covers not only the classic methyl ketone grouping, but also any structure that yields haloform ultimately with alkaline halogenation.

REFERENCES

1. Shilov, E.A., and Solodushenkov, S.M., The velocity of hydrolysis of chlorine, *J. Phys. Chem.* (USSR), *19*, 405–407, 1945.
2. Eigen, M., and Kustin, K., The kinetics of hydrolysis of Cl_2, Br_2, and I_2, *J. Am. Chem. Soc.*, *84*, 1355–1358, 1962.
3. Jakowkin, A.A., On the hydrolysis of chlorine, *Z. Physik. Chem.*, *29*, 613–657, 1899.

4. Connick, R.E., and Chia, Y., The hydrolysis of chlorine and its variation with temperature, *J. Am. Chem. Soc., 81,* 1280–1285, 1959.
5. Morris, J.C., The acid ionization constant of HOCl from 5 to 35°C, *J. Phys. Chem., 70,* 3798–3802, 1966.
6. Morris, J.C., The chemistry of aqueous chlorine in relation to water chlorination, in *Water Chlorination: Environmental Impact and Health Effects,* Vol. 1, Jolley, R.L., (ed.), Ann Arbor Science Publishers, Inc., 1978, pp. 21–33.
7. Anbar, M., and Taube, H., The exchange of hypochlorite and hypobromite ions with water, *J. Am. Chem. Soc., 80,* 1073–1079, 1958.
8. Halperin, J., and Taube, H., The reaction of halogenates with sulfite in aqueous solution, *J. Am. Chem. Soc., 74,* 375–379, 1952.
9. Morris, J.C., Kinetics of reactions between aqueous chlorine and nitrogenous compounds, in *Principles and Applications of Water Chemistry,* Faust, S.D., and Hunter, J.V., (eds.), John Wiley & Sons, Inc., New York, 1967, pp. 23–53.
10. Weil, I., and Morris, J.C., Kinetic studies on the chloramines. 1. The rates of formation of monochloramine, N-chloromethylamine and N-chlorodimethylamine, *J. Am. Chem. Soc., 71,* 1664–1670, 1949.
11. Palin, A.T., A study of the chloro derivatives of ammonia and related compounds, *Water and Water Engr., 54* (10), 151–200; (11), 189–200; (12), 248–258, 1950.
12. Wei, I.W., and Morris, J.C., Dynamics of breakpoint chlorination in *Chemistry of Water Supply, Treatment and Distribution,* Rubin, A.J. (ed.), Ann Arbor, MI; Ann Arbor Science Publ., Ann Arbor, MI, 1974, pp. 299–332.
13. Weil, I., Morris, J.C., and Culver, R.H., Kinetic studies on the breakpoint with ammonia and glycine, presented at Int'l Union for Pure and Appl. Chem. New York, NY, Sept. 20, 1952; see Morris, J.C., Reaction dynamics in water chlorination, in *Water Chlorination: Environmental Impacts and Health Effects,* Vol. 5, Jolley, R.L., et al. (eds.), Lewis Publishers. Inc., Chelsea, MI, 1985, pp. 701–712.
14. Brass, H.J., *J. Am. Water Works Assn., 74,* 107, 1982.
15. Cotruvo, J.A., *Environ. Sci. Technol., 15,* 268, 1981.
16. Helz, G.R., Dotson, D.A., and Sigleo, A.C., Chlorine demand: Studies concerning its chemical basis, in *Water Chlorination: Environmental Impact and Health Effects,* Vol. 4, Jolley, R.L., et al. (eds.), Ann Arbor Science Publ., Ann Arbor, MI, 1983, pp. 181–190.
17. Bean, R.L., Recent progress in the organic chemistry of water chlorination, in *Water Chlorination: Environmental Impact and Health Effects,* Vol. 4, Jolley, R.L., et al. (eds.), Ann Arbor Science Publ., Ann Arbor, MI, 1983, pp. 843–850.
18. Oliver, B.G., Chlorinated non-volatile organics produced by the reaction of chlorine with humic materials, *Can. Res. Dev., 6,* 21–22, 1978.
19. Morris, J.C., and Baum, B., Precursors and mechanisms of haloform formation in the chlorination of water supplies, in *Water Chlorination: Environmental Impacts and Health Effects,* Vol. 2, Jolley, R.L., Gorchev, H., and Hamilton, D.H. (eds.), Ann Arbor Science Publ., Inc., Ann Arbor, MI, 1977, pp. 29–48.
20. Ram, N.M., and Morris, J.C., Environmental significance of nitrogenous organic compounds in aquatic sources, *Environ. Int., 4,* 397–405, 1980.
21. Buelow, R.W., and Walton, G., Bacteriological quality versus residual chlorine, *J. Am. Water Works Assn. 63* (1), 28–35, 1971.
22. Morris, J.C., Disinfectant chemistry and biological activities, in *Disinfection,* Am. Soc. Civil Engrs., New York, 1971, pp. 609–633.
23. Hall, E.L., Quantitative assessment of disinfection interferences, *Water Treatment and Examination, 22,* 153–174, 1973.
24. *Drinking Water and Health,* Vol. 2, National Academy of Sciences, Washington, DC, 1980, pp. 5–138.
25. Sacks, R.S., Ann Arbor controls trihalomethanes, *J. Am. Water Works Assn., 76* (7), 105–108, 1984.

26. Norman T.S., et al., The use of chloramines to prevent trihalomethane formation, *J. Am. Water Works Assn., 71,* 87–90, 1979.
27. Babcock, D.B., and Singer, P.C., Chlorination and coagulation of humic and fulvic acids, *J. Am. Water Works Assn., 71,* 149–154, 1979.
28. Hoehn, R.C., Dixon, K.L., Malone, J.K., Novak, J.T., and Randall, C.W., Biologically induced variations in the nature and removability of THM precursors by alum treatment, *J. Am. Water Works Assn., 74* (4), 134–141, 1984.
29. World Health Organization, *Guidelines for Drinking Water Quality,* Vol. 1, Recommendations, WHO, Geneva, 1984, pp. 63, 76–77.
30. Craw, D.A., The kinetics of chlorohydrin formation. V. The reaction between hypochlorous acid and crotonic acid in buffered solutions at 25° and 35°, *J. Chem. Soc.,* 2510–2514, 1954; VI. The reaction between hypochlorous acid and tiglic acid at constant pH, *J. Chem. Soc.* 2515–2519, 1954.
31. Oliver, B.G., and Thurman, E.M., Influence of aquatic humic substance properties on trihalomethane potential, in *Water Chlorination: Environmental Impact and Health Effects,* Vol. 4, Jolley, R.L., et al. (eds), Ann Arbor Science Publ., Ann Arbor MI, 1983, pp. 231–241.
32. Sykes, P., A guidebook to mechanism in organic chemistry, John Wiley & Sons, New York, 1961, pp. 204–207.
33. Soper, F.G., and Smith, F.G., The halogenation of phenols, *J. Chem. Soc.,* 1582–1591, 1926.
34. Lee, G.F., and Morris, J.C., Kinetics of chlorination of phenol—chlorophenolic tastes and odors, *Int. J. Air Water Pollut., 6,* 419–431, 1962.
35. Ingols, R.S., and Ridenour, R.M., The elimination of phenolic tastes by chloro-oxidation, *Water and Sewage Works, 93,* 187–190, 1948.
36. Ingols, R.S., and Jacobs, G.M., BOD Reduction by chlorination of phenol and amino acids, *Sewage Ind. Wastes, 29,* 258–262, 1957.
37. Hopkins, C.Y., and Chisholm, M.J., Chlorination by aqueous sodium hypochlorite, *Can. J. Res., B24,* 208–210, 1946.
38. Rook, J.J., Chlorination reactions of fulvic acids in natural waters, *Environ. Sci. Technol., 11,* 478–482, 1977.
39. Boyce, S.D., Barefoot, A.C., Britton, D.R., and Hornig, J.F., Formation of trihalomethanes from the halogenation of 1,3-dihydroxybenzenes in dilute aqueous solution: Synthesis of 2-C^{13}-resorcinol and its reaction with chlorine and bromine, in *Water Chlorination: Environmental Impact and Health Effects,* Vol. 4, Jolley, R.L., et al. (eds.), Ann Arbor Science Publ., Ann Arbor, MI, 1983, pp. 253–268.
40. Fuson, R.C., and Bull, B.A., The haloform reaction, *Chem. Rev., 15,* 278–309, 1934.
41. Bartlett, P.D., and Vincent, J.R., The rate of the alkaline chlorination of ketones, *J. Am. Chem. Soc., 57,* 1596–1600, 1935.
42. Christman, R.F., et al., Identity and yields of major halogenated products of aquatic fulvic acid chlorination, *Environ. Sci. Technol., 17* (10), 625–631, 1983.
43. Miller, J.W., and Uden, P.C., Characterization of nonvolatile aqueous chlorination products of humic substances, *Environ. Sci. Technol., 17* (3), 150–155, 1983.
44. Reckhow, D.A., and Singer, P.C., The removal of organic halide precursors by preozonation and alum coagulation, *J. Am. Water Works Assn., 76* (4), 151–157, 1984.
45. Stanbro, W., and Smith, W., Kinetics and mechanism of the decomposition of N-chloroalanine in aqueous solution, *Environ. Sci. Technol., 13* (4) 446–451, 1979.
46. Culver, R.H., The reaction of chlorine with glycine in dilute aqueous solution, Ph.D. Dissertation, Harvard University, Cambridge, MA, 1955, 271 pp.
47. Trehy, M.L., and Bieber, T.I., Detection, identification and quantitative analysis of dihaloacetonitriles in chlorinated natural waters, in *Advances in the Identification and Analysis of Organic Pollutants in Water,* Keith, L.H. (ed.), Ann Arbor Science Publ., Ann Arbor, MI, 1981.

CHAPTER 3

Dissolved Organic Compounds in Natural Waters

E. M. Thurman

U.S. Geological Survey
Arvada, Colorado

3.1 INTRODUCTION

Over the past 10 years there has been an exponential increase in knowledge on the types and amount of dissolved organic substances in natural waters. Research on organic compounds in natural waters is the result of renewed interest in how natural dissolved substances are related to contaminants, as well as advances in analytical chemical analysis by gas chromatography/ mass spectrometry and liquid chromatography. Consequently, there is a need to summarize the new information on organic compounds in natural waters and to make it available to the expanding group of researchers interested in the biological, chemical, and ecological roles of dissolved organic substances. Two books that have summarized dissolved organic substances in natural waters are *Marine Organic Chemistry* (1) on marine waters and *Organic Geochemistry of Natural Waters* (2) on fresh waters. This chapter is a condensed summary of the amount of natural dissolved organic substances in fresh waters from Thurman (2). The chapter contains nine parts: terms for organic carbon, organic carbon continuum, amount of organic carbon in natural waters, histogram of dissolved organic carbon, aquatic humic substances, carboxylic acids, amino acids, carbohydrates, and hydrocarbons. Emphasis is placed on the amount of dissolved organic substances in fresh waters, with the main theme, "What are the compounds that constitute dissolved organic carbon?" Therefore, it is appropriate that the discussion begins with terms for organic carbon in natural waters.

3.2 TERMS FOR ORGANIC CARBON

The literature contains at least 13 terms for organic matter in water, as shown in Table 3.1. Of the 13 terms, this chapter will discuss four: dissolved organic carbon (DOC), suspended organic carbon (SOC), particulate organic carbon (POC), and total organic carbon (TOC). These measurements give the total amount of organic carbon in water. The other terms in Table 3.1, although sometimes used in engineering and aquatic ecology, are not as useful for natural waters, because they cannot be determined accurately by carbon analysis.

3.2.1 Dissolved Organic Carbon

DOC is the organic carbon passing through a 0.45-μm silver or glass-fiber filter and is the most important term used in the study of organic carbon. DOC quantifies the chemically reactive fraction and gives the mass of organic carbon dissolved in a water sample. It is a reliable measure of the many simple and complex organic molecules making up the dissolved organic load. In spite of the fact that DOC is organic carbon that passes a 0.45-μm filter, most of the DOC is dissolved and is smaller than the colloidal range of 0.45 μm to 1 nm. DOC is determined by oxidation to carbon dioxide and by measurement of carbon dioxide by infrared spectrometry (3,4).

TABLE 3.1
Acronyms of Commonly Used Terms
for Organic Matter in Water

Acronym	Meaning
DOC	Dissolved organic carbon
SOC	Suspended organic carbon
POC	Particulate organic carbon
FPOC	Fine particulate organic carbon
CPOC	Coarse particulate organic carbon
TOC	Total organic carbon
VOC	Volatile organic carbon
DOM	Dissolved organic matter
POM	Particulate organic matter
TOM	Total organic matter
COM	Colloidal organic matter
BOD	Biochemical oxygen demand
COD	Chemical oxygen demand

Source: Ref. 2.

3.2.2 Suspended Organic Carbon

SOC is the organic carbon retained on a 0.45-μm silver filter. The term suspended organic carbon is consistent with hydrologic terminology for inorganic constituents and contrasts with the term particulate organic carbon from the literature of limnology. These two terms are essentially identical, except that SOC refers to filtration through silver filters, and particulate organic carbon refers to filtration through glass-fiber filters. There are advantages in the use of silver filtration:

1. The size cutoff is between 0.1 and 0.45-μm for the silver filter, whereas the glass-fiber filter gives a slightly greater size cutoff of 0.5–2.0 μm. This is due to the fibrous nature of the glass filter, which allows particles to pass through the filter. These larger particles may consist of suspended organic carbon associated with clay.
2. The silver from the filter dissolves into the sample at a concentration of 50–100 μg/L. This acts as a preservative to prevent bacterial growth.

The disadvantages of silver filtration are that it is expensive and slow as compared to filtration through glass-fiber filters. In a comparison of glass and silver filters, Wangersky and Hincks (5) noted that particulate organic carbon was greater on the glass-fiber filter, which may be due to sorption of dissolved organic matter (6). Both SOC and POC are determined by wet combustion (4).

3.2.3 Particulate Organic Carbon

POC, the organic carbon retained on a 0.45-μm glass-fiber filter, is essentially identical to suspended organic carbon, as explained above. Limnologists further subdivide POC into fine and coarse organic carbon, called FPOC and CPOC. FPOC is from 0.45 μm to 1.0 mm, while CPOC is larger than 1.0 mm. The term, POC, rather than SOC, is most commonly used.

3.2.4 Total Organic Carbon

TOC is the sum of DOC and SOC or the sum of DOC and POC. Although TOC may be measured directly on a carbon analyzer, in itself it is not a useful term. Instead, separate measurements should be made for dissolved, particulate, and suspended organic carbon, for the following reasons:

1. DOC is chemically more reactive because it is a measure of individual organic compounds in the dissolved state, while POC and SOC are both discrete plant and animal organic matter and organic coatings on

silt and clay. Total organic carbon measured on the entire sample does not distinguish between these two important and different fractions.

2. SOC and POC increase dramatically with increasing discharge, while DOC varies less. Thus, TOC will reflect the increase of SOC and POC rather than DOC. If DOC is not measured separately, TOC will have little interpretative meaning.

3. In many lakes, small streams, and the open ocean, the concentration of suspended organic carbon is small, less than 10% of the total organic carbon. In these environments, TOC is almost identical to DOC. However, for rivers laden with sediment, SOC and POC may be larger than DOC. Thus, single measurements of TOC are not recommended, rather a measurement of DOC and POC, or DOC and SOC, are summed for the TOC load.

3.3   ORGANIC CARBON CONTINUUM

DOC is a mixture of hundreds of thousands of individual organic molecules. Many of these compounds have known structures and are biological molecules; they include: sugars, amino acids, fatty acids, and hydrocarbons. There are also compounds that are complex polyelectrolytes, which originate from the decomposition of soil and plant organic matter. These compounds are the humic substances and hydrophilic acids. Together these compounds contribute to the pool of organic carbon that constitutes DOC.

Figure 3.1 shows the continuum of organic carbon in natural waters. Filtration removes macroscopic particulate organic carbon, such as zooplankton, algae, bacteria, and detrital organic matter from soil and plants. Viruses (and some ultrasmall bacteria) are the only types of organisms that pass through a filter and enter the dissolved organic fraction. DOC is an operational definition and is the organic carbon smaller than 0.45 μm in diameter. The majority of the dissolved organic carbon that is present at the molecular level is polymeric organic acids, called humic substances. These yellow organic acids are of 1000–2000 molecular weight and are polyelectrolytes of carboxylic, hydroxyl, and phenolic functional groups. They comprise 50–75% of the dissolved organic carbon and are the major class of organic compounds in natural waters.

Dissolved molecules of fulvic acid are approximately 2 nm in diameter and at least 60 nm apart. Between each of these organic molecules, one would expect to see in any direction five inorganic ions: calcium, sodium, bicarbonate, chloride, and sulfate. If we imagine that an organic molecule is the height of an average man, it would be 10 m to his nearest neighbor and 70 m to his nearest relative. Thus, a great part of the dissolved organic carbon behaves as individual dissolved ions.

However, there is some colloidal organic matter in water. These are

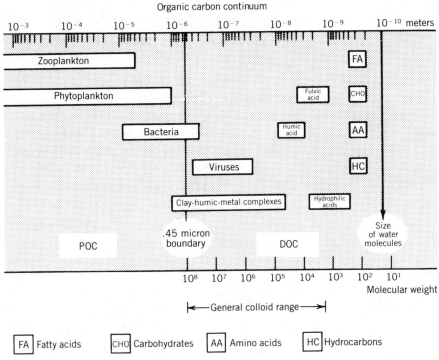

Figure 3.1. Continuum of particulate and dissolved organic carbon in natural waters. (Published with permission of Martinus Nijhoff-Dr. Junk Publishers after Ref. 2.)

large aggregates of humic acids, which are 2–50 nm in diameter. Commonly they are associated with clay minerals or oxides of iron and aluminum. In most natural waters the colloidal organic matter is approximately 10% of the dissolved organic carbon. This colloidal organic matter is the humic acid fraction of humic substances. This fraction is larger in molecular weight from 2000 to 100,000 (see Figure 3.1) and contains fewer carboxylic and hydroxyl functional groups than the fulvic acid fraction. Humic acid adsorbs and chemically bonds to the inorganic colloids modifying their surfaces.

Dissolved humic substances comprise 5–10% of all anions in streams and rivers. In fact, it is the anionic character of the organic matter that gives it aqueous solubility, binding sites for metals, buffer capacity, and other characteristics. The anionic character comes from the dissociation of carboxylic-acid functional groups:

$$R{-}COOH = R{-}COO^- + H^+$$

Carboxylic functional groups occur on aquatic humic substances with a frequency of 5–10 per molecule. At the pH of most natural waters, 6–8, all of these carboxylic groups are anionic or dissociated. These charged groups repulse one another and spread out the molecule. The counterions balancing the charge for these negative groups are mostly calcium and sodium. However, trace metals may be bound to some of these carboxylic groups that have a favorable steric location. That is, a carboxylic group in association with a phenolic group may form a chelate, or ring structure, and bind metal ions.

Besides the humic substances, what other types of organic compounds would we see at the molecular level? There are, of course, individual molecules that the organic chemist would recognize: simple sugars, amino acids, fatty acids, hydroxy acids, and hydrocarbons. These compounds account for 10–20% of the organic matter in water. These simple organic compounds come from decomposition of plant and soil organic matter and are in a constant state of flux because of chemical and biological activity. As one geochemist stated, "A sample of water for organic analysis is like a single frame of a moving picture."

The remainder of the DOC falls into a category called hydrophilic acids. These are polymeric molecules, which are a continuum of humic substances. Because they have not been isolated from water until recently, little is known of their chemistry or structure. Preliminary results show that they contain more carboxylic, hydroxyl, and carbohydrate character than humic substances, but are similar in molecular weight. Thus, they are called hydrophilic acids or hydrophilic humic substances, which is discussed in more detail in a later section.

3.4   AMOUNT OF ORGANIC CARBON
IN NATURAL WATERS

Figure 3.2 shows the "average" concentrations of dissolved and particulate organic carbon in surface and ground waters. Dissolved organic carbon varies with the type of water from approximately 0.5 mg/L for groundwater and seawater to over 30 mg/L for colored water from swamps.

Seawater has the lowest DOC with a median concentration of 0.5 mg/L for deep seawater and 1.0 mg/L for the upper 200 m. Coastal waters have DOCs of 1–3 mg/L because of carbon input from terrestrial sources and increased primary productivity in coastal waters. The concentration of POC is much less, ranging from 0.1 mg/L in shallow water to 0.01 mg/L below 300 m. A good summary of DOC in seawater is MacKinnon (6).

Dissolved organic carbon in groundwater ranges from 0.2 to 15 mg/L with a median concentration of 0.7 mg/L. The majority of all groundwaters have concentrations of DOC below 2 mg/L (7–11). However, there are exceptions; for example, in the southeastern United States, organic-rich surface waters recharge groundwater that contribute DOC in the form of humic substances. In these waters, DOCs range from 6 to 15 mg/L (2,12). Ground waters associated with coal deposits also may have larger than average concentrations of DOC, from 5 to 10 mg/L. These groundwaters contain

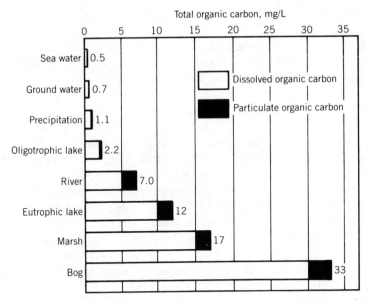

Figure 3.2. Approximate concentrations of dissolved and particulate organic carbon in natural waters. (Published with permission of Martinus Nijhoff-Dr. Junk Publishers, after Ref. 2.)

humic acid, which is only slightly soluble and is chemically different than fulvic acid (see later section of definition of fulvic and humic acid). Generally, groundwater does not contain POC; however, during sampling, sediment may be incorporated into the sample from aquifer sediments. This sediment should not be considered as POC.

The DOC of rivers varies with the size of a river, with climate, with the vegetation within a river's basin, and with the season of the year. For example, small streams in the arctic and alpine have concentrations of DOC from 1 to 5 mg/L with a median concentration of 2 mg/L. Streams in the taiga (a subarctic region) have a range from 8 to 25 mg/L and a median of 10 mg/L. In cool temperate climates the range of DOC is from 2 to 8 mg/L, with a median of 3 mg/L. In wet tropical climates, the range of DOC is from 2 to 15 mg/L, with a median of 6 mg/L. Rivers draining swamps and wetlands have the largest median DOC of 25 mg/L. These values hold even for large rivers, such as the Mississippi (DOC = 3.5 mg/L) and the Amazon (DOC = 5.0 mg/L).

Most small streams and rivers transport more DOC than POC. Because their discharge is small, their concentration of suspended sediment is low, and their concentration of POC is small (about 10% of DOC). There is an increase in POC relative to DOC with increasing discharge of a stream to a river. In the largest rivers, such as the Amazon, the concentrations of POC and DOC are equal at 5 mg/L. As a rule of thumb, POC will be 1–2% of the suspended matter of a stream; thus, a suspended load of 100 mg/L will contain 1–2 mg/L as POC. For in depth reviews of DOC and POC in rivers, see Meybeck (13,14).

The majority of organic carbon in lakes is dissolved, and particulate organic carbon contributes only about 10% of the total organic carbon. DOC varies with the productivity of the lake, increasing with trophic status of the lake. Lakes that are nonproductive, with little or no algal production, have concentrations of DOC from 1 to 3 mg/L. As algal productivity increases, so does the concentration of DOC. For instance, oligotrophic lakes (low productivity) have DOCs from 1 to 3 mg/L, eutrophic lakes (high productivity) have DOCs from 2 to 5 mg/L, and dystrophic lakes (low productivity with color from humic substances) have DOCs from 20 to 50 mg/L. The POC of lakes varies with trophic status also. Oligotrophic lakes have low concentrations of phytoplankton and low POC, usually less than 10% of the DOC. Eutrophic lakes on the other hand, have high concentrations of phytoplankton and POC and DOC may be equal. POC will vary throughout the water column as a function of plankton activity and accumulation. For reviews on DOC and POC in lakes, see a classic paper by Birge and Juday (15) and the chapter in Wetzel (16) on DOC and POC in lakes.

In wetlands (marshes, swamps, and bogs), the concentration of DOC is greater than in other types of aquatic environments. This is because of the decomposition of emergent plants in these environments, which contribute 1500–4000 g of organic matter per meter per year to the aquatic system. This organic matter amounts to 3–10 times more than forest and prairie areas. Not only is there increased productivity, but there is also anaerobic decomposition that commonly occurs in wetlands, which increases concentrations of DOC in water-logged soils. For these reasons, there is an increase in DOC of these waters compared to other natural waters (2). POC is also greater in concentration and amounts to about 10% of the DOC of these waters, except during heavy rains, when the concentration of particulate matter increases because of larger discharge. For reviews of organic carbon in wetlands, see Mulholland (17,18) and Thurman (2).

3.5 HISTOGRAM OF DISSOLVED ORGANIC CARBON

In natural waters there are many types of organic compounds, which will be classified by abundance into six major groups: humic substances, hydrophilic acids, carboxylic acids, amino acids, carbohydrates, and hydrocarbons. This simple grouping includes the majority of the organic compounds found in natural waters, which have been compiled from many studies cited in later sections and in Thurman (2). A simple way to display the abundance of these compounds is the histogram of dissolved organic carbon (Fig. 3.3).

Figure 3.3 shows that 50% of the DOC is aquatic fulvic and humic acids, the dominant group of natural organic compounds in water, and 30% of the DOC is hydrophilic acids, a relatively unknown yet large fraction of the

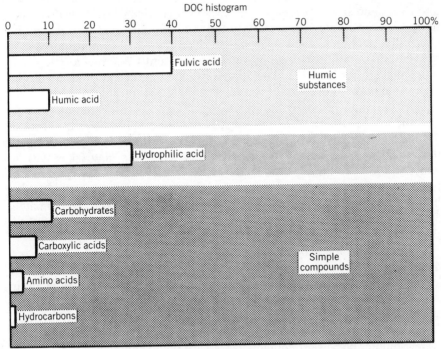

Figure 3.3. Dissolved organic carbon histogram for an average river water with a DOC of 5 mg/L. (Published with permission of Martinus Nijhoff-Dr. Junk Publishers, after Ref. 2.)

DOC. The remaining 20% of the DOC are identifiable compounds. The histogram in Figure 3.3 shows that of the identified compounds carbohydrates are 10%, carboxylic acids are 7%, amino acids are 3%, and hydrocarbons are less than 1%. This histogram is an approximation based on data compiled from the literature (2), and it is meant only as a guide to understand the nature of dissolved organic carbon. References to the original data are included in the following sections and Thurman (2). Before discussing each of these classes, let us look briefly at the nature of humic and hydrophilic acids, which is an operational classification based on adsorption chromatography and ion exchange. These substances are the key to the classification scheme used in this chapter.

Soil humic substances are the colored, polyelectrolytic acids that are operationally defined by their isolation from soil with 0.1 N NaOH (2,19). Aquatic humic substances are defined differently than soil humic substances. Aquatic humic substances are operationally defined as colored, polyelectrolytic acids isolated from water by sorption onto XAD resins, weak-base ion-exchange resins, or a comparable procedure. They are nonvolatile and range in molecular weight from 500 to 5000. They originate in plant and soil systems, where they are leached by interstitial water of soil into rivers and streams. Lakes and marine waters that contain algal productivity also contribute significantly to the pool of aquatic humic substances. Generally, aquatic humic substances have an elemental composition that is 50% carbon, 4–5% hydrogen, 35–40% oxygen, and 1% nitrogen. The major functional groups include: carboxylic acids, phenolic and alcoholic hydroxyl groups, and keto functional groups. The structure of aquatic humic substances is unknown (2).

The concentration of humic substances for various waters is shown in Table 3.2. Groundwaters and marine waters are lowest in concentration with 0.05–0.25 mg/L as humic substances. Streams, rivers, and lakes vary in concentration of humic substances from 0.5 to 4.0 mg/L. Tea-colored waters, such as marshes, bogs, and swamps vary from 10 to 30 mg/L DOC as humic substances. As a group humic substances account for approximately 30–50% of the DOC of most natural waters, except in colored waters, where they contribute 50–90% of the DOC.

Humic substances may be divided into fulvic and humic acid. This is a definition based on the soil literature, which uses precipitation to separate humic and fulvic acid (Fig. 3.4). The humic substances that precipitate in acid are humic acid, and those in solution are fulvic acid. In practice, this separation removes the larger aggregates of humic substances that originate from degradation of plant matter (20). Generally, the fulvic acid is more water soluble, because it contains more carboxylic and hydroxyl functional groups and is lower in molecular weight, from 800 to 2000. Humic acid is larger than 2000 molecular weight and is often colloidal in size (20). The humic acid fraction is sometimes associated with clay minerals and amorphous oxides of iron and aluminum. The combination of greater molecular weight, fewer carboxylic acid functional groups, and interaction with clay are the reasons that the humic acid precipitates. As Figure 3.3 shows, humic acid accounts for 10% of the DOC, and the combination of fulvic and humic acid account for 50% of the DOC of most natural waters. These data are the

TABLE 3.2
Concentration of Humic Substances
in Natural Waters

Water Type	DOC of Humic Substances (mg/L)
Groundwater	0.05–0.10
Seawater	0.10–0.25
Stream	0.5–2.0
Lake	1.0–4.0
River	1.0–4.0
Wetlands	10–30

Source: Compiled from references in Ref. 2.

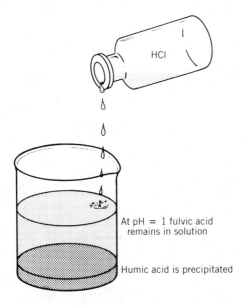

At pH = 1 fulvic acid
remains in solution

Humic acid is precipitated

Figure 3.4. Separation of humic and fulvic acids by acid precipitation. (Published with permission of Martinus Nijhoff-Dr. Junk Publishers, after Ref. 2.)

result of 50 analyses of fresh waters using an isolation procedure on XAD resin, which is a nonionic methylmethacrylate polymer that adsorbs organic matter from water by hydrophobic bonding (21,22). The pH of the water is lowered to 2.0, and the humic substances adsorb onto the XAD resin. Then, the humic substances are desorbed quantitatively with base and studied by various techniques. Numerous studies show that the colored organic acids are removed on the resins, approximately 85% or more of the visible absorbance of the sample at 400 nm and 50% of the DOC (Fig. 3.5). An interesting enigma arose from these basic studies on the nature of humic substances, which was: What is the remainder of the DOC? Part of the answer is the next bar in the dissolved organic carbon histogram (Fig. 3.3), the hydrophilic acids, or what may be called the hydrophilic humic substances.

Because approximately 30% of the dissolved organic carbon of a natural water are organic acids that are not retained by the XAD resin at pH 2, they have been called the "hydrophilic acids" (23). From the study of three samples of hydrophilic acids and specific organic compounds, Leenheer (24) postulated that the hydrophilic acids are a mixture of organic compounds that are both simple organic acids, such as volatile fatty acids and hydroxy acids, as well as complex polyelectrolytic acids that probably contain many hydroxyl and carboxyl functional groups. The hydrophilic acids probably contain sugar acids, such as uronic, aldonic, and polyuronic acids, but this is only speculation based on water solubility of these compounds and their abundance in nature. Because hydrophilic acids are difficult to isolate and purify, their study has only begun. This fraction will be an interesting part of the DOC to identify.

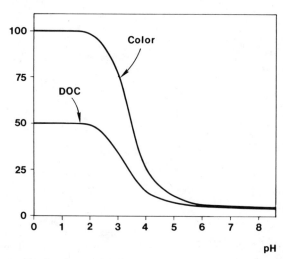

Figure 3.5. Removal of color and DOC from natural waters by XAD resins as a function of pH. (Data from Ref. 22.)

The hydrophilic acids have been isolated from several natural waters and separated from the humic substances by weak anion-exchange chromatography, which is shown in Figure 3.6. The first column contains XAD resin and sorbs humic substances from the solution, which is at pH 2. Next, amino acids and some polypeptides are adsorbed onto the cation exchange resin. All inorganic cations are exchanged onto the resin and hydrogen is released. At this point the solution contains the hydrogen form of the remaining organic acids and the hydrogen form of inorganic anions. The third column is a weak base ion-exchange resin that is in a free-base form. The hydrogen ion from the previous cation-exchange column protonates the weak-base resin. Now the weak-base resin is an anion-exchange column and retains all organic acids and all inorganic anions. Only the neutral, water-soluble organic species, such as simple sugars, two and three carbon alcohols, and ketones, will continue through all three columns (23,24).

Thus, column chromatography is the basis of understanding the fractions of natural DOC in water. The following sections are a synopsis of the important classes of compounds that constitute the DOC load. They include: aquatic humic substances, carbohydrates, amino acids, and hydrocarbons.

3.6 AQUATIC HUMIC SUBSTANCES

Aquatic humic substances constitute 40–60% of dissolved organic carbon and are the largest fraction of natural organic matter in water. They are called by many names: crenic and apocrenic acids (25,26), fulvic acid (27),

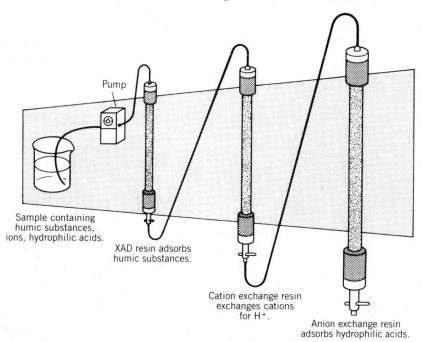

Pump

Sample containing
humic substances,
ions, hydrophilic acids.

XAD resin adsorbs
humic substances.

Cation exchange resin
exchanges cations
for H^+.

Anion exchange resin
adsorbs hydrophilic acids.

Figure 3.6. Separation of humic substances and hydrophilic acids by adsorption and ion-exchange chromatography. (Published with permission of Martinus Nijhoff-Dr. Junk Publishers, after Ref. 2.)

yellow organic acids (28), gelbstoff (29), and aquatic humus (30). What are these so-called "humic substances"? This is a difficult question. No reagent labeled aquatic humic substances exists, and there are no simple methods of analysis for aquatic humic substances. The problem is to define, in some limited yet useful way, what humic substances are. Thus, aquatic humic substances are defined with an operational definition, which was given in the previous section.

This operational definition of aquatic humic substances is related intimately to the isolation procedure using adsorption onto resins (22), a procedure that evolved over the last decade. Beginning in the late 1960s and early 1970s with the development of macroporous resins, Riley and Taylor (31) and Mantoura and Riley (32) successfully isolated and characterized aquatic humic substances by adsorption onto XAD resins (nonionic macroporous resins used for concentrating organic compounds and made by Rohm and Haas, Philadelphia). In the late 1970s, Aiken and others (33) studied the mechanisms of adsorption of humic substances onto various XAD resins and reached the conclusion that the acrylic-ester copolymer XAD-8 adsorbed and eluted aquatic humic substances most efficiently. Thurman and Malcolm (22) extended these studies and isolated gram quantities of aquatic humic substances with XAD-8 resin.

Consider how the definition of aquatic humic substances compares with the definition of humic substances from soil. Because the definition of humus originated in the soil literature, this is an important consideration. Humic substances from soil are: organic substances extracted from soil by sodium hydroxide (typically 0.1 N), the fraction that precipitates in acid is humic acid (pH 1–2), and the fraction remaining in solution is fulvic acid (19). This definition is different from the operational definition of aquatic humic substances. Therefore, the best comparison is between the general chemical characteristics of humic substances from soil and water. Both are polyelectrolytic, colored, organic acids with comparable molecular weights; their elemental composition is similar, and so is their functional group analysis. This does not prove that humic substances from soil and water are the same. Merely, it demonstrates that their general chemical characteristics are similar (22), and the definition of aquatic humic substances seems consistent with the definition of humic substances from soil.

The study of soil humic substances is a much older science than the study of aquatic humic substances. Because humus from soil holds water and nutrients, it is important to crop growth; therefore, it was a popular item of study during the eighteenth through the twentieth centuries. See the literature reviews of early work by Schnitzer and Khan (19) and Stevenson (34). Except for the study of Berzelius in 1806 (25), the literature of water chemistry contains little on aquatic humic substances until the 1900s, when studies began on the origin of color in water.

The concentration of humic substances varies for different natural waters, (see Table 3.2). The lowest concentrations of humic substances are in groundwater and seawater, where concentrations vary from 0.05 to 0.60 mgC/L. Streams, rivers, and lakes contain from 0.50 to 4.0 mgC/L, and colored rivers and lakes have much larger concentrations of humic substances, from 10 to 30 mgC/L. This section examines the amount and sources of humic substances in different aquatic environments.

3.6.1   Groundwater

As discussed earlier, the DOC of groundwater is low, from 0.1 to 2 mg/L, with an average value of 0.7 mg/L. Likewise, humic substances in ground water are low in concentration; for example, Thurman (35) isolated and characterized the aquatic humic substances in five groundwaters from dolomite, sandstone, and limestone aquifers. The concentration of the humic substances varied from 34 to 99 μgC/L for four aquifers. Only one of the aquifers, a shallow aquifer in Florida that received organic matter from the Everglades swamp, had a large concentration of DOC, 12 mgC/L. This is an unusually large DOC for a groundwater, and it was principally fulvic acid, which comprised 87%, and only 13% was humic acid. A mean value for humic substances in groundwater would be approximately 100 μgC/L, this value is only approximate due to the small number of samples analyzed.

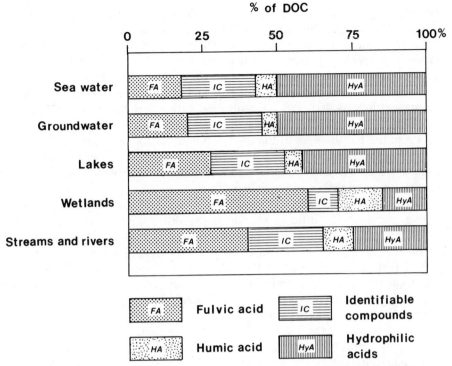

Figure 3.7. Distribution of humic substances in natural waters. (Published with permission of Martinus Nijhoff-Dr. Junk Publishers, after Ref. 2.)

Hydrophilic acids are an important part of the DOC of groundwater, as is shown in the bar diagram in Figure 3.7. Unfortunately, there are no data on the other fractions in groundwater, and Figure 3.7 is an estimate.

There are several sources of humic substances in groundwater, kerogen from the aquifer solids and organic matter carried in from soil and plant matter with precipitation. Thurman (35) found that the kerogen source is most important in deeper groundwaters (100 m and deeper) where dissolved organic carbon is less than 1 mg/L. In shallow groundwaters, 10–100 m deep, surface water laden with organic matter commonly contributes dissolved organic carbon to groundwater. Good examples of these types of groundwaters are in the southeastern United States, where the concentration of dissolved organic carbon is 5–10 mg/L from yellow organic acids, indicating organic carbon from soil and plant humic matter.

3.6.2 Seawater

The concentration of humic substances in seawater is 60–600 μgC/L and accounts for 10–30% of the DOC of seawater. These data are the results of many studies of seawater (29,36–45). The variation in concentration follows the concentration of dissolved organic carbon of coastal and open ocean waters. For example, Stuermer and Harvey (45) reported the concentration of humic substances in seawater from 60 to 200 μgC/L with greatest concentration near coastal regions. This is approximately 7–30% of the dissolved organic carbon in seawater, which is shown in Figure 3.7. Twenty-five percent is the average amount of humic substances (fulvic and humic acid) that are present in seawater, but concentration varies depending on the type of seawater.

From the bar diagram in Figure 3.7, humic substances account for 25% of the dissolved organic carbon in seawater, this is a lower percentage than in fresh waters. Because of the high salt content of seawater, humic acid from rivers precipitates in the estuary.

Hydrophilic acids account for a large fraction of the dissolved organic carbon, nearly 50%; little is known of the distribution of this dissolved organic carbon, and this is only an estimate. Identifiable compounds account for 10–15% of the DOC. The source of humic substances in seawater may be both terrestrial and aquatic. In the near-shore environments, terrestrial inputs from rivers are an important source of dissolved organic carbon. In open ocean and in lagoonal environments, algae are a major source of humic substances, but the exact pathway of origin is still an open question.

The majority of the humic substances in seawater are fulvic, approximately 85%, and 15% humic acid (36,45). The humic substances in seawater are considered by the majority of researchers to originate in the sea from the decomposition of phytoplankton. Sieburth and Jensen (37) noted seasonal variations in humic content of seawater off the coast of Norway, which they correlated with seasonal trends in algal productivity. In winter during dark months, the concentration of humus dropped from 0.2–0.8 mg/L to 0.01 mg/L, but increased during the spring and summer. The molecular weight of marine fulvic acid is generally small, less than 2000 (40,41,46), although some macromolecular material (greater than 100,000) has been found (40). The aliphatic content of humus from seawater is greater than that of terrestrial matter (44,47,48), and organic nitrogen is greater in humus from seawater (46). The functional group content of seawater humus is similar to terrestrial, with carboxyl and hydroxyl predominating, but the functional group content is different in that there is little or no phenolic hydroxyl in marine fulvic acid.

3.6.3   River Water

The amount of humic substances in mountain streams and melt waters from snow and ice varies from 0.05 to 0.5 mgC/L (49). For larger streams and rivers, the amount of humic substances increases to 0.5 to 4 mgC/L. For larger rivers and tropical rivers, the amount of humic substances increases from 2 to as much as 10 mgC/L. Humic substances are the chief or major component of dissolved organic carbon in rivers (Figure 3.7).

Humic substances vary in concentration from 10 to 30 mgC/L for the tea-colored rivers. These rivers are common in tropical regions and in southern temperate zones. Examples are the Suwannee River in southeastern Georgia, which has a concentration of humic substances of 30 mgC/L (50). An example of a tropical river is the Rio Negro in Brazil, a tributary of the Amazon, which has a concentration of aquatic humic substances of 10 mgC/L (51). In these colored waters, aquatic humic substances account for 70–90% of the dissolved organic carbon.

The concentration of humic substances in various rivers of the United States is shown in Table 3.3, but it must be recognized that concentrations vary seasonally, and these concentrations are average values. For example, seasonal variation is shown for an alpine stream in Colorado (Fig. 3.8). Concentration of fulvic acid in this stream peaked during the spring flush in early June, as melting snow flushed organic matter from the surrounding terrain. Small creeks and rivers show the largest seasonal input; commonly, their concentrations will increase two to three times during maximum discharge.

Humic substances have been studied in many rivers, including in the United States: Oyster River (52), Satilla River (53), Suwannee River (54), Williamson River (55), and the Ohio, Ogeechee, Bear, and Missouri Rivers (56), White Clay Creek (57), colored rivers in the northwestern United States (58), the Connetquot River (59), and a colored stream in the northwestern United States (60); in Germany the Rhine River (61); in England the Thames and Hull Rivers (36); in Norway the Nid River (37) and rivers and

TABLE 3.3
Concentration of Humic Substances in Various Rivers of the United States

River	Concentration (mgC/L)
Como Creek (Colorado)	1.00
Yampa River (Colorado)	0.80
Deer Creek (Colorado)	0.80
Snake River (Colorado)	0.30
South St. Vrain (Colorado)	1.20
South Platte River (Colorado)	1.50
Castle Creek (Oregon)	1.00
Colorado River (Arizona)	1.20
Missouri River (Iowa)	1.50
Ohio River (Ohio)	1.50
Ogeechee River (Georgia)	3.00
Suwannee River (Georgia)	30.00

Source: Thurman and Malcolm, unpublished data.

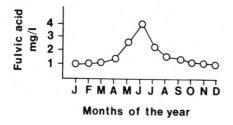

Figure 3.8. Seasonal variation in the concentration of humic substances for an alpine stream in Colorado. (Published with permission of Martinus Nijhoff-Dr. Junk Publishers, after Ref. 2.)

bogs of Norway (30); in Canada the Scoudouc River (62) and other Canadian rivers (63); In Brazil the Rio Negro and Amazon Rivers (51); in European rivers (64); in Russian rivers (65); in Swedish rivers (66); in Japanese rivers (67).

In general, humic substances from rivers and streams are of allochthonous origin from soil and plant matter; this is the consensus of many of the studies listed above. The molecular weight of humic substances is less than 2000 for fulvic acid and 2000–5000 or greater for humic acid (20). In general, fulvic acid makes up 85% of the humic substances, and humic acid makes up 15%. The major functional groups are carboxyl, hydroxyl, carbonyl, and lesser amounts of phenolic hydroxyl. They contain trace amounts of carbohydrates and amino acids (1–3%). These points are discussed again later in this chapter.

3.6.4   Lake Water

The concentration of humic substances in lakes varies with the concentration of dissolved organic carbon in the lake (Table 3.4). Generally, humic substances account for 40% of the dissolved organic carbon. The major deviation from this rule is colored waters, such as dystrophic lakes, where concentrations of dissolved organic carbon are 10 mg/L or greater. In lakes such as these, humic substances may contribute 80–90% of the dissolved organic carbon. The major studies of humic substances in lakes include: Shapiro (28,68), Ghassemi and Christman (58), Christman and Minear (69), Packham (36), Stewart and Wetzel (70), DeHaan (71–73), DeHaan and De-Boer (74,75), Hama and Handa (76), Tuschall and Brezonik (77), and Baccini and others (78).

Humic substances in lakes have both terrestrial and aquatic origins, and the origin is a function of the size of the lake and whether organic inputs are from land, a stream, or from algae within the lake. Generally, the larger the lake, the more important are algal inputs to the pool of organic carbon. Obviously, the source of organic matter has an important effect on the nature of humic material, and humic material from algal origin is different than humic material from terrestrial origin. In lake water, fulvic acid constitutes 85–90%, and humic acid comprises 10–15% of the total humic substances. Depending on the eutrophic state of the lake and the input of au-

TABLE 3.4
Concentration of Humic Substances
in Various Types of Lakes

Lake	Concentration (mgC/L)
Oligotrophic lake	0.50–1.00
Mesotrophic lake	1.00–1.50
Eutrophic lake	1.50–5.00
Dystrophic lake	10.0–30.0

Source: Ref. 2.

tochthonous carbon, the humic substances increase in aliphatic content and nitrogen (mostly algal carbon), while in dystrophic lakes the humic substances are aromatic and low in nitrogen. Finally, the dissolved organic carbon histogram for lake water is similar to that of river water (Fig. 3.7).

3.6.5 Wetland Water

Wetland areas, such as bogs, marshes, and swamps, contain considerable amounts of aquatic humic substances, from 10 to 50 mg/L, as shown in Table 3.5. These aquatic humic substances originate in wetlands from the decomposition of plant matter. Because the net primary productivity of emergent plants is large, from 1000 to 4000 g/m^2, the amount of decomposing plant matter is considerable.

Because decomposition in wetlands is slow, humic substances accumulate in the waters draining these areas. Decomposition is slow because of waterlogging of the soils, the lack of oxygen, and low pH (3–4). Because the pH of the water is acid in these environments, fungi are one of the major organisms of organic decomposition. If the water becomes anaerobic, fungal activity stops and organic carbon accumulates.

In wetlands the dissolved-organic-carbon histogram is different from river and lake waters, this difference being the increased percentage of humic and fulvic acid, which makes up 70–90% of the dissolved organic carbon (Fig. 3.7). Therefore, the larger concentrations of DOC of wetlands are chiefly from humic substances. The humic substances impose their chemistry upon the water, and commonly buffer pH and transport trace metals, such as iron and aluminum.

Studies on humic substances in wetlands include: Lake Celyn in North Wales (79), Suwannee and Okefenokee swamp (54), Norwegian bogs (30), Satilla River (53), in Rhode Island (80), in the eastern United States (81,82), Black Lake in North Carolina (83), Caine (84) on an alpine bog in Colorado, and McKnight and others (85) on Thoreau's Bog. On the basis of these studies, it is known that humic substances from bogs are the most aromatic

TABLE 3.5
Range of Concentration of Humic
Substances in Wetlands

Wetland	Concentration (mgC/L)
Marsh	10–30
Swamp	10–50
Bog	10–50

Source: Ref. 2.

of aquatic humic substances and contain the most humic acid, from 15–35%. They are depleted in carboxyl groups compared with other humic substances and have the least amount of aromatic carbon and the lowest atomic ratio of H/C. They generally have the largest amount of phenolic content and represent an aromatic end member in the suite of aquatic humic substances.

3.7 CARBOXYLIC ACIDS

In general, carboxylic acids are present at microgram-per-liter concentrations, but as a group may account for approximately 5–8% of the dissolved organic carbon of natural waters (Fig. 3.9). Nonvolatile fatty acids are the largest group of carboxylic acids found in natural waters, and they commonly account for approximately 4% of the dissolved organic carbon. Palmitic and stearic acids are the most abundant of the nonvolatile fatty acids and originate from the degradation of lipids and triacylglycerols, which are derived from algal and terrestrial plants. The next most important group of carboxylic acids are the volatile fatty acids, which account for approximately 2% of the DOC in waters containing oxygen. The most abundant of the volatile fatty acids is acetic acid. Acetic acid together with palmitic and stearic acids are the most common fatty acids found in all types of natural water. The volatile fatty acids originate from the microbiological degradation of both dissolved and particulate organic matter in aerobic and anaerobic environments. Generally, the volatile fatty acids will be much more abundant in anaerobic waters.

Hydroxy and dicarboxylic acids have not been studied because isolation is difficult. Weak ion-exchange chromatography may improve isolation, and this was discussed in a preceding section. Hydroxy and dicarboxylic acids originate from plant degradation products (86) and from biochemical cycles, such as the tricarboxylic acid cycle (87). It is now estimated that hydroxy and dicarboxylic acids account for 1–2% of the DOC of natural waters. Somewhat more is known of aromatic acids and phenols, because they are lignin degradation products and have been studied in conjunction with the

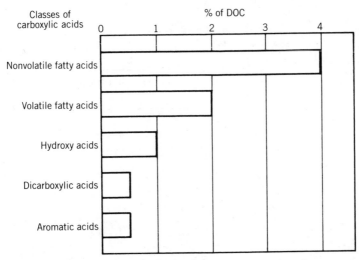

Figure 3.9. Contribution of carboxylic acids to the DOC of surface waters. (Published with the permission of Martinus Nijhoff-Dr. Junk Publishers, after Ref. 2.)

structure of humic substances. However, aromatic acids appear to constitute a small portion of the dissolved organic carbon, less than one percent.

Natural waters vary in concentration of carboxylic acids. Although literature on different waters is limited, it appears that carboxylic acids follow the concentration of dissolved organic carbon. That is, eutrophic lakes have the largest concentration of carboxylic acids followed by rivers, streams, seawater, and groundwater. No data are reported on the concentration of carboxylic acids in wetlands.

The factors controlling concentration include: origin, aqueous solubility, and biological availability. For example, long-chain fatty acids have many sources in plant and animal matter, but are of limited aqueous solubility and are less biologically available than volatile fatty acids, such as acetic acid, which is quite soluble and also biologically active. Thus, long-chain fatty acids are found in greater concentrations than volatile fatty acids, when oxidizing conditions exist, such as in surface waters. While in reducing conditions the volatile fatty acids are more abundant. Several good references on the origin of fatty acids in natural waters are: Lewis (88), Jeffries (89), and Poltz (92).

A general range of concentration for organic acids in fresh waters is shown in Table 3.6. The nonvolatile fatty acids are present at greatest concentrations from 100 to 600 µg/L, while the aromatic and dicarboxylic acids are present at the least concentrations, from 5 to 25 µg/L.

There are various procedures for the analysis of organic acids in water. An overview of identification of organic acids is that the organic acids are isolated from a water sample by liquid extraction into chloroform, or a similar solvent. The organic acids are methylated and analyzed by gas

TABLE 3.6
Concentration of Carboxylic Acids
in Fresh Waters

Type of Acid	Concentration Range (μg/L)
Volatile fatty	40–125
Nonvolatile fatty	100–600
Hydroxyl	10–250
Dicarboxylic	10–50
Aromatic	5–25

Source: Compiled from the following references: Hama and Handa (76), Barcelona and others (113), Mueller and others (114,115), Lamar and Goerlitz (116,117), and Hullett and Eisenreich (118).

chromatography and mass spectrometry. The most soluble organic acids may not be efficiently removed by solvent extraction with chloroform. For these compounds it is necessary to potassium-saturate the water sample, freeze-dry, and extract the salts with a crown ether/acetonitrile mixture. The samples are then esterified and analyzed by gas chromatography/mass spectrometry (GC/MS) (2).

3.8   AMINO ACIDS

This section discusses dissolved amino acids, which are the amino acids that pass through a 0.45-μm filter. All the amino acids that pass through are the total amino acids. Amino acids that are simple compounds are free amino acids. While those amino acids that are combined into polypeptides, proteins, and other structures (humic substances) are the combined amino acids. Concentrations are reported in micrograms per liter in order to agree with other sections of the chapter. However, concentrations of amino acids are generally reported in molar concentrations. For purposes of comparison, 1 μM is approximately 100 μg/L and 50 μgC/L.

The amino acid and peptide fraction is probably, next to humic substances, the most studied fraction in natural waters. Combined amino acids are found in dissolved aquatic humic substances and in dissolved polypeptides. The combined amino acids are four to five times as abundant as free amino acids. Because amino acids are present in proteinaceous matter in soils, plants, and aquatic organisms, they are ubiquitous in natural waters.

As a group, amino acids account for approximately 1–3% of the dissolved organic carbon of most natural waters. This is shown in Figure 3.10 and is based on a summary of the literature, which is cited Thurman (2). From

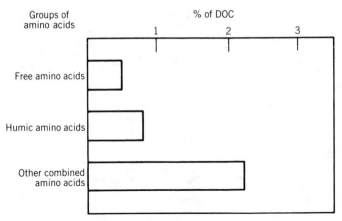

Figure 3.10. Histogram for free and combined amino acids in natural waters. (Published with permission of Martinus Nijhoff-Dr. Junk, after Ref. 2.)

the approximation shown in the histogram of Figure 3.10, one sees that combined amino acids account for the largest fraction of amino acids, followed by humic amino acids, and free amino acids.

Humic amino acids have been ignored in past studies because of either the isolation procedure, which does not isolate amino acids present in dissolved humic substances, or because of a total hydrolysis of all combined amino acids, which does not discriminate for humic amino acids. Humic substances behave as anions and are not part of the amino acid analysis, if the method uses preconcentration by cation exchange. However, recent methods of analysis use hydrolysis, formation of fluorescent derivatives, and direct injection onto a liquid chromatograph (91), and this procedure will include humic amino acids. Because humic carbon is 1–2 mg/L in rivers and because amino acids account for 1–3% of the humic carbon, humic amino acids are present at 10–60 µg/L. This is approximately the same concentration as dissolved free amino acids in river waters.

Besides the combined amino acids in humic substances, there are amino acids in dissolved polypeptides, proteinaceous compounds, and other organic substances. Although little is known of the distribution of the combined fraction, the types of amino acids in this fraction have been identified. This will be discussed in the following sections.

The amino acids that have been identified in both dissolved combined and free fractions are shown in Figure 3.11. Commonly the amino acids are grouped into neutral, secondary, aromatic, acidic, and basic amino acids. This grouping is based on the side chain present on the amino acid.

The concentration of total dissolved amino acids varies for different natural waters, and this is shown in Table 3.7. Seawater and groundwater have the smallest concentrations of amino acids, followed by oligotrophic lakes, rivers, eutrophic lakes, and marshes. Interstitial waters of soils and

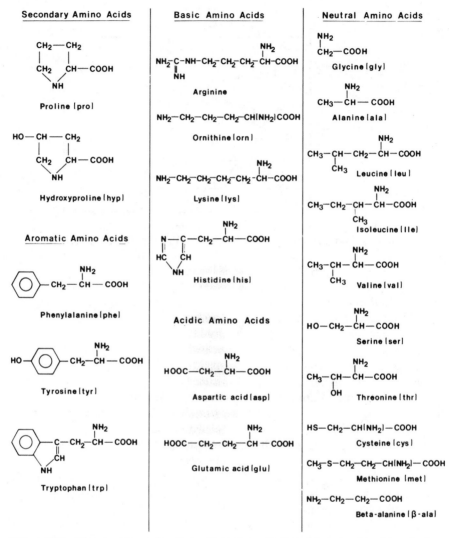

Figure 3.11. The structure of amino acids commonly found in natural waters. Amino acids are grouped into five categories: secondary amino acids, aromatic amino acids, basic amino acids, acidic amino acids, and neutral amino acids. The common abbreviation for each amino acid is also given.

sediments have the largest concentration of animo acids. Generally, it has been found that the greater the concentration of dissolved organic carbon, the greater the concentration of dissolved amino acids.

The origin of amino acids in water includes many sources, such as plant and soil organic matter, algae and aquatic plants, and bottom and suspended sediments. The importance of various sources has been studied most extensively in seawater, where algae are considered the major source of total

TABLE 3.7
**Median Concentration of Dissolved Total Amino
Acids in Various Natural Waters**

Water	Concentration (μg/L)	% DOC
Seawater	50 (20–250)	2–3
Groundwater	50 (20–350)	2–3
Rivers	300 (50–1000)	2–3
Oligotrophic lake	100 (30–300)	2–3
Eutrophic lake	600 (300–6000)	3–13
Marsh	600	4
Interstitial water	2000 (500–10000)	2–4

Approximate conversion is 100 μg/L equals 1 μM/L. A range
    is given in parentheses (2).

dissolved amino acids (see Chapter 6 in Ref. 1). In stream environments, terrestrial plants and soil organic matter make an important contribution to total dissolved amino acids. While in lakes, algae and terrestrial inputs vary with the eutrophic state of the lake, which is another way of showing that algae contribute to the pool of dissolved amino acids. In riverine environments, both aquatic and terrestrial sources may be important, with organic matter from sediment also contributing to the total pool of amino acids. However, little is known of these sources in rivers and is an obvious area for further research.

3.9   CARBOHYDRATES

Carbohydrates are polyhydroxy aldehydes, polyhydroxy ketones, or compounds that can be hydrolyzed to these compounds. Carbohydrates are an important, reactive fraction of organic matter in water, where they exist in several classes: monosaccharides, oligosaccharides, polysaccharides, and saccharides bound to humic substances (abbreviated as MS, OS, PS, and HS). Monosaccharides are simple one-unit sugars, such as glucose; oligosaccharides are 10 sugar units or less; polysaccharides are greater than 10 units; and humic saccharides are saccharides bound to humic substances. Carbohydrates may be modified by other functional groups, such as carboxyl groups, alcohols, amines and amino acids; these sugars are called sugar acids, sugar alcohols, and amino sugars (Fig. 3.12).

Monosaccharides may take two forms, an open-chain form and a ring form, including both six-membered rings for the hexoses and five-membered rings for the pentoses. Carbohydrates are the most abundant class of compounds produced in the biosphere. Generally, they are linked together into

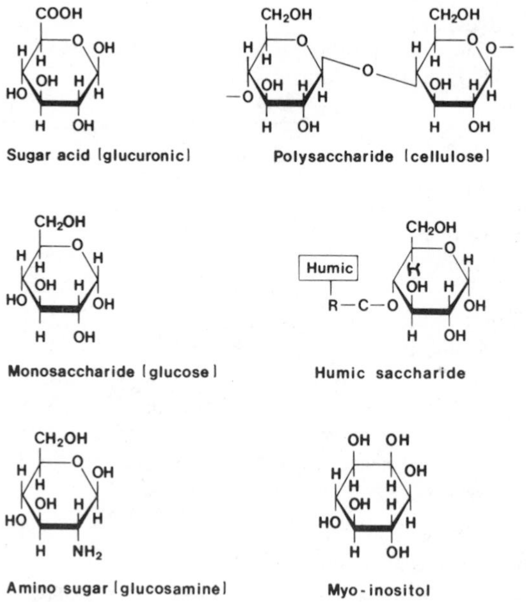

Figure 3.12. Different types of carbohydrates found in natural waters. (Published with permission of Martinus Nijhoff-Dr. Junk Publishers, after Ref. 2.)

polymers, and there are several important polymeric sugars that decompose and enter the aquatic system. Because cellulose comprises 30% of plant litter and may account for up to 15% of soil organic matter (34), it is probably the most important polysaccharide. Cellulose is a linear polymer of β-(1-4)-D-glucopyranose. Other important biopolymers include amylose, which is the water-soluble component of starch, and hyaluronic acid, which is a mixed polysaccharide of D-glucuronic acid and N-acetyl-D-glucosamine. This polymer is called an acid mucopolysaccharide, a jelly-like substance that provides intercellular lubrication and acts as a flexible cement for bacteria (87). Another polysaccharide that may be important in natural waters is alginic acid, which is a polymer of D-mannuronic acid and is a component of algae and kelp. All of these biopolymers are susceptible to degradation, both chemical and biochemical, and are important sources of MS and PS in aquatic environments.

The majority of the carbohydrates in fresh water originate in terrestrial systems. For example, plants, after death, dry out and may release 30% of their organic matter into water (84,92); half of this material is simple carbohydrates, probably monosaccharides and polysaccharides. The remaining half is organic acids rich in carbohydrates. Thus, water leachate of plant matter is an important source of carbohydrates in water.

Soils, on the other hand, contain carbohydrate-rich organic debris not readily soluble in water, and Stevenson (34) stated that as much as 5–25% of soil organic matter is carbohydrate, including amino sugars, uronic acids, hexoses, pentoses, cellulose, and its derivatives. The enzymatic hydrolysis of polysaccharides by soil microbes releases simple monosaccharides and oligosaccharides into soil solutions, which are flushed from soil during wet seasons into streams and rivers. Because simple sugars are easily utilized by soil organisms, such as bacteria, mold, and fungi, they are a reactive fraction and are continually used and released. Thus, plant and soil organic matter are important contributors to carbohydrates in water. In aquatic systems, such as large lakes and the ocean, algae are an important source of carbohydrates. Carbohydrate concentrations correlate closely to algal populations, and commonly concentrations of carbohydrate decrease with depth as algal populations decrease.

At least 14 different monosaccharides have been identified in natural waters; their structures are similar and include both five- and six-membered rings (Fig. 3.13). Carbohydrates are concentrated by rotoevaporation, hydrolyzed in acid, derivatized, and identified by gas or liquid chromatography. Several methods for the identification of sugars include: gas chromatography with alditol acetate and silyl derivatives (for applications to natural waters see 93–95), liquid chromatography (96–98), enzymatic methods (99,100), and colorimetric methods [phenol-sulfuric acid (101); periodate oxidation and colorimetric determination of formaldehyde (102)]. A review of methods for carbohydrate analysis in seawater is given by Dawson and Liebezeit (91).

The range of concentration of carbohydrates in natural waters is from 65 to 3000 μg/L with average concentrations of approximately 100–500 μg/L, depending on the type of natural water (see Table 3.8). This is the range of concentration for dissolved total carbohydrates, which are those carbohydrates present as simple monosaccharides, oligomers, and polysaccharides. It does not include the sugar acids and carbohydrates present in humic substances. These two categories contribute to the pool of dissolved organic carbon, which is shown in Figure 3.14.

Polysaccharides are the largest fraction of carbohydrate followed by sugars bound to humic substances. Monosaccharides constitute about 25% of the carbohydrates in water. There are various other types of sugars that account for the remaining 25%, including amino sugars, sugar alcohols, sugar acids, and methylated sugars. Little is known about their distribution and importance. Finally, sugar acids are postulated as being important types of dissolved carbohydrates, but little is known about their composition and distribution in natural waters.

3.10 HYDROCARBONS

As a group of compounds, hydrocarbons account for less than 1% of the DOC of most natural waters, but they are an important group of compounds for two reasons. First, the hydrocarbon, methane, contributes significantly

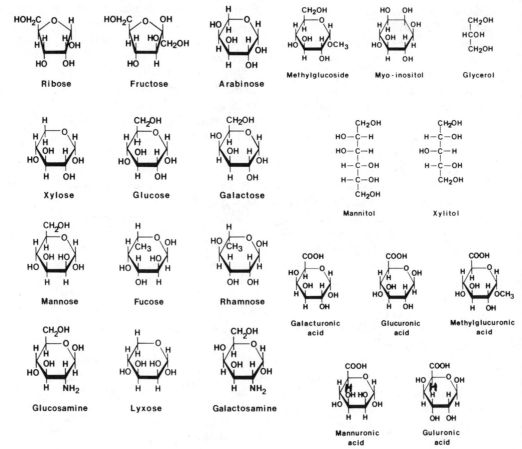

Figure 3.13. Carbohydrates identified in natural waters. (Published with permission of Martinus Nijhoff-Dr. Junk Publishers, after Ref. 2.)

to the carbon cycle of lakes, groundwaters, and interstitial waters of sediment. In fact the amount of methane in the atmosphere is estimated at 4.3 × 10^{15} g of carbon, which is nearly 10 times the total export of organic carbon by all the world's rivers in a single year (103–105). Second, many hydrocarbons enter water from urban sources, contaminating surface waters, and are a health hazard. Therefore, it is important to contrast the naturally occurring compounds with hydrocarbons from man's activities and to understand their source and occurrence.

Naturally occurring hydrocarbons in fresh water include saturated aliphatic hydrocarbons (alkanes and isoprenoids), unsaturated aliphatic hydrocarbons (*n*-alkenes and branched alkenes), saturated cyclic hydrocarbons, unsaturated cyclic hydrocarbons, and simple and fused ring aromatic hydrocarbons (Fig. 3.15).

Hydrocarbons originate both in the water and on the land and a good

TABLE 3.8
Concentration of Total Dissolved Carbohydrates in
Various Natural Waters (2)

Natural Water	Concentration (μg/L)	% DOC
Groundwater	100 (65–125)	1–4
Seawater	250 (100–1000)	5–10
River water	500 (100–2000)	5–10
Lake water	500 (100–3000)	8–12

Source: Ref. 2.

survey of the sources of hydrocarbons in marine environments is Saliot (106) and in sediments of fresh waters is Barnes and Barnes (107). Phytoplankton, benthic algae, zooplankton, and bacteria are important sources of hydrocarbons in the aquatic environment. Saliot (106) reviewed these sources for different types of hydrocarbons. For example, n-alkanes are a major type of hydrocarbon in the marine environment originating from phytoplankton, benthic and pelagic algae, and bacteria. The most common of the n-alkanes is heptadecane (n-C_{17}). Other hydrocarbons that are found include n-C_{15}, n-C_{21}, n-C_{23}, and n-C_{29}. Terrestrial alkanes are longer in chain length than the n-alkanes from algae; for example, cuticular waxes are common products of plants with a chain of n-C_{23} to n-C_{33} with a preference for odd-numbered chains with a dominance of n-C_{29} or n-C_{31} (108). Because the synthesis of alkanes in plants comes from even-numbered fatty acids and the subsequent

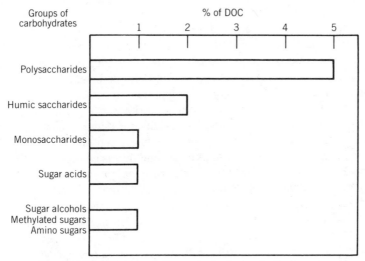

Figure 3.14. Histogram for carbohydrates in fresh waters.

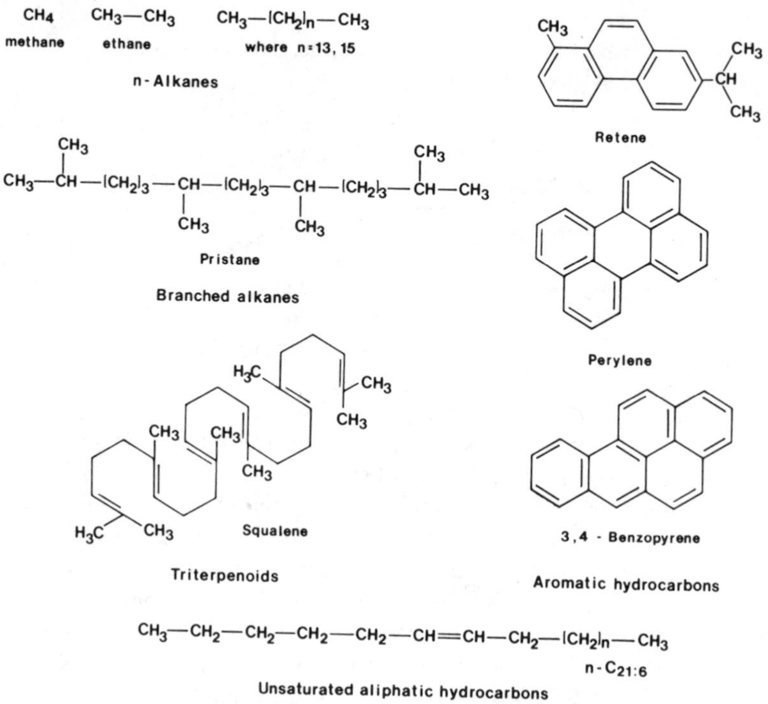

Figure 3.15. Types of natural hydrocarbons in water. (Summarized from data in Ref. 106.)

loss of CO_2 in the process (86), terrestrial alkanes are dominated by odd-numbered chain lengths. An important example of short-chain alkanes from higher plants is the hydrocarbon *n*-heptane (C_7), which is a major hydrocarbon in some species of pine (*Pinus jeffreyi* and *Pinus sabiniana*).

Important regular branched alkanes are isoprenoids, such as pristane and phytane. They are found in phytoplankton, benthic algae, zooplankton, and bacteria. The irregular branched alkanes, such as 7- and 8-methylheptadecane occur in algae and 8-methylheptadecane is an important branched alkane found in seawater (106).

The alkenes are found in phytoplankton, benthic algae and bacteria, and a typical alkene is 7-heptadecene (106). Unsaturated odd-carbon-numbered olefins, such as *n*-C_{21}:6, predominate in marine phytoplankton species (106). Squalene is an example of a branched alkene that is common in phytoplankton, bacteria, and zooplankton. Because squalene is presumed to be an intermediate in steroid biosynthesis, it must be made by all organisms that synthesize steroids (86). Saturated cyclic hydrocarbons that may occur in natural waters are the triterpenes, which are important products from higher plants. Hopane is a typical example. Finally, there are a few polynuclear aromatic hydrocarbons found in nature, such as retene, perylene, and 3,4-benzopyrene (106).

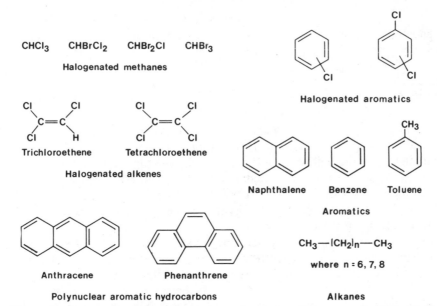

Figure 3.16. Structures of man-made hydrocarbons found in water. (Published with permission of Martinus Nijhoff-Dr. Junk Publishers, after Ref. 2.)

The common man-made hydrocarbons entering natural waters include chlorinated aliphatic and aromatic hydrocarbons (common solvents), polynuclear aromatic hydrocarbons (from combustion sources), saturated and unsaturated alkanes (from waste oil), and chlorinated methanes from chlorination of drinking water and wastewater (Fig. 3.16). In this chapter there is only a brief mention of these classes of compounds, and a review of their occurrence in water may be found elsewhere (references that would be useful are 109, 110).

Halogenated alkanes and alkenes are the result of chlorination of water and wastewater. Chlorinated solvents are common contaminants, they include volatile solvents used in industry as cleaning and degreasing solvents. Two compounds that seem to be common contaminants in groundwater are tetrachloroethene and trichloroethene. Alkanes and alkenes that result from petroleum products such as oil, diesel fuel, and gasoline products may enter many natural surface and groundwaters. Chlorinated aromatic compounds come from industrial use as solvents. Agricultural application of pesticides is an important source of these compounds in natural surface and groundwaters. Polynuclear aromatic hydrocarbons may come from combustion sources including automobiles, coal and oil burning, and natural sources of combustion such as forest fires.

The concentration of individual hydrocarbons in water is commonly at the nanogram per liter level (ppt), because of the insolubility and volatility of hydrocarbons. The short chain-length hydrocarbons are soluble, but because they are volatile they are commonly present at concentrations of nanograms per liter. The nonvolatile hydrocarbons are less soluble and are usually present at nanogram per liter levels also.

Generally, the solubility of hydrocarbons in water is a function of the size of the hydrocarbon and the number of aromatic rings and double bonds. The smaller the hydrocarbon is, the more water soluble it is. Also if hydrocarbons contain aromatic character or double bonds, this increases water solubility compared to a straight-chain hydrocarbon with the same number of carbon atoms. Volatility of a hydrocarbon is a function of its boiling point, with the lower the boiling point the greater the volatility of the hydrocarbon.

Although Saliot has tabulated data from seawater on hydrocarbon concentration and sufficient data are available to estimate concentrations in seawater, not enough is known of hydrocarbons in fresh waters to estimate concentrations precisely.

Hydrocarbons in seawater have been extensively studied (106) because of their importance in biosynthesis and transformation reactions that produce specific types of compounds with remarkable stability in water and sediment. Therefore, they are useful as biological markers. Another reason for the intense study of hydrocarbons in water and marine sediment is that hydrocarbons are source material for petroleum and are a key to the formation of oil. For this reason there have been many studies on hydrocarbons in seawater and marine sediments.

Because of the limited solubility of hydrocarbons in water, they may form a second phase, either as an emulsion or as a film on the water's surface. The concentration of hydrocarbons in water will vary depending on the location of the sample. If the surface film is sampled, concentrations of hydrocarbons may be considerably greater, as much as 50 µg/L. While in the bulk water phase, concentrations of total hydrocarbons will be 1–5 µg/L and individual hydrocarbons will be 1–20 ng/L.

Finally, hydrocarbons may be analyzed by several methods. Gas analysis by headspace has been used for methane (111) purge and trap has been used for volatile, chlorinated, organic compounds, closed-loop stripping for semi-volatile organics (112), and liquid extraction with methylene chloride and hexane (109).

3.11   CONCLUSIONS

The amount of dissolved organic compounds in natural waters is an interesting area of research in water chemistry. The advent of new instrumentation, such as gas and liquid chromatography coupled with mass spectrometry, allows us to determine structures of organic compounds with accuracy. The isolation of the organic matter and concentration to detectable limits is technologically available. Thus, there should be continued information on the types of natural organic compounds in water. One obstacle to our understanding of dissolved organic matter is how are the specific compounds present? For example, both carbohydrates and amino acids are present in more complex forms, which we call combined carbohydrates and amino acids. Studies should focus on where these substances come from and what are their structures and use in biological systems.

There is also a need for a compilation of anthropogenic compounds in waters. The literature has expanded considerably on organic contaminants in natural waters, and no compilation or explanation of pathways and geochemical processes are known. This topic has not been discussed in this chapter, and indeed, is a good subject for future books.

REFERENCES

1. Duursma, E.K., and Dawson, R., *Marine Organic Chemistry*, Elsevier, Amsterdam, 1981.
2. Thurman, E.M., *Organic Geochemistry of Natural Waters*, Martinus Nijhoff-Dr. Junk Publishers, Dordrecht, 1985.
3. Van Hall, C.E., Safranko, J., and Stenger, V.A., Rapid combustion method for the determination of organic substances in aqueous solutions, *Anal. Chem. 35*, 315–319, 1963.
4. Menzel, D.W., and Vaccaro, R.F., The measurement of dissolved organic and particulate carbon in seawater, *Limnol. Oceanogr., 9*, 138–142, 1964.
5. Wangersky, P.J., and Hincks, A.V., The shipboard intercalibration of filters used in the measurement of particulate organic carbon, National Research Council of Canada, Marine Analytical Chemistry Standards Progress Technical Report NRCC No. 16767, 1978.
6. MacKinnon, M.D., The measurement of organic carbon in sea water, in *Marine Organic Chemistry*, Duursma, E.K., and Dawson, R., (eds.), Elsevier, Amsterdam, pp. 415–444, 1981.
7. Leenheer, J.A., Malcolm, R.L., McKinley, P.W., and Eccles, L.A., Occurrence of dissolved organic carbon in selected groundwater samples in the United States, *U.S. Geol. Surv. J. Res., 2*, 361–369.
8. Robinson, L.R., Connor, J.T., and Engelbrecht, R.S., Organic materials in Illinois groundwaters, *Am. Water Works Assn. J., 59*, 227–236, 1967.
9. Maier, W.J., Gast, R.C., Anderson, C.T., and Nelson, W.W., Carbon contents of surface and underground waters in south-central Minnesota, *J. Environ. Qual., 5*, 124–128, 1976.
10. Spalding, R.F., Gormly, J.R., and Nash, K.G., Carbon contents and sources in ground waters of the central Platte region in Nebraska, *J. Environ. Qual., 7*, 428–434, 1978.
11. Barcelona, M.J., TOC determinations in ground water. *Groundwater, 22*, 18–24, 1984.
12. Feder, G.L., and Lee, R.W., Water-quality reconnaissance of Cretaceous aquifers in the southeastern coastal plain, *U.S. Geological Survey, Open-File Report* 81–696, 1981.
13. Meybeck, M., River transport of organic carbon to the ocean, in *Flux of Organic Carbon by Rivers to the Oceans*, Likens, G.E., (ed.), pp. 219–269, U.S. Department of Energy, NTIS Report # CONF-8009140, UC-11, Springfield, Virginia, 1981.
14. Meybeck, M., Carbon, nitrogen, and phosphorus transports by world rivers, *Oikos* (in review), 1983.
15. Birge, E.A., and Juday, C., Particulate and dissolved organic matter in inland lakes, *Ecol. Monogr., 40*, 440–474, 1934.
16. Wetzel, R.G., Organic carbon cycle and detritus, in *Limnology*, W.B. Saunders Company, Philadelphia, pp. 583–621, 1975.
17. Mulholland, P.J., Deposition of riverborne organic carbon in floodplain wetlands and deltas, in *Flux of Organic Carbon by Rivers to the Oceans*, Likens, G.E., (ed.), U.S.

Department of Energy, NTIS Report # CONF-8009140, UC-11, Springfield, Virginia, pp. 142–172, 1981a.

18. Mulholland, P.J., Formation of particulate organic carbon in water from a southeastern swamp-stream, *Limnol. Oceanogr., 26,* 790–795, 1981b.

19. Schnitzer, M., and Khan, S.U., *Humic Substances in the Environment,* Marcel Dekker, New York, 1972.

20. Thurman, E.M., Wershaw, R.L., Malcolm, R.L., and Pinckney, D.J., Molecular size of aquatic humic substances, *Organic Geochem., 4,* 27–35, 1982.

21. Thurman, E.M., Malcolm, R.L., and Aiken, G.R., Prediction of capacity factors for aqueous organic solutes adsorbed on a porous acrylic resin, *Anal. Chem. 50,* 775–779, 1978.

22. Thurman, E.M., and Malcolm, R.L., Preparative isolation of aquatic humic substances, *Environ. Sci. Technol., 15,* 463–466, 1981.

23. Leenheer, J.A., and Huffman, E.W.D. Jr., Classification of organic solutes in water by using macroreticular resins, *J. Res., U.S. Geol. Surv., 4,* 737–751, 1976.

24. Leenheer, J.A., Comprehensive approach to preparative isolation and fractionation of dissolved organic carbon from natural waters and wastewaters, *Environ. Sci. Technol., 15,* 578–587, 1981.

25. Berzelius, J.J., Undersokning af adolfsbergs brunnsvatten, Undersohning af Porlakallratten Berzelius och Hisinger's, *Afhandlingar i Physik, Kemi, och Mineralogi, 1,* 125–145, 1806.

26. Berzelius, J.J., Sur deux acides organiques qu'on trouve dans les eaux minerales, *Ann. Chim. Phys., 54,* 219–231, 1833.

27. Oden, S., The humic acids, studies in their chemistry, physics, and soil science, *Kolloidchemische Beihefte, 11,* 75–260, 1919.

28. Shapiro, J., Chemical and biological studies on the yellow organic acids of lake water, *Limnol. Oceanogr., 2,* 161–179, 1957.

29. Kalle, K., The problem of gelbstoff in the sea, in *Oceanographic Marine Biological Annual Reviews # 4,* Barnes, M., (ed.), Allen and Unwin, London, pp. 91–104, 1966.

30. Gjessing, E.T., *Physical and Chemical Characteristics of Aquatic Humus,* Ann Arbor Science, Ann Arbor, 1976.

31. Riley, J.P., and Taylor, D., The analytical concentration of traces of dissolved organic materials from seawater with Amberlite XAD-1 resin, *Anal. Chim. Acta, 46,* 307–309, 1969.

32. Mantoura, R.F.C., and Riley, J.P., The analytical concentration of humic substances from natural waters, *Anal. Chim. Acta, 76,* 97–106, 1975.

33. Aiken, G.R., Thurman, E.M., Malcolm, R.L., and Walton, H.F., Comparison of XAD macroporous resins for the concentration of fulvic acid from aqueous solution, *Anal. Chem., 51,* 1799–1803, 1979.

34. Stevenson, F.J., *Humus Chemistry,* John Wiley & Sons, New York, 1982.

35. Thurman, E.M., Isolation, characterization, and geochemical significance of humic substances from ground water, Ph.D. Thesis, University of Colorado, Boulder, 1979.

36. Packham, R.F., Studies of organic color in natural water, *Proc. Soc. Water Treat. Exam., 13,* 316–334, 1964.

37. Sieburth, J.M., and Jensen, A., Studies on algal substances in the sea, I. Gelbstoff interstitial and marine waters, *J. Explor. Marine Biol. Ecol., 2,* 174–189, 1963.

38. Khaylov, K.M., Dissolved organic macromolecules in seawater, *Geochem. Int., 5,* 497–503, 1968.

39. Skopintsev, B.A., Bakulina, A.G., Bikbulatova, E.M., Kudryavtseva, N.A., and Melnikova, N.I., Organic matter in the water of Volga and its reservoirs, *Trudy Instituta Biologii Vodokhranilishch Akademiya Nauk SSSR, 26,* 39–53, 1972.

40. Ogura, N., Molecular weight fractionation of dissolved organic matter in coastal seawater by ultrafiltration, *Marine Biol., 24,* 305–312, 1974.

41. Brown, M., High molecular-weight material in Baltic seawater, *Marine Chem., 3,* 253–258, 1975.
42. Kerr, R.A., and Quinn, J.G., Chemical studies on the dissolved organic matter in seawater. Isolation and fractionation, *Deep-Sea Research, 22,* 107–116, 1975.
43. Stuermer, D.H., The characterization of humic substances from sea water: PH.D. Thesis, Massachusetts Institute of Technology and Woods Hole Oceanographic Institution, Cambridge, 1975.
44. Stuermer, D.H., and Payne, J.R., Investigations of seawater and terrestrial humic substances with carbon-13 and proton nuclear magnetic resonance, *Geochim. Cosmochim. Acta, 40,* 1109–1114, 1976.
45. Stuermer, D.H., and Harvey, G.R., The isolation of humic substances and alcohol-soluble organic matter from seawater, *Deep-Sea Res., 24,* 303–309, 1977.
46. Stuermer, D.H., and Harvey, G.R., Humic substances from sea water, *Nature, 250,* 480–481, 1974.
47. Stuermer, D.H., and Harvey, G.R., Structural studies on marine humus. A new reduction sequence for carbon skeleton determination, *Marine Chem., 6,* 55–70, 1978.
48. Stuermer, D.H., Peters, K.E., and Kaplan, I.R., Source indicators of humic substances and protokerogen. Stable isotope ratios, elemental compositions, and electron spin resonance spectra, *Geochim. Cosmochim. Acta, 42,* 989–997, 1978.
49. Thurman, E.M., Origin and amount of humic substances in an alpine-subalpine watershed, *Arctic Alpine Res.,* (in review), 1984.
50. Malcolm, R.L., and Durum, W.H., Organic carbon and nitrogen concentrations and annual organic carbon load of six selected rivers of the United States, *U.S. Geological Survey Water-Supply Paper,* 1817-F, 1976.
51. Leenheer, J.A., Origin and nature of humic substances in the waters of the Amazon River basin, *Acta Amazonica, 10,* 513–526, 1980.
52. Weber, J.H., and Wilson, S.A., The isolation and characterization of fulvic acid and humic acid from river water, *Water Res., 9,* 1079–1084, 1975.
53. Beck, K.C., Reuter, J.H., and Perdue, E.M., Organic and inorganic geochemistry of some coastal plain rivers of the southeastern United States, *Geochim. Cosmochim. Acta, 38,* 341–364, 1974.
54. Thurman, E.M., and Malcolm, R.L., Structural study of humic substances: New approaches and methods, in *Aquatic and Terrestrial Humic Materials,* Christman, R.F. and Gjessing, E.T., (eds.), Ann Arbor Science, Ann Arbor, pp. 1–23, 1983.
55. Perdue, E.M., Lytle, C.R., Sweet, M.S., and Sweet, J.W., The chemical and biological impact of Klamath Marsh on the Williamson River, Oregon, *Water Resources Research Investigation,* #71, 1981.
56. Malcolm, R.L., Humic substances in rivers and streams, in *Humic Substances I. Geochemistry, Characterization, and Isolation,* Aiken, G.R., MacCarthy, P., McKnight, D., and Wershaw, R.L., (eds.), John Wiley Inc., New York, 1985.
57. Larson, R.A., and Rockwell, A.L., Fluorescence spectra of water-soluble humic materials and some potential precursors, *Arch. Hydrobiol., 89,* 416–425, 1980.
58. Ghassemi, M., and Christman, R.F., Properties of the yellow organic acids of natural waters, *Limno. Oceanogr, 13,* 583–597, 1968.
59. Hair, M.E., and Bassett, C.R., Dissolved and particulate humic acids in an east coast estuary, *Estuarine and Coastal Marine Science, 1,* 107–111, 1973.
60. Dawson, H.J., Hrutfiord, B.F., Zasoski, R.J., and Ugolini, F.C., The molecular weight and origin of yellow organic acids, *Soil Sci., 132,* 191–199, 1981.
61. Eberle, S.H., and Schweer, K.H., Bestimmung von huminsaure und ligninsulfonsaure im wasser durch flussig-flussig estraktion, *Vom Wasser, 41,* 27–44, 1974.
62. Rashid, M.A., and Prakash, A., Chemical characteristics of humic compounds isolated from some decomposed marine algae, *J. Fish. Res. Board Canada, 29,* 55–60, 1972.
63. Oliver, B.G., and Visser, S.A., Chloroform production from the chlorination of aquatic

humic material, The effect of molecular weight, environment, and season, *Water Res., 14,* 1137–1141, 1980.

64. Buffle, J., Deladoey, J.P., and Haerdi, W., The use of ultrafiltration for the separation and fractionation of organic ligands in freshwaters, *Anal. Chim. Acta, 101,* 339–357, 1978.

65. Fotiyev, A.V., The nature of aqueous humus, *Doklady Akademii Nauk SSSR, 199,* 198–201, 1971.

66. Wilander, A., A study on the fractionation of organic matter in natural water by ultrafiltration techniques, *Schweizerische Zeitschrift fuer Hydrologie, 34,* 190–200, 1972.

67. Ishiwatari, R., Hanana, H., and Machehana, T., Isolation and characterization of polymeric organic materials in a polluted river water, *Water Res., 14,* 1257–1262, 1980.

68. Shapiro, J., Yellow acid-cation complexes in lake water, *Science, 127,* 702–704, 1958.

69. Christman, R.F., and Minear, R.A., Organics in lakes, in *Organic Compounds in Aquatic Environments,* Faust, H., (ed.), Marcel Dekker, New York, pp. 119–141, 1971.

70. Stewart, A.J., and Wetzel, R.G., Dissolved humic materials: Photodegradation, sediment effects, and reactivity with phosphate and calcium carbonate precipitation, *Arch. Hydrobiol., 92,* 265–286, 1981.

71. DeHaan, H., Some structural and ecological studies on soluble humic compounds from the Tjeukemeer, *Internationale Vereinigung fuer Theoretische und Angewandte Limnologie, Verhandlungen, 18,* 685–695, 1972a.

72. DeHaan, H., Molecule-size distribution of soluble humic compounds from different natural waters, *Freshwater Biol., 2,* 235–241, 1972b.

73. DeHaan, H., Limnological aspects of humic compounds in Lake Tjeukemeer, Ph.D. Thesis, University of Groningen, The Netherlands, 1975.

74. DeHaan, H., and DeBoer, T., A study of the possible interactions between fulvic acids, amino acids, and carbohydrates from Tjeukemeer, based on gel filtration at pH 7.0, *Water Res., 12,* 1035–1040, 1978.

75. DeHaan, H., and DeBoer, T., Seasonal variations of fulvic acid, amino acids, and sugars in the Tjeukemeer, The Netherlands, *Archiv fur Hydrobiologie, 85,* 30–40, 1979.

76. Hama, T., and Handa, N., Molecular weight distribution and characterization of dissolved organic matter from lake waters, *Arch. Hydrobiol., 90,* 106–120, 1980.

77. Tuschall, J.R., and Brezonik, P.L., Characterization of organic nitrogen in natural waters: Its molecular size, protein content, and interactions with heavy metals, *Limnol. Oceanogr., 25,* 495–504, 1980.

78. Baccini, P., Grieder, E., Stierli, R., and Goldberg, S., The influence of natural organic matter on the adsorption properties of mineral particles in lake water, *Schweizerische Zeitschrift fuer Hydrologie, 44,* 99–116, 1982.

79. Wilson, M.A., Barron, P.F., and Gillam, A.H., The structure of freshwater humic substances as revealed by ^{13}C-NMR spectroscopy, *Geochim. Cosmochim. Acta, 45,* 1743–1750, 1981.

80. Midwood, R.B., and Felbeck, G.T. Jr., Analysis of yellow organic matter from fresh water, *J. Am. Water Works Assn., 60,* 357–366, 1968.

81. Black, A.P., and Christman, R.F., Characteristics of colored surface waters, *J. Am. Water Works Assn., 55,* 753–770, 1963a.

82. Black, A.P., and Christman, R.F., Chemical characteristics of fulvic acids, *J. Am. Water Works Assn., 55,* 897–912, 1963b.

83. Christman, R.F., Johnson, J.D., Pfaender, F.K., Norwood, D.L., Webb, M.R., Haas, J.R., and Babenrieth, M.J., Chemical identification of aquatic humic chlorination, in *Water Chlorination Environmental Impact and Health Effects,* Vol. 3, Jolley, R.L., Brungs, W.A., and Cumming, R.B., (eds.), Ann Arbor Science, Ann Arbor, pp. 75–84, 1980.

84. Caine, J., Sources of dissolved humic substances of a subalpine bog in the Boulder Watershed, Colorado, Master's Thesis, Department of Geology, University of Colorado, 1982.

85. McKnight, D.M., Thurman, E.M., Wershaw, R.L., and Hemond, H., Biogeochemistry of dissolved organic material in Thoreau's Bog, Concord, Massachusetts, *Ecology,* (in press), 1984.

86. Robinson, T., *The Organic Constituents of Higher Plants,* 4th edition, Cordus Press, North Amherst, Massachusetts, 1980.
87. Lehninger, A.L., *Biochemistry,* Worth Publishers, New York, 1970.
88. Lewis, R.W., The fatty acid composition of arctic marine phytoplankton and zooplankton with special reference to minor acids, *Limnol. Oceanogr., 14,* 35–40, 1969.
89. Jeffries, H.P., Fatty-acid ecology of a tidal marsh, *Limnol. Oceanogr., 17,* 433–440, 1972.
90. Poltz, V.J., Untersuchungen uber das Vorkommen und den Abbau von Fetten und Fett-sauren in Seen, *Arch. Hydrobiol., 4,* Supplement 40, 315–399, 1972.
91. Dawson, R., and Liebezeit, G., The analytical methods for the characterization of organics in seawater, in *Marine Organic Chemistry,* Duursma, E.K., and Dawson, R., (eds.), Elsevier, Amsterdam, pp. 445–496, 1981.
92. Dahm, C.N., Pathways and mechanisms for removal of dissolved organic carbon from leaf leachate in streams, *Can. J. Fish. Aquat. Sci., 38,* 68–76, 1981.
93. Sweet, M.S., and Perdue, E.M., Concentration and speciation of dissolved sugars in river water, *Environ. Sci. Technol., 16,* 692–698, 1982.
94. Cowie, G.L., and Hedges, J.I., Determination of neutral sugars in plankton, sediments, and wood by capillary gas chromatography of equilibrated isomeric mixtures. *Anal. Chem., 56,* 497–504, 1984a.
95. Cowie, G.L., and Hedges, J.I., Carbohydrate sources in a coastal marine environment, *Geochim. Cosmochim. Acta, 48,* 2075–2088, 1984b.
96. Mopper, K., Sugars and uronic acids in sediment and water from the Black Sea and North Sea with emphasis on analytical techniques, *Marine Chem., 5,* 585–603, 1977.
97. Mopper, K., Dawson, R., Liebezeit, G., and Ittekkot, V., The monosaccharide spectra of natural waters, *Marine Chem., 10,* 55–66, 1980.
98. Mopper, K., and Johnson, L., Reversed-phase liquid chromatographic analysis of Dns-sugars Optimization of derivatization and chromatographic procedures and applications to natural samples, *J. Chromatogr., 256,* 27–38, 1983.
99. Hicks, S.E., and Carey, F.G., Glucose determination in natural waters, *Limnol. Oceanogr., 13,* 361–363, 1968.
100. Cavari, B.Z., and Phelps, G., Sensitive enzymatic assay for glucose determination in natural waters, *Appl. Environ. Microbiol., 33,* 1237–1243.
101. Handa, N., Examination on the applicability of the phenol sulfuric acid method on the determination of dissolved carbohydrate in sea water, *J. Oceanogr. Soc. Jpn., 22,* 81–86, 1966.
102. Johnson, K.M., and Sieburth, J.McN., Dissolved carbohydrates in seawater. I. A precise spectrophotometric analysis for monosaccharides, *Marine Chem., 5,* 1–13, 1977.
103. Ehhalt, D.H., Methane in the atmosphere, *J. Air Pollut. Control Assn., 17,* 518–519, 1967.
104. Ehhalt, D.H., Methane in the atmosphere, *Atomic Energy Comm. Symp. Ser., 30,* 144–158, 1973.
105. Khalil, M.A.K., and Rasmussen, R.A., Secular trends of atmospheric methane (CH_4), *Chemosphere, 11,* 877–883, 1982.
106. Saliot, A., Natural hydrocarbons in sea water, in *Marine Organic Chemistry,* Duursma, E.K. and Dawson, R., (eds.), Elsevier, Amsterdam, pp. 327–374, 1967.
107. Barnes, M.A., and Barnes, W.C., Organic compounds in lake sediments, in *Lakes: Chemistry Geology Physics,* Lerman, J. (ed.), Springer-Verlag, New York, pp. 127–152, 1978.
108. Kolattukudy, P.E., Plant waxes, *Lipids, 5,* 259–275, 1970.
109. Keith, L.H., *Identification and Analysis of Organic Pollutants in Water,* Ann Arbor Science, Ann Arbor, 1976.
110. Keith, L.H., *Advances in the Identification and Analysis of Organic Pollutants in Water,* Volume 1 and 2, Ann Arbor Science, Ann Arbor, 1981.
111. Strayer, R.F., and Tiedje, J.M., In situ methane production in a small, hypereutrophic, hard-water lake: Loss of methane from sediments by vertical diffusion and ebullition, *Limnol. Oceanogr., 23,* 1201–1206, 1978.

112. Grob, K., and Grob, G., Organic substances in potable water and in its precursor, Part II. Applications in the area of Zurich, *J. Chromatogr., 90,* 303–313, 1974.
113. Barcelona, M.J., Liljestrand, H.M., and Morgan, J.J., Determination of low molecular weight volatile fatty acids in aqueous samples, *Anal. Chem., 52,* 321–325, 1980.
114. Mueller, H.F., Larson, T.E., and Ferretti, M., Chromatographic separation and identification of organic acids, *Anal. Chem., 32,* 687–690, 1960.
115. Mueller, H.F., Larson, T.E., and Lennarz, W.J., Chromatographic identification and determination of organic acids in water, *Anal. Chem., 30,* 41–44, 1958.
116. Lamar, W.L., and Goerlitz, D.F., Characterization of carboxylic acids in unpolluted streams by gas chromatography, *J. Am. Water Works Assn., 55,* 797–802, 1963.
117. Lamar, W.L., and Goerlitz, D.F., Organic acids in naturally colored surface waters, *U.S. Geological Survey Water-Supply Paper* 1817 A, 1966.
118. Hullett, D.A., and Eisenreich, S.J., Determination of free and bound fatty acids in river water by high performance liquid chromatography, *Anal. Chem., 51,* 1953–1960, 1979.

PART TWO

Identification Methods

CHAPTER 4

Detection of Organic Carcinogens in Drinking Water: A Review of Concentration/Isolation Methods

Charles D. Hertz
I. H. (Mel) Suffet

Environmental Studies Institute
Drexel University
Philadelphia, Pennsylvania

4.1 SAMPLING THE ENVIRONMENT

It has become a cliché that: "The results of the analysis are only as good as the sampling step." Yet, this is a research area that has received the least amount of attention by analysts. This is especially true for the sampling and concentration of potential organic carcinogens in water. The objective of the analysis should control the method of sampling.

This chapter is designed to review methods used for the sampling and concentration of organic carcinogens in water that are present at levels that are too low to be detected by direct determination. Therefore, the purpose for concentrating these organics is to achieve detectable levels with state of the art instruments. The level of concentration that is considered is ng/L to μg/L.

The choice of grab versus composite sampling is a function of the objective of the sampling. This has been reviewed by Huibregtse and Moser (1) along with the means to complete both types of sampling procedures. In the case of trace organics of potential health concern, natural water samples show variability of concentration and type of compounds present. This has been shown in specific studies exemplified for surface water by Suffet et al. (2) and for groundwater by Suffet et al. (3) and thus variability must be taken into consideration when sampling.

Composite sampling is best utilized (1) when:

1. An average concentration is the objective of the study.
2. A dose per unit time is considered.
3. The history of water quality is to be determined based upon long time intervals.

Grab sampling is best utilized when:

1. The composition of the chemicals is constant.
2. The peak concentration is being examined.
3. The maximum variability is to be determined.
4. The history of water quality is to be determined based upon short time intervals.

If the evaluation of long-term health effects is a major objective, then composite sampling based upon volume or time is optimum. An example would be the case of sampling organic carcinogens that are present at low concentrations. Too many short-term grab samples would be needed to obtain average values desired for this type of evaluation and the expense of obtaining and analyzing the samples would be prohibitive. Therefore, for drinking water evaluations of potential carcinogenic concern, composite sampling is recommended.

Drinking water standards are set as maximum contaminant levels (MCLs). However, since grab sampling can only evaluate MCLs, perhaps using composite sampling and the term average contaminant level (ACL) would be more fitting. This was suggested by Yohe et al. (4) when the granular activated carbon process was being evaluated for the removal of trace organic contaminants from drinking water. In summary, it should be stated again that the objective of the sampling program is the guide for its application and not the ease of completing an analysis.

This review is methodology based. Methods were classified into nine categories based on the major processes involved in isolating and concentrating an analyte. Any such classification is difficult because of the overlap of processes in most of the methods. Methodology-based reviews have been written by Jolley (5) and Kopfler (6) for concentrating organics for biological testing. Other recent reviews have been written by Jennings and Rapp (7), Karasek et al. (7a), Khromchenko and Rudenko (8), McElroy et al. (9), the National Research Council (10), and Trussell and Umphres (11).

Leenheer's (12) review of concentration methods was somewhat unique as it focused on the analytical techniques from the perspective of the analyte. In his excellent review, Leenheer was able to:

1. Define various methods for the same analyte.
2. Explore the principles and mechanisms of the methods.

3. Describe important sampling and storage considerations.
4. Propose a comprehensive analytical scheme.

The authors of the current review do not attempt to review concentration methods to the same extent as Leenheer (12). This chapter is an overview of concentration and isolation methods with an emphasis on organic carcinogens in drinking water.

In the current review, several topics are emphasized:

1. Analytical approach. Selection of broad spectrum or specific compound analysis.
2. Quality assurance supporting analytical limit of detection. Quality control procedures are not emphasized as these are best left to manuals and protocols (13,14).
3. Recognizing that carcinogenicity is not a physical property, both carcinogenic and noncarcinogenic compounds will be concentrated by a single method. Compounds considered carcinogenic or potentially carcinogenic are used for examples of recovery and limit of detection. Other organics are used to illustrate the capability or limitation of a particular method.

4.2 ANALYTICAL APPROACH

Two different analytical approaches can be used to determine specific organic compounds: (1) analysis of individual compounds and (2) broad spectrum analysis. In the first approach, individual compounds are selected for analysis (e.g., because of health implications). Traditional methods are used for an absolute quantitative analysis required to meet a standard. One example is the haloform methods used to meet an MCL (13,15–17).

Broad spectrum analysis is a method of screening many compounds. Compounds of interest are selected from the sample matrix by the choice of the concentration or isolation method. Such a method investigates the largest possible number of compounds in a single sample. This analytical approach is based upon the ability to compare entire chromatograms for the purpose of observing changes in water quality. Simultaneous comparison of many chromatograms becomes manageable by using computer reconstructed chromatograms (profiles) (18,19).

The characteristics of samples for which broad spectrum analysis is best applied are:

1. Samples that require a minimum of pretreatment.
2. Samples that contain a large molecular weight range of organic compounds.

Possible applications of broad spectrum analysis which observe changes between chromatograms are:

1. Sampling before and after a unit process, to observe the efficiency of a process.
2. Temporal sampling program; sampling at one location over a time interval.
3. Spatial sampling program; sampling different locations at the same time.

Examples of the broad spectrum analysis of water include the semiquantitative changes between samples by evaluating:

1. Peak to peak changes in GC detector response (2,20,21).
2. Visual impressions of the overall numbers of peaks and their areas (22).
3. Peaks that have different concentrations (23).
4. Changes in peak response in different parts of the chromatogram (2,3,20–23).
5. Comparison of overall chromatograms (3,24,25).

One major instrumental advancement that has aided broad spectrum analysis is the availability of capillary columns for the gas chromatograph (GC). These columns make possible the separation of extremely complex mixtures of organic compounds at the ng/L level. With the use of capillary columns, the number of peaks observed in a drinking water sample (Delaware River at Philadelphia, PA) increased from 60 with packed column GC to over 250 peaks with capillary column (20). The increased ability of capillary GC to resolve organic compounds also brought about disadvantages. The precision of the analysis became more sensitive to the sample matrix than when packed column GC was used. This reduction of precision occurred because the amount of stationary phase on a capillary column is much less than on the stationary phase of a packed column. This, in turn, increases the likelihood of observing interferences originating from the sample matrix (26).

Coyle et al. (20) used chromatographic profiles generated by a desk top computer to do a broad spectrum analysis. They used this approach to study the removal of trace organic compounds in a water treatment plant. Organic compounds were isolated by adsorption onto XAD-2 resin. Compounds were eluted with four portions of diethyl ether which was then concentrated to 1 mL by Kuderna-Danish evaporation. All samples were analyzed by GC/FID to generate broad spectrum profiles. Selected samples were analyzed by GC/mass spectrometry (MS) in order to identify the most prominent peaks in the profile.

In summary, broad spectrum analysis is one approach to water analysis that can determine overall trends and semiquantitative changes between samples. When coupled with specific compound analysis, this approach can be a valuable method of assessing general water quality. The decision to apply broad spectrum and/or specific compound analysis depends on the objectives of the analysis.

4.2.1   Quality Assurance/Quality Control (QA/QC)

Analytical objectives for environmental samples are different from many other types of samples because analyte concentrations are usually very low, i.e., ng/L to μg/L. These concentrations can represent small responses from analyte detectors and a critical issue is the reliability of the determination. The question is: "Is the analyst confident that the result is not a random fluctuation of the instrument's baseline?" Often the confidence with which a result is reported is a function of the quality assurance program that has supported the analytical measurement. Therefore, to report low analyte concentrations with high confidence, there must be in place a good quality assurance plan. Quality control is the mechanism established to assure reliable results (27,28).

An example of excellent quality assurance is in the EPA's Master Analytical Scheme (MAS) (14,29–31). Each MAS protocol includes extensive quality control procedures, such as sampling considerations, sample preservation, blanks, standards, and instrument checks. Considerable attention is given to the performance of the GC/MS instrument.

An important aspect of trace analysis is the concept of limit of detection (LOD). The American Chemical Society (ACS) has defined LOD as "the lowest concentration that can be determined to be statistically different from a blank" (27,28). One of the major purposes of concentrating and/or isolating an analyte is to improve the overall limit of detection. The LOD is associated with many phases of an analysis and is dependent upon:

1. Analyte.
2. Analyte recovery.
3. Sample size; original and final.
4. Analytical procedure, including concentration/isolation.
5. Instruments; detector and associated equipment.
6. Interferences from compounds similar to analyte.
7. Matrix effects.

Over the years, limits of detection have been calculated differently by many groups (32–35). Recently, several organizations and individuals have attempted to standardize this calculation process (27,28,36–38). Some of these definitions of LOD are shown in Table 4.1. A general similarity among the definitions is that the LOD is located 3 standard deviations above the

TABLE 4.1
Methods Used to Determine Limit of Detection

Method	Terminology	Comments	References
ACS	Limit of detection (LOD) Limit of quantitation (LOQ) Instrument detection limit	1. LOD = 3σ + blank 2. LOQ = 10σ + blank 3. Values < LOD should be reported as "Not Detected"	27,28
ASTM	Criterion of detection Limit of detection T code W code	1. For labs with established quality control programs 2. Negative results are reported rather than censoring data 3. LOD = $2\times$ criterion of detection	36
EPA	Method detection limit (MDL)	1. Emphasis on operational characteristics of MDL 2. Analyte must be present for MDL to be meaningful 3. Analyte can be quantified below MDL 4. As blank $\rightarrow$ 0, MDL = LOD	37
Classical graphic method	LOD	1. Subject to errors in estimation of slope and intercept 2. Can approximate lower detection limit	38
IUPAC	LOD	1. Acceptable when slope is well defined and intercept is zero 2. Often estimates low values for detection limit	38
Propagation of errors method	LOD	1. Considers error in the intercept 2. Accounts for nonzero intercept and nonlinear calibration curve	38

blank response. The authors recommend use of the procedures set forth by the American Chemical Society because it is a clear and well-defined approach.

The concepts of LOD apply to all methods of concentration and isolation of "trace" amounts of analytes. In many cases, reporting an analyte as being present in "trace amounts" is unacceptable. An important factor in deciding how much QC is needed is to determine the consequences of reporting a result that is "wrong" (28). If the objectives of the analysis include a screening program, looking for "hot spots" or very general trends, the objectives may be satisfied with a minimum amount of QC. However, if the objective of the analysis is compliance with Federal regulations, a comprehensive QA plan is desirable—if not mandatory.

Snyder and van der Wal (39) studied the error involved in a typical liquid chromatographic analysis using a liquid–liquid extraction procedure for individual compounds. Error in the determination was divided into contributions proportional to or independent of the measurement technique. A general error model is developed where the total source of variation is the sum of the squares for the variation for each step in an analysis. Sources of error are discussed for each operation in an analysis. Error associated with the concentration or isolation technique include:

1. Measurement of solvent volumes and subsequent changes in volume due to evaporation.
2. Variation of analyte extraction.
3. Measurement of extract volume.
4. Handling losses of analyte in extract.
5. Loss of analyte by evaporation.
6. Incomplete dissolution of analyte.
7. Measurement of aliquot of extract.

Snyder and van der Wal (39) also mention that many of these errors can be accounted for by the use of internal standards. Gibs et al. (26) recently completed a similar evaluation of error for broad spectrum analysis of trace organics by the macroreticular resin isolation method. These two error analyses are models of what should be done for every individual compound and broad spectrum analysis.

Problems of sample integrity and artifacts are common to all concentration methods. Field blanks and spiked blanks are usually the best way to identify problems. Artifacts or losses can be produced at any point in the analytical method. Possible sources of artifacts are shown in Table 4.2. The list of artifacts is not intended to point out a hopeless situation; rather it should remind the analyst of potential problems within the analytical process.

TABLE 4.2
Sources of Artifacts in Analytical Methods

Source	Selected References
Sampling	
Sample vessels (adsorption and/or leaching)	55,140
Tubing	
Sampling technique	
Storage	
Microbiological changes	56
degradation of analyte	
metabolic products	
Photosensitive changes	
Analytical procedure	
Solvent impurities	141
Solvent preservatives	142
Adsorbent impurities	26,65,94
Membrane impurities or preservatives	143
Reagents	
derivatizing or drying of glassware	
reaction of analytes with:	
solvent, resin, membrane	
impurities in solvent, resin, membrane	144
preservatives in solvent, membrane	
other solutes, e.g., humic substances	109
Heated reactions:	
during distillation	80,145
during Kuderna-Danish evaporation	
within instrument, i.e., hot injection port of GC	
Impurities:	
from laboratory air, e.g., CLSA	24,64
from carrier gas, e.g., GC	

4.2.2 Concentration/Isolation Methods for the Analysis of Organics in Water

The concentration or isolation of an organic compound from water depends primarily on the movement of the analyte into or onto a second phase. This second phase can be in the form of a gas, a solid, or another liquid. Methods can be classified according to the type of compound that is isolated and boiling point range (see Fig 4.1) or the phases through which an analyte must transfer:

Liquid/gas: static headspace, purge and trap, closed loop stripping analysis (CLSA), distillation, residue (evaporative) methods.

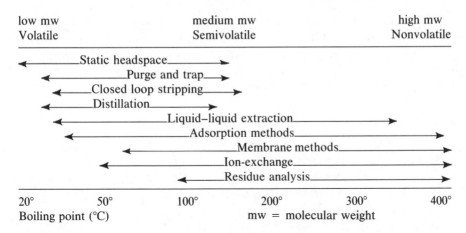

Figure 4.1. Concentration/isolation methods for the analysis of organics in water.

Liquid/solid: adsorption resins, ion-exchange, membranes.
Liquid/liquid: liquid–liquid extraction.

The latter classification method is more complicated because several of the methods involve more than one phase transfer. For example, the CLSA method involves three distinct phase transfers, that is, liquid/gas; gas/solid; solid/liquid. This is discussed in more detail in each section of this chapter.

The distribution coefficient is a common thermodynamic parameter used to describe the equilibrium condition between two phases. The Gibb's phase rule is the basis of the distribution coefficient (40,41).

$$F = C + 2 - P \tag{4.1}$$

The phase rule shows the relationship between the number of components (C), the number of phases (P), and the number of independent variables (F) necessary to characterize a system. For example, Gibb's phase rule describes the behavior of a two-phase system of three components: water, organic solvent, and analyte. In such a system there are three independent variables (degrees of freedom). If two degrees of freedom—pressure and temperature—are kept constant, then only one degree of freedom—concentration—needs to be specified to define the system. This holds true for any single analyte distributed between two phases.

The distribution coefficient (D) and the partition coefficient (K) are often used as synonyms yet the theoretical D should be used as the thermodynamic parameter and the K reserved for the partition coefficient determined experimentally.

$$D = \frac{\gamma_a}{\gamma_b} \times \frac{C_a}{C_b} = K \frac{\gamma_a}{\gamma_b} \tag{4.2}$$

$$K = \frac{C_a}{C_b} \tag{4.3}$$

In dilute solution, the activity coefficients (γ) approach unity and the ratio of concentrations (C) approach the distribution coefficient. The partition coefficient should remain constant, under a given set of conditions, regardless of solute concentration. The partition coefficient will be utilized to describe many of the isolation methods in this chapter.

4.2.3 Static Headspace Analysis

Application: Volatile compounds that are, in general, nonpolar, low molecular weight, and have a low water solubility.

Principles/Processes: Liquid/gas partition that may be aided by addition of heat and/or salt. Partitioning is proportional to Henry's law constants.

General Sensitivity: ppb to ppm.

In addition to drinking water analysis, this method can also be used for samples that might cause problems with other techniques. Examples would be soils or sludges. Static headspace analysis can avoid some of the matrix effects that might lead to problems with these analyses. For aqueous samples, solute partitioning is proportional to its Henry's law constant.

Static headspace equilibrium can be interpreted thermodynamically through either Raoult's law or Henry's law. Raoult's law is concerned with the behavior of the solvent whereas Henry's law models the equilibrium behavior of the solute (40,42). Headspace analysis is a solute isolation technique, therefore it concerns the transfer of an organic compound from the aqueous phase to the vapor phase. The activity coefficient of a solute can be expressed as:

$$\gamma_i = \frac{P_i}{K_H X_i} \tag{4.4}$$

where γ_i = activity coefficient of organic solute i
   P_i = partial pressure of i
   K_H = Henry's law constant
   X_i = mole fraction of i

In dilute solution where ideality is approached, $X_i \to 0$ and $\gamma_i \to 1$ therefore:

$$K_H = P_i \tag{4.5}$$

When salt is added to a water sample, the activity coefficient increases and the solubility of i decreases. This is salting out and when it occurs, the partial pressure of i increases and more solute partitions into the vapor phase.

In terms of analytical measurements, a partition coefficient can be used to describe headspace analysis. In dilute solution it will be assumed that each solute molecule partitions independently.

$$K = \frac{C_v}{C_w} \times \frac{\gamma_v}{\gamma_w} \tag{4.6}$$

where K = partition coefficient
             C_v = concentration of solute in vapor phase
             C_w = concentration of solute in water phase
   γ_v and γ_w = the corresponding activity coefficients

At equilibrium, the activity coefficients are equal. Writing equilibrium concentrations in terms of weight per volume,

$$K = \frac{W_v/V_v}{W_w/V_w} = \frac{W_v}{W_w} \times \frac{V_w}{V_v} \tag{4.7}$$

where the variables are the weights of the solute (W_v, W_w) and volumes (V_v, V_w) in the vapor and water phases, respectively. If α is defined as V_v/V_w and the total amount of solute added to the system is W_t, then $W_w + W_v = W_t$. Replacing W_w with ($W_t - W_v$) and plugging α into Eq. 4.7, the partition coefficient is:

$$K = \frac{W_v}{\alpha(W_t - W_v)} \tag{4.8}$$

Using this equation, Friant and Suffet (43) calculated detection limits of selected polar organics ranging from 50 µg/L for methylethyl ketone to 740 µg/L for dioxane. Other groups (44–46) have also applied static headspace analysis to the isolation of volatiles in drinking water although dynamic headspace methods have generally replaced static methods. The sensitivity is usually not high enough for low ppb quantification for broad spectrum analysis, yet for specific compound analysis, static headspace and purge and trap analysis may be equivalent.

The sensitivity of headspace analysis can be improved by the addition of salt and/or heat. Friant and Suffet (43) optimized such a system for the determination of volatile trace organics. The optimum conditions were an aqueous pH of 7.1, 50°C, and 3.5M Na$_2$SO$_4$. Enrichment factors of up to 66 times were calculated for the optimized system over that of the reference state (pH 7.1, 30°C, without added salt). The combined effects of the salt and temperature were shown to be greater than additive (synergystic) by an ANOVA statistical design.

Otson et al. (47) compared static headspace and purge and trap methods for the determination of trihalomethanes in water. Using a Hall detector, the limit of detection with the purge and trap method was 0.5 ppb for both chloroform and bromoform. Static headspace analysis was able to detect 10 ppb and 8 ppb for chloroform and bromoform, respectively. Both methods are able to determine these compounds at low ppb concentrations, although the purge and trap method is generally more sensitive.

Static headspace analysis can be useful as a low-cost alternative to the dynamic headspace techniques. This is especially true for aqueous samples that tend to form emulsions or foam during sparging. The method is also used in the Master Analytical Scheme (14) as a scouting technique to determine if more elaborate/expensive methods are required. Reviews on the subject have been written recently by McNally and Grob (48) and Ioffe and Vitenberg (46).

4.2.4 Purge and Trap

Application: Volatile to semivolatile compounds; they are relatively nonpolar, low molecular weight, and have a low water solubility.

Principles/Processes: Dynamic headspace technique coupled with sorption and thermal desorption.

Analyte Transfers: Liquid/gas; gas/solid; solid/gas by thermal desorption.

General Sensitivity: ppt to ppm.

The purge and trap technique is most applicable to nonpolar, low molecular weight, volatile compounds. This technique was first used to determine hydrocarbons in seawater (50). The procedure was modified by Bellar and Lichtenberg (51) and has formed the basis for standard protocols adopted by the U.S. Environmental Protection Agency (EPA) Methods 502, 503, 601, and 602 (13,16). In addition to the EPA, other organizations have adopted similar protocols (52). These methods have the ability to determine 32 purgeable priority pollutants in one sample, and can be used for specific analysis, e.g., trihalomethanes (THMs) (52) or for broad spectrum analysis with a flame ionization detector [EPA Method 624 (13)].

The purge and trap method consists of several laboratory unit operations. First, the organic solute must partition from the water sample to the gaseous phase. The ease with which a compound is purged from solution is proportional to the compounds Henry's law constant. Once purged from aqueous solution, the analyte must be adsorbed on resin particles of the trap. Then, the analyte is desorbed thermally from the trap onto the head of a chromatographic column. Both the addition of salt and heat (14,53,54) have been used successfully to improve the transfer of organic solutes from the water to the gas phase. For example ppb amounts of acetone were analyzed by Tai (54) with addition of salt and heat, where the EPA Methods 601, 602 could not analyze it with good precision.

The recovery of purgeable organics is dependent not only on the particular purge and trap/chromatographic system but also on the aqueous matrix. The recovery ± standard deviation of several suspected carcinogens is shown below.

	% Recovery ± SD		
Compound	Distilled Water	Municipal Effluent	Energy Effluent
Benzene	76 ± 7	104 ± 5	111 ± 8
Chloroform	79 ± 11	105 ± 18	95 ± 6
Trichloroethylene	87 ± 2	84 ± 10	96 ± 7

Details of the recovery study are reported elsewhere (30). Overall mean precision for the purge and trap method has been reported to be ± 12% (31).

Sampling considerations may be the limiting factor in an accurate analysis (1,55). Samples must be taken so that no headspace is present when the sample is ready for analysis. When determining oxidant reaction products, such as THMs, the reaction process must be stopped so that the compounds and concentrations determined reflect those present during sampling. A reducing agent, often sodium sulfite or sodium thiosulfate, is used to quench the chlorination reaction for drinking water samples.

Other methods must also be considered to preserve the integrity of the sample. Changes in sample composition can result from biological processes; therefore it is wise to refrigerate samples to minimize microbiological activity. Even at low temperatures, however, microbiological degradation can occur (56) and a storage time of less than 7 days has been recommended (13).

Specific problems may be encountered depending on the type of sample. Typical problems include suspended solids, the formation of emulsions, and foaming. The design of the sparge vessel has been modified to accommodate a variety of samples (57,58), yet usually the procedures of the standard protocols are applicable to the isolation of volatile organics from drinking water.

Recent advances in purge and trap analysis include the use of capillary columns. The mismatch in flow rates required for trap desorption and capillary columns has been addressed so that ppt concentrations of closely eluting solvents can be determined. Using this system Pankow et al. (59) found 80–90% recovery for selected chlorinated organics.

4.2.5   Closed Loop Stripping Analysis (CLSA)

Application: Compounds with midrange volatility, water solubility, and molecular weight.

Principles/Processes: Dynamic headspace technique coupled with sorption on activated carbon and solvent desorption.

Analyte transfers: Liquid/gas; gas/solid; solid/liquid by solvent desorption.

General Sensitivity: ppt to ppb.

In 1973, Grob (60) reported a closed loop stripping technique for the analysis of semivolatile organics in drinking water. An improved and standardized procedure was later published (61) where almost a one million-fold concentration was realized without evaporating the extract. This CLSA procedure used a 1-liter sample through which, 500 mL headspace gas was recycled. Organics were stripped from solution and adsorbed onto a 1.5-mg trap of activated carbon. The organics were later desorbed from the trap with 12 μL of carbon disulfide.

There are several aspects of the CLSA method that make it a good procedure to complement the purge and trap method. The sensitivity of CLSA is high; detection limits are often in the low ng/L range (62). Complex mixtures can often be separated because the procedure calls for the use of capillary column GC. On one hand, this method allows the determination of compounds with higher water solubility and higher molecular weight than can be determined by the standard purge and trap method (63), yet on the other hand, polar compounds are minimized as carbon is the adsorbent.

System blanks can present a problem although with extreme care very clean blanks are obtainable (63). Contamination is usually caused by leaks in the system or from the carbon filter. Laboratory air that leaks into the system can contain much higher amounts of organics than the water sample (24,64). A recently published procedure addresses the problem of background contamination (65). The carbon filter can be cleaned with H_2O_2 and 5% NH_4OH. Filters are then rinsed with methanol, methylene chloride, and carbon disulfide. This procedure has proven successful in reducing background contamination from fatty acids and alkanes.

Some disadvantages of the method may be laboratory specific. In the past, most CLSA units were homemade, therefore it took a good deal of time to set up and become proficient with its operation. The apparatus is now available commercially. With highly polluted waters, there is also the possibility of overloading the trap. This problem may be resolved by using a 5-mg carbon trap. Both the 1.5-mg and 5-mg traps are also available commercially.

Coleman et al. (62) reported recoveries of many organics using CLSA. The recovery of selected organic compounds is shown below.

Analyte	% Recovery		% RSD
Dibromochloromethane	56.5	±	12.6
2-Chlorotoluene	90.0	±	7.7
bis-(2-Chloroethyl) ether	11.6	±	10.7
1,4-Dichlorobenzene	92.5	±	8.2
Hexachlorobenzene	31.1	±	12.9
2,2',4,4',6,6' Hexachlorobiphenyl	104.3	±	7.1

Recoveries were determined by spiking low ppb concentrations of analytes in reagent water. The 1-liter water sample was maintained at 40°C during CLSA. As is the case with other headspace methods, salt can be added to increase the sensitivity of CLSA. Hwang et al. (66) reported that the addition of Na_2SO_4 to a water sample resulted in a doubling of the sensitivity in the analysis of taste and odor compounds in drinking water. In addition to decreasing stripping time, this salting out effect improved the limit of detection to 0.8 ng/L for six compounds.

The MAS (14) contains no protocol that uses the CLSA technique. The range of compounds isolated by CLSA methods is covered by several methods including heated purge and trap, distillation, liquid–liquid extraction, and resin adsorption. The CLSA seems to be preferable for a single broad spectrum analysis whereas the MAS protocols tend to determine a wide range of organics through a series of more specific procedures. Advantages of one approach over the other depend on the objectives of the analysis.

4.2.6 Distillation Methods

Application: Volatile, low molecular weight, polar compounds.

Principles/Processes: Phase changes: liquid/gas; gas/liquid via condensation; important factors are boiling point and vapor pressure.

General Sensitivity: ppb to ppm.

Polar compounds having a high water solubility have been separated from water by distillation. Methods have been approved to determine both general classes of organics and specific compounds (67). Types of distillation and distillation theory are discussed elsewhere (68).

The MAS (14) contains a protocol that uses anion-exchange resin and distillation to determine volatile strong acids in water. A drinking water sample analyzed by this MAS protocol would have a typical detection limit of 1 ppb.

Richard and Junk (69) used steam distillation as one of three methods to determine polar organic compounds in shale process waters. Using synthetic process water, they found good recoveries for most organic acids tested. Only acetic acid was recovered with less than 90% efficiency. These results compared favorably with liquid–liquid extraction with methylene chloride. Carboxylic acids (C_8–C_{10}) were concentrated by the distillation method yet were not detected by liquid–liquid extraction. Concentrations used in this study were in the ppm range.

Azeotropic distillation has been used to concentrate volatile polar organics such as acrolein and acrylonitrile (14,70,71). Concentration factors of almost 300 were reported for these compounds in the ppb to ppm range. Recovery of several low-molecular-weight alcohols, ketones, and nitriles was approximately 80%.

A variation on simple distillation is vapor/vapor extraction. A special distillation column (72,73) allows for simultaneous condensation of distillate and extraction solvent. Schultz et al. (74) have used a modified design of the simultaneous distillation and extraction apparatus. Recovery depended on extraction solvent, pH of aqueous solution, analyte, and pressure in the apparatus. In yet another design, Godefroot et al. (75) demonstrated very high solute recovery in a short amount of time. Over 99% recovery was observed for 12 chlorinated pesticides at low ppb concentrations. Individual peaks of a polychlorinated biphenyl mixture (Arochlor 1260) were recovered with 95% efficiency. Vapor/vapor extraction is a method that deserves further consideration for specific and broad spectrum analysis.

4.2.7   Liquid–Liquid Extraction

Application: Semivolatile compounds of a relatively wide range of polarity and molecular weight.

Principles/Processes: Transfer of solutes between two liquid phases. Can be run in batch or continuous mode.

General Sensitivity: ppt to ppm.

Liquid–liquid extraction has been used by many generations of analytical chemists because of its simplicity and wide range of application. All types of liquid–liquid extraction (LLE) are based on the distribution of solutes between two immiscible liquids. There are a number of ways to express the equilibrium distribution of solutes such as percent extracted, distribution coefficient, partition coefficient, and *p*-value (76–80).

The *p*-value method is a convenient way of expressing the distribution of a solute between two phases at equilibrium at a 1:1 water-to-solvent ratio. That fraction of solute that partitions into the organic phase is called the *E*-value. It is defined here as:

$$E = \frac{A_n}{A_s} \times \frac{V_n}{V_s}$$ (4.9)

where A_n = amount of analyte in organic phase
A_s = amount of analyte in water phase
V_n = final volume of organic solvent at equilibrium
V_s = initial volume of organic solvent
V_n/V_s = fractional recovery of organic solvent after the phases have been mixed and equilibrium is obtained

The *E*-value is calculated under any set of LLE conditions. The extraction is normalized back to the 1:1 ratio and expressed as the *p*-value.

$$p = \frac{E}{\alpha - [E(\alpha - 1)]}$$ (4.10)

where α is the equilibrated solvent-to-water ratio

$$\alpha = \frac{V_n}{V_p}$$ (4.11)

where V_p = final volume of water phase at equilibrium.

The *p*-value is the partition coefficient (*K*) on a fractional basis.

$$K = \frac{E}{\alpha(1 - E)} = \frac{p}{q}$$ (4.12)

where p = fractional amount of analyte in organic phase
q = fractional amount of solute in aqueous phase

If $p = 0.85$ this indicates that 85% of the solute is in the organic phase and 15% is in the aqueous phase, since $p + q = 1$. The partition coefficient

$K = 0.85/0.15 = 5.7$ and $p = 0.85$ represent the same degree of partitioning yet the p-value allows a more convenient way of expressing the result.

It must be noted that small changes in p-value can reflect large changes in the fractional amount recovered (E-value). This is true for typical liquid–liquid extractions where α values are low. Typically, a LLE is operated at water-to-solvent ratios ranging from 10:1 to 100:1. Water-to-solvent ratios outside these subjective limits can cause problems such as in solvent concentration or solvent recovery. Microextraction procedures have been reported that use very high water-to-solvent ratios (81–84). The p-value is useful because it aids in the comparison of LLE systems operated at different water-to-solvent ratios.

There are many variations of continuous extractors such as flow under extractor, flow over extractor (14,85), flow through extractor (86,87), Teflon helix extractor (80,88), mixer-settler type (89), and others. The ease of use of any LLE is a function of the analyte, its expected concentration and the sample matrix. Continuous liquid–liquid extractors have allowed larger volumes of water to be sampled, thus giving a more representative composite sample as well as increasing the sensitivity of the analysis.

For the determination of organics in drinking water, a simple batch operation has been used. For protocols of the MAS, detection limits ranged from 5 to 10 ppb, depending on type of water and final volume of concentrate (14).

The extraction of more water-soluble, polar organics can pose certain problems. Nonpolar solvents have been used with much success in the determination of relatively nonpolar solutes, however, when the objective is to determine polar organic compounds, the selection of solvents becomes more complex. The solubility of one liquid phase in the other is negligible for nonpolar solvents, yet the issue becomes critical when using polar organic solvents. This problem of mutual solubility is especially important in continuous liquid–liquid extractors and in microextraction methods. This is an area in which there needs to be more research.

In most of the LLE procedures there is the need to concentrate the organic phase after the extraction step has occurred. This often requires the use of a Kuderna-Danish (K-D) evaporator along with a Snyder condensation column. Evaporation to near the final volume of the concentrate usually entails a nitrogen blowdown along with a micro-Snyder column. The loss of highly volatile compounds during the extraction and the K-D procedure limits most LLE methods to the determination of semivolatile to nonvolatile compounds that boil $> 100°C$ (80).

The recovery and limit of detection can vary tremendously depending on the circumstances of the liquid–liquid extraction. The protocol for the batch extraction of semivolatile, strong acids (ESSA), found in the MAS (14) is summarized here. After a preliminary extraction with methylene chloride, an acidified 1-liter sample is extracted twice with methyl-t-butyl ether. The analytes are derivatized and the solvent is concentrated to 0.5 ml. A typical ESSA analyte would have a limit of detection of 5 ppb for a drinking water sample. Recovery ± SD of model compounds included pentachlorophenol at 100% ± 19 and 2,4-dichlorophenoxyacetic acid at 110% ± 10. Oleic acid had a mean recovery of 73% ± 7.

An example of specific compound analysis is the colorimetric determination of benzidine in water (90). Benzidine is isolated from water by ethyl acetate extraction. Interferences are minimized by adjustment of pH and multiple extractions. Benzidine had a 69% recovery from river water spiked at low ppb concentrations. An example of semispecific analysis is the liquid–liquid extraction of trihalomethanes. Although a wide range of compounds can be isolated from water in a single LLE, the use of specific detectors can minimize interference from nonanalytes. Using a halogen-specific Hall detector, Dressman et al. (15) reported limits of detection between 0.1 and 0.8 ppb for various trihalomethanes in water. Electron capture detectors have also found considerable use in the detection of ng/L concentrations of specific compounds (90a).

Broad spectrum analysis can be useful to determine trends in a wide range of organic compounds. With this goal in mind, both batch and continuous liquid–liquid extraction procedures have been used in conjunction with flame ionization detectors and mass spectrometers (21,80).

4.2.8   Adsorption Methods

Application: Semivolatile to nonvolatile compounds of a wide range of polarity and molecular weight.

Principles/Processes: Compounds are adsorbed onto a surface then desorbed with an elution solvent (liquid/solid; solid/liquid).

General Sensitivity: ppt to ppm.

A variety of adsorbent materials have been used to concentrate organic compounds from water. Common adsorbents include activated carbon (91), resins (92–95), and polyurethane foam (96,97). Group selectivity can be achieved by careful selection of adsorbent and elution solvents. However, within the group selected, for example, acidic, basic, polar, or nonpolar compounds, a large range of compounds within a group can be sorbed for any given adsorbent or elution solvent. Thus, these methods are well suited to broad spectrum analysis. Competition between organic solutes for sites on the sorbent is a subject that must be addressed. Some solutes will interact more strongly with the adsorbent and therefore be adsorbed preferentially. This can present a problem during broad spectrum analysis because the results will be biased toward those compounds that fared well in the competition for adsorption sites. Also, when doing a specific compound analysis, there will be a low recovery for compounds that were desorbed.

Other problems with adsorption methods are artifacts and samples with high concentrations of suspended solids. Artifacts can be formed from either the resin or the elution solvent. Considerable attention must be paid to the cleaning of sorbents (especially resins). Cleaning usually involves successive soxhlet extractions with at least two organic solvents (94). Solvent purity must be evaluated in conjunction with adsorbent cleaning procedures. To insure good data, it is important to run system blanks as part of a quality control program (14). Tenax is a resin that is usually purer than other resins yet has a high price. In addition, because of its thermal stability Tenax is often used as the bulk of the adsorbent trap for the purge and trap methods (14,95).

Currently, activated carbon is used more for water treatment (91) than for water analysis, although the CLSA method uses a small, highly purified charcoal filter to adsorb organics that have been sparged from water. Synthetic resins have received more attention for sampling water because they allow for more consistent solvent desorption than from activated carbon. Junk et al. (95) have reported procedures for the concentration of a range of specific organic compounds using XAD resins. The sensitivity of these isolation procedures is at the ppb level. Broad spectrum, semiquantitative analyses have been used for the evaluation of water treatment processes at ppt to ppb levels (2,3,5,6,20,26).

Polyurethane foams have been used as an alternative to synthetic resins (96,97) but because of their selectivity they are not recommended for broad-spectrum analysis. Foam sorbents have been used for the isolation of nonpolar, semivolatile compounds, such as polycyclic aromatic hydrocarbons (96). Because large volumes of water can be sampled, sensitivity is often good; ppt levels have been reported (96). Gough and Gesser (97) evaluated polyurethane foams for the extraction of phthalate esters from water. Excellent recoveries were reported for di-n-butyl phthalate and diisobutyl phthalate yet poor recoveries were reported for dimethyl phthalate and di-n-hexyl phthalate.

The MAS (14) contains a procedure that uses XAD-4 resin to isolate weak organic acids, neutrals, and bases. Detection limits for model compounds were estimated to be 0.5 ppb using 11.5 L samples. Because macroreticular resin columns allow the sampling of large volumes of water, detection limits could be improved by increasing the sample size up to 100 L. For such large volumes, it was recommended that resin columns be taken into the field rather than transporting samples to the laboratory.

Another group of methods termed "direct injection techniques" fall under the category of adsorption. Organic compounds have been isolated (adsorbed) on both liquid and gas chromatographic columns (98–100). Tetrachloroethylene was determined in natural waters by direct injection of up to 10 mL into a reverse-phase LC column (101). A detection limit was reported as 10 ppb using UV detection and a methanol/water elution. Another form of direct injection has been used by Pankow (102). This method involves the cryogenic trapping of volatile organics in a capillary column.

4.2.9   Ion-Exchange Methods

Application: Polar organics that are often nonvolatile.
Principle/Processes: Substitution of ionic analyte for ion of the substrate.
General Sensitivity: ppb to ppm.

Both natural and synthetic ion-exchange materials have found many applications in the laboratory. Important applications include water softening, deionization, and in analytical methods (103,104).

The mass-action law describes the equilibrium that exists during the process of ion-exchange (105).

$$A + BR \rightleftharpoons AR + B \tag{4.13}$$

A and B represent ions in solution and AR and BR represent ions bound to ion-exchange resin. The constant (K_{eq}) for this system can be written as

$$K_{eq} = \frac{(AR)\,(B)}{(BR)\,(A)} = \frac{a_{AR}a_B}{a_{BR}a_A} \tag{4.14}$$

where a = activity of corresponding species.

By replacing the activities of the ion/resin terms with the product of the activity coefficient (f) and the mole fraction (X) the equilibrium constant becomes:

$$K_{eq} = \frac{(X_{AR}f_{AR})a_B}{(X_{BR}f_{BR})a_A} \tag{4.15}$$

An apparant distribution coefficient (K_D) can be defined as:

$$K_D = K_{eq}\frac{f_{BR}}{f_{AR}} = \frac{X_{AR}a_B}{X_{BR}a_A} \tag{4.16}$$

This distribution coefficient can be calculated by determining the concentration of ions on the resin (AR and BR) and the activities of A and B in solution. During typical ion-exchange conditions, one of the ions is in large excess and a partition coefficient (K_p) can be calculated.

$$K_p = K_D\frac{X_{BR}}{a_B} = \frac{X_{AR}}{a_A} \tag{4.17}$$

The value of K_p represents the affinity of a particular ion-exchange resin for an ion (A) relative to another ion (B) (105). Note the similarity of this partition coefficient to partition coefficients of other two phase equilibrium systems.

Synthetic ion-exchange resins can be divided into at least four categories (104,105).

Type of Resin	Functional Group
1. Strong acid cation exchange	Sulfonic acid (R SO_3H)
2. Weak acid cation exchange	Carboxylic acid (RCOOH)
3. Strong base anion exchange	Quaternary amine (RN $(CH_3)_3^+$ OH^-)
4. Weak base anion exchange	Secondary amine (R_2NH)
	Tertiary amine (R_3N)

Some ion-exchange resins are available commercially such as those used in the MAS (14). Others have been prepared from synthetic adsorption resins such as the XAD resins, for the purpose of retaining both adsorption and ion-exchange sites for more broad spectrum analysis (106–108).

The MAS contains three protocols that use ion-exchange methods. The first is a combination anion exchange/distillation method for the determination of volatile strong acids (VOSA). Using Biorad AG1-X8 resin limits of detection are reported to be 1 ppb. The same resin is used for the isolation of nonvolatile organic acids (NOVA). This protocol calls for the elution of organic acids with HCl in methanol. Analytes are then concentrated from the eluant by rotary evaporation. The reported LOD is 2 ppb for these nonvolatile analytes in drinking water. Interferences can result from humic acids and artifacts from the resin.

The third protocol of the MAS calls for the use of Biorad AG50 W-X8 resin for the determination of strong amines (SAM) by cation exchange. When drinking water was spiked at 35 ppb the following recoveries were reported (14).

Analyte	% Recovery ± SD
Allylamine	80 ± 8
D-n-butylamine	112 ± 9
Piperidine	21 ± 9
t-Butylamine	101 ± 7

Although interferences from derivatizing reagents are possible, the overall LOD was reported to be 1 ppb in a 2.5-L drinking water sample.

XAD resins have formed the basis of both cation and anion exchange resins. Kaczvinsky et al. (107) prepared strong cation exchange resin by sulfonating XAD-4 resin. Organic bases in the H^+ form, were exchanged for Na^+ and eluted with ammonia, methanol, and diethyl ether. Recoveries of at least 85% were reported for most of the analytes tested at 50 ppb. Richard and Fritz (108) prepared an anion exchange resin from XAD resins. Organic acids were eluted with diethyl ether saturated with gaseous HCl. Recovery of selected analytes is listed below (108).

Analyte	% Recovery @ 100 ppb
2,4-Dichlorophenoxy acetic acid	93
2,4,5-Trichlorophenoxy acetic acid	102

Analyte	*% Recovery @ 100 ppb*
2,4-Dinitrophenol	75
Pentachlorophenol	91
o-Phthalic acid	65

Problems with the ion-exchange methods include contaminants from the resin and artifacts from derivatizing reagents. Although some methods are available for volatile compounds, most analytes in ion-exchange methods will be nonvolatile. This can present problems after the isolation step has occurred, that is, analytes may have to be derivatized for separation or chromatographic detection. In addition, the use of certain reagents, particularly diazomethane, require specific safety precautions. Humic substances and suspended solids can also interfere with efficient isolation of analytes (109).

While ion-exchange methods in general can isolate a wide range of organic compounds, they are not often considered applicable to broadspectrum analysis. These methods usually have a more specific set of target analytes, that is, ionizable solutes above or below a certain pK_a. The use of macroreticular resins (such as XAD resins) that have been prepared as ionexchange resins are exceptions. If both adsorption and ion-exchange resin sites are available on these resins, broad spectrum analysis becomes possible. Junk and Richard (106) reported simultaneous isolation of neutral and ionic organics although these groups of compounds were eluted in separate fractions.

4.2.10   Membrane Methods

Application: Wide range of polar compounds; semivolatile to nonvolatile compounds of mid to high molecular weight.

Principles/Processes: Transfer of solvent through membrane; retention of solutes based on size, charge, and affinity for the membrane.

General Sensitivity: ppb to ppm.

Analytical methods using membranes have been used to concentrate organics of a wide range of molecular weight. Often these compounds are polar, ionic, and relatively nonvolatile. Membrane methods to be discussed in this section include reverse osmosis, ultrafiltration, and dialysis.

Reverse osmosis (RO) will be emphasized because this type of membrane has received the most attention in drinking water analysis. Common RO membranes are made of cellulose acetate or an aromatic polyamide. Typical pore diameters are 4–11 Angstroms. Ultrafiltration membranes (UF) differ from RO membranes in the lower pressure in the system and the larger pore size, typically 15–500 Angstroms (110). Dialysis membranes take advantage of the differences in diffusion rates of solutes through a membrane. The driving force in dialysis is the solute's concentration gradient as well as applied forces such as pressure or electric potential. Ionic strength, pH, and chlorine residual can be important operating parameters for all membrane methods.

Osmosis can be defined as the diffusion of water through a semipermeable membrane. The driving force in osmosis is the concentration gradient of water. The van't Hoff equation describes this phenomenon (111).

$$\pi_i = RTC_i \tag{4.18}$$

where π_i = osmotic pressure
R = gas constant
T = absolute temperature
C_i = concentration of solute i

For reverse osmosis to occur, a pressure greater than π_i must be applied to the solution so that the solvent is transported against its concentration gradient. As water diffuses through a membrane, the solutes, both organic and inorganic, are concentrated in the retentate. The permeate is that solution that diffuses through the membrane.

Many theories exist which attempt to explain the rejection and transport of solute and solvent molecules. Several of the popular theories are listed below (112). Reverse osmosis is thought to depend upon:

1. Preferential sorption/capillary flow.
2. Solution/diffusion mechanism.
3. Solute/membrane partition coefficient.
4. Size, charge, and dissociation of hydrated solutes.

Transport mechanisms through RO membranes are discussed in detail elsewhere (112,113).

Much of the drinking water research in which RO membranes were used had the objective of concentrating large volumes of water for biological testing (114–116).

Although the best recoveries (a function of rejection) of analytes occur when their molecular weight is over 200, smaller molecules have also been concentrated. Fang and Chian (117) reported the separation of 13 low-molecular-weight organics by three RO membranes. Their general results are shown below.

Membrane Type	Overall Separation
Aromatic polyamide	50%
Cross-linked polyethylenimine	75%
Cellulose acetate	13–27%

The model compounds included acetone, aniline, methanol, phenol, and urea. While cellulose acetate membranes reject inorganic ions with high efficiency, other membranes are usually required to obtain high recoveries of organic compounds. The relative polarity of the solute and the membrane are important because some compounds may adsorb to the membrane and appear to be rejected yet not recovered for analytical purposes (117,118). Since free chlorine degrades many nylon membranes, a dechlorination step is usually required when doing drinking water studies. New membranes are being developed that are more resistant to chlorine and changes in pH.

Combination systems have also been set up using cellulose acetate and aliphatic nylon membranes in series (113). Tap water was concentrated in such a system. Some of the organics identified in the retentate were 2,4,6-trichlorophenol, diethyl phthalate, and a tetrachlorobiphenyl isomer. The recovery of these organics varied with the degree to which the sample was concentrated. Kopfler et al. (119) also used a combination of membranes. In this study, Donnan dialysis was used to exchange Na^+ for Ca^{2+} prior to RO concentration by two types of membranes.

Ultrafiltration differs from RO primarily in the size of molecules separated and the mechanism of solute transport. UF membranes often are reported to reject compounds with molecular weights above 1000 (111), although some membranes can retain solutes as small as 500 daltons (49). These membranes have found most application in the concentration of macromolecules or colloids.

Aiken (120) recently identified errors associated with the use of UF membranes for molecular weight determinations of fulvic acids. Molecular weight data can be in error because of:

1. Nonuniform membrane pore size.
2. Breakthrough of higher-molecular-weight solutes.
3. Reactivity of humic and fulvic acids with each other as well as with the membrane.

This last source of error highlights one of the problems of membranes in general. Solutes that do not permeate accumulate along the membrane and cause membrane fouling and concentration polarization. The result is decreased flux through the membrane; reduced driving force for permeable species and increased driving force for less permeable molecules (49,111). This can often be controlled by agitation of the solution. The concentration of inorganic salts along with organics may or may not cause problems. Another disadvantage is the loss of volatile compounds.

Advantages of membrane methods include the following observations.

1. It is a gentle concentration method.
2. There is the potential to process large volumes of water quickly.
3. Selectivity and fractionation are possible.

While membrane methods are, in general, used to concentrate solutes, membranes have also been used to isolate organics from water. Kurtz (118) has used cellulose triacetate membrane filters as an isolation device to determine PCBs and DDTs.

The MAS (14) contains no protocols that use membrane methods; however, what is applicable is the sampling and storage considerations. If large volumes of water are to be processed, concentration of organics in the field may be desirable. If samples are to be stored, microbiological degradation should be minimized and, as always, samples should be analyzed as soon as possible.

4.2.11   Residue Analysis (evaporation and freezing techniques)

Application: Nonvolatile compounds that are often polar and have a moderate to high molecular weight. Sample preparation for LC systems.

Principles/Processes: Sample matrix is removed, residue is analyzed.

General Sensitivity: ppb to ppm.

This set of concentration methods includes evaporation to dryness, freeze drying, vacuum distillation, and freeze concentration. The temperatures to which analytes are subjected will define the specific compounds that can be determined by these techniques. In its simplest form, a sample could be evaporated over a steam bath. The water matrix is removed and the residue remaining in the sample vessel is analyzed. Therefore, analytes must have boiling points greater than 100°C to be determined. Vacuum distillation is a technique that can produce a residue at about 35°C. Freeze drying (lyophilization) is a similar technique yet operated at lower temperatures and lower pressures. A freeze dryer will first freeze the sample then subject the sample to high vacuum. Heat is then added and the water matrix is removed by sublimation, that is, direct transfer from the solid to the gas phase. The residue must be extracted with a solvent prior to further isolation techniques or direct determination of constituents.

Crathorne et al. (121) have used rotary evaporation (vacuum distillation) and freeze-drying to remove water from water samples for broad spectrum analysis of polar organics. Methanol was used to extract the residue prior to separation of components by either high-performance liquid chromatography (HPLC) or GC. Compounds identified in drinking water included 5-chlorouracil, n-alkanes (C_{24}–C_{31}), cholesterol, dioctyl phthalate, and many others. In another study (122), the detection limit for 5-chlorouracil was 0.1 ppb for a 1-L sample using HPLC with UV detection.

Freeze concentration (123–127) is a related technique in that the sample matrix is removed, although not to the same extent as some of the evaporative techniques. This method is applicable to volatile, polar organics—especially to those compounds that are heat liable. With a 15-fold concentration, recoveries greater than 90% have been demonstrated for compounds such as ethyl formate, 4-heptanone, and ethanol (127). Kammerer and Lee (124) reported 88–101% recovery of μg/L levels of lindane from distilled water. The method is often limited by the presence of inorganic ions because salts will be concentrated along with the organic analytes. Mechanical stirring of the sample is essential during freezing out (126), because the formation of supercooled layers of ice tend to decrease the recovery of organics. Recovery of organic solutes was found to be inversely proportional to ionic strength (123).

Sampling for nonvolatile organic compounds in water need not be as difficult as for methods used to determine volatile compounds. There is generally no need to worry about an analyte volatilizing or adhering to the walls of the sample vessel because these solutes prefer to be in aqueous solution. However, care should be taken to prevent biological degradation and oxidation of analytes. Crathorne et al. (122) stored water samples in the dark at 4°C. Samples were extracted as soon as possible. Buffle et al. (128) also suggested that samples should be analyzed as soon as possible yet found that samples concentrated by ultrafiltration or freeze concentration were stable for 1 year. Stability was measured by change in UV absorbance.

One concern regarding the evaporative techniques is that of matrix interference. Inorganic salts will be concentrated along with the nonvolatile organics therefore the solvent used to extract the residue must be selected carefully. Ideally, the solvent will dissolve the organic analytes, leaving behind the inorganic salts, therefore a compromise must be reached so that interference is minimized. Also, the presence of nonvolatile acids or bases, that is, H_2SO_4, H_3PO_4, or NaOH can result in hydrolyzed and charred residues (12). Buffle et al. (128) cautioned against concentrating hard waters more than a factor of 10 because as the $CaCO_3$ precipitates, some organic compounds may be adsorbed to the inorganic salt. Decarbonation of the sample allowed concentration factors greater than 10.

The identification of nonvolatile compounds may not be as straightforward as with other classes of compounds. Several techniques that have been used in conjunction with residue analysis are shown below. This is likely to be an important area of research especially in improved methods for LC/MS (129), LC/FTIR (130,131), and tandem MS (132–134). The use of these additional detection techniques will provide much needed information about this group of organics.

Identification Technique	Selected References
UV absorbance	122,135
Fluorescence detection	122,136
Flame ionization detection	137
GC/MS	137
LC/MS	121

It is important that the complex subject of nonvolatile, polar organics is the thrust of present and future research. These compounds are among the least understood of any solutes yet comprise at least 80% of the dissolved organic carbon in water.

4.3 DISCUSSION

Two analytical approaches have been discussed: broad spectrum and specific compound analysis. Neither approach can serve the objective of the analysis without support of a quality assurance program. The amount of quality control is a function of the objective of the analysis. Broad spectrum analysis can be applied to any concentration or isolation method as long as the limitations of the analysis are known. Broad spectrum analysis can assess the overall water quality within the range of compounds concentrated by a particular method. In the more traditional approach, specific compound analysis, the objective is to do quantitative analysis on selected compounds. Both approaches have their merits as well as drawbacks. A combination of the two approaches might simply be to do specific compound analysis with respect to other peaks in a chromatogram. Analysts must not be satisfied with just knowing the concentration of selected compounds; they must know the relationship of these specific compounds to other compounds present in a sample.

Seven isolation and two concentration methods have been discussed in this chapter. They are described in terms of the typical analyte which is determined by each method. They are listed in order of application to analytes of increasing molecular weight and decreasing volatility. Figure 4.1 outlines the molecular weight, boiling point, and volatility ranges that are applicable to the specific techniques.

4.3.1 Static Headspace Analysis

Application: Volatile compounds that are, in general, nonpolar, low molecular weight, and have a low water solubility.

Principles/Processes: Liquid/gas partition that may be aided by addition of heat and/or salt. Partitioning is proportional to Henry's law constants.

General Sensitivity: ppb to ppm.

4.3.2 Purge and Trap

Application: Volatile to semivolatile compounds; they are relatively nonpolar, low molecular weight, and have a low water solubility.

Principles/Processes: Dynamic headspace technique coupled with sorption and thermal desorption.

Analyte Transfers: Liquid/gas; gas/solid; solid/gas by thermal desorption.

General Sensitivity: ppt to ppm.

4.3.3 Closed Loop Stripping Analysis (CLSA)

Application: Compounds with midrange volatility, water solubility, and molecular weight.

Principles/Processes: Dynamic headspace technique coupled with sorption on activated carbon and solvent desorption.

Analyte transfers: Liquid/gas, gas/solid; solid/liquid by solvent desorption.

General Sensitivity: ppt to ppb.

4.3.4   Distillation Methods

Application: Volatile, low-molecular-weight, polar compounds.

Principles/Processes: Phase changes: liquid/gas, gas/liquid via condensation; important factors are boiling point and vapor pressure.

General Sensitivity: ppb to ppm.

4.3.5   Liquid–Liquid Extraction

Application: Semivolatile compounds of a relatively wide range of polarity and molecular weight.

Principles/Processes: Transfer of solutes between two liquid phases. Can be run in batch or continuous mode.

General Sensitivity: ppt to ppm.

4.3.6   Adsorption Methods

Application: Semivolatile to nonvolatile compounds of a wide range of polarity and molecular weight.

Principles/Processes: Compounds are adsorbed onto a surface (liquid/solid) then desorbed with an elution solvent (solid/liquid).

General Sensitivity: ppt to ppm.

4.3.7   Ion-Exchange Methods

Application: Polar organics that are often nonvolatile.

Principles/Processes: Substitution of ionic analyte for ion of the substrate.

General Sensitivity: ppb to ppm.

4.3.8   Membrane Methods

Application: Wide range of polar compounds; semivolatile to nonvolatile compounds of mid to high molecular weight.

Principles/Processes: Transfer of solvent through membrane; retention of solutes based on size, charge, and affinity for the membrane.

General Sensitivity: ppb to ppm.

4.3.9 Residue Analysis (evaporation and freezing techniques)

Application: Nonvolatile compounds that are often polar and have a moderate to high molecular weight. Sample preparation for LC systems.
Principles/Processes: Sample matrix is removed, residue is analyzed.
General Sensitivity: ppb to ppm.

A good quality assurance program is at the heart of many studies of trace organics in water. The analyst should recognize the limitations of each step of an analysis. This is especially true when determining volatile organics. Modern instruments can produce results of high accuracy and precision yet the result may be meaningless if the sample was taken, preserved, transported, or stored incorrectly. Therefore, the weak link in the analytical process can lie outside of the laboratory, but not outside of the laboratory's responsibility. A good quality assurance program can minimize errors throughout the entire process.

Great strides have been made in analytical methods, of which the concentration/isolation methods are a critical part. Advances are especially needed in the concentration, separation, and detection of high-molecular-weight organics. It is this group of compounds that comprise most of the organic matter in water and have been shown to contribute substantially to the carcinogenic activity of drinking water (138,139).

REFERENCES

1. Huibregtse, K.R., and Moser, J.H., Handbook for Sampling and Sample Preservation of Water and Wastewater. EPA-600/4-76-049. Environmental Monitoring and Support Laboratory. Cincinnati, OH, 1976.
2. Suffet, I.H., Brenner, L., Coyle, J.T., and Cairo, P.R., Evaluation of the capability of granular activated carbons and XAD-2 resins to remove trace organics from treated drinking water. *Environ. Sci. Technol., 12,* 1315–1322, 1978.
3. Suffet, I.H., Gibs, J., Coyle, J.A., Chrobak, R.S., and Yohe, T.L., Applying analytical techniques to solve groundwater contamination problems, *J. Am. Water Works Assoc., 77,* 65–72, 1985.
4. Yohe, T.L., Suffet, I.H., and Cairo, P.R., Specific organic removals by granular activated carbon adsorption, *J. Am. Water Works Assoc., 73,* 402–410, 1981.
5. Jolley, R.L., Concentrating organics in water for biological testing, *Environ. Sci. Technol., 15,* 874–880, 1981.
6. Kopfler, F.C., Alternative strategies and methods for concentrating chemicals from water, in *Short-term Bioassays in the Analysis of Complex Environmental Mixtures,* Vol. 2, Waters, M.D., Sandhu, S.S., Huisingh, J.L., Claxton, L., and Nesnow, S. (eds.), Plenum Press, New York, 1981, pp. 141–153.
7. Jennings, W.G. and Rapp, A., *Sample Preparation for Gas Chromatographic Analysis.* Dr. Alfred Huthig Verlag, Heidelberg. 1983, pp. 33–71.
7a. Karasek, F.W., Clement, R.E., and Sweetman, J.A., Preconcentration for trace analysis of organic compounds, *Anal. Chem., 53,* 1050A–1058A, 1981.
8. Khromchenko, Y.L., and Rudenko, B.A., Determination of trace amounts of polluting organic substances in drinking and natural waters and in effluents by gas chromatography, *Soviet J. Water Chem. Technol., 3,* 26–63, 1981.

9. McElroy, F.C., Searl, T.D., and Brown, R.A., Sample preparation for environmental trace analysis, in *Trace Organic Analysis: A New Frontier in Analytical Chemistry*. Hertz, H.S., and Chester, S.N., (eds.), National Bureau of Standards, Special Publication 519, U.S. Govt. Printing Office, Washington, DC, 1979, pp. 7–18.

10. Jolley, R.L., and Suffet, I.H., Concentration techniques for isolating organic constituents in environmental water samples, in *Sampling and Analysis of Organic Pollutants from Water*, Suffet, I.H., and Malaiyandi, M., (eds.), ACS, Advances in Chemistry, No. 214, American Chemical Society, Washington, DC (in press).

11. Trussell, A.R., and Umphres, M.D., An overview of the analysis of trace organics in water, *J. Am. Water Works Assoc., 70*, 595–603, 1978.

12. Leenheer, J.A., Concentration, partitioning, and isolation techniques, in *Water Analysis*, Vol. 3, Minear, R.A., Keith, L.H., (eds.), Academic Press, Orlando, FL, 1984, pp. 83–166.

13. Longbottom, J.E., and Lichtenberg, J.J., (eds.), *Methods for Organic Chemical Analysis of Municipal and Industrial Wastewater*, EPA-600/4-82-057, U.S. Environmental Protection Agency, Environmental Monitoring and Support Laboratory, Cincinnati, OH, 1982.

14. Pellizzari, E.D., Sheldon, L.S., Bursey, J.T., Michael, L.C., and Zweidinger, R.A., Master analytical scheme for organic compounds in water. Part I: Protocols. U.S. Environmental Protection Agency, Office of Research and Development, Athens, GA, 1984.

15. Dressman, R.C., Stevens, A.A., Fair, J., and Smith, B., Comparison of methods for determination of trihalomethanes in drinking water, *J. Am. Water Works Assoc., 71*, 393–396, 1979.

16. U.S. Environmental Protection Agency, The Analysis of Trihalomethanes in Drinking Water by Liquid/Liquid Extraction. Method 501.2. Environmental Monitoring and Support Laboratory, Cincinnati, OH, 1979.

17. Varma, M.M., Siddique, M.R., Doty, K.T., and Machis, A., Analysis of trihalomethanes in aqueous solutions: A comparative study. *J. Am. Water Works Assoc., 71*, 389–391, 1979.

18. Glaser, E.R., Silver, B., and Suffet, I.H., Computer plots for the comparison of chromatographic profiles, *J. Chromatogr. Sci., 15*, 22–28, 1977.

19. Suffet, I.H., and Glaser, E.R., A rapid gas chromatographic profile/computer data handling system for qualitative screening of organic compounds in waters at the part-per-billion level. *J. Chromatogr. Sci., 18*, 12–18, 1978.

20. Coyle, T.G., Maloney, S.W., Gibs, J., and Suffet, I.H., Broad-spectrum analysis of the removal of trace organics in ozone-granular activated carbon potable water pilot plant study, in *Water Chlorination: Environmental Impact and Health Effects*, Vol. 4, Ann Arbor Science Ann Arbor, MI, Jolley, R.L. et al., (eds.), 1983, pp. 421–443.

21. Yohe, T.L., Suffet, I.H., and Coyle, J.T., Monitoring and analysis of aqueous chlorine effects of GAC pilot contactors, in *Activated Carbon Adsorption of Organics from the Aqueous Solution*, Vol. 2, McGuire, M.J., and Suffet, I.H., (eds.), Ann Arbor Science, Ann Arbor, MI, 1980, pp. 27–69.

22. Stevens, A.A., Seeger, D.R., Slocum, C.J., and Domino, M.M., Gas chromatographic techniques for controlling organics removal processes, *J. Am. Water Works Assoc., 73*, 548–554, 1981.

23. Van Rensburg, J.F.J., Hassett, A., Theron, S., and Wiecher, S.C., The fate of organic micropollutants through an integrated waste water treatment/water reclamation system, *Progr. Water Technol., 12*, 537–552, 1980.

24. Boren, H., Grimvall, A., and Savenhed, R., Modified stripping technique for the analysis of trace organics in water, *J. Chromatogr., 252*, 139–146, 1982.

25. Grimvall, A., Savenhed, R., and Boren, H., Statistical analysis of chromatographic data obtained by stripping techniques, *Water Sci. Tech., 15*, 169–179, 1983.

26. Gibs, J., Najar, B., and Suffet, I.H., Broad spectrum analysis of organics in drinking water using macroreticular resins–A quality assurance evaluation, in *Water Chlorination Environmental Impact and Health Effects*, Vol. 5, Jolley, R.L. et al. (eds.), Lewis Publ., Chelsea, MI, 1985, p. 1099.

27. American Chemical Society, Committee on Environmental Improvement, Guidelines for data acquisition and data quality evaluation in environmental chemistry, *Anal. Chem., 52,* 2242–2249, 1980.

28. American Chemical Society, Committee on Environmental Improvement, Principles of environmental analysis, *Anal. Chem., 55,* 2210–2218, 1983.

29. Garrison, A.W., Application of the master analytical scheme to polar organics in drinking water, in *Sampling and Analysis of Organic Pollutants from Water,* Suffet, I.H., and Malaiyandi, M., (eds.), ACS, Advances in Chemistry, No. 214, American Chemical Society, Washington, DC (in press).

30. Michael, L.C. et al., Quality of master analytical scheme data: Purgeables and volatile organic acids, in *Advances in the Identification & Analysis of Organic Pollutants in Water,* Vol. 1, Keith, L.H., (ed.), Ann Arbor Science, Ann Arbor, MI, 1981, pp. 87–114.

31. Tomer, K.B., Bursey, J.T., Michael, L.C., Sheldon, L.S., Pellizzari, E.D., Alford, A.L., Pope, J.D., and Garrison, A.W., Quantitative aspects of the master analytical scheme for organics in water, in *Advances in the Identification & Analysis of Organic Pollutants in Water,* Vol. 1, Keith, L.H., (ed.), Ann Arbor Science, Ann Arbor, MI, 1981, pp. 67–85.

32. Garfield, F.M., *Quality Assurance Principles for Analytical Laboratories,* Association of Official Analytical Chemists, 1984.

33. Gilliom, R.J., Hirsch, R.M., and Gilroy, E.J., Effect of censoring trace-level water-quality data on trend-detection capability, *Environ. Sci. Technol., 18,* 530–535, 1984.

34. Harris, T.D., and Williams, A.M., Fundamental detection limits in spectrophotometric analysis, *Appl. Spectros., 39,* 28–32, 1985.

35. Hubaux, A., and Vos, G., Decision and detection limits for linear calibration curves, *Anal. Chem., 42,* 849–855, 1970.

36. American Society of Testing & Materials, *Annual Book of ASTM Standards, Water and Environmental Technology,* Vol. 11.01, D4210-83, pp. 7–16, 1983.

37. Glaser, J.A., Foerst, D.L., McKee, G.D., Quave, S.A., and Budde, W.L., Trace analysis for wastewaters, *Environ. Sci. Technol., 15,* 1426–1434, 1981.

38. Long, G.L., and Winefordner, J.D., Limit of detection—A closer look at the IUPAC definition, *Anal. Chem., 55,* 712A–724A, 1983.

39. Snyder, L.R., and van der Wal, Sj., Precision of assays based on liquid chromatography with prior solvent extraction of the sample. *Anal. Chem., 53,* 877–884, 1981.

40. Stumm, W., and Morgan, J.J., *Aquatic Chemistry,* 2nd ed., John Wiley & Sons, New York, 1981, pp. 40–44, 103–104.

41. Suffet, I.H., and Radziul, J.V., Guidelines for the quantitative and qualitative screening of organic pollutants in water supplies, *J. Am. Water Works Assoc., 68,* 520–524, 1976; Addendum, *69,* 174, 1977.

42. Nicholson, B.C., Maguire, B.P., and Bursill, D.B., Henry's Law constants for the trihalomethanes: Effects of water composition and temperature, *Environ. Sci. Technol. 18,* 518–521, 1984.

43. Friant, S.L., and Suffet, I.H., Interactive effects of temperature, salt concentration, and pH on headspace analysis for isolating volatile trace organics in aqueous environmental samples, *Anal. Chem., 51,* 2167–2172, 1979.

44. Drozd, J., and Novak, J., Headspace determination of benzene in gas-aqueous liquid systems by the standard additions method, *J. Chromatogr., 152,* 55–61, 1978.

45. McAullife, C., GC determination of solutes by multiple phase equilibration, *CHEMTECH, 1,* 46–51, 1971.

46. Ioffe, B.V., and Vitenberg, A.G., *Headspace Analysis and Related Methods in Gas Chromatography,* John Wiley & Sons, New York, 1984.

47. Otson, R., Williams, D.,T., and Bothwell, P.D., A comparison of dynamic and static head space and solvent extraction techniques for the determination of trihalomethanes in water, *Environ. Sci. Technol., 13,* 936–939, 1979.

48. McNally, M.E., and Grob, R.L., A review: Current applications of static and dynamic headspace analysis: Part one: Environmental applications, *Am. Lab., 17,* 20–33, 1985.

49. Michaels, A. S., Ultrafiltration: An adolescent technology, *CHEMTECH.*, Jan., 36–43, 1980.

50. Swinnerton, J.W., and Linnenbom, V.J., Determination of C_1 to C_4 hydrocarbons in sea water by gas chromatography, *J. Gas Chromatogr., 5,* 570–573, 1967.

51. Bellar, T.A., and Lichtenberg, J.J., Determining volatile organics at microgram-per-litre levels by gas chromatography, *J. Am. Water Works Assoc., 66,* 739–744, 1974.

52. American Public Health Association, *Standard Methods for the Examination of Water and Wastewater,* 16th ed., Washington, DC, pp. 591–602, 1985.

53. Spraggins, R.L., Oldham, R.G., Prescott, C.L., and Baughman, K.J., Organic analysis using high-temperature purge-and-trap techniques, in *Advances in the Identification & Analysis of Organic Pollutants in Water,* Vol. 2, Keith, L.H., (ed.), Ann Arbor Science, Ann Arbor, MI, 1981, pp. 747–761.

54. Tai, D.Y., The determination of acetone and methyl ethyl ketone in water, U.S. Geological Survey, Water Resources Investigation 78–123, NTIS Report PB 291151, 1981.

55. Grob, R.L., and Kaiser, M.A., Environmental problem solving using gas and liquid chomatography, *J. Chromatogr. Library,* Vol. 21, Elsevier Scientific Publishing Co., Amsterdam, 1982, pp. 7–136.

56. Ogawa, I., Junk, G.A., and Svec, H.J., Degradation of aromatic compounds in groundwater and methods of sample preservation, *Talanta, 28,* 725–729, 1981.

57. Ellington, J., Analysis of volatile organics on sediments and in associated water, in *Advances in the Identification & Analysis of Organic Pollutants in Water,* Vol. 2, Keith, L.H., (ed.), Ann Arbor Science, Ann Arbor, MI, 1981, pp. 729–746.

58. Haile, C.L., Shan, Y.A., Malone, L.S., and Northcutt, R.V., Development of methods for the analysis of purgeable organic priority pollutants in municipal and industrial wastewater treatment sludges, in *Advances in the Analysis & Identification of Organic Pollutants in Water,* Vol. 2, Keith, L.H., (ed.), Ann Arbor Science, Ann Arbor, MI, 1981, pp. 763–791.

59. Pankow, J.F., Isabelle, L.M., and Kristensen, T.J., Tenax—GC cartridge for interfacing capillary column gas chromatography with adsorption/thermal desorption for determination of trace organics, *Anal. Chem., 54,* 1815–1819, 1982.

60. Grob, K., Organic substances in potable water and in its precursors. Part 1. Methods for their determination by gas-liquid chromatography, *J. Chromatogr., 84,* 255–273, 1973.

61. Grob, K., and Zurcher, F., Stripping of trace organic substances from water, equipment and procedure, *J. Chromatogr., 117,* 285–294, 1976.

62. Coleman, W.E., Munch, J.W., Slater, R.W., Melton, R.G., and Kopfler, F.C., Optimization of purging efficiency and quantification of organic contaminants from water using a 1-L closed-loop-stripping apparatus and computerized capillary column GC/MS, *Environ. Sci. Technol., 17,* 571–576, 1983.

63. Melton, R.G. et al., Comparison of Grob closed-loop stripping analysis with other trace organic methods, in *Advances in the Identification & Analysis of Organic Pollutants in Water,* Vol. 2, Keith, L.H., (ed.), Ann Arbor Science, Ann Arbor, MI, 1981, pp. 597–673.

64. Grob, K., Organic substances in potable water and in its precursors, Part 1. Methods for their determination by gas-liquid chromatography, *J. Chromatogr., 84,* 255–273, 1973.

65. Grob, K., Grob, G., and Habich, A., Overcoming background contamination in closed loop stripping analysis (CLSA), *J. High Resol. Chromatogr. Chromatogr. Commun., 7,* 340–342, 1984.

66. Hwang, C.J., Krasner, S.W., McGuire, M.J., Moylan, M.S., and Dale, M.S., Determination of subnanogram per liter levels of earthy-musty odorants in water by salted closed-loop stripping method, *Environ. Sci. Technol., 18,* 535–539, 1984.

67. American Public Health Association, *Standard Methods for the Examination of Water and Wastewater,* 14th ed., Washington, DC, 1976, pp. 529–531.

68. Karger, B.L., Snyder, L.R., and Horvath, C., *An Introduction to Separation Science,* John Wiley & Sons, New York, 1973, pp. 181–210.

69. Richard, J.J., and Junk, G.A., Steam distillation, solvent extraction, and ion-exchange for determining polar organics in shale process waters, *Anal. Chem., 56,* 1625–1628, 1984.

70. Kuo, P.P.K., Chian, E.S.K., and DeWalle, F.B., Determination of trace low-molecular-weight volatile polar organics in water by gas chromatography using distillation method, *Water Res., 11,* 1005–1011, 1977.
71. Peters, T.L., Steam distillation apparatus for concentration of trace water soluble organics. *Anal. Chem., 52,* 211–213, 1980.
72. Likens, S.T., and Nickerson, G.B., Detection of certain hop oil constituents in brewing products, *Proc. Am. Soc. Brew. Chem.,* 5–11, 1964.
73. Nickerson, G.B., and Likens, S.T., Gas chromatographic evidence for the occurrence of hop oil components in beer, *J. Chromatogr., 21,* 1–5, 1966.
74. Schultz, T.H., Flath, R.A., Mon, T.R., Eggling, S.B., and Teranishi, R., Isolation of volatile components from a model system, *J. Agric. Food Chem., 25,* 446–449, 1977.
75. Godefroot, M., Stechele, M., Sandra, P., and Verzele, M., A new method for the quantitative analysis of organochlorine pesticides and polychlorinated biphenyls, *J. High Resol. Chromatogr. Chromatogr. Commun., 5,* 75–79, 1982.
76. Beroza, M., and Bowman, M., Extraction of insecticides for cleanup and identification, *J. Assoc. Offic. Agric. Chem., 48,* 358–370, 1965.
77. Beroza, M., and Bowman, M.C., Apparatus and procedure for rapid extraction and identification of pesticides by single and multiple distribution in binary solvent systems, *Anal. Chem., 38,* 837–841, 1966.
78. Suffet, I.H., The p-value approach to quantitative liquid-liquid extraction of pesticides and herbicides from water. 3. Liquid-liquid extraction of phenoxy acid herbicides from water, *J. Agric. Food Chem., 21,* 591–598, 1973.
79. Suffet, I.H., and Faust, S.D., The p-value approach to quantitative liquid-liquid extraction of pesticides from water. 1. Organophosphates: Choice of pH and solvent, *J. Agric. Food Chem., 20,* 52–56, 1972.
80. Yohe, T.L., Suffet, I.H., and Grochowski, R.J., Development of a Teflon helix continuous liquid-liquid extraction apparatus and its application for the analysis of organic pollutants in drinking water, ASTM STP 686, American Society for Testing and Materials, 1979, pp. 47–67.
81. Grob, K., Grob, K., and Grob, G., Organic substances in potable water III. The closed-loop stripping procedure compared with rapid liquid extraction, *J. Chromatogr., 106,* 299–315, 1975.
82. Keith, L.H., (ed.), *Advances in the Identification & Analysis of Organic Pollutants in Water,* Vol. 1, Section 4: Microextraction, 1981, pp. 241–281.
83. Murray, D.A.J., Rapid micro extraction procedure for analyses of trace amounts of organic compounds in water by gas chromatography and comparisons with macro extraction methods, *J. Chromatogr, 177,* 135–140, 1979.
84. Van Rensburg, J.F.J., and Hassett, A.J., A low-volume liquid-liquid extraction technique, *J. High Resol. Chromatogr. Chromatogr. Commun. 5,* 574–576, 1982.
85. Peters, T.L., Comparison of continuous extractors for the extraction and concentration of trace organics from water, *Anal. Chem.* 54, 1913–1914, 1982.
86. Ahnoff, M., and Josefsson, B., Simple apparatus for on-site continuous liquid–liquid extraction of organic compounds from natural waters, *Anal. Chem., 46,* 658–663, 1974.
87. Ahnoff, M., and Josefsson, B., Apparatus for in situ solvent extraction of nonpolar organic compounds in sea and river water, *Anal. Chem., 48,* 1268–1269, 1976.
88. Wu, C., and Suffet, I.H., Design and optimization of a Teflon helix continuous liquid-liquid extraction apparatus and its application for the analysis of organophosphate pesticides in water, *Anal. Chem., 49,* 231–237, 1977.
89. Bruchet, A., Cognet, L., and Mallevialle, J., Continuous composite sampling and analysis of pesticides in water, *Water Res., 18,* 1401–1409, 1984.
90. American Public Health Association, Supplement to the Fifteenth Edition of *Standard Methods for the Examination of Water and Wastewater,* pp. S48–S50, 1981.
90a. Mieure, J.A., A rapid and sensitive method for determining volatile organohalides in water, *J. Am. Water Works Assoc., 69,* 60–62, 1977.

91. Suffet, I.H., and McGuire, M.J., (eds.), *Treatment of Water by Granular Activated Carbon*, American Chemical Society, Advances in Chemistry Series, No. 202, Washington, DC, 1983.

92. Aiken, G.R., Thurman, E.M., Malcolm, R.L., and Walton, H.F., Comparison of XAD macroporous resins for the concentration of fulvic acid from aqueous solution, *Anal. Chem., 51,* 1799–1803, 1979.

93. Dressler, M., Extraction of trace amounts of organic compounds from water with porous organic polymers, *J. Chromatogr., 165,* 167–206, 1979.

94. James, H.A., Steel, C.P., and Wilson, I., Impurities arising from the use of XAD-2 resin for the extraction of organic pollutants in drinking water, *J. Chromatogr., 208,* 89–95, 1981.

95. Junk, G.A., Synthetic polymers for accumulating organic compounds from water, in *Sampling and Analysis of Organic Pollutants from Water,* Suffet, I.H., and Malaiyandi, M., (eds.), ACS, Advances in Chemistry, No. 214, American Chemical Society, Washington, DC (in press).

96. Saxena, J., Kozuchoski, J., and Basu, D.K., Monitoring of polynuclear aromatic hydrocarbons in water 1. Extraction and recovery of benzo(*a*)pyrene with porous polyurethane foam, *Environ. Sci. Technol., 11,* 682–685, 1977.

97. Gough, K.M., and Gesser, H.D., The extraction and recovery of phthalate esters from water using porous polyurethane foam, *J. Chromatogr., 115,* 383–390, 1975.

98. Cramer, P.H., Drinkwine, A.D., Going, J.E., and Carey, A.E., Determination of carbofuran and its metabolites by high performance liquid chromatography using on-line trace enrichment, *J. Chromatogr., 235,* 489–500, 1982.

99. Fuill, T., Trace determination of vinyl chloride in water by direct aqueous injection gas chromatography-mass spectrometry, *Anal. Chem., 49,* 1985–1987, 1977.

100. Zlatkis, A., Wang, F.S., and Shanfield, H., Direct gas chromatographic analysis of aqueous samples at the part-per-billion and part-per-trillion levels, *Anal. Chem., 55,* 1848–1852, 1983.

101. Kummert, R., Molnar-Kubica, E., and Giger, W., Trace determination of tetrachloroethylene in natural waters by direct aqueous injection high-pressure liquid chromatography, *Anal. Chem., 50,* 1637–1639, 1978.

102. Pankow, J.F., Cold trapping of volatile organic compounds of fused silica capillary columns, *J. High Resol. Chromatogr. Chromatogr. Commun., 6,* 292–299, 1983.

103. Rieman, W., and Walton, H.F., *Ion Exchange in Analytical Chemistry,* Pergamon Press, Oxford, 1970, 295 pp.

104. Rothbart, H.L., Ion-exchange separation processes, in *An Introduction to Separation Science,* Karger, B.L., Snyder, L.R., Horvath, C., (eds.), John Wiley & Sons, New York, 1973, pp. 337–373.

105. Skoog, D.A., and West, D.M., *Fundamentals of Analytical Chemistry,* 3rd ed., Holt, Rinehart & Winston, New York, 1976, pp. 664–668.

106. Junk, G.A., and Richard, J.J., Anionic and neutral organic components in water by anion exchange, in *Advances in the Identification & Analysis of Organic Pollutants in Water,* Vol. 1, Keith, L.H., (ed.), Ann Arbor Science, Ann Arbor, MI, 1981, pp. 295–315.

107. Kaczvinsky Jr., J.R., Saltoh, K., and Fritz, J.S., Cation-exchange concentration of basic organic compounds from aqueous solution, *Anal. Chem., 55,* 1210–1215, 1983.

108. Richard, J.J., and Fritz, J.S., The concentration, isolation, and determination of acidic material from aqueous solution, *J. Chromatogr. Sci., 18,* 35–38, 1980.

109. Carlberg, G.E., and Martinsen, K., Adsorption/complexation of organic micropollutants to aquatic humus—Influence of aquatic humus with time on organic pollutants and comparison of two analytical methods for analysing organic pollutants in humus water, *Sci. Total Environ., 25,* 245–254, 1982.

110. Josephson, J., Crossflow filtration, *Environ. Sci. Technol., 18,* 375A–376A, 1984.

111. Hwang, S.T., and Kammermeyer, K., Membrane processes, in *Treatise on Analytical Chemistry,* 2nd ed. Vol. 5, Elving, P.J., Grushka, E., and Kolthoff, I.M., (eds.), John Wiley & Sons, New York, 1982, pp. 185–238.

112. Sourirajan, S., and Matsuura, T., Physiochemical Criteria for Reverse Osmosis Separations, in *Reverse Osmosis and Synthetic Membranes. Theory—Technology—Engineering*, Sourirajan, S., (ed.), Nat'l. Res. Counc., Ottawa, Canada, 1977, pp. 5–43.

113. Deinzer, M., Melton, R., and Mitchell, D. Trace organic contaminants in drinking water: Their concentration by reverse osmosis, *Water Res., 9*, 799–805, 1975.

114. Coleman, W.E., Melton, R.G., Kopfler, F.C., Barone, K.A., Aurand, T.A., and Jellison, M.G., Identification of organic compounds in a mutagenic extract of a surface drinking water by a computerized gas chromatography/mass spectrometry system (GC/MS/COM). *Environ. Sci. Technol., 14*, 576–588, 1980.

115. Lynch, S.C., Smith, J.K., Rando, L.C., and Yauger, W.L., Isolation or Concentration of Organic Substances from Water: An Evaluation of Reverse Osmosis Concentration. U.S. Environmental Protection Agency, Contract No. 68-03-2999, Health Effects Research Laboratory, Cincinnati, OH, January 1983.

116. Malaiyandi, M., Wightman, R.J., and LaFerriere, C., Concentration of selected organic pollutants: Comparison of adsorption versus reverse osmosis techniques, in *Sampling and Analysis of Organic Pollutants from Water*, Suffet, I.H., and Malaiyandi, M., (eds.), ACS, Advances in Chemistry, No. 214, American Chemical Society, Washington, DC, (in press).

117. Fang, H.H.P., and Chian, E.S.K., Reverse osmosis separation of polar organic compounds in aqueous solution, *Environ. Sci. Technol., 10*, 364–369, 1976.

118. Kurtz, D.A., Adsorption of PCB's and DDT's on membrane filters—A new analysis method, *Bull. Environ. Contam. Toxicol. 17*, 391–398, 1977.

119. Kopfler, F.C., Coleman, W.E., Melton, R.G., Tardiff, R.G., Lynch, S.C., and Smith, J.K., Extraction and identification of organic micropollutants: Reverse osmosis method, *Ann. N.Y. Acad. Sci., 298*, 20–30, 1977.

120. Aiken, G.R., Evaluation of ultrafiltration for determining molecular weight of fulvic acid, *Environ. Sci. Technol., 18*, 978–981, 1984.

121. Crathorne, B., Fielding, M., Steel, C.P., and Watts, C.D., Organic compounds in water: Analysis using coupled-column high performance liquid chromatography and soft-ionization mass spectrometry, *Environ. Sci. Technol., 18*, 797–802, 1984.

122. Crathorne, B., Watts, C.D., and Fielding, M., Analysis of non-volatile organic compounds in water by high performance liquid chromatography, *J. Chromatogr., 185*, 671–690, 1979.

123. Baker, R.A., Trace organic contaminant concentration by freezing-IV. Ionic effects, *Water Res., 4*, 559–573, 1970.

124. Kammerer, P.A., and Lee, G.F., Freeze concentration of organic compounds in dilute aqueous solutions, *Environ. Sci. Technol., 3*, 276–278, 1969.

125. Kobayashi, W., and Lee, G.F., Freeze-concentration of dilute aqueous solutions, *Anal. Chem., 36*, 2197–2198, 1964.

126. Shapiro, J., Freezing-out, a safe technique for concentration of dilute solutions, *Science, 133*, 2063–2064, 1961.

127. Kepner, R.E., Von Straten, S., and Weurman, C., Freeze concentration of volatile components in dilute aqueous solutions, *J. Agric. Food Chem., 17*, 1123–1127, 1969.

128. Buffle, J., Deladoey, P., Zumstein, J., and Haerdi, W., Analysis and characterization of natural organic matters in freshwaters, I. Study of analytical techniques, *Schweiz. Z. Hydrol., 44*, 325–362, 1982.

129. Bombaugh, K.L., The use of HPLC for water analysis, in *Water Analysis*, Vol. 3., Minear, R.A., and Keith, L.H., (eds.), Academic Press, Orlando, FL, 1984, pp. 317–379.

130. Kawahara, F.K., Infrared spectrophotometry of pollutants in water systems, in *Water Analysis*, Vol. 3, Minear, R.A. and Keith, L.H., (eds.), Academic Press, Orlando, FL, 1984, pp. 381–443.

131. Sadowski, L.H., and Harris, J.C., Analysis of Water-Soluble Organics, EPA-600/2-84-012, U.S. Environmental Protection Agency, Industrial Environmental Research Laboratory, Research Triangle Park, NC, Jan., 1984.

132. Hunt, D.F., Shabanowitz, J., Harvey, M., and Coates, M.L., Analysis of organics in the

environment by functional group using a triple quadrupole mass spectrometer, *J. Chromatogr., 271,* 93–105, 1983.

133. Hunt, D.F., Shabanowitz, J., Harvey, T.M., and Coates, M., Scheme for the direct analysis of organics in the environment by tandem mass spectrometry, *Anal. Chem., 57,* 525–537, 1985.

134. McLafferty, F.W., Tandem mass spectrometry, *Science, 214,* 280–286, 1981.

135. Katz, S., Pitt, W.W., and Scott, C.D., The determination of stable organic compounds in waste effluents at microgram per liter levels by automatic high-resolution ion exchange chromatography, *Water Res., 6,* 1029–1037, 1972.

136. Pitt, W.W., Jolley, R.L., and Katz, S., Automated analysis of individual refractory organics in polluted water. EPA-660/2-74-076, Office of Research and Development, U.S. EPA Washington, DC, August 1974.

137. Watts, C.D., Crathorne, B., Fielding, M., and Killops, S.D., Nonvolatile organic compounds in treated waters, *Environ. Health Persp., 46,* 87–99, 1982.

138. Loper, J.C., Mutagenic effects of organic compounds in drinking water, *Mutat. Res., Rev. Genet. Toxicol., 76,* 241–268, 1980.

139. Loper, J.C., Tabor, M.W., and Miles, S.K., Mutagenic subfractions from nonvolatile organics of drinking water, in *Water Chlorination: Environmental Impact and Health Effects,* Vol. 4, Jolley, R.L. et al., (eds.), Ann Arbor Science, Ann Arbor, MI, 1983, pp. 1199–1210.

140. Gabler, R., Hegde, R., and Hughes, D., Degradation of high purity water on storage, *J. Liq. Chromatogr., 6,* 2565–2570, 1983.

141. Bowers, W.D., Parsons, M.L., Clement, R.E., Eiceman, G.A., and Karasek, F.W., Trace impurities in solvents commonly used for gas chromatographic analysis of environmental samples, *J. Chromatogr., 206,* 279–288, 1981.

142. Keith, L.H., Lee, K.W., Provost, L.P., and Present, D.L., Methods for gas chromatographic monitoring of the environmental protection agency's concent decree priority pollutants, in *Measurement of Organic Pollutants in Water and Wastewater,* Van Hall, C.E., (ed.), ASTM STP 686, American Society for Testing and Materials, 1979, pp. 85–107.

143. Jay, P.C., Anion contamination of environmental water samples introduced by filter media, *Anal. Chem., 57,* 780–782, 1985.

144. Ligocki, M.E., and Pankow, J.F., Oxidative losses of anthracene, acenaphthylene, and benzo[a]pyrene during florisil and silica gel cleanup using diethyl ether containing peroxides, *Anal. Chem., 56,* 2984–2987, 1984.

145. Jeos, I.J., Reineccivs, G.A., and Thomas, E.L., Artifacts in flavor isolates produced by steam vacuum distillation and solvent extraction of distillate, *J. Agric. Food Chem., 24,* 433–434, 1976.

CHAPTER 5

Application of Combined Gas Chromatography/Mass Spectrometry to the Identification and Quantitative Analysis of Trace Organic Contaminants

D.S. Millington*
D.L. Norwood

Department of Environmental Sciences and Engineering
School of Public Health
University of North Carolina at Chapel Hill
Chapel Hill, North Carolina

5.1 INTRODUCTION

Combined gas chromatography-mass spectrometry (GC/MS) is arguably the most powerful technique available for the analysis of complex organic mixtures (1). It unites the separating power of gas-liquid chromatography, which has received a considerable boost since the introduction of open-tubular capillary columns, with mass spectrometry, the most versatile and sensitive method for the detection and analysis of organic compounds. Not surprisingly, GC/MS has found widespread application in all the life sciences, including studies of organic pollution in the environment. There is no doubt that progress in this field would be severely restricted without the formidable capabilities of GC/MS for qualitative and quantitative analysis of specific organic contaminants in the bulk terrestrial matrices, air, water, and soil.

Owing to their low concentration, the analysis of such compounds in drinking water requires, as an essential preliminary step, the enrichment of organic solutes to levels detectable by GC/MS. The major techniques for concentration are reviewed in the preceding chapter, which also discusses the difficulties arising from the introduction of additional matrices with

* Current address: Division of Genetics and Metabolism, Department of Pediatrics, Duke University Medical Center, Durham, North Carolina.

which the organic material will be associated before analysis. Not the least of these problems is the introduction of artifacts, derived mainly from contaminated solvents and adsorbents. Some specialized concentration techniques have been developed to overcome or minimize this problem. The best example of this is the closed loop stripping apparatus (CLSA) developed by Grob (2), which relies on scaling down the quantities of the secondary matrices. Other techniques avoid solvents altogether, as in the adsorption-thermal desorption approach (3). It seems certain that more new methods will emerge as research continues in this area. In any event, the concentration step is a necessary complication that imposes limitations on the analysis, including the significant element of selectivity. There is no guarantee that all organic contaminants in a water sample that are amenable to GC/MS analysis will be recovered by any or all of the existing methods of concentration. GC/MS has its own practical limitations, which are reviewed later (Section 5.3.3).

Another issue of greater concern has arisen from the rapid growth of GC/MS applications in water analysis. Recent technological advances in electronics and computers have reduced the controlling functions of a sophisticated apparatus to simple keyboard operation. The hitherto awesome and time-consuming procedures of data reduction and interpretation have also been greatly simplified. Such simplification and cost reduction have contributed much to making GC/MS available to a large additional number of research laboratories in recent years. These instruments, which can be and often are operated by persons with little previous knowledge or training in mass spectrometry, produce results in prodigious quantities that are contributing to a rapidly expanding data base in the literature. There also appears to be a popular but quite unjustified belief in the integrity of all data generated by GC/MS, which is considered by many to be a fundamentally valid and accurate technique. Meanwhile, the validity and meaning of much of the recent GC/MS literature has been challenged (4) owing to the lack of accompanying information required for the independent assessment of technical quality. This applies particularly to the data in nonrefereed publications that comprise the so-called "gray" literature (5). Since fundamental policy decisions may be affected by GC/MS data from such sources, it is reasonable to expect that the methodology used should be open to independent evaluation and peer review. This requires that evidence of scientific rigor should accompany the data in GC/MS publications, an element that has too often been overlooked.

The purpose of this chapter is to outline the problems that exist owing to limitations and pitfalls of computerized GC/MS, which the authors feel have been underpublicized. Systematic guidelines are presented for the qualitative and quantitative analysis of mixtures by GC/MS, designed to encourage analysts to critique their own data and present it in a form suitable for objective assessment by other scientists. These procedures are not necessarily difficult or time-consuming, but should represent a common sense solution to a problem of increasing magnitude. Failure to recognize this need will jeopardize future decisions and policies based on GC/MS data, and the technique itself could needlessly fall into disrepute. Auxiliary techniques for increasing the information content of GC/MS analysis are also discussed, using examples drawn from the authors' experiences.

5.2 TERMINOLOGY

A number of terms appear frequently in GC/MS publications that either are used inaccurately or have no universally accepted standard definition, leading to confusion for both experts and newcomers to the field. Some of these terms have been selected for special consideration in this section. Attention is also drawn to several recent reviews on the subject of principles, definitions, and nomenclature in environmental analysis. One of the most useful is that by Keith et al. (6), which defines commonly used terms such as precision, accuracy, validation, verification, limits of detection, and quantitation. Other useful information pertinent to good project planning and execution is also included. More detailed articles on validation (7), quality control (8), and limit of detection (9,10) provide additional reference material.

5.2.1 Identification

This term generally implies the definition of an exact chemical structure for a sample constituent, based on the interpretation of physico-chemical data and the matching comparison of such data with a known compound when possible. Unfortunately, there is no standard set of criteria for what constitutes an "identification." Some organic chemists may consider rigorous identification to consist of the matching comparison of all commonly available physico-chemical characteristics, such as melting point, chromatographic behavior, infrared, ultraviolet, nuclear magnetic resonance, and mass spectra for an isolated and purified unknown with those of authentic reference material. The more complex the molecular structure, the more rigorous the comparison should be. In many GC/MS publications, a component is said to have been identified if its mass spectrum matches one of those in a library of mass spectral data according to criteria determined by a computer program. Between these unpalatable extremes is a range of options, which implies that the term "identification" should be qualified to indicate a degree of confidence or uncertainty. It must be understood that in any analysis of an unknown mixture by GC/MS, there will be a range of confidence or uncertainty levels associated with component identifications. While specific suggestions for a satisfactory nomenclature have been described previously (4,11) and are further elaborated later in this chapter, more discussion of this important and much neglected subject is warranted.

5.2.2   Internal Standard

This is another term without a standard definition that implies different meanings to scientists with different backgrounds. Confusion can, therefore, arise when the term is not defined in the context of its application. To most analytical mass spectroscopists, an internal standard is a substance added in known amount *to the original sample matrix* for the purpose of quantifying the amount of one or more previously identified analytes. This is accomplished by measuring the ratio of detector responses for characteristic mass spectral ions of the analyte and internal standard by GC/MS. The response ratio is calibrated by an appropriate procedure. The selection of internal standards for quantification is arguably the most important criterion affecting the accuracy. This subject is discussed in detail later in the chapter (Section 5.4.2).

Compounds added to the recovered or concentrated sample for the purpose of evaluating chromatographic performance or determination of retention indices are often called internal standards. More appropriately, perhaps, they should be referred to as chromatography standards. In some procedures, an internal standard for quantitation is added to the recovered or concentrated sample just before GC/MS analysis. In this case, the recovery of each analyte from the original sample matrix must be determined separately. Another classification of internal standard, often referred to as a "recovery" or "surrogate" standard, may be added to the matrix in known concentration for this purpose. It can be argued that the use of surrogate standards in GC/MS analysis of water samples is an unnecessary complication that can be avoided by appropriate selection of internal standards and a valid calibration procedure (Section 5.4.3).

5.2.3   Reconstructed Ion-Current Profile (or Chromatogram)

This term refers to the computer technique of recovering selected data from a completed GC/MS analysis when the mass spectrometer was used in a full cyclic scanning mode. A total ion-current (TIC) profile is the sum of all the individual ion intensities plotted as a function of scan number or time and is the nearest GC/MS equivalent to a flame ionization detector (FID) trace. These TIC profiles are often computer-enhanced to improve signal-to-noise ratio, either by summing ion intensities over a limited mass range or by use of automated background subtraction. Mass-selected ion-current profiles, also called "mass chromatograms" are the intensities of individual ions plotted as a function of time. These are often used to search a complex mixture for components of a certain structural class. They can also be used for quantitative analysis of particular components when suitable internal standards are present. When used in cyclic scanning mode, the mass spectrometer can provide the maximum amount of information, both qualitative and quantitative, and is the method of choice for multicomponent analysis.

5.2.4 Selected Ion Monitoring

This term is synonymous with "multiple ion detection," and numerous other permutations of multiple or selected with ion detection, recording, or monitoring. It was also referred to as "mass fragmentography" by the original pioneers (12). In this technique, mass-selected ion-current profiles are generated and stored during the GC/MS analysis, instead of full-scan data (13). The mass spectrometer is then operated as a *selective* detector, being made to jump rapidly between different preselected positions in the mass scale in a cyclic fashion. Sensitivity is 100–1000 times greater in this mode than in full-scan mode because the detector is employed most of the time collecting useful data. Selected ion monitoring is therefore the method of choice for the quantitative analysis of specific trace components in a biological sample.

5.2.5 Response Ratio and Response Factor

These terms are appearing more often in the context of quantitative GC/MS analysis. A response ratio is the ratio of the detector response for a characteristic mass spectral ion in an analyte, integrated for the duration of its elution from the GC, to that for a corresponding ion in an internal standard. A response factor is a calibrated response ratio. Often, it is simply the response ratio corresponding to equimolar concentrations of the analyte and internal standard. Single response factors derived in this way are often used, erroneously, as the basis for quantitation in environmental samples. There is no justification for assuming either the linearity of response ratio with analyte concentration or that response factors are physical constants. The correct procedure for calibration of response ratio is to determine "response factors" over a range of analyte concentration to derive a standard curve.

5.3 QUALITATIVE ANALYSIS

The purpose of qualitative analysis is to identify as many of the components of a mixture as possible. The following steps are involved: presentation of the sample in a form suitable for GC/MS analysis, efficient separation of the mixture components, generation of good-quality mass spectra, interpretation, rejection of artifactual components, and confirmation of component identification. The first of these procedures is discussed in Chapter 4 and the remaining steps are covered in the following paragraphs.

5.3.1 Principles and Application of Capillary Columns in GC/MS

The theory and techniques of capillary GC are well documented (14) and only certain aspects pertinent to GC/MS are discussed here. The advent of glass capillary columns has certainly revolutionized the art of GC/MS. Their primary advantages for GC/MS over packed GC columns are higher resolving power and enhanced absolute sensitivity, owing to the elution of solutes in very narrow, sharp peaks. Also, the low flow rates (1–2 ml per min) are compatible with direct coupling to MS without the need for a molecular separator. The recently introduced "fused-silica" capillary columns with chemically bonded liquid phases offer still more advantages when coupled to MS (15). These include high thermal stability, resulting in very low column "bleed," and flexibility, enabling their direct insertion into the ion source of the MS. The avoidance of sample losses due to molecular separators or exposure to untreated surfaces at the GC/MS interface strongly justifies the use of these columns. The authors routinely use and recommend commercially available 30 m × 0.24 mm i.d. columns with chemically bonded methyl silicone liquid phase of 1 μm thickness for the routine analysis of mixtures containing acidic, basic, and neutral components. The performance of these columns can easily be permanently degraded by allowing air to enter them when hot or by injecting samples containing water, mineral acids, or refractory material. By carefully avoiding these problems and not overloading the column, satisfactory performance can be maintained for long periods. It is stressed, however, that all GC columns are somewhat compound-selective and become more selective with use, owing to generation of active sites.

The useful dynamic range of capillary columns for full-scan GC/MS is 1–200 ng per GC component. For routine analysis, the sample is split by the injector so that only 1–3% is actually analyzed. For very small, uncontaminated samples, the on-column technique is preferred, but this cannot be automated without difficulty (16). Before performing GC/MS analysis on any new sample, experience has shown that it is prudent to analyze the sample first on an off-line GC/FID system equipped with an identical capillary column. Thus, the optimum injection technique, sample volume, and GC program conditions can be determined before the more costly and time-consuming GC/MS analysis is undertaken.

The presentation of many very sharp GC peaks, often only a few seconds wide, imposes certain constraints on the scanning mass spectrometer as a detector. Unlike flame ionization or electron capture detectors, the MS is a discontinuous detector that samples the column effluent at regular intervals. To preserve chromatographic integrity, the total scan cycle time must be short enough to collect at least five to seven scans on each GC peak (see Fig. 5.1). Until recently, only quadrupole MS were capable of achieving the necessary scan rates, but revolutionary technological advances have given some magnetic sector instruments an almost equivalent capability (17,18). The computer's interface, data processing speed, and storage capacity now impose the practical limitation of scan speed. For most purposes, a scan cycle time of about 1–2 s is adequate. With the help of modern computer

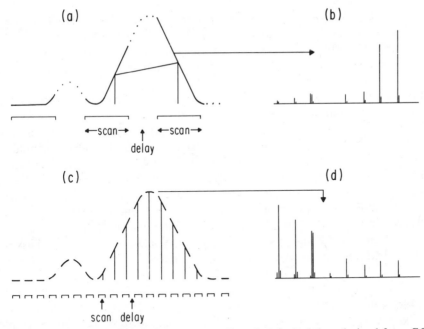

Figure 5.1. Effect of scan speed on the quality of analytical data derived from GC/MS. (*a*) Scan time is too long compared with elution profile of GC peaks, hence too few data points to achieve accurate quantitation based on peak area measurement. (*b*) Mass spectrum from tailing side of GC peak when scan speed too slow is grossly distorted owing to change in sample concentration during the scan. (*c*) Scan cycle time increased by a factor of 3, enabling much more accurate quantitation. (*d*) Mass spectrum taken at top of peak or at most places on the sides does not exhibit significant distortion.

programs, the quantity of data generated under these conditions (1800–3600 mass spectra/h) can be processed in a reasonable amount of time.

5.3.2 Computer Analysis of GC/MS Data

For the best chances of successful interpretation of their mass spectra, GC components should be presented to the MS in the purest form possible, which justifies the use of the best available capillary columns for separation. Even then, some quite sophisticated computer treatment of the acquired GC/MS data is usually necessary. When used in the conventional electron ionization (EI) mode, the MS ionizes all vaporized materials in the ion source at all times. Consequently, there is always a "background" of detectable signals due to air, carrier gas, contaminating residue from earlier analyses, and GC column bleed. The relative contribution to the background from these sources changes during a GC/MS analysis. In addition, there is often a carry-over or "memory" effect from one GC component eluting just before another, especially if the first component has a high relative concentration. Thus, the first computer task applied to the raw GC/MS data is to remove the interfering background signals from the true analyte signals. This process, which is usually referred to as "enhancement," can be carried out in one of two principal modes—either continuously updated background subtraction or a modification of the Biller-Biemann algorithm (19) for recognition of simultaneous maxima of ion intensities. As some manufacturers provide both forms of enhancement, it is important that the analyst understands their differences and knows how to apply them. It is also essential to preserve at least one complete copy of the raw, unadulterated data for storage, because enhancement always significantly alters the original data and may irretrievably subtract some valuable information.

The authors' experience has indicated that one variant of the first method, in which the mass spectrum taken just before the elution of a GC peak is subtracted from all scans until the next peak, and so on, is relatively safe and effective for providing good quality mass spectra for most of the GC components, especially in a complex chromatogram. The alternative enhancement procedure is much more subtle. Success depends on such factors as the scan rate and GC peak width, and on differences between the manufacturer's version and the original Biller-Biemann program. Its principle advantage is to indicate multiplicity in a single GC peak, unless two or more components exactly coelute, that may otherwise be undetected. In our hands, the Biller-Biemann type of enhancement is best invoked only to generate enhanced TIC chromatograms. The "enhanced" mass spectra have often exhibited erroneous features that could lead to misinterpretation.

Another option, best performed on the raw data, is to display the spectrum of an analyte and an operator-selected background scan, subtract one from the other and compare the result with the unsubtracted spectrum. This is the most reliable method of enhancement but is is also the most time-consuming.

After enhancement, the operator or computer selects "scans of interest," typically those corresponding to the GC-peak maxima, and may store them with reconstructed ion-current profiles as a reduced form of the original data. Each scan of interest is then compared with a library of mass spectra, compiled, for example, from the EPA/NIH data base (20). This automated library search attempts to match the sample spectra with those of known compounds in the data base and reports successful matches as computer "identifications." Some of the computer programs now available can execute the enhance, interest-mark, and library search functions as the GC/MS analysis is proceeding by using simultaneous foreground/background processing features. Thus, modern GC/MS systems can automatically perform multiple sample analyses without operator intervention. It is very tempting to accept the computer's results without contest in the belief that the elimination of human error and, therefore, human interaction is always a positive benefit. Unfortunately, computerized analysis of mixtures by GC/

MS is very far from a state of total reliability and judicious human interaction is absolutely vital to the credibility of GC/MS data.

5.3.3 Pitfalls of Automated GC/MS Analysis

The major weakness of computerized GC/MS analysis is the library search procedure for component identification, which compares one imperfect, complex set of data with another using an imperfect algorithm. The result is a high potential for misinterpretation, where false negative and false positive "identifications" can occur with unpredictable frequency. To promote an understanding of this important limitation, there follows a brief review of the most common sources of error.

A mass spectrum may be too severely distorted for computer matching if the scan rate is too low, owing to a significant change in sample concentration during the scan (see Fig. 5.1). Distortion may also result from inaccurate background subtraction and by poor calibration or mass discrimination in a badly tuned instrument. This latter is an inherent problem of the small quadrupole mass analyzers commonly used in GC/MS systems. Most of the mass spectra in the commercially available libraries were originally recorded on various instruments of the magnetic sector type. Owing to an inherent relative mass discrimination (18), the same spectra from quadrupole spectrometers may appear substantially different. Some quadrupole instrument manufacturers modify the spectra in the data base in an attempt to compensate for these differences. The original data are already known to be partly corrupted by mistakes and artifacts due to impurities in the reference compounds. Furthermore, a typical library only contains a paltry 25,000–30,000 entries compared with over five million known synthetic organic structures. When one adds to this the fact that usually only the main features of the unknown spectra are compared with those in the library, the skepticism of experienced mass spectroscopists for compound identifications based solely on computerized library search procedures can be appreciated.

It is true that many compounds that occur ubiquitously in environmental samples exhibit highly characteristic, sometimes unique mass spectra that are not likely to be missed by a library search. Chlorobenzene, tetrachloroethylene, and haloforms are in this category, for example. A compound identified as 2,3,7,8-tetrachlorodibenzo-p-dioxin, on the other hand, could be one of 22 possible isomers. For all its limitations, the library search is a very useful interpretative aid but it should only be considered as one step in a rigorous identification procedure.

5.3.4 Recommended Procedure for Qualitative Analysis

Good planning and preparation are essential to rigorous GC/MS analysis. The recommended procedure, summarized in Figure 5.2, includes the rejection of artifactual components that often plague GC/MS experiments (21).

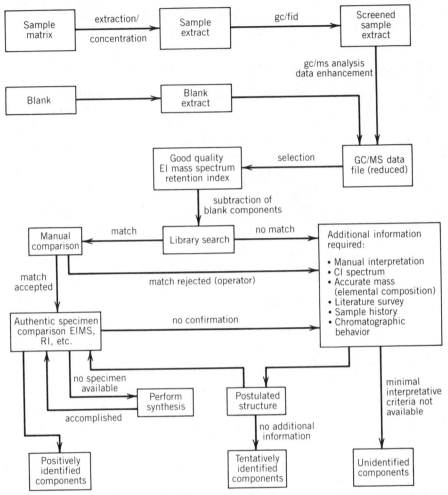

Figure 5.2. Flow diagram summarizing the recommended procedure for qualitative analysis of trace organic compounds in water samples using GC/MS. Human interaction is an essential ingredient both for interpretation and for assessment of confidence in the component identifications.

Their origins can be glassware, solvents, reagents, laboratory air, the GC injector, and the GC column itself. Trace amounts of organic material from some of these sources can become very significant after concentration. Therefore, it is most important to include the analysis of appropriate blank and control samples, even when the utmost precautions are taken to avoid contamination. The more steps involved prior to GC/MS, the more essential is the role of the blank in the analytical scheme. In fact, GC/MS data in publications that fail to mention blanks or controls should be considered invalid a priori.

Before conducting GC/MS analysis, an adequate quality control proce-

dure is required to ensure correct tuning and testing of the equipment especially for quadrupole-based systems (22). The spectrometer's scan cycle
time should be adequate to prevent severe distortion of the mass spectra.
The sample should have been preanalyzed by off-line GC/FID to establish
the correct sample concentration, injection volume, and analytical conditions.

After acquiring the GC/MS data and generating a condensed file of
background-subtracted spectra of the salient components, the identification
procedure (Fig. 5.2) can be implemented. When the library search appears to
be successful and provides one or more matches, the analyst must visually
compare the mass spectrum of the unknown with the library data and make
the final decision on acceptance. The human eye is a very good pattern-
matching device. An acceptable match of the EI spectra should be followed
with an acceptable match of retention index (RI) on the same capillary GC
column to complete the identification. The RI is easily determined by coin-
jection of a mixture of *n*-alkanes (23) and is a very reproducible parameter.
Unfortunately, it is not included in mass spectral libraries because there is
no universally accepted standard GC column or set of conditions. Therefore,
the analyst must obtain or synthesize an authentic specimen of the postulated compound for this comparison. It is timely to suggest that a common
effort should be made by laboratories that specialize in GC/MS analysis to
compile RI information, standardized on the use of one or more of the
commercially produced chemically bonded phase columns alluded to earlier.
Otherwise, each laboratory faces the prospect of having to custom build its
own extended library of MS and RI data.

5.3.5 Use of Supplementary MS Information
to Improve Interpretation

To illustrate further use of the flow diagram in Figure 5.2, examples from the
author's experiences are now cited. The degradation products of aquatic
humic acid after digestion with oxidizing agents, such as potassium permanganate (24) and chlorine (11,25), are of interest owing to their potential
health hazard, since they may be formed during the treatment of drinking
water. Analysis of the lower-molecular-weight degradation products was
accomplished by GC/MS after ether extraction and methylation of carboxylic acids using diazomethane. These analyses presented difficulties in interpretation owing mostly to the paucity of comparable data in the mass spectral libraries and in other literature. For example, the library search
"identified" four peaks in the permanganate degradation mixture as "dimethyl phthalate." Visual comparison showed that the spectra were indeed
very similar and that three of them contained an ion at m/z 194, corresponding to the molecular ion of dimethyl phthalate. Thus it was readily
deduced that the three possible benzenedicarboxylic acid isomers must all
be present. This was subsequently confirmed by obtaining specimens of

these readily available compounds and showing that after methylation, the retention index and mass spectrum of each corresponded exactly with one of the degradation products. The fourth component in the mixture, which had to be very similar in structure to the others, nevertheless could not be identified without further information.

In this case, the chemical ionization (CI) mass spectrum provided the vital structural clue. This was obtained by reanalyzing the sample by GC/MS in the "CI" mode, in which ionization takes place mostly by transfer of protons from a reactant gas, usually isobutane (26). Protonated molecular ions are generated from any compound whose proton affinity is greater than that of isobutane, thus permitting the determination of molecular weight. The CI spectrum of the fourth component (Fig. 5.3b) indicated a molecular weight (MW) of 222, which is 28 mass units higher than that of dimethyl phthalate (Fig. 5.3a). We reasoned that only an additional carbonyl group between the benzene ring and one of the carboxyl groups satisfactorily accounts for the similarity of the EI spectra. Cleavage at the weak bond between the functional groups (Fig. 5.3a and b) produces the same base ion (m/z 163) as dimethyl phthalate. Since only one MW 222 product was formed in the oxidation, the most likely structure was dimethylphthalonate, which is known to be formed in high yield as phthalonic acid from naphthalene and permanganate (27). This tentative structural identification was later confirmed by synthesis and comparison of EI, CI, and RI data with those of the authentic sample.

Further analysis of the humic acid/KMnO$_4$ degradation products revealed more structures of this type, containing up to five carboxyl groups but only one α-carbonyl group (24). Furthermore, the same new oxidation products were observed from terrestrial humic acid (28). Earlier GC/MS studies had failed to identify these structures. This example illustrates how a simple piece of supplementary MS information can greatly enhance structure elucidation and lead to interesting discoveries; in this case, strong evidence for the presence of fused-ring structures in aquatic and terrestrial humic acid. It is noteworthy that although almost every GC/MS system produced commercially during the past 5 years is equipped with a combined EI/CI ion source, very little use has been made of CI in water analysis.

5.3.6 Use of CI and Low-Resolution Accurate Mass Data as Interpretative Aids

Recent publications have indicated that the major products of humic acid chlorination are chloroform and trichloroacetic acid (11,29,30). A large number of additional products were formed, which could not readily be identified on the basis of their EI spectra owing to the general lack of molecular ions and comparative spectral information. In this case, the supplementary information required for interpretation included both CI spectra and, more important, accurate mass measurements (11). Briefly stated, the measurement

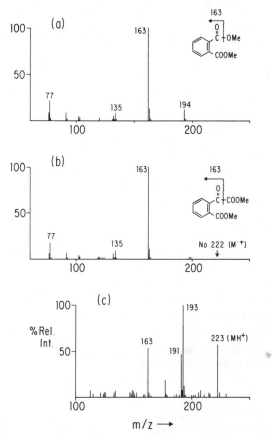

Figure 5.3. Example of the use of supplementary chemical ionization (CI) data for detection of phenylglyoxylic acid residues in permanganate degradation product of aquatic humic acid. (*a*) EI mass spectrum of dimethylphthalate. (*b*) EI mass spectrum of related degradation product, later shown to be that of phthalonic acid dimethyl ester after interpretation of (*c*) the CI mass spectrum of the same product, which indicated a MW of 222.

of the mass of an ion to within an accuracy of 15 ppm facilitates the determination of its elemental composition, since only a few combinations of the common elements (C, H, O, N, and Cl) are possible up to a mass of about 400. This information is not routinely available to most GC/MS laboratories, since specialized computer programs and instrumentation are necessary (18). The authors have routinely performed such analyses in the EI mode, however, and have chosen to include examples to indicate the great value and future potential of this analytical capability. In Figure 5.4, the EI and CI spectra of a chlorination product of humic acid are presented with the possible elemental compositions of two major EI fragment ions, whose measured masses were 140.952 ± 0.002 and 155.977 ± 0.002 daltons. The CI spectrum (Fig. 5.4*b*) exhibits a protonated molecular ion at *m/z* 201 with an isotope

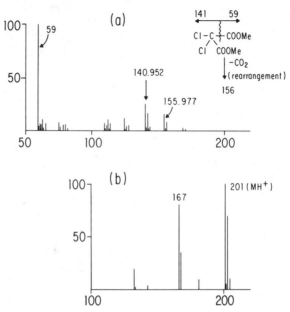

Figure 5.4. Interpretation of the structure of a methylated component from the chlorination of aquatic humic acid, using (a) low-resolution accurate mass data to supplement the EI spectrum and (b) the isobutane-CI spectrum to determine molecular weight and extent of chlorine substitution.

cluster consistent with the presence of two chlorine atoms. The same isotope cluster accompanies the m/z 141 and 156 ions, whose elemental compositions can only be $C_3H_3O_2Cl_2$ and $C_4H_6O_2Cl_2$ according to their accurate masses. From these data, the postulated structure of this compound was $Cl_2C(CO_2CH_3)_2$ (dimethyl dichloromalonate, $C_5H_6O_4Cl_2$), which would generate the m/z 141 ion by loss of a CO_2CH_3 radical. The m/z 156 ion presumably arises from the loss of CO_2 by means of a skeletal rearrangement (Fig. 5.4a). Because this compound has not been found in the literature and no synthetic material is yet available, the identification is considered incomplete. However, the CI and accurate mass data have at least provided a structure in which there is a high level of confidence. Such structural assignments were made for many other components of the original mixture (11,25), which would not have been feasible without the supplementary mass spectral data.

When two or more compounds coelute from a GC column, the mass spectrum is a mixture that is usually very difficult to interpret by either manual or computer methods. This problem recently arose during capillary GC/MS analysis of an organic solvent extract from a sample of spent granular activated carbon (GAC), taken from one of the U.S. locations that employ GAC filter beds in the treatment of drinking water (31). The mixed EI and CI spectra (Fig. 5.5a and b) are confusing, because at first sight the data

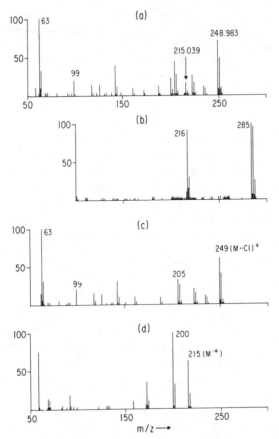

Figure 5.5. Example of use of EI-accurate mass and CI data to interpret a coeluting mixture during analysis of a GAC filter extract. The EI spectrum (*a*) is a mixture of at least two components as shown by the CI spectrum (*b*), which indicates one component having 3 × Cl atoms, MW 284 and another having 1 × Cl, MW 215. The reference EI mass spectra of tris(2-chloroethyl) phosphate (*c*) and atrazine (*d*), when superimposed in the ratio of about 5:1, generate (*a*). The retention indices of each standard and the mixture component are the same.

seem to be consistent with a single, chlorinated species of MW 284. The isotope cluster for the protonated molecular species is consistent with a content of three chlorine atoms, but the m/z 216 ion, which contains one Cl atom, is not a logical fragment of 285, since 285 − 216 = 69 and 2 × ^{35}Cl = 70. The correct interpretation of the CI spectrum is that two species are present with molecular weights of 215 and 284. Further analysis of this mixture was possible only with the EI accurate mass data. The accurate mass determined for the m/z 215 molecular ion in the EI spectrum was 215.039, for which the only possible elemental composition is $C_8H_{14}N_5Cl$ (215.040). A literature survey for compounds of this molecular formula suggested atrazine, a common pesticide, as a likely candidate. Indeed, all the principal ions in the standard mass spectrum (Fig. 5.5d) were observed in the mixed EI spectrum. Although the atrazine spectrum was also in the mass spectral library, the computer search failed owing to the presence of the second, coeluting component. This is a good example of a "false negative," previously alluded to as a potential failing of the automated library search procedure. Interpretation of the remaining information was straightforward after it was realized that m/z 99 is commonly observed in the spectra of organic phosphates and that the molecular formula of the second component must be $C_6H_{12}O_4Cl_3P$, since m/z 249 (measured mass 248.983) corresponds to $C_6H_{12}O_4Cl_2P$ (calculated mass 248.985). The postulated structure, tris(chloroethyl)phosphate($[C_2H_4ClO]_3P{=}O$), has the possibility of structural isomerism. Tris(2-chloroethyl)phosphate, a known proprietary compound, is used as a flame retardant. An authentic sample gave the spectrum shown in Figure 5.5c and was almost exactly coincident in retention index with pure atrazine. Thus, a difficult interpretative problem was again solved by the systematic application of supplementary MS techniques. Several other components in the GAC extract, for which no library or literature MS information were available, were identified in this manner (31).

In reporting component identifications derived by GC/MS in the preceding studies, the authors have employed the nomenclature suggested in Figure 5.2 to express the level of confidence in their data. There may be simpler and more effective ways to indicate confidence levels, but at least an important issue that deserves more open debate has been promoted.

5.4   QUANTITATIVE ANALYSIS BY GC/MS

In 1978, the National Bureau of Standards (NBS) distributed a blood sample containing 25 ppm of the drug diphenylhydantoin to 107 laboratories claiming to perform quantitative GC/MS analysis (32). The mean value reported was 24.8 ppm, but the coefficient of variation was 41% and the range of values was from 0 to 380 ppm. One conclusion from such a study would be that GC/MS is an unreliable method for quantitative analysis. However, the NBS itself develops and promotes GC/MS assays for biologically important compounds such as cholesterol (33) and glucose (34) as "definitive," meaning the most specific and accurate available—in fact, the benchmark for all other methods. This apparent anomaly is explained by the fact that it is the analytical procedure, not the GC/MS technique per se, that determines the analytical precision and accuracy.

The analysis of trace components in water is complicated by the number of steps involved in the procedure, especially the concentration step. The pitfalls and difficulties associated with automated qualitative analysis are relatively minor compared with those of quantitative analysis, yet there is ever-increasing interest and desire to obtain total quantitative data from the organic analysis of water. The oversimplification of procedures in this area for the sake of data production is an established and profoundly disturbing trend.

5.4.1 The Principles of Quantitative GC/MS

Procedures for quantitative analysis must take into account matrix effects, analyte losses during extraction and concentration, variations in behavior of the GC injector, the GC column, and the detector. The analysis of many components in the same matrix (multicomponent analysis) is a tall order and it is not easy to accomplish. First, it is appropriate to consider the steps required to assay a single mixture component by GC/MS.

The detector response, usually the area of a characteristic ion current profile, is related directly to quantity, but it is both unnecessary and unwise to calibrate this response by external standardization. Therefore, an internal standard (IS) is employed and a response ratio, which is also related to the quantity of analyte, is determined. The response ratio is calibrated by analysis of solutions containing various known amounts of the analyte with a fixed amount of the IS. A plot of response ratio against the quantity or concentration of the analyte—the standard curve—is derived. The purpose of this procedure is to eliminate or minimize errors associated with variations in the GC injector, column, and detector. If the IS is added to the extract just prior to GC/MS, then the quantitation refers only to the concentrated extract. If the objective is to quantify the analyte in the original sample matrix, it will be necessary to determine the recovery of the analyte from the matrix. The determination of recovery is tedious, difficult, and error-prone. Therefore, it makes very good sense to avoid it altogether if possible. For water analysis, the solution is a very simple one, namely, the addition and equilibration of the IS to the original water sample. This establishes, at the outset, a ratio between the analyte and IS that is carried through all the stages of extraction, concentration, and analysis and is ultimately determined by the mass spectrometer.

The recommended procedure, outlined in Figure 5.6, is highly tolerant of inconsistencies in recovery, losses during concentration, and even of accidental spillage and inferior laboratory technique. To validate the entire procedure, it is necessary only to show that the ratio of analyte to IS remains unaltered. Fortunately, using mass spectrometry, this is easily accomplished. There is, in fact, no strategy for quantification that is superior to GC/MS with internal standardization, as outlined in Figure 5.6. Because the choice of the internal standard is a critical factor, the criteria for its selection are reviewed.

5.4.2 Selection of an Internal Standard

Bearing in mind that the purpose of the internal standard (IS) is to establish a concentration ratio with the analyte in the sample matrix that is accurately

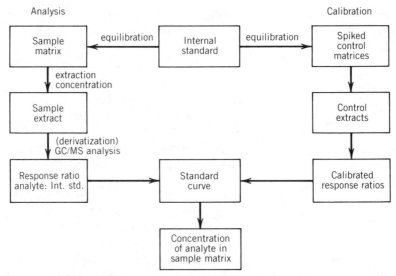

Figure 5.6. Flow diagram summarizing recommended procedure for quantitative analysis by GC/MS. Calibration and analysis are performed in the same manner, using a suitable internal standard for each analyte.

reflected by the response ratio of selected characteristic ions determined at the GC/MS stage, the following factors determine the choices.

1. It must not be a component of the sample matrix.
2. It must have physical and chemical properties very similar to those of the analyte.
3. The GC retention index must be very similar to that of the analyte.
4. The characteristic ion should have a similar m/z value to that of the analyte, especially if a quadrupole MS is employed.

The first point is self-explanatory. The second and third are vitally important. The analyte and its internal standard should equilibrate equally between various potential binding sites in the sample matrix. Such sites may include dissolved or suspended humic substances, also the walls of the sample container. If liquid–liquid extraction is involved, the partition coefficients between aqueous and organic phases must be equal or, if adsorption on a solid phase is involved, the adsorption and recovery properties must be the same. There is a potential for large errors if these conditions are not fulfilled. Equally important, at the critical GC/MS stage, the analyte and IS must also exhibit very similar behavior in the injector and on the GC column.

The types of IS that fulfill the desired criteria can be summarized as follows:

1. A very closely related positional isomer—must separate from the analyte on the GC column since both have same characteristic m/z value.

2. A closely related homolog or analog—may have the same or a different RI as the analyte because the two ions being monitored are usually different (but preferably by not more than 20 amu).

There is documented evidence to show that large errors can result from the use of an IS that is not structurally closely related to the analyte (35,36). Although numerous possible choices for a suitable IS may exist for any particular analyte, the only type that is virtually guaranteed to fulfill all the required conditions is a stable isotope-labeled analog. The alteration in chemical structure—hence in physical and chemical behavior—caused by the introduction of a few atoms of isotopically pure ^{13}C or ^{2}H into a molecule are so slight that mass spectrometry is, in fact, the only technique that can detect the difference unequivocally at the required sensitivity level.

The technique of adding a labeled analog for quantification is known as isotope dilution, which is the basis of many definitive assays for biologically important compounds including those previously cited (33,34). Isotope dilution has the advantage that calibration of response ratios is straightforward and exhibits a linear relationship over at least 2 orders of magnitude in analyte concentration when the ion abundances are corrected for any overlap. Furthermore, validation is easily accomplished by demonstrating that neither the sample matrix nor the GC system exerts a selective effect on the recovery of the analyte and IS. Most important, the precision and accuracy (37) of the method are readily determined and are usually excellent. These are important criteria for judging the acceptability of a method, which is based on an estimation of the total error (38).

These principles are adequately demonstrated in the recent publication (30) of a definitive assay for trichloroacetic acid (TCA), a ubiquitous by-product of water chlorination (11,25). The internal standard was ^{13}C-1 trichloroacetic acid, whose mass spectrum after methylation exhibits an intense m/z 60 ion corresponding to $[^{13}COOCH_3]^+$, instead of m/z 59 as in the unlabeled compound. The assay was based on the determination of the m/z 59/60 ion area ratio for the GC peak corresponding to TCA.

^{13}C is a very stable label, but ^{2}H and ^{18}O can potentially exchange with protons or oxygen atoms in the sample matrix. Thus, the chemical stability of such an isotopically labeled IS, both in the sample matrix and in stored standard solutions, must be checked before use.

5.4.3 Multicomponent Analysis

On the premise that the complete analysis of a mixture, such as the trace organic compounds in a water sample, requires qualitative and quantitative analysis, then the principles of quantitative GC/MS must be applied to each component in the mixture.

Large volumes of quantitative data have been reported on waste water analysis methods published by the United States Environmental Protection Agency, in which GC/MS quantitation can be performed with either an internal or external standard (39). A recent evaluation of these methods by Kirchmer et al. (35) has revealed that both accuracy and precision are greatly affected by the method of calibration and by the selection of the internal standard. Colby and Rosecrance (36) also suggest that the recently introduced isotope dilution methods for priority pollutant analysis (40) overcome all the major difficulties.

The inescapable facts are that each mixture component must have its own internal standard, ideally but not necessarily an isotope-labeled analog, calibration and analysis must be carried out in the same manner, and a validation of each assay must be performed. The supreme advantage of GC/MS is that these procedures can be carried out simultaneously for all components to be assayed. It is just as easy to add 40 internal standards to a water sample as it is to add one. The computer can easily determine 40 different response ratios in one GC/MS experiment. Furthermore, 40 calibration curves can be derived as easily as one. All of these procedures readily lend themselves to computer automation, which removes the tedium of manual calculations and plots.

5.4.4   Recommended Procedure for Multicomponent Quantitative Analysis

The procedure is outlined in Figure 5.6, the essential steps being as follows:

1. Analyze sample qualitatively using recommended procedure (Section 5.3.4).
2. Prepare list of compounds to be quantified (analytes).
3. Select an appropriate internal standard for each analyte (Section 5.4.2).
4. Devise an acceptable "control matrix" for calibration/validation.
5. Derive calibration curves for each analyte.
6. Determine response ratios for sample analytes using GC/MS.
7. Obtain concentrations of analytes by reference to standard curves.

Obviously, if there is no reference material or valid internal standard for a particular analyte, then that component cannot be quantified. The availability of isotopically labeled compounds is very limited at present, but it is increasing. The EPA's priority pollutants, for example, can now be purchased in labeled form as a kit (41). When lack of availability or expense prohibits the use of a labeled IS, a closely related homolog or analog is the best alternative.

We would suggest that the term "quantitation" should apply only to validated GC/MS assays using an internal standard, otherwise, the term "estimation" is more acceptable. General adoption of the guidelines suggested here (or similar ones) for qualitative and quantitative GC/MS analysis of water samples will make it easier to assess the quality of published data and will preserve the standing of GC/MS as the finest method available.

REFERENCES

1. W.H. McFadden, *Techniques of Combined Gas Chromatography/Mass Spectrometry,* Wiley, New York, 1973, 463 pp.
2. K. Grob and F. Zurcher, *J. Chromatogr. 117,* 285, 1976.
3. J.F. Pankow, L.M. Isabelle, and T.J. Kristensen, *Anal. Chem., 54,* 1815, 1982.
4. R.F. Christman, *Environ. Sci. Technol., 16,* 143A, 1982.
5. M.A. Shiffman, *Environ. Sci. Technol., 16,* 85A, 1982.
6. L.H. Keith, W. Crummett, J. Deegan, R.A. Libby, J.K. Taylor, and G. Wentler, *Anal. Chem., 55,* 2210, 1983.
7. J.K. Taylor, *Anal. Chem., 55,* 600A, 1983.
8. C.J. Kirchmer, *Environ. Sci. Technol., 17,* 174A, 1983.
9. J.A. Glaser, D.L. Foerst, G.D. McKee, S.A. Quave, and W.L. Budde, *Environ. Sci. Technol., 15,* 1427, 1981.
10. G.L. Long, and J.D. Winefordner, *Anal. Chem., 55,* 713A, 1983.
11. J.D. Johnson, R.F. Christman, D.L. Norwood, and D.S. Millington, *Environ. Health Persp., 46,* 63, 1982.
12. C.G. Hammar, B. Holmstedt, and R. Ryhage, *Anal. Biochem., 25,* 532, 1968.
13. J.T. Watson, *Introduction to Mass Spectrometry: Biomedical, Environmental and Forensic Applications,* Raven Press, New York, 1976, pp. 199–221.
14. W. Jennings, *Gas Chromatography with Glass Capillary Columns,* Academic Press, New York, 1980, 320 pp.
15. J. Settlage and H. Jaeger, *J. Chromatogr. Sci., 22,* 192, 1984.
16. G. Schomburg, H. Behlau, R. Doélmann, F. Weeke, and H. Husmann, *J. Chromatogr., 142,* 87, 1977.
17. P. Burns, B.N. Green, and D.S. Millington, *Proceedings of the 28th Annual Conference on Mass Spectrometry and Allied Topics,* ASMS, New York, 1980, pp. 345–346.
18. R.D. Craig, R.H. Bateman, B.N. Green, and D.S. Millington, *Phil. Trans. Roy. Soc. London A, 293,* 135, 1979.
19. J.W. Biller and K. Biemann, *Anal. Lett., 7,* 515, 1974.
20. National Bureau of Standards, *EPA/NIH Mass Spectral Data Base,* National Standard Reference Data Service Publication No. 63 (1978).
21. M. Ende and G. Spiteller, *Mass Spectrom. Rev., 1,* 29, 1982.
22. J.W. Eichelberger, L.E. Harris, and W.L. Budde, *Anal. Chem., 47,* 995, 1975.
23. E. Kovats, *Adv. Chromatogr. 1,* 229, 1965.
24. R.F. Christman, W.T. Liao, D.S. Millington, and J.D. Johnson, in *Advances in the Identification & Analysis of Organic Pollutants in Water,* Vol. 2, L.H. Keith (ed.), Ann Arbor Science, Michigan, 1982, p. 979.
25. D.L. Norwood, J.D. Johnson, R.F. Christman, and D.S. Millington, in *Water Chlorination: Environmental Impact and Health Effects,* Vol. 4, Book 1, R.L. Jolley et al. (eds.), Ann Arbor Science, Michigan, 1981, p. 191.
26. F.H. Field, in *MTP International Review of Science,* Physical Chemistry, Vol. 5, A. McColl (ed.), Butterworth & Co., Baltimore, MD, 1972, Chapter 5; M.S.B. Munson, *Anal. Chem., 48,* 28A, 1971.
27. J.H. Gardner and C.H. Naylor, *Org. Syn. Coll., 2,* 523, 1943.
28. D.S. Millington and A. Stevens, unpublished.
29. J.W. Miller, P.C. Uden, and R.M. Barnes, *Anal. Chem., 54,* 485, 1982.
30. R.F. Christman, D.L. Norwood, D.S. Millington, and J.D. Johnson, *Environ. Sci. Technol., 17,* 625, 1983.
31. D.S. Millington, D.J. Bertino, T. Kamei, and R.F. Christman, in *Water Chlorination, Environmental Impact and Health Effects,* Vol. 4, R.L. Jolley et al. (eds.), Book 1, 1981, p. 445.
32. D.J. Reeder, D. Enagonio, R.G. Christensen, and R. Schaffer, Symposium on Trace Organic Analysis, Nat. Bureau of Standards, Gaithersburg, MD, April 1978.

33. A. Cohen, H.S. Hertz, J. Mandel, R.C. Paule, R. Schaffer, L.T. Sniegoski, T. Sun, M.J. Welch, and E. White, *Clin. Chem., 26,* 854, 1980.
34. E. White, M.J. Welch, T. Sun, L.T. Sniegoski, R. Schaffer, H.S. Hertz, and A. Cohen, *Biomed. Mass Spectrom, 9,* 395, 1982.
35. C.J. Kirchmer, M.C. Winter, and B.A Kelly, *Environ. Sci. Technol., 17,* 396, 1983.
36. B.N. Colby, and A.E. Rosecrance, in *Advances in the Identification and Analysis of Organic Pollutants in Water,* Vol. 1, L.H. Keith (ed.), Ann Arbor Sci., Michigan, 1981, p. 221.
37. Guides for measures of precision and accuracy, *Anal. Chem. 40,* 2271, 1968.
38. E.F. McFarren, R.J. Lishka, and J.H. Parker, *Anal. Chem. 42,* 358, 1970.
39. EPA Methods 624 and 625, Federal Register, *44,* 69532–69552, 1979.
40. Methods 1624 and 1625, U.S. EPA (WH-552), Washington, DC.
41. Available from KOR Isotopes, Cambridge, Massachusetts.

CHAPTER 6

Alternatives to Gas Chromatography/Mass Spectrometry for the Analysis of Organics in Drinking Water

William H. Glaze

Environmental Science and Engineering Program
University of California
Los Angeles, California

6.1 INTRODUCTION

Alternatives to gas chromatography/mass spectrometry (GC/MS) may be classified into three categories. First, there are methods that use GC with some detector other than a mass spectrometer. These include the traditional detectors of environmental analysis, especially the electron capture detector, and also those that are sensitive to specific elements such as the halogens, nitrogen, phosphorous, and sulfur. Although these detectors do not offer confirmation of structure, they continue to be attractive because of their relatively low cost and ease of operation. Recently, they have been complemented by new spectrometric detectors such as the atmospheric pressure microwave-induced plasma detector (MIPD) with multielement capability, and, more importantly, by advanced design fourier transform infrared (FTIR) detectors.

Second, there are methods that transcend the foremost limitation of gas chromatography by using separations procedures that do not require volatilization of the sample. These consist principally of variants of liquid chromatography, including what is now referred to as high-performance liquid chromatography (HPLC). HPLC with mass spectroscopic detection (HPLC/MS) and with FTIR detection (HPLC/FTIR) are of increasing interest, although they are still in developmental stages.

Finally, there are the analytical methods that do not focus on specific contaminants. These are in two subcategories: those that measure a group of

153

compounds with some common property, such as a phenolic functionality, halogen bound to carbon, etc.; and, those that seek to measure a biochemical effect of a total concentrate such as toxicological or toxicologically mimetic methods.

In this chapter we will discuss GC methods with detectors other than the mass spectrometer, HPLC-based methods, and some of the group parameter methods, particularly total organic halogen (TOX). Toxicological methods are discussed in following chapters.

6.2   GAS CHROMATOGRAPHIC METHODS

6.2.1   GC Methods with Conventional Detectors

Under certain conditions, GC with conventional detectors is an acceptable method for routine monitoring of specific organic compounds in water. The most familiar example is the analysis of trihalomethanes and other volatile organic halides. EPA-approved methods, now in Standard Methods, include a purge and trap method (1) and a liquid–liquid extraction method (2) with a halogen-sensitive detector, namely, the Hall electrolytic conductivity detector (HECD) and the electron capture detector (ECD), respectively. In general, GC methods with conventional detectors should be used on a routine basis for samples that are well known, where gross changes in contaminants on a sample-by-sample basis are not expected, and where interfering contaminants are not likely to be present. Under these conditions the methods are usually reliable, although in the absence of spectrometric confirmation misidentification of a GC peaks remains a possibility.

6.2.2   GC with Spectrometric Detectors

Spectrometric detection of GC fractions includes, in addition to mass spectrometric detection, the observation of absorption and emission characteristics of the fractions, either in the molecular state or as dissociated atoms. The atmospheric pressure microwave-induced plasma detector (MIPD) dissociates the molecular fractions into atoms in a plasma and focuses on characteristic emission frequencies of the constituent atoms. The MIPD is highly sensitive and selective for molecules containing certain types of atoms, depending on the emission wavelength being monitored. Detection limits of 3–20 pg for Br and Cl and 1.4 ng for S have been reported (3). Quimby et al. (4) and more recently, Chiba and Haraguchi (5) have coupled an MIPD to a conventional purge and trap apparatus for the determination of halogenated organics in water. Detection limits for chlorides, bromides, and iodides were reported to be in the sub-ppb range with a linear range of four orders of magnitude. Quimby et al. have also used GC/MIPD to investigate the chlorination products of soil humic and fulvic acids (6). Figure 6.1 shows

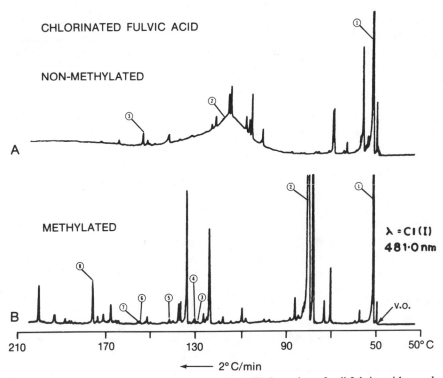

Figure 6.1. Capillary gas chromatogram with MIP detection of soil fulvic acid sample after chlorination. Peak identities: (above) (1) chloroform, (2) trichloroacetic acid, (3) 1-chlorophenol; (below) (1) chloroform, (2) trichloroacetic acid methyl ester, (3) 2,4,6-trichlorophenol methyl ester, (4) 2-chlorobenzoic acid methyl ester, (5) 3,5-dichlorobenzoic acid methyl ester, (6) 1-chlorophenol methyl ester, (7) 2-chlorophenol, (8) pentachlorophenol methyl ether. (From Ref. 6, with permission.)

a capillary GC/MIPD chromatogram indicating the formation of relatively large amounts of trichloroacetic acid, as well as other chlorinated products. The figure also shows the beneficial effect of sample methylation on the quality of the GC separation. While not as definitive as the GC/MS studies reported by Christman et al. (7), these studies do show the power of the MIPD detector, a very important feature of which is the ability to discriminate between the different halogens.

A variety of other spectrometric detectors for GC have been reviewed (8), but in the main these do not now appear to have outstanding potential for water analysis. The exception is the FTIR detector which is discussed in more detail below.

6.2.3 GC with FTIR Detection

Recent developments in FTIR spectroscopy have made this technique a viable alternative to mass spectrometry for the identification of organic contaminants in water. Both on-line and off-line FTIR measurements are feasible. For off-line measurement the fractions are condensed and collected, then their infrared (IR) spectra are obtained in a separate procedure. Collection procedures include the use of cooled KBr disks of a size such that they can be inserted directly in the FTIR unit, or gold collector disks in which the GC fractions are collected in a frozen argon matrix. On-line GC/FTIR is similar in principle to GC/MS, with the essential accessory being a digital data storage and control system of substantial capacity.

In principle, on-line FTIR should provide valuable spectral information for the identification of GC fractions, but two deficiencies have so far prevented the realization of this potential. These include, first, a rather high operational detection limit for FTIR when used in conjunction with capillary GC systems, and second, a rather small data base of standard reference spectra. Azarraga was among the first to recognize the potential of GC/FTIR for environmental analysis (9). He and his colleagues have been instrumental in the development of a data base of vapor phase spectra of organic compounds of environmental significance, but at the present time the conveniently usable data base is far smaller than typical MS data bases (3300 versus 40,000 spectra). Undoubtedly this deficiency is only temporary and FTIR data bases will probably rival MS data bases in the near future. Moreover, it has been proposed that capillary GC/FTIR detection limits will be lowered to the subnanogram level soon (10). It is clear from the data in Table 6.1 that such low detection limits were not available only a short time ago. The table, taken from Gurka et al. (11), lists minimum identifiable quantities (MIQ) in the microgram range for packed and capillary columns using an FTIR instrument with two IR scans per second. In this case MIQ is defined

TABLE 6.1

Minimum Identifiable Quantities of Selected Compounds by On-Line Packed and Fused Silica Capillary GC Columns with FTIR Detection

	MIQ	
Compound	Packed (μg)	FSCC (μg)
Nitrobenzene	1.0	0.5
Dichlorobenzenes	2.0–6.0	0.5
Dinitrotoluenes	2.0–6.0	0.5
N-Nitrosoamines	2.0–6.0	0.5–1.0
β-Chloroethers	6.0	0.5–1.0
Isophorone	—	0.5
Hexachlorobutadiene	7.5	1.0
Fluorene	10.0	2.2

Source: Ref. 38.

conservatively as the minimum level required to exceed the peak locator threshold of the FTIR software. Instruments with 20 scans per second and improved interfaces are now available which have lowered MIQ values to approximately 100 ng, but detection limits comparable to GC/MS remain a goal at this time.

GC/FTIR and GC/MS often provide complementary information on the structure of unknown substances. For example, FTIR is superior for identification of structural isomers of aromatic hydrocarbons, while not very sensitive to changes in the structure of aliphatic isomers. In view of this complementarity there is currently a great deal of interest in combined GC/FTIR/MS systems (12). In principle such systems should allow one to identify a much larger fraction of "untargeted" compounds, that is, unknowns in environmental samples. In a recent review of GC/FTIR Griffiths et al. (10) have noted the work of Shafer et al. (13) in which more peaks could be identified in a hazardous waste extract by GC/FTIR than by GC/MS, the difference in the size of the two data bases notwithstanding. Gurka et al. (14) have pointed out that neither method is presently capable of identifying the majority of compounds in a hazardous waste extract, but that the two methods combined should be successful at least in classifying the majority of volatile compounds according to molecular type, such as aromatic or aliphatic hydrocarbons, aromatic amines, etc. Molecular type information could then be combined with short-term bioassay results and structure-activity relationships to predict the toxicity of a hazardous waste sample (14). Presumably the same could be said for drinking water samples. It has been suggested that commercial GC/FTIR/MS systems will probably be available in the near future at costs of less than $150,000 (10). If so, and if the required sensitivity is achieved for FTIR, these systems should be extremely valuable in the identification of the large number of nontargeted compounds that occur at low levels in drinking water.

6.3 LIQUID CHROMATOGRAPHIC TECHNIQUES

It is well known that the organic compounds that are amenable to gas chromatography represent only a small fraction of the total organic carbon in aquatic samples. Most natural organic compounds, and many of synthetic origin, are too polar, too involatile, or otherwise too intractable to pass through a GC system. HPLC offers great potential as an analytical method amenable to these types of compounds.

Several monographs and review articles have described the principles of HPLC, and this is beyond the scope of the present review. Rather, we shall examine here some of the applications of HPLC to the study of organic compounds in water and discuss the potential of some new HPLC techniques for future development.

6.3.1 HPLC Analysis of Low-Molecular-Weight Compounds

HPLC is well suited for the analysis of certain types of low-molecular-weight organic compounds that are difficult to analyze at the trace level by GC. These include the very polar organics which can be extracted from water only with great difficulty, such as the chloroacetic acids formed during water chlorination (5,7). Also, HPLC is the method of choice in the analysis of labile substances such as carbamate pesticides. Unfortunately, HPLC has suffered from the lack of a detector capable of low level detection of a wide variety of organic compounds. Presently, HPLC is most applicable to analytes that have strong electronic absorption or emission characteristics. This makes the technique very valuable for solutes such as polynuclear aromatic hydrocarbons, which has been extensively exploited (15, 16), but it is of marginal value for materials without a strong chromophore. [See, however, the recent paper by van der Wal and Snyder (17) on the use of an absorption photometric detector operating at 184 nm for enhancing sensitivity of HPLC with UV detection.] As a result much effort has gone into detector development, with environmental trace analysis being one of the principal goals. Of particular interest are the new and improved electrochemical detectors, halogen-sensitive detectors, total organic carbon detectors (useful only with aqueous solvents), and spectrophotometric detectors such as FTIR (18).

Jolley was among the first to use liquid chromatography for the examination of organics in water (19). In a landmark study of the effects of chlorination on waste waters, he employed ion-exchange chromatography to demonstrate the formation of new chlorinated substances during the disinfection process. Figure 6.2 shows a typical chromatogram taken from recent work (20). Note that the time scale covers a period of many hours, a severe limitation of the technique for routine application. (It is significant that Jolley identified compounds by off-line MS studies on collected fractions, and it could be argued that most could have been determined by GC/MS after preconcentration and derivatization.) Prior to the development of fused silica capillary GC columns, phenols could not be determined conveniently due to poor resolution on packed columns or adsorption effects on glass capillaries. Also MS was unable to differentiate between poorly resolved isomers. As a result there was stimulus to develop alternative techniques for the analysis of phenols. Haeberer and Scott (21) have reported an HPLC procedure which utilized XAD-4/8 resins for sample preconcentration. A cesium silicate cleanup step isolated the phenolic fraction from carboxylic acids. Figure 6.3 shows the liquid chromatogram using reverse-phase separation of municipal effluent with and without phenol fortification. By the standards of present-day methods, the resolution is not remarkable, but the method is still useful, especially for intractable nitrophenols. Recently, Shoup and Mayer (22) used an amperometric detector to determine environmental phenols in the range of 100 ppt to 500 ppm. At the lower levels, solute enrichment was achieved by passing the sample through an octadecylsilane-

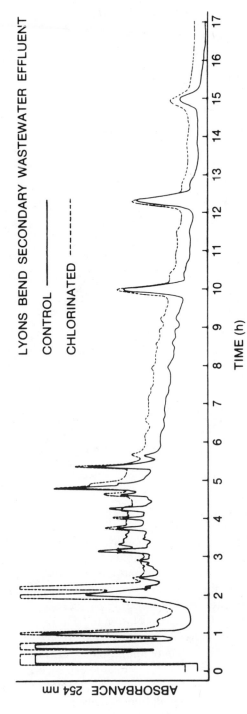

Figure 6.2. Analytical-scale ion-exchange HPLC chromatogram of municipal waste water before and after chlorination. (From Ref. 19, with permission.)

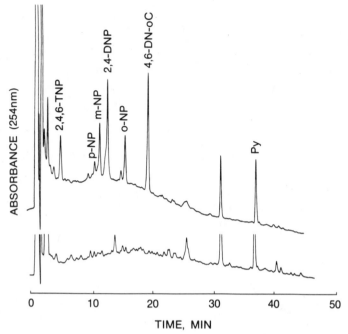

Figure 6.3. HPLC analysis of a municipal effluent with (*upper curve*) and without (*lower curve*) phenol fortification at 10 ppm; reverse-phase separation. (From Ref. 20, with permission.)

bonded silica cartridge that comprised one loop of a multiport injection value. The analytes were then flushed from the cartridge by changing the solvent from 0 to 40% acetonitrile. This technique, with variations in the type of cartridge and solvents used, is widely applicable for preconcentration and analysis of water samples.

6.3.2 HPLC Analysis of High-Molecular-Weight Organics

Liquid chromatography (LC) may also be used for the characterization of high-molecular-weight materials. Size exclusion chromatography has been extensively employed for the study of natural organics in water for many years (23). However, the most common size-exclusion column materials are subject to adsorption effects by polyelectrolytes such as those that constitute natural organics. The effect causes elution to be delayed, leading one to conclude that the fractions are of lower molecular weight than they are, in fact. In addition, natural organics may ionize and/or aggregate, which tends to distort molecular size measurements. Nonetheless, size-exclusion HPLC is useful for studying organics in water, especially for determining the effect of treatment processes, e.g., coagulation and disinfection. For example, Glaze and co-workers (24–26) used several LC methods including size exclusion to study natural organics in surface water sources, before and after disinfection with hypochlorite and ozone. The results showed a slight lowering of apparent molecular weight upon mild chlorination, but more significantly, a nearly uniform ratio of organic-bound chlorine to organic carbon throughout the polymer. Three fractions of the raw water showed trihalomethane formation potentials [per unit total organic carbon (TOC)] which were all about the same level. Veenstra and Schnoor (27) fractionated surface water organics with Sephadex gels and chlorinated the fractions to determine trihalomethane potentials. They showed that about 88% of the TOC had an apparent molecular weight of less than 3000 daltons and that 87% of the total trihalomethanes were produced from this fraction. Watts and co-workers (28) used a reverse-phase separation method to show that organic halogen was present in all fractions collected with especially high levels in two peaks eluting early, which were inflated due to inorganic halide.

Investigations of the production of by-products from ozonation of natural waters have not been as fruitful as those that have examined chlorinated by-products. This is due, in part, to the absence of a convenient "tag" on the by-products to aid in the identification process. Reverse-phase and size-exclusion HPLC studies of a Canadian surface water source before and after ozonation (25) illustrates one of the deficiencies of conventional HPLC detectors. Figure 6.4 is a reverse-phase chromatogram using fluorescence detection with no sample preconcentration. Chromatogram B, taken after ozonation, has essentially the same features as chromatogram A, but with diminution of intensity. What this illustrates is not that ozonation produces no by-products, but rather that ozonation destroys aromatic sites that have high fluorescencing properties. Other types of separations using electrochemical and UV absorption detectors showed similar results, including the observation that the apparent molecular weight of the organics did not appear to shift significantly upon ozonation (25). On the other hand, Gloor and co-workers (29) used a size-exclusion HPLC procedure with total organic carbon detection to show the formation of new low-molecular-weight substances after ozonation and other changes which are not observed with spectrometric detectors (Fig. 6.5). It is clear that ozonation does produce low-molecular-weight by-products, but the methods used so far have not elucidated the structures nor the toxicological hazards of these by-products.

The work of Watts et al. (28) illustrates another puzzling feature of studies on aquatic organics. Analysis of the fractions collected from the HPLC column by mass spectroscopy, including field desorption ionization MS, have yielded very little definitive structural information. At the time of this writing, no papers have appeared that have successfully applied MS to the analysis of aquatic organics. Although molecular size studies indicate that a substantial fraction of aquatic organics are in the range of 1000–2000 daltons, they are apparently too polar to undergo ionization without decomposition. Degradation techniques in combination with GC/MS (30) have yielded useful information on the structure of aquatic organics, but our

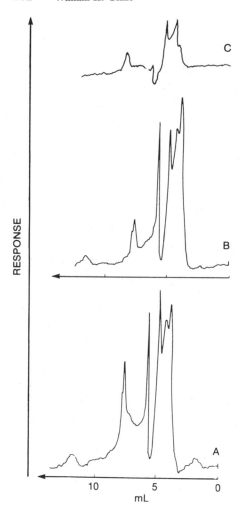

Figure 6.4. Reverse-phase HPLC chromatogram of a Canadian surface water source before (*A*) and after treatment with ozone (*B:* dose = 0.2 × TOC, *C:* 2.5 × TOC). (From Ref. 24, with permission.)

understanding of these materials is still primitive. This situation may be improved in the future with the development of new MS methods and with the refinement of HPLS/MS and HPLC/FTIR.

6.3.3   HPLC with MS, FTIR, and Nuclear Magnetic Resonance Detectors

The primary obstacle to the coupling of HPLC and MS systems is the large volume of HPLC solvent to be disposed of so as not to flood the MS ionization chamber. There are presently two types of interfaces that are competing for application in the area of HPLC/MS: direct liquid inlet (DLI) and moving band or wire. DLI introduces the HPLC fraction directly into the ionization chamber, whereas in the other method the effluent is applied to a moving

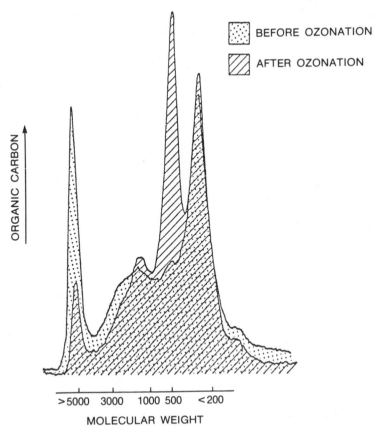

Figure 6.5. Sephadex gel chromatogram with total organic carbon detection before and after ozone treatment. (From Ref. 28, with permission.)

band or wire where a majority of the solvent evaporates before the sample is introduced into the ionization chamber. DLI has become more attractive recently as a result of the development of new microcolumns that use solvent flow rates below 1 mL/min, and by the invention of the thermospray interface by Vestal and co-workers (31).

Hayes et al. (32) applied a moving belt with a spray deposition interface for the analysis of polynuclear aromatic hydrocarbons using normal and reverse phase HPLC. Figure 6.6 is a comparison of the UV and MS traces observed in one case. The deficiencies of the method are its rather high detection limits (> 40 ng) caused in part by loss of analyte in the belt interface, and the difficulty of using aqueous solvents.

Apffel and co-workers (33) have designed a DLI LC/MS interface and applied it for the analysis of 15 phenylurea herbicides and polynuclear aromatic hydrocarbons. The device gives somewhat lower detection limits than other DLI devices that require effluent splitting (34).

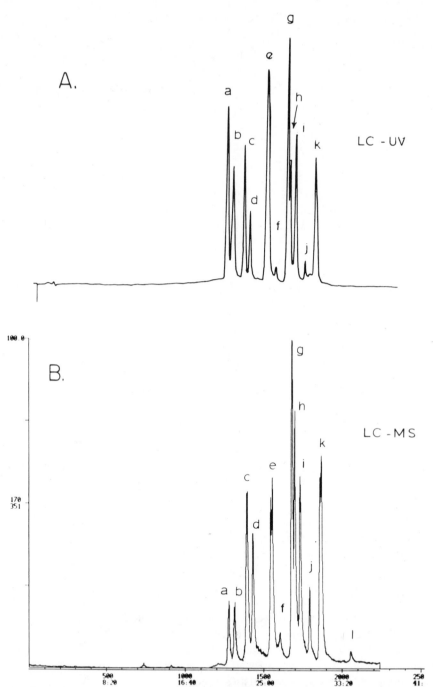

Figure 6.6. Comparison of UV and MS profiles for analysis of polynuclear aromatic hydrocarbons. MS detection by spray deposition on moving belt. (*a*) phenanthrene; (*b*) anthracene; (*c*) fluoranthene; (*d*) pyrene; (*e*) chrysene; (*f*) unknown; (*g*) benzo-[*b*]fluoranthene; (*h*) benzo[*k*]fluoranthene; (*i*) benzo[*a*]pyrene; (*j*) benzo[*a*]anthracene; (*k*) indeno[1,2,3-cd]pyrene; (*l*) unknown. (From Ref. 32 by permission.)

Perhaps the most exciting recent development in LC/MS interfaces is the thermospray developed by Vestal and co-workers (35). In the presence of certain ionic substances, such as ammonium acetate in the solvent, no ionization of the sample is required and flow rates up to 2 mL/min of aqueous mobile phase can be handled. Subnanogram detection limits are reported with selected ion monitoring.

So far, there have been few applications of advanced HPLC/MS methods in the area of drinking water analysis. This is due in part to the rapid emergence of the field, and to the emphasis so far on substances amenable to GC analysis. This situation will no doubt change in the future as more research workers become familiar with HPLC/MS methods and as more attention is given to solutes that resist GC identification.

HPLC with FTIR detection is still in a much more primitive state of development than HPLC/MS. The principal difficulty in obtaining usable IR spectra of HPLC fractions is the interference of the common solvents used in the HPLC separation. This can be alleviated by deposition of the fractions onto KBr disks with evaporation of solvent; however, the procedure is laborious and time consuming. Perhaps the most promising prospect is to utilize supercritical solvents with large IR transparent windows, such as carbon dioxide (36). However, carbon dioxide alone is a poor solvent for polar compounds and therefore of marginal value as an HPLC solvent for environmental samples. Another development is the use of microbore HPLC columns that minimize the amount of solvent interference. Brown and Taylor have recently used this technique to detect aromatic amines of interest in synthetic fuel waste waters (37). Figure 6.7 shows the IR spectrum of 6 μg of indole after HPLC separation and a standard spectrum of indole for comparison. In general, minimum detectable quantities are in this range, making the technique of value for drinking water analysis only after sample preconcentration, and then only for components of high concentration.

On-line observation of HPLC fractions by nuclear magnetic resonance (NMR) spectroscopy has also been limited by high detection limits (38). Commonly available 200-MHz instruments have been able to detect about 10–20 μg of compounds by proton NMR, that is, about the same sensitivity as a refractive index detector. However, design modifications could make HPLC/NMR more sensitive by a factor of at least 10, making this nondestructive technique potentially useful for preconcentrated samples.

In summary, HPLC instruments with MS, FTIR, and NMR detectors may someday provide adequate structural information for the identification of water contaminants that have so far eluded attention. Developments in these areas will be interesting to watch in the coming years.

6.4 NONCHROMATOGRAPHIC METHODS

Chromatographic methods for the analysis of the organic constituents in water require a substantial capital investment and relatively high skill levels

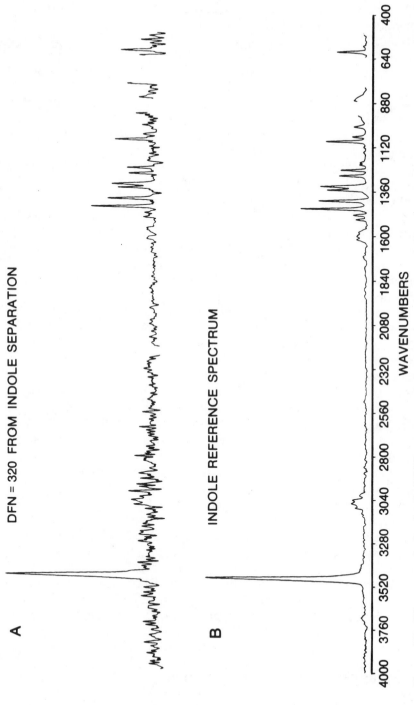

Figure 6.7. (Upper trace): On-the-fly FTIR spectrum of component of mixture of amines separated on microbore (1-mm diam.) polar amino cyano HPLC column. (Lower trace): Indole reference spectrum. (From Ref. 36, with permission.)

for the personnel involved. This is especially true when one wishes to confirm the identity of the chromatographic fractions by some spectroscopic technique. Moreover, chromatographic methods are time consuming, often requiring hours for the accumulation of data, and perhaps days for its interpretation. Most significantly, although the major components in a water sample will probably yield to some form of chromatographic assay, chromatographic methods cannot identify the majority of constituents in a water sample, at least on a routine, cost-effective basis. These considerations have led to a variety of efforts to develop rapid, low-cost assays that will allow one to judge the safety of a water without having to resort to complete chromatographic/spectrometric characterization. These efforts have taken several approaches, including the following:

1. Development of tests to determine the toxicity of the sample or a sample concentrate, without preseparation of its components, if possible.
2. Development of tests to determine organic constituents of interest without preseparation.

Only the latter category will be covered in this chapter.

6.4.1   Tests that Measure Gross Organic Levels

The traditional measures of water quality include tests for gross organic content such as, biological oxygen demand (BOD), chemical oxygen demand (COD), and total organic carbon (TOC). For each of these parameters, one may distinguish between "dissolved" and "suspended" organic matter by some prefiltration step, but it is recognized that the split will be determined by factors such as pore size distribution and the efficiency of the filter. In any case, BOD, COD, and TOC, while valuable for certain purposes, have not proven to be very useful in ascertaining the safety of water samples. In the case of drinking waters, there may be a correlation between TOC and trihalomethane formation potential (39), but fluctuations of weather and reservoir conditions may change this correlation even on a local basis.

The failure of gross parameters such as TOC to be useful as indicators of toxicity is apparently due to the fact that most of the organic content in a natural water, even after treatment, is due to innocuous background organics. These background organics may fluctuate in concentration due to natural causes, changing the TOC level and thus obscuring any change in levels of potent toxicants at the micro-level. It is for this reason that research has focused lately on more toxicologically significant gross parameters.

The ideal gross parameter would satisfy the following requirements:

1. Be correlatable to one or more toxic responses.
2. Be capable of rapid measurement, at low cost, and by personnel with relatively low skill levels.
3. Require a modest capital investment and low maintenance costs.

Unfortunately, no such ideal parameter has yet been discovered; however, one that represents the closest approximation at the present time is the parameter total organic halogen (TOX or TOH).

TOX was first developed by Kuhn and Sontheimer (40), who were seeking a method to monitor chlorinated organic levels in West German water supplies, such as the Rhine River. The method developed consists of slurrying a water sample with powdered activated carbon (PAC), filtering the PAC, and heating the filter cake at 900°C in a stream of oxygen and steam. The organic compounds adsorbed onto the PAC are oxidized in the furnace and any organic halogen is converted into hydrogen halide, which is carried over with the steam condensate. The halide may be titrated potentiometrically or, more sensitively, by microcoulometry. To avoid interferences by inorganic halide ions in the sample, the filter cake is washed by nitrate solution, which competitively desorbs halide ion from the PAC. Recent developments in the TOX method, which more properly should be called "carbon adsorable organic halogen" (CAOX), have been made by Takahashi (41) at Dohrmann Envirotech (now Dohrmann-Xertex Corp). The tentative standard EPA method for TOX (42) is the Takahashi method, in which the PAC slurry is replaced by a cartridge of approximately 40 mg of PAC, through which the water sample is passed under pressure. The cartridge is then washed with nitrate solution and pyrolyzed in a specially manufactured furnace module coupled to a microcoulometer.

The Dohrmann procedure also allows one to determine purgeable organic halides (POX) by a step prior to PAC adsorption, in which the sample is purged with inert gas. Purgeable organics are carried in the gas stream directly into the furnace, and the hydrogen halide formed there is subsequently trapped and titrated in the microcoulometer.

Other attempts to measure organic halogen have utilized resin accumulators (43) and solvent extraction (44). Each method has some advantages over the others, but it is the PAC method that appears to be most favored at the present time (45).

Sorrell et al. (46) have recently evaluated the Dohrmann POX and TOX procedures on a variety of different water samples. The average accuracy of the POX and TOX methods for volatile organic halieds was found to be 64 and 79%, respectively, while projected costs for duplicate analyses were $77 and $99, respectively. In screening a variety of groundwaters, POX appeared to be reliable method for halogenated-VOCs at low concentrations, and afforded no false positives. TOX gave false positives for six of seven groundwaters with no GC-quantifiable halo-VOCs. The precise nature of this background is not known. Table 6.2 illustrates the background effect noted above.

Kruithof and co-workers (47) have provided one of the most striking examples of the possible usefulness of combined organic halogen measurements as a surrogate for specific compound determinations. XAD-4 extractable organic halogen (XOX) was measured on samples taken from a pilot

TABLE 6.2
Comparison of TOX and POX Levels and GC/HECD
Concentrations of Volatile Organic Halides for Selected
GroundWater Samples (μg/L as Cl)

City Code No.	TOX[a]	POX[a]	POX by GC/HECD[a]
1	21(21%)	<2	ND
2	10(56%)	<2	0.3(4%)
3	18(46%)	<2	0.6
4	27(10%)	21(5%)	20(7%)
5	30(15%)	<2	ND
6	20(12%)	<2	ND
7	25(27%)	<2	ND
8	9(50%)	<2	ND
9	10(5%)	2(50%)	6(6%)
10	<5	<2	ND
11	300(2%)	300(2%)	33
12	>2000	2300(5%)	1800(5%)
13	80(50%)	110(11%)	75(5%)
14	37(4%)	28(4%)	52(4%)
15	9(56%)	<2	ND

Source: Ref. 46.
[a] Values in parentheses refer to percent relative standard deviation for
POX and TOX and percent difference for GC/HECD); ND = not
detected.

plant starting with Rhine River water. Figure 6.8 shows XOX values on
these samples plotted versus the number of induced *Salmonella* revertants
using the Ames TA100 strain without S9 liver homogenate. There appears to
be a correlation between XOX values and the Ames test results, with a
different correlation for extracts obtained at pH 7 and 2. Further investigation of the use of TOX as a surrogate for mutagenicity and other forms of
toxicity is certainly in order in view of these results.

6.5   CONCLUSIONS AND FUTURE DEVELOPMENTS

Although GC/MS is presently the method of choice for the analysis of many
important types of organic compounds in water, scientists and public health
authorities continue to be interested in alternatives to GC/MS for several
important reasons. GCs with conventional detectors will continue to be
popular because of their relatively low cost and ease of operation, although
errors in identification are possible in complex samples. GC with FTIR

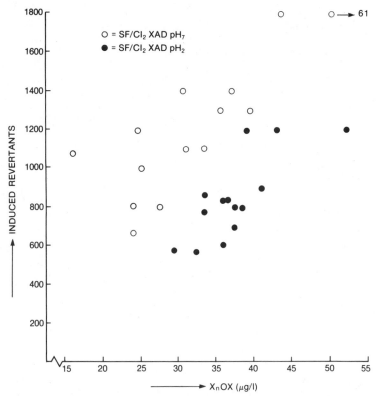

Figure 6.8. TA 100 revertants (without S9 liver homogenate) of XAD-4 extracts of chlorinated, rapid sand-filtered water (SF/Cl₂) from the Rhine River, plotted against organic halogen levels in the extracts determined by combustion/microcoulometry. (From Ref. 46, with permission.)

detection appears to be complementary to GC/MS, and commercial GC/FTIR/MS instruments are expected on the market shortly.

The power of HPLC has not been exploited in the analysis of drinking water, partly because of the limitations of the method, and partly because of the current regulatory emphasis on compounds that can be determined conveniently by GC methods. With the advent of higher-resolution columns and with the prospect of breakthroughs in HPLC/MS, HPLC/FTIR, and HPLC/NMR methodologies, more attention will be given to these techniques in the future. Of particular interest will be compounds whose lability prevents their derivatization and detection by GC or GC/MS.

There will continue to be great interest in the development of group and surrogate parameters that are capable of rapid and low-cost measurement by personnel with relatively low skill levels, and that correlate well with toxicological effects. Among the group parameters now available, TOX and POX appear to be the most promising candidates. Future developments in this area may have to wait for a more thorough understanding of the biochemical mechanisms of organic carcinogens. When that understanding is available, it may be possible to design assays that measure biochemical processes in vitro or in vivo that are directly related to or mimic the toxic effect of interest (48).

ACKNOWLEDGMENT

Until September of 1984, the author was a member of the faculty of the University of Texas at Dallas. He is grateful to that university and to the staff of the Graduate Program in Environmental Sciences, particularly Shirley Price, for help in the preparation of this manuscript.

REFERENCES

1. *Standard Methods for the Examination of Water and Wastewater,* 15th ed., American Public Health Association, Washington, DC, 1981, p. 538.
2. *Supplement to the Fifteenth Edition of Standard Methods for the Examination of Water and Wastewater,* American Public Health Association, Washington, DC, 1981, p. S92.
3. Cerbus, C.S., and Gluck, S.J., *Spectrochim. Acta, Part B, 38B,* 387, 1983.
4. Quimby, B.D., Delaney, M.F., Uden, P.C., and Barnes, R.M., *Anal. Chem., 51,* 875, 1979.
5. Chiba, K., and Haraguchi, H., *Anal. Chem., 55,* 1504, 1983.
6. Quimby, B.D., Delaney, M.F., Uden, P.C., and Barnes, R.M., *Anal. Chem., 52,* 259, 1980.
7. Christman, R.F., Johnson, J.D., Pfaender, F.K., Norwood, D.L., Webb, M.R., Hass, J.R., and Bobenrieth, M.J., in Jolley, R.L., et al. (eds.), *Water Chlorination: Environmental Impact and Health Effects,* Vol. 3, Ann Arbor Science, Ann Arbor, MI, 1980, p. 75.
8. Karasek, F.W., Onuska, F.I., Clement, R.E., and Yang, F.J., *Anal. Chem., 56,* 174R, 1984.
9. Azarraga, L.V., *Appl. Spectrosc., 34,* 224, 1980.
10. Griffiths, P.R., de Haeth, J.A., and Azarraga, L.V., *Anal. Chem., 55,* 1361A, 1983.
11. Gurka, D.F., Laska, P.R., and Titus, R., *J. Chromatogr. Sci., 20,* 145, 1982.
12. Williams, S.S., Lam, R.B., Sparks, D.T., Isenhour, T.L., and Hass, J.R., *Anal. Chem. Acta, 138,* 1, 1982.
13. Shafer, K.H., Hayes, T.L., Brasch, J.W., and Jakobsen, R.J., *Anal. Chem., 56,* 237, 1984.
14. Gurka, D.F., Hiatt, M., and Titus, R., *Anal. Chem., 56,* 1102, 1984.
15. Furuta, N., and Otsuki, A., *Anal. Chem., 55,* 2407, 1983.
16. Grzybowski, J., Radecki, A., and Rewkowska, G., *Environ. Sci. Technol., 17,* 44, 1983.
17. Van der Wal, S., and Snyder, L.R., *J. Chromatogr., 255,* 463, 1983.
18. Majors, R.E., Barth, H.G., and Lochmuller, C.H., *Anal. Chem., 56,* 300R, 1984.
19. Jolley, R.L., *Chlorination Effects on Organic Constituents in Effluents from Domestic Sanitary Sewage Treatment Plants,* ORNL-TM-4290, Oak Ridge National Laboratory, Oak Ridge, TN, 1973.
20. Jolley, R.L., Cumming, R.B., Lee, N.E., Thompson, J.E., and Mashni, C.I., in Jolley, R.L., et al. (ed.), *Water Chlorination: Environmental Impact and Health Effects,* Vol. 4, Book 1, Ann Arbor Science, Ann Arbor, MI, 1983, p. 499.
21. Haeberer, A.F., and Scott, T.A., in Keith, L.F., (ed.), *Advances in the Identification and Analysis of Organic Pollutants in Water,* Vol. 1, Ann Arbor Science, Ann Arbor, MI, 1981, p. 359.
22. Shoup, R.E., and Mayer, G.S., *Anal. Chem., 54,* 1164, 1982.

23. Schnitzer, M., and Khan, S.U., *Humic Substances in the Environment*, Marcel Dekker, Inc., New York, 1972, p. 106.

24. Glaze, W.H., Peyton, G.R., Saleh, F.Y., and Huang, F.Y., *Int. J. Environ. Anal. Chem.*, 7, 143, 1979.

25. Glaze, W.H., Jones, P.C., and Saleh, F.Y., in Keith, L.H., (ed.), *Advances in the Identification and Analysis of Organic Pollutants in Water*, Vol. 1, Ann Arbor Science, Ann Arbor, MI, 1981, p. 371.

26. Glaze, W.H., Saleh, F.Y., and Kinstley, W., in Jolley, R.L., et al. (eds.), *Water Chlorination: Environmental Impacts and Health Effects*, Vol. 3, Ann Arbor Science, Ann Arbor, MI, 1980, p. 99.

27. Veenstra, J.N., and Schnoor, J.L., in Jolley, R.L., et al. (eds.), *Water Chlorination: Environmental Impacts and Health Effects*, Vol. 3, Ann Arbor Science, Ann Arbor, MI, 1980, p. 109.

28. Watts, C.D., Crathorne, B., Crane, R.I., and Fielding, M., in Keith, L.H. (ed.), *Advances in the Identification and Analysis of Organic Pollutants in Water*, Vol. 1, Ann Arbor Science, Ann Arbor, MI, 1981, p. 383.

29. Gloor, R., Leidner, H., and Wuhrmann, K., *Water Research, 15*, 457, 1981.

30. Liao, W., Christman, R.F., Johnson, J.D., Millington, D.S., and Hass, J.R., *Environ. Sci. Technol., 16*, 403, 1982.

31. Borman, S.A., "Focus," *Anal. Chem., 56*, 1031A, 1984.

32. Hayes, M.J., Lankmayer, E.P., Vouros, P., Karger, B.L., and McGuire, J.L., *Anal. Chem., 55*, 1745, 1983.

33. Apffel, J.A., Brinkman, U.A.T., Frei, R.W., and Evers, E.A.I.M., *Anal. Chem., 55*, 2280, 1983.

34. Voyksner, R.D., and Bursey, J.R., *Anal. Chem., 56*, 1582, 1984.

35. Blakley, C.R., and Vestal, M.L., *Anal. Chem., 55*, 750, 1983.

36. Shafer, K.H., and Griffiths, P.R., *Anal. Chem., 55*, 1939, 1983.

37. Brown, R.S., and Taylor, L.T., *Anal. Chem., 55*, 1492, 1983.

38. Dorn, H.C., *Anal. Chem., 56*, 747A, 1984.

39. Symons, J.M., et al., *Treatment Techniques for Controlling Trihalomethanes in Drinking Water*, EPA-600/2-81-156, U. S. Environmental Protection Agency, Municipal Environmental Research Laboratory, Cincinnati, OH, 1981, p. 9.

40. Kuhn, W., and Sontheimer, H., *Vom Wasser, 41*, 65, 1973.

41. Takahashi, Y., and Moore, R., "Measurement of Total Organic Halides (TOX) in Water by Carbon Adsorption/Microcoulometric Determination," presented at 177th National Meeting, American Chemical Society, Div. of Environ. Chem., Honolulu, HI, 1979.

42. Billets, S., and Lichtenberg, J.J., "Total Organic Halide, Method 450.1-Interim," EPA-600/4-81-056, U.S. Environmental Protection Agency, Environmental Monitoring and Support Laboratory, Cincinnati, OH, 1981.

43. Glaze, W.H., Peyton, G.R., and Rawley, R., *Environ. Sci. Technol., 11*, 685, 1977.

44. Wegman, R., and Greve, P., *The Sci. of the Total Environ., 7*, 235, 1977.

45. Dressman, R., and Stevens, A., *J. AWWA, 75*, 431, 1983.

46. Sorrell, R.K., Daly, E., Boyer, L., and Brass, H.J., "Monitoring for Volatile Organohalides Using Pureable and Total Organic Halide as Surrogates," U. S. Environmental Protection Agency, Technical Support Division, Office of Drinking Water, Cincinnati, OH, 1984; presented at Water Chlorination Conference, Williamsburg, VA, June, 1984.

47. Kruithof, J.C., Noordsij, A., Puijker, L.M., and van der Gaag, M.A., "The Influence of Water Treatment Processes on the Formation of Organic Halogens and Mutagenic Activity by Post Chlorination," presented at Water Chlorination Conference, Williamsburg, VA, June, 1984.

48. Christensen, R.G., Hertz, H.S., Reeder, D.J., and White V, E., in Keith, L.H. (ed.), *Advances in the Identification and Analysis of Organic Pollutants in Water*, Vol. 1, Ann Arbor Science, Ann Arbor, MI, 1981, p. 447.

CHAPTER 7

Quality Assurance Programs in the Analysis of Trace Organic Contaminants

Herbert J. Brass

U.S. Environmental Protection Agency
Office of Drinking Water
Technical Support Division
Cincinnati, Ohio

Barbara A. Kingsley

SRI International
Menlo Park, California

The objective of this chapter is to describe the role and importance of quality assurance (QA) in gathering environmental data. While always significant, QA becomes especially critical in trace analysis since frequently state-of-the-art measurement technology is used to identify and quantify contaminants at or near detection levels. An objective assessment, based on available data, must always define the bounds of scientific observations, so that the data gathered can be fully and properly evaluated. Quality assurance is discussed in the subsequent sections as follows: (1) definition of QA and quality control (QC) and their importance; (2) levels of QA and QC required based on project objectives; (3) elements of QA and QC planning; (4) incorporation of QA and QC into analytical methodology; and (5) examples of QA programs using specific case histories. Owing to the background of the authors, much of this discussion will focus on the procedures and policies of the U.S. Environmental Protection Agency (USEPA) and related experiences.

7.1 WHY QUALITY ASSURANCE AND QUALITY CONTROL?

Quality assurance is, in fact, a management function that is based on establishing documentation, for example, by the use of standard operating proce-

173

dures, of environmental data gathering activities, and then evaluating the quality of the results that are obtained (1). Quality control is a technical function that provides the scientific data on which QA assessments and reports are based. It assists in evaluating the performance of the measurement systems and defines the quality of the generated data. It includes the determination of such parameters as precision, accuracy, detection limit, and method bias. Prior to the initiation of an environmental data gathering activity, a QA project plan should be prepared that presents the policies, organization, objectives, functional activities, and specific QA and QC activities designed to achieve the data quality objectives of this specific activity (2). The QA plan is primarily a management tool in which supervision, evaluation and review, summarization, and documentation are vital.

Included in the QA and QC program is not just the analytical operation. It also includes the project as a whole, focusing on those elements that affect overall QA requirements. Thus, the statistical design of a national survey can dictate the design of the sampling program, which in turn affects the required analytical accuracy, precision, and detection limits. This in turn can determine which analytical methods are selected for the measurement activity. The QA and QC program extends to the collection of field samples as well, since generated analytical data are useless unless the sample is stabilized until analysis and is representative of the source sampled.

A large component of a successful QA and QC program is to establish the credibility of organizations performing measurement activities (1). While the approach to, and application of, QA and QC must be objective, credibility is subjective. When sampling and analytical results generated in an organization are suspect, credibility is lost, and future data produced will be viewed as suspect irrespective of the quality of subsequent data. An instance arose in which review of a data base for a series of analytes was required. The reviewer was perplexed and perturbed since the validity of the data in question could not be determined due to inadequate QA and QC procedures. He remarked that "the data was as likely to be correct as incorrect"—flip a coin, if you will. Nothing more needs to be said about the consequences of such a review.

7.1.1   Drinking Water Laboratory Certification Program

Recognizing the importance of quality assurance and quality control, the Congress saw fit to include in the Safe Drinking Water Act (SDWA) a provision that defines a primary drinking water regulation as including "criteria and procedures . . . including quality control and testing procedures to insure . . . compliance . . ." (for regulated contaminants) (3). The National Interim Primary Drinking Water Regulations (4) "require that all testing for compliance purposes, except for turbidity, free chlorine residual, temperature, and pH, be performed by laboratories approved by either the USEPA or those states with primary enforcement responsibilities" (5). This is called primacy, which means the individual states must assume responsibilities for managing the drinking water program.

From these requirements, the Laboratory Certification Program (LCP) has been developed (5) which requires that all laboratories performing measurements for compliance with regulations be certified. Procedurally, the USEPA through its Environmental Monitoring and Support Laboratory (EMSL) located in Cincinnati, OH, is responsible for determining the certification status of USEPA Regional laboratories for chemistry and microbiology. The EMSL laboratory located in Las Vegas, NV, has the responsibility for radionuclides. Certification officers within the Regions subsequently determine the certification status of principal state laboratories in primacy states and local laboratories in states that have not assumed responsibility for the drinking water program. Each primacy state must have a certification program for local laboratories if all analyses are not performed in state laboratories. These programs must be at least as stringent as that of the USEPA.

The SDWA was the first act specifying the need for QA and QC in a USEPA program. The need was essential and the Laboratory Certification Program managed by the Office of Drinking Water provides the management function for carrying out this activity. In addition, the LCP takes into account the directives of the Quality Assurance Management Staff, which was established to implement the USEPA mandatory QA program (6).

7.2 LEVELS OF REQUIRED QUALITY ASSURANCE AND QUALITY CONTROL

QA programs are required for all environmentally related measurements, that is, all field and laboratory investigations that generate data. Since this definition spans a range of activities from exploratory investigations to compliance monitoring, no single QA protocol will be applicable to all. The first requirement is to identify the intended use to which the generated data will be applied and then to design a suitable QA program specific for the particular investigation.

One broad category includes those data generated as a result of various types of environmental data gathering activities. This category can be further subdivided on the basis of scope.

7.2.1 National Surveys

Large-scale, data base generating activities, such as national surveys, require QA procedures that carefully monitor all elements of the study. Such surveys are used to provide data on the frequency and magnitude of occurrence of specific types of chemical pollutants and may require collection and analysis of several thousand samples. A comprehensive QA plan is necessary as an integral part of such surveys. A sampling plan must be formulated to ensure that samples are representative of nationwide occurrence according to the design of the survey. Sampling procedures must be suitable for performance by state personnel, personnel at individual utilities, or other staff required to collect the samples. Preservation, shipping, and storage conditions must be specified and monitored. Sample custody and tracking procedures must ensure that no sample gets lost and that all required analyses are performed within allowable holding times for each subsample. Analytical methods must be specified. The analytical laboratory must demonstrate competency in analyte measurement before samples are received. Criteria for accuracy, precision, and detection limits must be established. Internal and external checks of accuracy and precision must be performed routinely. High sample volume and limited sample storage time mandate measures to assure preventive maintenance of instrumentation to minimize downtime. The availability of back-up equipment is desirable. Calibration may encompass a large number of parameters over a wide range of concentrations. There must be provisions for checking data reduction and reporting procedures for errors. All of these elements require careful planning and complete documentation. Systematic procedures for corrective action should be planned. Regular formal QA reports must be submitted to project management.

7.2.2   Limited-Scope Activities

Data gathering activities of more limited scope may require more emphasis on some QA elements than on others. A utility may need to monitor routinely a specific water quality parameter. For example, wells near a source of potential ground water pollution (e.g., an underground storage tank) may be checked routinely for the suspected pollutant to ensure that the water remains safe from contamination. In other utilities, the effluent from activated carbon beds is monitored routinely for total organic carbon to determine when replacement or regeneration is required. In such cases, QA elements such as ensuring correct sample preservation, storage, and custody procedures may be very simple if the sampler and analyst are the same person and the sample is analyzed immediately after collection. Calibration procedures may be limited to a single parameter over a narrow concentration range. However, there would be little point in producing data unless all of the QA elements are addressed in some form, no matter how simplified.

7.2.3   Exploratory Activities

Other types of environmental measurements are those generated as a result of exploratory investigations. These types of activities have their own QA emphasis. If the research requires analysis of environmental samples, each of the cited QA elements must be addressed, taking into account the intended use of the data. For example, many studies seek to determine the effect of changes in a treatment process on some measured water quality parameter. In these cases, the requirements for analytical accuracy and precision and for day-to-day reproducibility may well be more stringent than those acceptable for other types of data gathering activities. Further, in these types of investigations, additional control procedures may be required to ensure that all elements of the treatment process (other than the experimental variable) remain constant or, if this is not possible, that the experimental design ensures that effects of the uncontrollable variables can be eliminated or accounted for.

To be accepted, the results of any research activity must be reproducible by other investigators following the same procedures. A thoroughly documented and carefully followed QA program can greatly facilitate verification of research findings.

7.2.4   Compliance Monitoring

The activity that requires the most stringent QA and QC program is compliance monitoring. The objective is to gather information that violation of an enforceable standard has taken place. In this case, the sampling and analytical methodology used, whether simple or complex, must be fully verified by extensive testing and must be approved by the regulating agency. An example is the maximum contaminant level for total trihalomethanes (TTHMs) in drinking water, where measurement for each THM and for TTHMs must be accomplished with an accuracy of $\pm 20\%$. A rigorous and mandatory QA and QC program is required to insure the credibility of the organization generating the data and the validity of the data itself. The Laboratory Certification Program for drinking water (5) is intended to deal with this situation.

7.3   ASSESSING QUALITY ASSURANCE AND QUALITY CONTROL

The subsequent comments on QA and QC procedures are based on USEPA guidance and experience and are intended to serve as examples. Obviously, there are many acceptable alternatives to document and implement an acceptable QA program.

7.3.1.   Data Quality Objectives

As previously noted, an assessment of the objectives of an activity is required to develop an appropriate QA and QC program. At the time decisions are reached regarding the scope of a particular project, data quality objectives (DQOs) should be determined (7), which serve as the initial criteria for the preparation of QA project plans. Data quality objectives are "qualitative and quantitative statements of the data quality that is required to support specific decisions" (7). The QA and QC program applied depends on the specific intended use of the data. Both management and technical staffs should interact to develop DQOs that are based on a realistic and a balanced assessment of information that is required.

For example, consider two projects with very different objectives. The first relates to a compliance monitoring activity, where data are required to determine whether contaminants are present above an established regulatory standard. Here, the QA planning must be designed to assure data quality within narrow bounds defined by a regulation to evaluate whether a potential legally enforceable violation has occurred. The second example is a study, limited in scope, conducted to determine qualitatively whether a class of chemicals occurs in water. Loose bounds are required on the absolute values of the concentrations that are measured (e.g., $\pm 100\%$). A totally different set of QA requirements are needed in this case.

7.3.2   Quality Assurance Project Plans

After DQOs have been determined, a QA project plan should be developed. According to EPA guidance (2), 16 elements or sections are required for the QA plan as detailed below. As appropriate, comments are offered for some of these elements.

Section 1. Title page
Section 2. Table of contents
Section 3. Project description. This section includes a general description of the project, including the experimental design, and a summary statement of the DQOs.
Section 4. Project organization
Section 5. QA objectives for measurement data in terms of precision, accuracy, completeness, representativeness, and comparability. This is a key section where data quality objectives for measurements are defined in numerical terms and criteria for method selection are given. Substantial detail is required to document how measurement data will be obtained and used.
Section 6. Sampling procedures. This section includes specifications used to select sampling sites and to conduct the sampling program.
Section 7. Sample custody
Section 8. Calibration procedures and frequency
Section 9. Analytical methods
Section 10. Data reduction, validation, and reporting. This section includes criteria used to validate data, details of the reduction scheme to be used for collected data, and the mechanism for data reporting.
Section 11. Internal quality control checks. Examples of items to be in-

cluded are replicate samples, spiked samples, split samples, control charts, blanks, internal standards, and reference samples.

Section 12. Performance and system audits. This is a description of the internal and external checks and audits that will be required to monitor the performance of the total measurement operation.

Section 13. Preventive maintenance

Section 14. Specific routine procedures used to assess data precision, accuracy, and completeness. This section includes the specifics of the data to be assessed and the details of the assessment.

Section 15. Corrective action. This section includes defining when corrective action is to be taken and the specifics of its implementation.

Section 16. Quality assurance reports to management

The preparation of a document of this magnitude requires substantial effort. However, its merit lies in documenting the quality of the data that are generated. It also assists in establishing credibility. Arguably, the effort in preparing a QA plan can be unreasonable for small projects where a limited amount of data are to be gathered. For this situation, it is suggested that an abbreviated version of a QA plan be prepared that delineates the objective of the activity and the measurement systems that will be used.

7.3.3 Standard Operating Procedures

Inherent in a QA program is QC, which should be an ongoing preventive function rather than a response to an emergency situation. It should be in place from the outset to the completion of a project (1). QC not only pertains to inanimate objects such as equipment and supplies, but equally, or perhaps more importantly, to the people who must be in charge of all functional activities. If control is maintained continuously, then one will always be in a position to be aware immediately of problems that have arisen, and, hopefully, be able to correct them quickly. It is important to realize that if a problem cannot be rectified quickly, this could mean halting data gathering activities until a solution has been found and implemented. A mistake that many organizations make is to continue measurement activities when a problem has been identified but not resolved. This leads to generating potentially suspect information.

One way to maintain QC is to prepare and regularly use written standard operating procedures (SOPs) which specify in detail each key task that is undertaken to ensure that the results are obtained and interpreted correctly. For example, analytical methods often do not deal with all the required tasks, therefore the project designer, sample collector, analyst, and others must provide their own documentation (SOP) of the individual steps in the data gathering process.

Another function of an SOP is to provide continuity, so that a clear record of procedures used can be understood by others who subsequently follow them. SOPs also provide a reference for information about how the measurement activity was conducted. The project planner, project manager, and quality assurance staff are then in a position to interpret the data fairly and accurately in the context of the overall project.

SOPs can be developed in many ways. In the following discussion, a brief description will be presented of the way in which SOPs are developed by the Office of Drinking Water, Technical Support Division (TSD), in Cincinnati, OH. The approach is to link information contained in SOPs with the requirements of QA project plans. For each analytical operation, an SOP is prepared that addresses pertinent elements of the QA plan (Sections 5, 6, 8, 9, 10, 11, and 13). These sections contain much detail, and when the QA plan is prepared, appropriate SOPs are referenced (without inclusion), thereby minimizing the amount of material needed for the plan.

An example of guidance provided by TSD for the preparation of one section of an SOP is presented below.

Calibration Procedures and Frequency
   1. Calibration solutions. This information may be in the "Reagent" section of a method. If so, include by reference to a method section (MS). The preparation procedure and storage information can probably be referenced to a MS, but information about concentrations will probably need to be supplied. For most methods, the following applies, if not, adjust as required.
      a. Stock standard solutions
           Preparation procedure-solvent
           Number of components
           Final concentration (can be a range)
           Storage conditions
           Stability
      b. Secondary dilution standard solution(s)
           Preparation procedure-solvent
           Number of components
           Range and basis for final concentration, e.g., 1–50 µg/mL to make calibration standards that produce a response close ($\pm 20\%$) to that of the unknowns
           Storage conditions
           Stability
      c. Calibration solutions
      d. Internal standard spiking solution(s)
      e. Surrogate compound spiking solution(s)
      f. Other solutions
           As required, for procedures specific to a method or for performance checks after system modifications, etc. (see 4 and 5 below)
           For c, d, e, and f, same kind of information as listed for secondary dilution standard solution(s)

2. Initial Calibration(s)
   Frequency
   Procedure(s) and solutions used
   Outcome(s), for example, calibration curve; quantitation limits; and/or criteria for acceptability of results; action if criteria not met
   As applicable; records kept and their location
3. Daily calibration(s)
   Same kind of information as listed for initial calibration(s)
4. Other procedures
   Specific to a method, for example, a performance test for a MS detector
   Same kind of information as listed for initial calibration(s)
5. Equipment performance tests
   Performance checks after modifications of the system or after extended instrument inactivity; system or method efficiency tests, and so forth
   Same kind of information as listed for initial calibration(s)

7.4 ANALYTICAL METHODS

Important to the determination of an analyte or a series of analytes is a carefully described sampling and analytical method, which sets forth the detailed specifics of conducting the measurements. Contents generally include the following sections, depending on the particular method and the organization that develops and publishes it. Although one section of a method is devoted to overall QC, most of the individual elements require the application of specific QC procedures. These are usually described in standard operating procedures.

- Scope and application
- Discussion of interferences
- Equipment and apparatus
- Reagents and materials
- Sample collection, preservation, and storage
- Calibration and standardization
- Conducting the procedure, generally including an illustrative example
- Quality Control (internal and external)
- Calculations and reporting results

7.4.1 Levels of Method Validation

The applicability of a method to a particular measurement activity depends to a large extent on the intended use of the data. Just as there are levels of QA and QC required for environmentally related measurements, there are also levels of validation required for methods. At one extreme is the case where very limited documentation of validity and broad applicability are available. Such a method may be applied to a preliminary investigation, where the objective is to determine the presence of an analyte and its approximate concentration. Further detailed investigation requires additional method development. The other extreme is a fully documented and verified method, including round-robin testing, needed for evaluating compliance with a regulatory enforceable standard.

Within the USEPA, methods are developed in phases and are applied to the measurement of contaminants in water (8). Two phases involve establishing the level of method performance, while two serve to enhance and evaluate laboratory performance using the method. These phases are described below in the sequential order in which they take place.

Single operator accuracy and precision (validates method performance by one analyst). This is a case where the method has been developed in one or more laboratories and has generally been used in several laboratories. It has been fully evaluated in a single laboratory by a single analyst. A complete method write-up is generated and includes all method elements. Also included is an assessment of accuracy, precision and method detection limit by the analyst.

Availability of QC samples (enhances laboratory performance). For the method just described, concentrates of the analytes are available which contain known concentrations. These are prepared by a laboratory other than the laboratory performing the analysis and are intended to serve as an independent check on the ability to identify qualitatively and measure quantitatively the analytes of concern. For organic, inorganic, and microbiological contaminants, the Environmental Monitoring and Support Laboratory located in Cincinnati, provides these QC samples. The EMSL laboratory in Las Vegas, provides QC samples for radionuclides.

Availability of performance evaluation (PE) samples (evaluates laboratory performance). Concentrates are available as described for QC samples. However, these are unknown and are distributed to a number of laboratories to evaluate their performance for specified analytes. In the Laboratory Certification Program, in order for a laboratory to be certified to analyze for regulated contaminants in drinking water, it must successfully analyze PE samples once each year within prescribed limits of accuracy.

Method validation study (validates method performance by multiple analysts and laboratories). This is equivalent to round-robin testing, and involves the analysis of unknown concentrations of a single or a group of analytes using the specified method. The overall performance of the method is thereby statistically established.

7.4.2 Quality Control Provisions

Within a method, QC provisions are explained in detail. Their nature and the amount of detail depend on the method itself as well as on the requirements of the program. The following is a listing of frequently used QC provisions in addition to the use of QC and PE samples.

- External standard
- Internal standard
- Surrogate compound spike
- Field blank
- Method blank
- Reagent blank and/or solvent blank
- Laboratory control standard
- Spiked sample
- Field duplicate
- Laboratory duplicate

7.5   SPECIFIC CASE HISTORIES

Sampling and analytical QA and QC programs used for actual cases of environmental measurements will help illustrate some of the principles discussed so far in this chapter.

7.5.1   National Surveys for Volatile Organic Compounds

Two national surveys of purgeable volatile organic compounds (VOCs) in drinking water have been conducted by the USEPA within the past 8 years. These surveys were conducted by the Office of Drinking Water, Technical Support Division (TSD), in Cincinnati. The purpose of the first of these, the Community Water Supply Survey (CWSS), was to gather occurrence data for VOCs in water systems that are supplied by ground and surface water sources and that serve from 25 to 100,000 persons (9). These data were to be used in the formulation of drinking water regulations. For this survey, 1189 field samples of raw and finished water were collected at 452 individual water supplies. The second of these national surveys was the Ground Water Supply Survey (GWSS) (10). In addition to building a data base representative of VOC occurrence in the nation's public ground water systems, this survey also sought to provide the states with information on systems suspected of being contaminated with purgeable VOCs. In Phase 1 of the GWSS, approximately 1000 ground water systems were sampled. About half were selected randomly to provide national occurrence data; the other half were selected by the states in an attempt to encourage them to identify supplies suspected of being contaminated. In the second phase of this survey, selected sites, in which contaminants were identified in the initial analyses, were resampled at the original sampling point and, often at a number of additional locations including the distribution systems and well heads. Increasing interest in ground water quality expanded this phase to include a number of smaller subsurveys. Approximately 1400 field samples were collected and analyzed in Phase 2.

Each of these surveys required a substantial commitment of time and resources. To ensure that they produced meaningful data, detailed QA and QC protocols were established as integral parts of the sampling and analytical schemes. A formal written QA project plan was prepared for the GWSS. The QA programs were similar for both surveys. Since the CWSS QA program has been reported elsewhere in detail (11), examples from the QA program for the GWSS will be discussed here (12).

Samples for the GWSS were analyzed for a target list of 37 purgeable organic compounds and for total organic carbon. The target compounds included those halogenated aliphatic and aromatic compounds such as vinyl chloride, trichloroethylene, tetrachloroethylene, benzene, xylene, etc., that had been reported in ground water. Attempts were made to identify additional compounds of this class that were observed occasionally. Specific quality control procedures were in place to monitor each step of the program, from sampling to final data reporting and confirmation. In fact, QA issues were addressed before the first sample was collected; laboratories bidding on the analytical contract were required to demonstrate competence in the measurement methods by submitting the results of duplicate analyses of performance evaluation samples containing 15 of the target compounds.

Complete sampling kits were prepared by TSD for each sampling site. These kits included sample bottles already containing a preservative and labeled with preprinted, fill-in labels stamped with a unique sample identification number for each sampling point. Replicate bottles were provided for each point. The bottles were packed in a custom-molded polystyrene foam container designed to fit the number of bottles to be collected. A field blank (organic-free water plus preservative) was placed in the kit before shipment to the sampling site and accompanied the samples at all times thereafter. This field blank was analyzed to monitor for contamination introduced during shipping and storage. Detailed sampling and shipping instructions were included in each kit.

Samples were refrigerated as soon as possible after collection, then packed in ice and shipped by overnight air express to TSD. There, they were unpacked, logged in, inspected, and placed in a cold room (4°C) that was monitored for organic vapor contamination. Approximately half of the replicate samples collected at each location were then repacked in ice and shipped (overnight air express) to the contract analytical laboratory (SRI International, Menlo Park, CA). The remaining samples were retained in cold storage at TSD. These samples served as backups in case a shipment went astray, and also provided samples for independent verification of results obtained by the contract laboratory.

On arrival at SRI the samples were again inspected and recorded in a logbook, then placed in a locked walk-in refrigerator (4°C) used exclusively for storage of water samples. Bottles of organic-free water were also stored there to monitor for possible contamination introduced during storage. The sample logbook would later be used to record the dates and types of analyses performed for each sample, and to indicate the laboratory notebooks in which the analytical results were recorded. (The results of this survey filled 57 standard laboratory notebooks, and six file drawers of chromatograms.)

The VOC analyses were performed using USEPA approved methods: the purge and trap technique of analyte concentration, followed by gas chromatographic separation, and then detection and quantification using photoionization and electrolytic conductivity detectors. For this survey, the procedures for purgeable aromatics (EPA Method 503.1) (13) and for halocarbons (EPA Method 502.1) (14) were combined in a single analysis by using serially coupled detectors (15). Before samples were analyzed, the equivalency of this modification to the individual approved methods was demonstrated by analyses of quality control and environmental samples using both techniques.

The analytical QA program began before the first sample was analyzed. Each analytical instrument used was initially calibrated over a range of concentrations for each target compound using standard solutions prepared in-house from pure chemicals. This calibration was verified by analyses of quality control samples (discussed below) before beginning sample analysis. Calibration standards were analyzed daily to verify the calibration, monitor changes in sensitivity, and so forth.

The analytical QA protocol for this survey consisted of the following elements:

- Analysis of EPA-supplied quality control samples
- Analysis of duplicate samples
- Confirmation of qualitative and quantitative results by use of alternative chromatographic columns
- Analysis of selected samples by gas chromatography/mass spectrometry (GC/MS)
- Analysis of blind samples
- Analysis of split samples

Results of all QA procedures were submitted monthly to the EPA project officer in written reports. Examples from each element will be discussed separately.

QC samples, provided by the contract project officer, were concentrates containing known quantities of selected target compounds. These samples were prepared by diluting the concentrates with blank water according to instructions. Analysis of quality control samples provided a measure of both accuracy and precision for many of the commonly observed purgeable organics.

The protocol initially specified biweekly duplicate analyses of two concentration levels for purgeable halocarbons, and one concentration level for purgeable aromatics using each analytical instrument used for sample analysis during that period. Precision was to be calculated as the difference between duplicate measurements, divided by their average (converted to percentage). The minimum acceptable level of precision was specified as 20% for concentrations >5 μg/L and 40% for lower levels. To allow more frequent monitoring, the QA protocol was changed to allow weekly, single analyses of each reference sample type. Since the data were reported monthly, a more appropriate measure of precision for the four weekly measurements was the coefficient of variation (CV), the standard deviation divided by the mean, expressed as a percentage.

Accuracy for QC sample analyses was reported as the difference between the mean concentration found for the four weekly analyses for each parameter and the true concentration, divided by the true concentration (also converted to percentage). The acceptable error was limited to 20% for high (>5 μg/L) concentrations and 40% for lower levels.

A summary of the precision and accuracy data obtained for halocarbon QC sample analyses using the primary analytical system during the first phase of the GWSS is shown in Table 7.1. These data show that the mean precision (CV) was 13% for all compounds at concentrations <5 μg/L and 11% for those at higher concentrations. The accuracy of these analyses was also well within QA specifications.

Duplicate analyses were specified for every tenth sample. The precision between duplicate samples was required to be 20% for concentrations >5 μg/L and 40% for lower concentrations. Results of duplicate analyses are shown in Table 7.2 for all duplicate samples in which a target compound (other than trihalomethanes) was found during the first phase of the GWSS. This table shows the range of precision values obtained for all duplicate pairs for a given compound and the mean precision for that compound. For concentrations <5 μg/L, 79 of the 84 pairs of observations (94%) fell within the 40% precision limit. For higher concentrations 14 of 18 pairs (78%) fell within the tighter 20% precision requirement.

Occurrences of compounds were confirmed both qualitatively and quantitatively by a second analysis using an alternative chromatographic column according to methods 502.1 and 503.1. Columns were selected to provide a different elution order for the target compounds. The original QA protocol specified second column confirmation for 10% of the samples. In Phase I of the GWSS, however, it was possible to perform second column confirmatory analyses for *all* samples in which a target compound (other than trihalomethanes) was observed—about 35% of the samples. Since these analyses also provided quantitative data, QC samples were analyzed regularly using the confirmatory system and the results transmitted in the monthly QA reports.

Selected samples were, in addition, analyzed by GC/MS. A USEPA GC/MS method was not available at the time the survey was initiated. However, prior methods development at TSD provided a procedure for VOC analyses by purge and trap GC/MS (16). Subsequently USEPA Method 524 was is-

TABLE 7.1.
Halocarbon Quality Control Sample Analyses—Primary Column (Ground Water Supply Survey)

Compound	Expected Conc. (μg/L)	Low Level[a] Concentration Found (ppb)				Expected Conc. (μg/L)	High Level[b] Concentration Found (ppb)			
		Range	Mean	CV[c]	% Error[d]		Range	Mean	CV[c]	% Error[d]
Chloroform	8.2	6.1–9.2	7.2	12	−12	34	25–36	31	9	−8.8
1,2-Dichloroethane	3.3	2.4–3.6	2.9	10	−12	14	9.9–15	13	11	−7.1
1,1,1-Trichloroethane	1.3	0.85–2.0	1.1	22	−15	5.6	3.7–8.0	5.0	20	−11
Carbon tetrachloride	1.5	1.2–1.8	1.4	10	−6.7	6.2	5.1–7.3	6.4	8	3.2
Bromodichloromethane	1.4	0.96–1.6	1.4	11	0	6.0	5.1–7.5	6.4	9	6.7
Trichloroethylene	2.3	1.7–2.6	2.0	9	−13	9.1	7.4–10	8.7	6	−4.4
Dibromochloromethane	2.1	1.0–2.5	1.7	17	−19	8.5	5.3–8.9	7.1	11	−16
Bromoform	1.7	0.59–2.0	1.6	18	−5.9	7.0	5.1–8.2	6.9	10	−1.4
Tetrachloroethylene	1.1	0.82–1.4	1.0	11	−9.1	4.4	3.6–5.4	4.3	8	−2.3

[a] 48 analyses.
[b] 47 analyses.
[c] Coefficient of variation: 100 times the standard deviation divided by the mean value.
[d] Error expressed as 100 times the difference between the expected and mean measured concentrations, divided by the expected concentration.

187

TABLE 7.2
Precision of Duplicate Analyses[a] (Ground Water Supply Survey)

Compound	Concentration <5 ppb			Concentration > 5 ppb		
	Number Duplicate Pairs[a]	Range of Precision Values[b]	Mean Precision[b]	Number Duplicate Pairs[a]	Range of Precision Values[b]	Mean Precision[b]
Vinyl chloride	1	36	—	1	34	—
1,1-Dichloroethylene	6	4.6–51	29	0	—	—
1,1-Dichloroethane	11	0–53	17	0	—	—
cis- or trans-1,2-Dichloroethylene	14	0–43	13	6	0–22	11
1,2-Dichloroethane	1	0	—	0	—	—
1,1,1-Trichloroethane	12	0–41	14	2	5.4–5.7	5.5
Carbon tetrachloride	8	2.0–38	17	0	—	—
1,2-Dichloropropane	2	0–1.2	0.6	0	—	—
Trichloroethylene	8	0–37	22	8	2.5–24	13
Tetrachloroethylene	8	0–27	13	1	17	—
Chlorobenzene	2	11–23	17	0	—	—
Bromobenzene	1	67	—	0	—	—
Toluene	2	8–20	14	0	—	—
m-Xylene	3	3.3–35	17	0	—	—
o-, p-Xylenes	3	13–20	17	0	—	—
o-Dichlorobenzene	2	3.8–20	12	0	—	—

[a] Number of times compound found at or above the quantification limit in both analyses, separated into high and low ranges.
[b] For each pair, precision calculated at 100 times the absolute value of their difference, divided by their average. The range of the precision values and mean precision value are shown for each parameter.

sued (17). Analytical procedures followed were similar to those detailed in method 524.

Selection of samples for analysis by GC/MS emphasized those samples in which nontarget compounds were observed. In these analyses, 16 additional compounds were identified tentatively. (Identification was limited to spectral matching in most cases. Authentic standards of the tentatively identified compound were analyzed for only four of these compounds.) Since the samples generally also contained compounds on the target list, GC/MS analyses were structured to yield quantitative data for these compounds. Thus, QC samples were analyzed in duplicate each time the GC/MS was used for sample analyses. In addition to the QA requirements listed for the overall survey, GC/MS analytical procedures have their own QA protocol. Many of the QA elements outlined in method 524 were used.

All of the QA elements discussed so far were accomplished exclusively within the analytical laboratory. The final two provisions of the QA protocol, split and blind sample analyses, provided interlaboratory checks of the data quality. Split samples were water samples collected at the same time in separate bottles that were analyzed both by the contract laboratory and by TSD. Blind samples were formulated water standards (i.e., blank water with or without addition of one or more target compounds). The blind samples were shipped from TSD to the analytical laboratory and intermixed with real water samples to prevent their identification. The blind samples were also analyzed by TSD.

In Phase 1 of the GWSS, 30 of the water samples analyzed by both TSD and SRI contained target compounds other than trihalomethanes. The results obtained by both laboratories are shown graphically in Figure 7.1. The concentrations found by TSD are plotted against those found by SRI. A log–log representation is used for ease of visual presentation of the data at low concentrations. Sixty-eight pairs of data are included for 12 halogenated and aromatic VOCs at concentrations ranging from 0.2 to 120 ppb. The slope of 0.99 and correlation coefficient of 0.99 of the linear correlation demonstrate good agreement between the laboratories. The spread in the data points is expected at low concentrations as interlaboratory precision becomes poorer.

Five blind samples were submitted to the analytical laboratory within the first month of the contract period. These samples in every way resembled authentic samples. TSD and SRI results obtained for these blind samples are shown in Table 7.3. These data provided early assurance that the sampling and analytical procedures were under control.

The results obtained as part of the QA protocol were reviewed by the SRI laboratory project manager and transmitted in monthly reports to the contract project officer at TSD. The actual sample data, however, were directly entered by the analytical laboratory into a computer file accessible to both the laboratory and TSD. The data were then reviewed by the TSD contract project officer and any questions resolved by consultation between

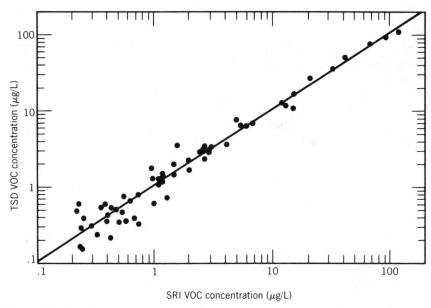

Figure 7.1. Log-log plot of concentrations determined by TSD and SRI for the analyses of split samples in the GWSS.

TSD and the SRI laboratory project manager. Results were considered final only after all confirmatory analyses were complete and all problems were resolved.

Even with all of the internal and external cross-checking of results, the massive amounts of data generated provided ample opportunity for reporting errors. To determine if this was a problem, at the end of each phase of the GWSS the files for every tenth sample were pulled and the data followed from sample log through chromatographic trace, to final computer printout for each analysis performed for that sample. Only minor transcription errors were found.

It is useful to estimate the level of effort required to maintain this type of QA program. Opinions will differ as to exactly what sorts of tasks should be included as part of a QA and QC effort. For simplification, consider the following breakdown: A sample is analyzed once, yielding a set of data. All other analyses performed are solely for the purpose of verifying that the data are correct and defining the limits of precision and accuracy. Taking this definition, the actual breakdown below of analyses performed during 23 months of the GWSS contract is illustrative. Note that these totals include VOC analyses that were performed as part of several small projects separate from the GWSS but using identical QA protocols.

TABLE 7.3
Comparison of Concentrations (μg/L) Determined from Blind Sample Analyses[a] (Ground Water Supply Survey)

Compound	Sample 1	Sample 2	Sample 3	Sample 4	Sample 5
cis- or trans-1,2-Dichloroethylene					
Chloroform	(<0.5)[b]/0.27	61/49		7.5/9.0	1.4/1.5
1,1,1-Trichloroethane		7.7/10		1.7/1.6	
Carbon tetrachloride		1.7/1.2			9.6/12
Bromodichloromethane		3.5/2.2			0.9/1.2
Trichloroethylene				1.7/1.7	
Dibromochloromethane		1.8/1.6			0.77/0.92
Dichloroiodomethane	1.6/1.6				
Bromoform		2.1/1.5			1.1/1.3
1,1,2-Tetrachloroethane			2.2/2.3		
Tetrachloroethylene	(<0.5)[b]/0.50	3.9/3.6		1.3/1.4	
Chlorobenzene			5.6/5.0		
Benzene	1.4/1.3	0.97/1.2	1.1/1.1		
Toluene	13/13	6.4/5.2	5.5/5.2		
Ethylbenzene	1.6/1.2	1.5/1.6	0.94/1.0		
m-Xylene	11/11	5.1/4.6	17/21		
p-Dichlorobenzene		4.6/4.7			

[a] The first number given was determined from analysis at TSD and the second reported by SRI.
[b] Not reported below quantification limit.

Analysis Type		Number of Analyses	Percent of Total
Primary analyses		3556	70
Quality control analyses		1422	30
Quality control samples	319		
Duplicates	477		
Second column	560		
GC/MS	66		
Total		4978	100

Other QA and QC practices not shown include such things as daily analysis of calibration check standards, sample tracking, monitoring of storage facilities, spot checking of data, preparation of QA reports, and analyses as well as blind sample preparation performed by TSD.

7.5.2 National Survey for Inorganic and Radionuclide Contaminants

The USEPA is currently conducting a National Inorganics and Radionuclides Survey (NIRS). Although not involving the measurement of organics, the quality assurance program for this survey is extensive and is presented here to illustrate the application of QA and QC policies discussed in this chapter. The NIRS involves the use of primary and secondary laboratories all located within USEPA. Primary laboratories analyze samples from each sampling location included in the study (~1000). In addition, they analyze samples for QA and QC purposes. Secondary laboratories only analyze samples related to QA and QC. A summary of the QC sample analyses for the survey is presented in Table 7.4. In addition to the analyses of these "external samples," each laboratory has implemented a vigorous QA and QC program with regular reports provided to managers of the survey. This QA and QC program includes:

- Preparing SOPs and providing them for use in the survey QA plan
- Determining precision, accuracy, and minimum detection limits for each analyte prior to the initiation of the study, and at regular intervals as the study is being conducted
- Using spiked samples for internal quality control
- Analyzing quality control samples on a predetermined schedule
- Participating in USEPA's performance evaluation studies, where samples of unknown concentration are analyzed

7.5.3 Smaller-Scale Projects

A large amount of environmental data gathered by the USEPA is obtained from the analyses of small batches of samples submitted to various contract

TABLE 7.4
Summary of Quality Control Sample Analyses[a]
(National Inorganic and Radionuclide Survey)

Analyses	Laboratory	Duplicates	Blinds	Field Blanks
Inductively coupled plasma	Primary	55	55	15
(32 elements)	Secondary	110	55	15
Cold vapor atomic absorption	Primary	55	55	15
spectroscopy (mercury)	Secondary	55	55	15
Furnace atomic absorption	Primary	55	55	15
(cadmium, lead, selenium, arsenic)	Secondary	55	55	15
Radionuclides (gross alpha,	Primary	55	55	15
gross beta, radon-222, radium-226, radium-228, uranium	Secondary	110	55	15

[a] Based on the analyses of 1000 sampling locations.

laboratories. A substantial portion of data gathered by USEPA Regional, state, and local bodies derive from such small-scale efforts. The QA and QC objectives for these types of studies are the same as for the large surveys, but the methods used to achieve them must be modified to accommodate small sample numbers. For USEPA contractors used by TSD, only approved or widely accepted analytical methods are used where possible. As described for national surveys, quality control and blind samples are routinely utilized to evaluate the capabilities of potential contractors, and to monitor the performance of current contractors. A single laboratory may contract to perform analyses for VOCs, pesticides, metals, radionuclides, and other analytes, and must, therefore, at its own expense demonstrate competency in each method. In addition, duplicates, splits, blanks, confirmations, spikes, as well as quality control and blind samples, are all used to validate and document the analytical results after contract award.

Samples submitted to a contractor for analyses are routinely accompanied by instructions specifying methods to be used and what QC analyses must be performed. For example, as few as four water samples of a given type may require analysis of one of the samples in duplicate, a shipping blank, a quality control sample, and a spiked sample. Clearly, small batches of samples require heavy QA "overhead." For larger sample sets (e.g., greater than 10 samples), the QA burden is reduced, since relatively fewer blanks are required, duplicates can be limited to 10% of the samples and so forth. For laboratories continuously engaged in analyses of small sets of samples of a given type, routine QC sample analyses (e.g., twice monthly) further reduce the QA burden for any given sample set.

Samples submitted for nonroutine types of analyses present unique requirements. For example, when ethylene dibromide (EDB) was discovered in the ground water in Hawaii and California, there was no "approved" analytical method that would allow detection of this compound at the low levels required (20 ng/L). QC samples were not available for use by contract laboratories. For EDB measurement, an analytical method that had been developed by state and Regional laboratories (though not an "official" EPA method) was specified. Blind samples of EDB in water were specially prepared by TSD and sent to the contractor. In addition, a capillary purge and trap GC/MS procedure was developed for in-house qualitative and quantitative confirmation of contract laboratory results (18). These cases required much more involvement of the contracting agency than would be necessary for more routine types of analyses.

ACKNOWLEDGMENTS

The authors wish to thank Audrey D. Kroner for her assistance in developing the QA program at TSD and for her review of this paper; Mr. Robert F. Thomas (TSD) for helping to implement the QA provisions of the Ground Water Supply Survey analytical contract; Marianne Feige for implementing QA provisions in other TSD analytical contracts; and Cynthia Bultman and Dorothy Davis (TSD) for their assistance in the preparation of this manuscript. Lowell A. Van Den Berg, James J. Westrick (TSD), and Dale M. Coulsen (SRI) are thanked for their overall management commitment to quality assurance.

REFERENCES

1. King, D.E., Credibility: The Consequence of Quality Assurance, Laboratory Services Branch, Ontario Ministry of the Environment, May, 1982.
2. U.S. Environmental Protection Agency, Interim Guidelines and Specifications for Preparing Quality Assurance Project Plans, Quality Assurance Management Staff, Office of Research and Development, Publication 005/80, Washington, DC, December, 1980.
3. U.S. Congress, The Safe Drinking Water Act; As Amended Through November 1977, U.S. Government Printing Office, Serial No. 95-10, Washington, DC, November, 1977.
4. U.S. Environmental Protection Agency, National Interim Primary Drinking Water Regulations, *Fed. Reg., 40*, 59566, December 24, 1975.
5. U.S. Environmental Protection Agency, Manual for the Certification of Laboratories Analyzing Drinking Water, Office of Drinking Water, "Publication EPA-570/9-82-002, Washington, DC, October, 1982.
6. U.S. Environmental Protection Agency, Policy and Program Requirements to Implement the Mandatory Quality Assurance Program, Office of Research and Development, EPA Order 5360.1, Washington, DC, April, 1984.
7. U.S. Environmental Protection Agency, Guidelines and Specifications for Preparing Quality Assurance Program Plans for National and Program Offices, ORD Headquarters and ORD Laboratories, Quality Assurance Management Staff, Office of Research and Development, Appendix B, May, 1985.

8. U.S. Environmental Protection Agency, Environmental Monitoring and Support Laboratory, Cincinnati, OH.
9. Brass, H.J., Weisner, M.J., and Kingsley, B.A., Community Water Supply Survey: Sampling and Analysis for Purgeable Organics and Total Organic Carbon, American Water Works Association Annual Conference, St. Louis, MO, June, 1981.
10. Westrick, J.J., Mello, J.W., and Thomas, R.F., *J. Am. Water Works Assoc., 76,* 52, 1984.
11. Kingsley, B.A., Gin, C., Peifer, W.R., Stivers, D.F., Allen, S.H., Brass, H.J., Glick, E.M., and Weisner, M.J., in Keith, L.H. (ed.), *Advances in the Identification and Analysis of Organic Pollutants in Water,* Vol. 2, Ann Arbor Science, Ann Arbor, MI, 1981.
12. Kingsley, B.A., Quality Assurance in a Contract Laboratory, Proc. Amer. Water Works Assoc, Water Quality Technology Conference, 69, 1982.
13. U.S. Environmental Protection Agency, The Analysis of Aromatic Chemicals in Water by the Purge and Trap Method, Publication EPA-600/4-81-057, Cincinnati, OH, April, 1981.
14. U.S. Environmental Protection Agency, The Determination of Halogenated Chemicals in Water by the Purge and Trap Method, Publication EPA 600/4-81-059, Cincinnati, OH, April, 1981.
15. Kingsley, B.A., Gin, C., Coulson, D.M., and Thomas, R.F., in Jolley, R.L., et. al. (eds.), *Water Chlorination: Environmental Impact and Health Effects,* Vol. 4, Book 2, Ann Arbor Science, Ann Arbor, MI, 1983.
16. Munch, D.J., Mello, J.W., Feige, M.A., Glick, E.M., and Brass, H.J., in Keith, L.H., (ed.), *Advances in the Identification and Analysis of Organic Pollutants in Water,* Vol. 2, Ann Arbor Science, Ann Arbor, MI, 1981.
17. U.S. Environmental Protection Agency, Purgeable Organic Compounds in Water by Gas Chromatography/Mass Spectrometry, Cincinnati, OH, January, 1983.
18. Madding, C., and Brass, H.J., Analysis of Volatile Organic Chemicals in Drinking Water by Capillary Column Gas Chromatography/Mass Spectrometry, Amer. Water Works Assoc., Water Quality Technology Conference, Poster Session, Denver, CO, December, 1984.

PART THREE

Water Treatment Processes That Prevent or Remove Trihalomethanes and Other Organic Contaminants in Drinking Water Supplies

CHAPTER 8

Conventional Water Treatment and Direct Filtration: Treatment and Removal of Total Organic Carbon and Trihalomethane Precursors

James K. Edzwald

Department of Civil Engineering
University of Massachusetts
Amherst, Massachusetts

8.1 INTRODUCTION

Conventional water treatment plants are used to remove suspended and colloidal particles from water supplies. These plants employ a series of processes consisting of coagulation-flocculation, sedimentation, and filtration as shown in Figure 8.1a. A variety of coagulants and coagulant aids are used. Alum is the most common primary coagulant, and organic polymers are frequently used as coagulant aids. Their function is to destabilize particles and produce large aggregates or flocs that settle readily. An objective then in a conventional water plant is to remove most suspended solids by sedimentation leaving filtration as a polishing step. Direct filtration, on the other hand, is a treatment scheme whereby all of the particulate solids must be removed in the filters; there is no sedimentation step as shown in Figure 8.1b. Direct filtration may include a flocculation tank or the water may bypass flocculation and go directly to the filters. The latter scheme is called in-line filtration. In either case it is necessary to add coagulants before filtration to destabilize the particles in the raw water. Primary coagulants used include alum, iron salts, and cationic polyelectrolytes.

Conventional plants are used to treat both turbid waters and waters containing natural color (humic substances). For the latter case, alum is most often used as the coagulant. Direct filtration plants in North America

199

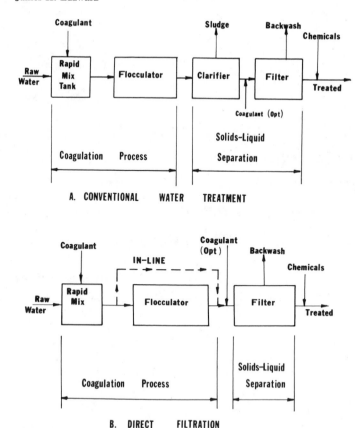

Figure 8.1. Schematic diagrams of conventional water treatment plant (*a*) and direct filtration plant (*b*).

have been designed primarily on the basis of turbidity removal with best application to low-turbidity raw waters. It was discovered in the 1970s that trihalomethanes (THMs) are produced within water treatment plants as by-products of chlorination (1,2,3). A variety of organic compounds present in natural waters can react with chlorine to form volatile chlorinated by-products (mostly, the THMs) and nonvolatile chlorinated by-products [frequently measured as a group parameter, as nonpurgeable organic halide (NPOX)]. Humic substances that cause natural color are both the largest fraction of organic matter and THM precursors in most natural waters. The Environmental Protection Agency (EPA) in 1979 amended the National Interim Primary Drinking Water Regulations (4) and established a maximum contaminant level (MCL) of 0.10 mg/L for total trihalomethanes (TTHMs). Water treatment plants must now be concerned about control of THMs and organic matter as well as the traditional parameters of color and turbidity.

There are several ways to control THM formation and to remove THM precursors. EPA (5) has identified several methods as "Best Generally Available." One of these is improved clarification for reduction of THM precursors. Improved clarification means improved coagulation and applies to both conventional and direct filtration plants. The major goal of this chapter is to describe what can be accomplished by conventional water treatment plants and direct filtration plants in removing organic matter and THM precursors. To accomplish this goal the chapter is divided as follows. First, fundamentals of coagulation of humic substances are described for two different coagulants, alum and cationic polyelectrolytes. Second, the results of field studies of two conventional-type water treatment plants are described and discussed. THM formation through the plants is examined and removals of TOC and THM precursors are summarized. To insure general application of the results, the discussion focuses on important water quality, plant operating, and plant design variables that affect THM formation control and THM precursor removal. Finally, direct filtration is examined as a treatment scheme in terms of traditional measures of performance (filtered water turbidity and color, and head loss development) and in terms of performance for removals of TOC and THM precursors.

8.2   COAGULATION OF HUMIC SUBSTANCES

8.2.1   Background

Prior to the early 1970s and the discovery of THMs in drinking water, the primary interest in the coagulation and removal of humic substances was for the removal of natural color. Research by Black and co-workers (6,7), Hall and Packham (8), and Mangravite et al. (9) demonstrated the importance of pH and coagulation stoichiometry. An optimum pH range of 5 to 6 was cited for alum coagulation of humic substances. Optimum pH means those pH conditions at which the largest color reduction can be achieved with the lowest alum dosage. Their research also showed a stoichiometric dependence between the initial concentration of humic matter (color) and the alum dosage needed for good coagulation.

Since the discovery of THMs, research interests in the coagulation of humic substances shifted from color reduction to the removals of total organic carbon (TOC) and THM precursors (e.g., 10–15). More recently, coagulation studies have examined the removals of nonvolatile chlorinated precursors (16,17). These studies have verified the older work on color removal regarding optimum pH conditions and the stoichiometric alum coagulant requirements.

In practice, a wide range in removal efficiencies for TOC and THM precursors have been reported (see Ref. 18). Some of the variation is due to differences in the ability of coagulants to remove a variety of precursors from various waters. However, some of the variation in removals is due to differences in coagulants tested—alum versus cationic polymers. Other important factors explaining variation in the reported removals are differences in the nature of the THM precursors, in treatment conditions, and in the experimental protocol. Data have been reported for humic acids, fulvic acids, and uncharacterized organic matter from real waters. The pH of coagulation and the experimental protocol for measuring THM precursors varies widely, especially where field data are reported. THM precursors are measured indirectly by measuring the total trihalomethane formation potential (TTHMFP). Experimental methods for TTHMFP measurements have not been standardized. Different investigators may use different pH, temperature, or formation period conditions. All of the above contribute to discrepancies in reported data on the efficiences of coagulation for removals of TOC and THM precursors.

The goal of Section 8.2 is to present principles on the removal of humic substances by coagulation for alum and cationic polyelectrolytes. The removals of TOC and TTHMFPs for waters containing naturally occurring organic matter, particularly humic substances, are reported for well-controlled coagulation experiments. First, the procedures for properly evaluating the effectiveness of coagulants via jar tests (bench-scale batch studies) are explained. Next alum coagulation is explained. Finally, cationic polyelectrolytes are examined as sole coagulants. The material in this section serves as a foundation for subsequent sections.

8.2.2 Jar Test Procedures

Jar tests are bench-scale batch experiments used by environmental engineers and water plant operators. They are used to evaluate the destabilizing ability of various coagulants, to choose coagulant dosages, and to evaluate coagulant aids. The jar test data are commonly judged in terms of the ability of the coagulant to produce a low residual turbidity following coagulation (rapid mix), flocculation (slow mix), and sedimentation. To evaluate coagulation for THM precursor removal, the traditional turbidity and color measurements are not adequate. Jar test experiments must include additional measurements.

THM precursor removal by coagulation can be determined by measuring the difference between the TTHMFPs of the raw and treated waters. Herein TTHMFP measurements are based on 7-day formation periods at pH 7.5 and 20°C. It is critical that sufficient chlorine be added to the samples in the TTHMFP test so that a free chlorine residual exists for the entire 7-day formation period. A rule of thumb is an initial dose based on 3–5 mg Cl_2 per milligram of TOC is usually adequate. Standardizing the pH of the TTHMFP test is also critical to separate the influence of pH on coagulation from pH effects on THM formation.

pH is an important variable in coagulation experiments that must be controlled, especially when using hydrolyzing metal coagulants such as alum. pH affects the chemistry of aluminum and the stability of colloids and organic matter present in the raw water. The proper way to control pH in jar test experiments is described below.

The pH of the raw water is first adjusted to the desired study value with strong acid (HCl) or base (NaOH). Alum is then added, but with the simultaneous equivalent addition of $NaHCO_3$ to neutralize the acidity of the alum, and thus maintain a constant pH in the critical rapid mixing stage of coagulation. This method has been used successfully in the author's laboratory. In coagulation experiments examining a range of pH conditions from 5.5 to 7.5, measured pH values at the conclusion of the jar tests (after the settling period) are usually within 0.2 pH units of the initial pH. Addition of alum to the samples followed by adjustment of pH to a target study value is not a good procedure because the kinetics of aluminum reactions with water and organic matter are so rapid that these reactions will have taken place prior to pH adjustment.

In summary, pH must be controlled in coagulation experiments and standardized for all TTHMFP measurements. TTHMFP measurements are, however, time consuming. It may be a week or more before the THM data are available to evaluate the effectiveness of coagulation. TOC measurements may be used as a nonspecific indicator of the effectiveness of coagulation for removal of organic matter. However, this test also takes time and requires specialized, expensive equipment. A good surrogate parameter for TOC and THM precursors (TTHMFPs) is UV (254 nm) absorbance. It is especially good in jar test studies because, like turbidity, it can be measured easily, rapidly, and inexpensively. Measurements on filtered (glass fiber filters, Whatman 934-AH) and unfiltered samples at a fixed pH are recommended (e.g., pH 7.5). In this chapter unfiltered measurements are referred to as total UV (TOTL-UV) and filtered measurements are referred to as soluble UV (SOL-UV). The usefulness of UV is demonstrated in the material that follows.

8.2.3 Alum Coagulation

Table 8.1 compares the main features of the coagulation of humic substances with alum. It is used to guide us through this subject in which data are presented to illustrate pH effects, stoichiometric considerations, and removals of TOC and THM precursors.

8.2.3.1 pH

The literature indicates that optimum pH conditions exist between pH 5 to 6. Figure 8.2 compares coagulation at pH 5.5 with pH 7.2 for Grasse River water, a highly colored water supply of low alkalinity and turbidity. A detailed description of this supply is provided in Section 8.3.1. The data clearly

TABLE 8.1
Comparison of Alum Coagulation of Humic Substances
at Different pH Conditions

	pH 5–6	pH 6–8.5
Literature	Optimum	Not optimum
Stoichiometry	Coagulant dosage increases with increasing humic concentration	Coagulant dosage increases with increasing humic concentration
Flocculation kinetics	Could be limiting due to dilute humic concentrations	Should be more rapid due to precipitation of $Al(OH)_3(s)$ particles
Mechanism	(1) Low TOC: precipitation of aluminum humates (2) High TOC: adsorption of humic substances on aluminum hydroxide particles	Adsorption of humic substances on aluminum hydroxide particles

demonstrate that lower alum dosages are needed at pH 5.5 than at the higher pH for good coagulation. Optimum pH conditions of 5–6 not only apply to the coagulation of TOC and THM precursors, but also organic halide precursors (measured as TOXFP). Reckhow and Singer (17) have reported on the coagulation of fulvic acid showing optimum pH conditions of 5 to 6 for removals of TOXFP. Consequently, optimum pH conditions for coagulation applies to the total pool of organic matter (TOC), to that fraction of the organics that produces volatile chlorinated organics (THMs), and to that fraction that produces nonvolatile chlorinated organics (NPOX).

Figure 8.2 shows that no restabilization of floc particles occurs at pH 7.2. Furthermore, alum doses above that first needed for good coagulation produces only a small increase in the removals of NPTOC and TTHMFP. There is evidence at pH 5.5 of slight restabilization of flocs at aluminum dosages of 4–6.5 mg/L (alum dosages of 50–75 mg/L). The residual turbidity and unfiltered UV data show poorer settling of the floc in that alum dosage region, followed by good coagulation and settling at higher aluminum doses.

The data also show that UV (254 nm) absorbance is a good surrogate parameter for evaluating removals of NPTOC and TTHMFP in jar test experiments. Data collected in the author's laboratory have demonstrated excellent correlations between UV and both NPTOC and TTHMFP (19,20). Its use in practice is addressed in Sections 8.3 and 8.4.

8.2.3.2 Stoichiometry

Figure 8.3 presents coagulation data at pH 5.5 for two different systems—the Grasse River and a synthetic water containing fulvic acid (Contech)—at

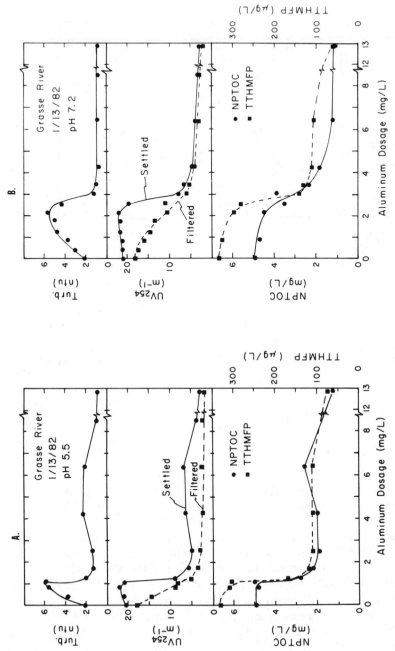

Figure 8.2. Jar test results using alum for pH 5.5 (*a*) and pH 7.2 (*b*). Raw water quality: 2.1 ntu turbidity, 20.9 m^{-1} (0.209 cm^{-1}) UV (254 nm), 4.9 mg/L NPTOC, and 332 µg/L TTHMFP. 1 mg/L Al equals 11 mg/L Al$_2$(SO$_4$)$_3$;14 H$_2$O.

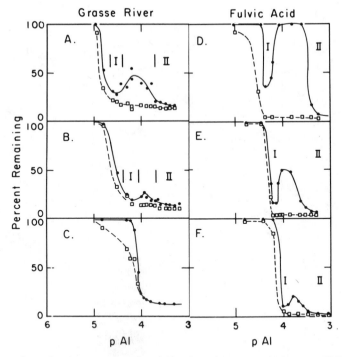

Figure 8.3. Organic matter remaining following alum coagulation at pH 5.5. Initial NPTOC conditions for the Grasse River: (*a*) 2.1 mg/L; (*b*) 4.5 mg/L; (*c*) 6.5 mg/L. Initial NPTOC conditions for the fulvic acid system: (*d*) 2.5 mg/L; (*e*) 3.4 mg/L; (*f*) 4.9 mg/L. (*Solid lines*) UV (254 nm) on settled samples; (*dashed lines*) UV (254 nm) on settled plus glass fiber filtered samples.

different initial concentrations of TOC. At the lowest concentrations of TOC (Fig. 8.3, *a* and *d*), there are two coagulation zones labeled I and II. The area of restabilization separating these coagulation zones is characterized as an area in which the particles are stable and do not settle well. Zone I coagulation occurs at relatively low total aluminum concentrations (Al_T), while Zone II occurs at pAl of about 3.5 (Al_T of 3.2×10^{-4} M) or greater.

Increasing the initial TOC of the two waters requires additional aluminum for Zone I coagulation. The data in Figure 8.3 also show that as the concentration of TOC increases, Zone I shifts to the right (higher Al requirements) and apparently merges with Zone II. For the Grasse River, no restabilization is observed at the highest TOC water of 6.5 mg/L (Fig. 8.3*c*), and only a small degree of restabilization is observed at 4.5mg/L (Fig. 8.3*b*). For the fulvic acid system, the area of restabilization essentially disappears at a TOC of 4.9 mg/L (Fig. 8.3*f*).

The data show a stoichiometry between the alum dose and the raw water TOC concentration. This is measured by an Al_T/TOC ratio—the Al_T dose required for good coagulation divided by the raw water TOC. At pH 5.5 this

TABLE 8.2
Stoichiometric Ratios and Removals of TOC and TTHMFPs

	pH 5.5			pH 7.2		
	Al/TOC	Percent Removals		Al/TOC	Percent Removals	
System	(mg/mg)	NPTOC	TTHMFP	(mg/mg)	NPTOC	TTHMFP
Fulvic acid	0.5–0.6	70–85	65–85	0.9–1.6	70–80	75–80
Grasse River	0.4–0.6	50–70	60–80	0.9–1.3	50–70	60–75

ratio is about 0.5 for both raw water systems as shown by the data in Table 8.2. If coagulation is practiced at higher pH, there is also a stoichiometric dependence, but the Al_T/TOC ratio is higher. At pH 7.2 the ratio is about twice that at pH 5.5. The Al_T/TOC ratio is significant and practical. If the raw water TOC is known, it can be used to estimate alum dosages for humic-type waters. The TOC can be measured directly or estimated from TOC-UV correlations. The use of UV absorbance at 254 nm as a surrogate parameter for organic matter and THM precursors is discussed in Sections 8.3 and 8.4.

8.2.3.3   Coagulation Mechanisms

Two mechanisms are described and illustrated in Figure 8.4. They follow concepts of particle formation for aluminum in the presence of humic matter presented by Snodgrass et al. (21), and coagulation mechanisms for humic substances presented by Dempsey et al. (15). The coagulation mechanism for alum depends on the pH of coagulation, the Al_T dose, and the raw water concentration of humic matter represented by the TOC concentration.

At the optimum pH conditions of pH 5–6, coagulation can occur by two mechanisms depending on the TOC concentration. At low TOC concentration the required Al_T dose is relatively low, and coagulation is achieved by precipitation of aluminum-humate particles produced by the chemical interaction of positively charged aluminum hydrolysis species with negatively charged humic macromolecules (see Fig. 8.4). The humic matter precipitates out of solution when the negative charge is satisfied by an equivalent positive charge from the dissolved aluminum species. It is proposed that Zone I in Figure 8.3 is due to this mechanism. Using data in Figure 8.3 and Table 8.2, it is proposed that this mechanism is applicable when raw water TOC concentrations do not exceed 4–5 mg/L. The required Al_T dose would be 10^{-4} M or less (alum dose of 30 mg/L or less). If alum is added above the dosage required for precipitation of aluminum-humate, the aluminum-humate particles can adsorb positively charged soluble aluminum from solution, resulting in stable, positively charged aluminum-humate floc.

The second mechanism involves the precipitation of aluminum hydrox-

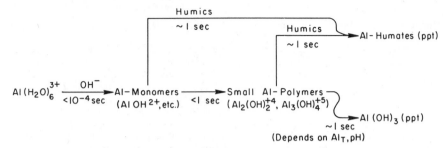

Figure 8.4. Proposed mechanisms for alum coagulation of humic substances. Pathways at top resulting in Al-humate precipitation favored by low TOC and pH < 6. Pathway resulting in $Al(OH)_3$ precipitation favored by high TOC or pH above approximately 6.

ide and the removal of organic matter by adsorption on aluminum hydroxide particles. This mechanism operates at pH 5.5 when high Al_T doses are needed due to high TOC concentrations (above 4–5 mg/L), or at higher pH (pH 6–8) conditions. At Al_T dosages of about 3.2×10^{-4} M or higher (alum of 100 mg/L or higher), aluminum hydroxide precipitation occurs. In Figure 8.3 this is shown as Zone II. For low TOC waters (less than 4–5 mg/L), it is shown as a second zone of coagulation at pH 5.5 following restabilization. For TOC concentrations greater than 4–5 mg/L, the Al_T doses are so high (about 3.2×10^{-4} M or higher) that aluminum hydroxide precipitation is favored kinetically over precipitation of aluminum humates (see Fig. 8.4). It is proposed then that removal of humic substances occurs by adsorption on aluminum hydroxide particles. The adsorption process may involve both soluble aluminum-humate complexes and uncomplexed humic matter.

8.2.3.4   Removals of NPTOC and THM Precursors

Table 8.2 summarizes the removals of NPTOC and TTHMFP achieved by alum coagulation with flocculation and settling periods. No filtration of the samples took place. Alum coagulation is effective in removing TOC and THM precursors at both pH conditions; however, coagulation at pH 5.5 requires lower alum dosages—about one-half the dosage needed at the higher pH condition. The data show that removals of 70% can be achieved for the Grasse River. In Section 8.3.4 the results of these laboratory experiments are compared with full-scale treatment plant data for the Grasse River.

For the Grasse River, the TOC concentration frequently exceeds 4 mg/L so that the mechanism of removal of dissolved organic matter is by adsorption on aluminum hydroxide. The lower alum dosage at pH 5–6 is consistent with the observations that adsorption of organic matter on aluminum oxides is pH dependent with maximum adsorption occurring at pH 5–6 (22).

8.2.4 Cationic Polyelectrolytes

Cationic polymers can coagulate humic substances (23–25). Below, jar test data are used to describe the coagulation model. Important features of the model are then outlined. The application of cationic polymers in direct filtration is presented in Section 8.4.

8.2.4.1 Coagulation Model

Figure 8.5 shows jar test data for a synthetic water containing 10 mg/L humic acid buffered at pH 6.3 with $NaHCO_3$. A high charge density, cationic polyelectrolyte (a quaternized polyamine) was used as the coagulant over a wide range of dosages.

In this jar test, electrophoretic mobility (EPM) measurements were also made and are shown at the bottom of the figure. At low dosages (0–6 mg/L), underdosing is observed whereby an insufficient amount of positively charged polymer has been added to achieve particle destabilization. This is

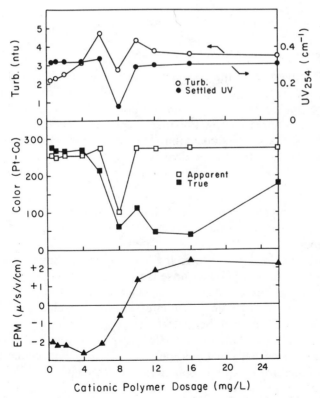

Figure 8.5. Jar test results for the coagulation of 10 mg/L humic acid at pH 6.3 using a cationic polymer.

reflected in the UV and color data as well as the negative EPM values. At a dosage of 8 mg/L, the jar test optimum for UV and color, the mobility is close to zero with a slight negative charge. At this point, enough cationic polymer has been added to neutralize the charge of the humic matter. Charge neutralization is important, but the mechanism is different from that traditionally described for cationic polymers adsorbing on colloidal solid surfaces. Aggregates (precipitates of humic matter and cationic polymer) are formed as a result of chemical cross-linking between the negatively charged humic macromolecules and the cationic polymers (23). Overdosing is observed at dosages greater than 8 mg/L. Here, the suspension of particles is restabilized as indicated by the increase in UV and apparent color. Charge reversal occurs with excess polymer addition as indicated by the positive EPM values.

The name given to this model is the Charge Neutralization/Precipitation model (26). Coagulation occurs at the point that the negatively charged humic macromolecules are neutralized by the addition of cationic polyelectrolyte causing precipitation of a cationic polymer-humic solid phase. The mechanism is analogous to the alum coagulation model for low TOC waters at pH 5–6.

8.2.4.2 Features of the Model

Polymer effectiveness in the coagulation of humic substances depends on polymer properties, the humic substances (concentration, chemical structure, charge, molecular weight), and the bulk water chemistry (pH, Ca^{2+} concentration, ionic strength). Polymer properties include charge density, molecular weight, and functional group chemistry; however, the single most important property is the charge density.

A summary of important features follows:

- Cationic polymers can coagulate humic substances. The dose depends on the raw water humic concentration and the charge density of the polymer.
- The pH of the water affects the total negative charge of the humic matter; therefore, polymer dose is pH dependent.
- Overdosing with polymer will produce a stable (positively charged) precipitate. This stable precipitate will not flocculate rapidly nor be removed well by deep bed filtration.
- Kinetics of flocculation can be slow due to the low floc volume fraction produced when cationic polymers form polymer-humic precipitates. Therefore, cationic polymers as sole coagulants would be more applicable in direct filtration rather than conventional treatment plants.

8.3 CONVENTIONAL WATER TREATMENT PLANT PERFORMANCE

8.3.1 Description of Water Supplies

Two representative water supplies were selected for study. Both supplies are treated by conventional-type water treatment plants that were evaluated for their ability to control THM formation and to remove TOC and THM precursors.

The Grasse River (Canton, New York) is a low alkalinity, highly colored supply typical of many water supplies found in many regions of the United States such as New York, New England, and the Pacific Northwest. The Glenmore Reservoir (Oneida, New York) is a protected upland water supply of low turbidity. The concentrations of organic matter and THM precursors in both raw water supplies undergo a well-defined seasonal variation (19,20). Figure 8.6 shows this for the Grasse River. Highest concentrations occur in the summer, particularly late in the summer, and lowest concentrations occur in the winter during river ice-cover. The observed seasonal variation is not unique. It occurs for other surface waters and should be recognized by water supply utilities, consulting engineering firms, and regulatory agencies

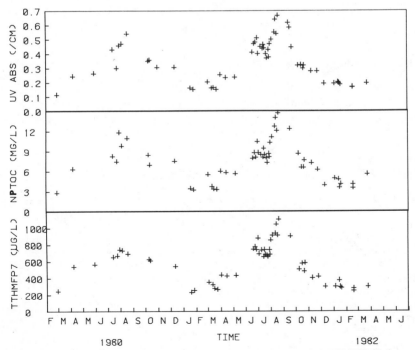

Figure 8.6. Seasonal variation in UV (254 nm), NPTOC, and TTHMFP (7-day, pH 7.5, 20°C) for the Grasse River.

concerned with pilot plant studies or full-scale water plant investigations. The poorest water quality with regard to the highest concentration of natural organic matter does not occur during autumn leaf fall or during spring runoff conditions, but in the summer. The higher water temperatures in the summer coupled with high raw water THM precursor concentrations means the potential is greatest for THM formation in water plants and the distribution systems during this period.

The Grasse River and Glenmore Reservoir are good water supply systems to study. The organic matter in both supplies is derived from natural sources. The Glenmore Reservoir is a protected potable supply, and there are no significant upstream wastewater dischargers on the Grasse River. They are also contrasting systems. The Grasse River is highly colored due to humic substances leached from the upland watershed and bogs whereas a major source of organic matter for the Glenmore Reservoir is that produced within the reservoir from primary biological production—algal cells, algal by-products, and algal decomposition products.

The full-scale water plants at Canton and Oneida were each monitored on 16 dates between February 1980 and June 1982. The monitoring dates were selected to assess any effect of seasonal variation in water quality on plant performance. Raw water quality data are summarized in Table 8.3. These data show that the Grasse River contains higher amounts of color, NPTOC, THM precursors (TTHMFP), and higher UV (254 nm) absorbance values than the Glenmore Reservoir.

8.3.2   Description of Water Plants

The Canton treatment plant normally operates at a flow of 11.4 ML/day (3 MGD), and is a high-rate conventional-type plant. Treatment processes consist of coagulation-flocculation, sedimentation (tube settlers), and

TABLE 8.3
Raw Water Data for Canton and Oneida Water Plant Monitoring

Raw Water Parameter	Canton ($n = 16$)		Oneida ($n = 16$)	
	Mean	Range	Mean	Range
Temperature (°C)	8	0.5–23	11	3–21
pH	7.1	6.9–7.4	6.9	6.5–7.4
Turbidity (ntu)	2.1	1.2–3.2	1.4	0.54–3.5
True Color, Pt-Co	125	73–276	65	24–140
TOTL-UV (cm^{-1})	0.278	0.113–0.666	0.179	0.084–0.309
SOL-UV (cm^{-1})	0.279	0.137–0.640	0.165	0.085–0.317
NPTOC (mg/L)	6.1	2.75–14.7	4.7	2.0–10.65
TTHMFP (μg/L)	473	245–1108	320	152–597

TABLE 8.4
Treatment Conditions and Inst TTHM Data for the Canton and Oneida Plants

	Canton (Mean)	Oneida (Mean)
Alum dosage (mg/L)	42.5 (35–50)[a]	22.7 (12–44)
pH		
raw water	7.1 (6.9–7.4)	6.9 (6.5–7.4)
after settling	6.4 (5–6.9)	6.2 (5.5–6.6)
clearwell	6.5 (5.7–7.0)	7.5 (7.0–9.2)
Prechlorination (mg/L)		
rapid mix tank	1.1 (0–2.5)	See below[b]
top of filter		1.1 (0.7–2.9)
Postchlorination (mg/L)	1.7 (0–4)	3.4 (2.3–6.5)
Chlorine residual (mg/L)		
clearwell	1.1 (0.23–1.7)	2.5 (2.0–3.2)
Inst TTHM (µg/L)		
settled	10.2 (ND[c]–33.5)	9.2 (ND–87.5)
after filtration	10.2 (ND–43.6)	19.3 (ND–97.5)
clearwell	36.6 (ND–126.8)	47 (10.1–148.5)

[a] Range.
[b] Two occasions at 4.2 and 4.9 mg/L.
[c] ND, Not detectable.

filtration (mixed media filters). Summaries of treatment conditions for both plants are presented in Table 8.4. Alum coagulation under acidic pH conditions is practiced at Canton in treating the highly colored Grasse River. Prechlorination is normally practiced (rapid mix tank), but low chlorine dosages are used—average of 1.1 mg/L. The postchlorination dosage is also low, and is applied at a point between the filters and the clearwell.

The Oneida plant has a rated process capacity of 15.1 ML/day (4 MGD). It is a conventional plant using coagulation-flocculation, sedimentation (rectangular settling basins), and filtration (dual media filters). As indicated in Table 8.4, alum is used to treat this low turbidity supply. Currently, the chlorination practice includes a small dose on top of the filters followed by postchlorination. On two of the monitoring dates, the plant used high prechlorination doses because of reduced Fe and Mn in their raw water supply.

Samples were collected at several locations through the plants—raw, settled, filtered, and finished water (pumped water from clearwell to the distribution system). The analyses included temperature, pH, turbidity, apparent color, true color, total UV (254 nm) absorbance, soluble UV (254 nm) absorbance, NPTOC, instantaneous TTHM (Inst TTHM), TTHMFP, terminal TTHM (Term TTHM), and free residual Cl_2. For the raw waters, the 7-

day TTHMFPs (pH 7.5, 20°C) were used as the measure of raw water THM precursor concentration. Within the water plants, Term TTHM samples were buffered at pH 7.5 and spiked with chlorine to maintain a residual for 7 days—that is, Term TTHMs (7-day, pH 7.5, 20°C). THM precursor removal by treatment is the difference between the raw water 7-day TTHMFP and the 7-day Term TTHM at the sampling point chosen to evaluate treatment.

8.3.3 Illustrations of Plant Monitorings

The results of three plant monitorings are presented first to illustrate the effects of raw water THM precursor concentration, temperature, pH, and chlorination practice. Next the removals of NPTOC, THM precursors, and UV achieved by the plants are summarized and discussed. Finally, THM formation in the treatment plants is evaluated.

8.3.3.1 *Canton Plant: Low TOC and Low TTHMFP*

Plant monitoring results are shown in Figure 8.7 for winter water quality conditions. The raw water temperature (1°C) and concentrations of TTHMFP (255 μg/L), NPTOC (3.2 mg/L), and UV (0.151 cm^{-1}) were low. Process detention times are shown at the top of the figure. Note that this is a high-rate plant with a short hydraulic detention time—only 60 min ahead of the clearwell. Overall plant performance for removals of organic matter and THM precursors is based on raw water data and filtered water data. Sampling after a specific filter is preferred to sampling after the clearwell because the water quality in the clearwell reflects a composite of several filters. Sampling ahead of the clearwell permits sampling from a common train of flocculation and sedimentation tanks leading to designated filters.

The plant performs well with good removals of TOC and THM precursors: 69% for Term TTHM (i.e., THM precursors), 60% for NPTOC, and 67% for UV absorbance. The turbidity and NPTOC data show that the filters do most of the work under cold water conditions. Removals prior to filtration are poor. The effect of water temperature on performance will be discussed later. THM formation through the plant as indicated by the Inst TTHM data is low, only 25 μg/L in the water following the clearwell. Several factors contribute to this, including the low raw water precursor concentration, the chlorination practice (low prechlorination and postchlorination doses; no free chlorine residual measured ahead of the clearwell), and pH conditions through the plant (pH 6.5–7.3), but the single most important factor is the low water temperature, which slows the rate of THM formation.

8.3.3.2 *Canton Plant: High TOC and High TTHMFP*

Figure 8.8 illustrates monitoring results under early summer water quality conditions when the concentrations of TTHMFP and TOC are relatively

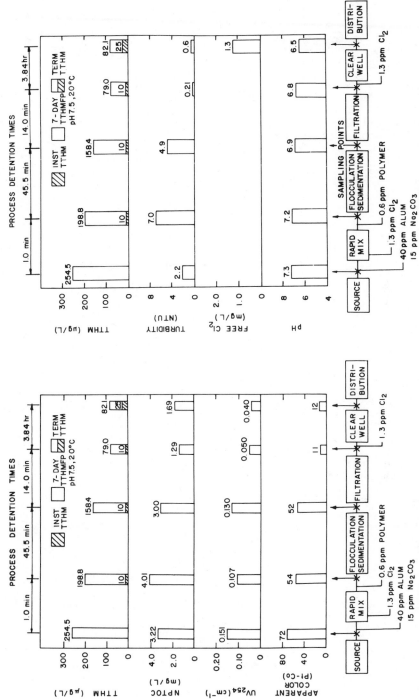

Figure 8.7. Canton water plant monitoring results during winter: Low raw water TOC and TTHMFP, water temperature at 1–2.5°C.

215

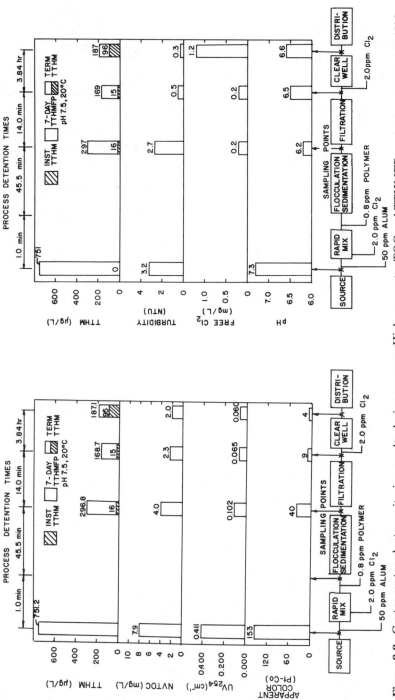

Figure 8.8. Canton water plant monitoring results during summer: High raw water TOC and TTHMFP, water temperature at 19°C.

high and the water temperature is 19°C. Again overall plant performance for removals of Term TTHM, NPTOC, and UV is good. In contrast to the prior case, substantial removals are accomplished prior to filtration—that is, by coagulation-flocculation and sedimentation.

THM formation through the plant prior to the clearwell is low—Inst TTHM of 15 μg/L after the filter—considering the high raw water precursor concentration. This is attributed to the pH conditions of treatment (less than pH 6.5 after prechlorination), the short detention period, and the prechlorination practice. The raw water contains a high concentration of TOC and would have a high chlorine demand, but a relatively low prechlorination dose of 2 mg/L was applied. A free chlorine residual of 0.2 mg/L was measured through the plant, but this free chlorine concentration is too low to drive the THM formation reaction and, therefore, limits THM formation. An Inst TTHM concentration of 95 μg/L is measured after the clearwell due to the postchlorination dose of 2 mg/L (free residual of 1.2 mg/L) and the nearly 4-h detention period in the clearwell.

8.3.3.3 Oneida Plant: Effect of Preoxidant

During the summer stratification period for the Glenmore Reservoir, reduced Fe and Mn present in the raw water require oxidation and removal by the Oneida treatment plant. Monitoring results presented in Figure 8.9 contrast the use of two preoxidants, chlorine and potassium permanganate ($KMnO_4$). The plant is a conventional one with long rectangular sedimentation basins. Process detention times listed at the top of the figure show a long hydraulic detention period of 6 h in the plant prior to the clearwell.

Figure 8.9a shows that when chlorine is used as a preoxidant, removal of THM precursors is poor and THM formation is high. The Inst TTHM concentration of 134 μg/L after the clearwell exceeds the MCL for TTHMs of 0.10 mg/L. Additional THM formation occurs in the distribution system. Two samples collected from the distribution system had concentrations of 211 and 221 μg/L. The THM formation problem is caused largely by using a high prechlorination dose of 4.2 mg/L in the rapid mix basin. The need for a high postchlorination dose at a high pH (pH 8.6) also contributes to the high THMs. The postchlorination practice of relatively high chlorine addition is due to a 20-mile transmission main to the city in which a free chlorine residual must be maintained. The high pH was due to their corrosion control practice used during this monitoring period. Their corrosion control practice was later changed.

The following summer Oneida switched to potassium permanganate as a preoxidant. Figure 8.9b shows the results for a date in which the raw water THM precursor concentration was its highest over a 30-month period. The benefit of using potassium permanganate is shown by the relatively low THM formation through the plant—60 μg/L Inst TTHM after the clearwell. The plant also benefited from a change in corrosion control practice to

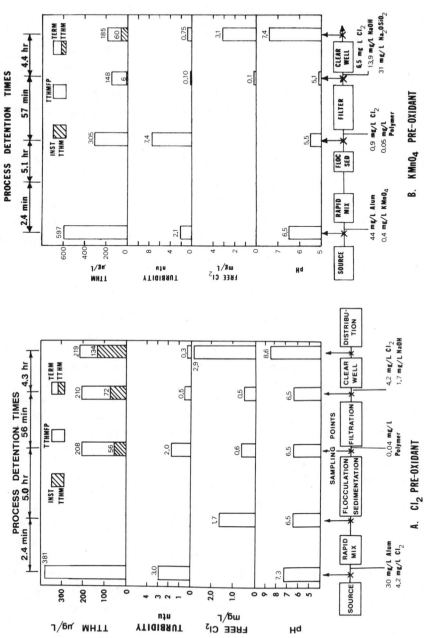

Figure 8.9. Oneida water plant monitoring results: Effect of preoxidants prechlorination (*a*) and potassium permanganate (*b*).

sodium silicate, which allows finished water pH at a full pH unit lower than prior practice. Elimination of prechlorination permitted good plant performance for removing THM precursors, 75% on this date.

8.3.4   Discussion of Performance

8.3.4.1   Summary of Removals

The Grasse River and Glenmore Reservoir water supplies are different in terms of type, origin of organic matter, and concentrations of organic matter and THM precursors. In spite of these differences the data in Table 8.5 show that both water plants achieve excellent and similar removals of NPTOC, UV (254 nm), and Term TTHMs (THM precursors).

The excellent performance of the Canton plant in both removing THM precursors and controlling THM formation through the water plant is due to the effectiveness of the coagulation process, pH conditions during treatment, chlorination practice, the short hydraulic detention time of the plant, and the relatively low water temperatures. The Canton plant uses alum coagulation with the addition of a nonionic polymer to aid floc formation. Normally coagulation is practiced at pH 5.5–6 (rapid mix tank) which is optimum according to the principles presented in Section 8.2. In spite of the high THM precursor concentrations in the Grasse River, the average Inst TTHM concentration of the finished water was only 37 μg/L (see Table 8.4). Although prechlorination is practiced, the dosages are low at 1–2 mg/L. The plant has tube settlers so that the total hydraulic detention time through the plant (excluding the clearwell) is approximately 1 h. The prechlorination practice and the short hydraulic detention time allow removal of THM precursors before large concentrations of THMs are formed.

The Oneida plant also achieves excellent removals of NPTOC and THM precursors. It also practices alum coagulation at pH conditions near op-

TABLE 8.5
Performance Summaries for the Canton and Oneida Water Plants

	Mean Percent Removals[a]	
Parameter	Canton ($n = 16$)	Oneida ($n = 12$)
NPTOC	69 (8)[b]	71 (14)
Total UV (254 nm)	80 (7)	82 (4)
Soluble UV (254 nm)	84 (6)	81 (7)
Term TTHM[c]	72 (5)	68 (7)

[a] Excludes performance data for Oneida during the first year when chlorine was used at the plant as a preoxidant.
[b] Standard deviation.
[c] 7-day, free Cl_2 present, pH 7.5, 20°C.

timum. The detention time through the Oneida plant (excluding the clear-well) is approximately 6 h. Seasonal problems with Fe and Mn in the Glen-more Reservoir require use of a preoxidant. The type and dosage of preoxidant has a significant effect on Inst TTHM concentrations in the finished water. The data in Table 8.6 show that when chlorine was used as a preoxidant, the Inst TTHM concentration exceeded the MCL of 0.10 mg/L. The plant substituted potassium permanganate for chlorine as a preoxidant, and they changed their corrosion control practice so that the treated water has a pH of 7–7.5. These changes have controlled THM formation in the plant, as shown by the data in Tables 8.4 and 8.6. Chlorine is still added, but at relatively low doses to the top of the filters and to the clearwell.

8.3.4.2 Effect of Temperature

Temperature effects on removals of THM precursors and on Inst TTHMs are discussed using data from Canton. The data in Table 8.7 show that overall treatment plant efficiences for removal of NPTOC, UV (254 nm), and THM precursors are not significantly affected by temperature; however, water temperature does affect where in the treatment plant the removals take place. For cold water conditions, the organic matter is coagulated, but settling of the floc is poor so that a large portion of the removal takes place in the filters during the winter months. Similar results were observed for the Oneida plant.

Water temperature also influences the Inst TTHM concentration of the treated water. The data in Table 8.8 demonstrate that at the higher water temperatures, the mean Inst TTHM concentration and variation are much higher than at colder water temperature conditions.

8.3.5 UV as a Surrogate Parameter

8.3.5.1 Plant Monitoring of TOC and THM Precursors

UV absorbance at 254 nm is explored as a surrogate parameter for NPTOC and THM precursors (7-day TTHMFP, pH 7.5, 20°C). Linear least-squares regression analysis was used to develop predictive equations from the data. Two sampling locations were used for the data base: raw water and filtered water (filter effluent).

Figure 8.10 illustrates the correlation between NPTOC and UV for the Canton water plant. A summary of the predictive equations for both the Canton and Oneida water plants is given in Table 8.9. The statistical measures in the table indicate how much of the variation (r^2) in the dependent variable is explained by the predictive equation and the significance of the regression (α). These statistical measures show that UV is an excellent parameter for estimating TOC and THM precursors. The predictive equations can be used to estimate raw water and treated water quality, and

**TABLE 8.6
Effect of Preoxidant Practice on THM Formation
in the Oneida Water Plant**

Preoxidant	Dosage Range (mg/L)	Number of Monitorings	Clearwell Inst. TTHMs (μg/L)
Cl_2	4.2–4.9	2	141
$KMnO_4$[a]	0.12–0.7	6	35
None[a]	0	6	20

[a] 0.7–1.1 mg/L Cl_2 on top of filters.

**TABLE 8.7
Effect of Water Temperature on Performance at the Canton Water Plant**

	Mean Percent Removals			
	Temp. 0.5–5°C ($n = 9$)		Temp. 9–23°C ($n = 7$)	
	Sed.[a]	Filt.[b]	Sed.[a]	Filt.[b]
NPTOC	25 (14)[c]	65 (8)	43 (15)	73 (4)
Total UV (254 nm)	44 (17)	79 (7)	67 (8)	81 (7)
Term TTHM[d]	42 (9)	70 (5)	53 (8)	74 (4)

[a] Cumulative removal by coagulation-flocculation and sedimentation.
[b] Cumulative removal by coagulation-flocculation, sedimentation, and filtration.
[c] Standard deviation.
[d] 7-day, free Cl_2 present, pH 7.5, 20°C.

**TABLE 8.8
Effects of Temperature on Clearwell Water Inst TTHMs
at the Canton Water Plant**

Case	n	Inst. TTHM (μg/L)
All data	16	37 (33)[a]
Temperature		
0.5–5°C	9	23 (14)
9–23°C	7	54 (44)

[a] Standard deviation.

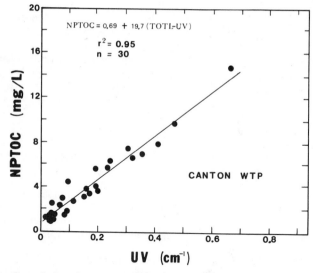

Figure 8.10. Correlation between NPTOC and TOTL-UV (254 nm) for raw and finished waters for the Canton water plant.

therefore monitor overall plant performance. UV absorbance has certain attributes that make it a good surrogate parameter for water plant monitoring and operational control. It can be measured easily, rapidly, and inexpensively. The equipment (UV spectrophotometer) could be installed on-line providing real time, continuous measurements.

   The selection of sampling locations and sample handling procedures are important factors when considering UV as a surrogate parameter (19). UV absorbance measurements are affected by pH, turbidity, and oxidants. Sample handling procedures should consider pH adjustment of all samples prior

TABLE 8.9
UV Predictive Equations for NPTOC and TTHMFP for the Canton and Oneida Water Plants

Equation	r^2	α	n
Canton[a]			
NPTOC = 19.7 (TOTL-UV) + 0.69	0.95	0.0001	30
TTHMFP = 1629.3 (TOTL-UV) + 37.8	0.97	0.0001	31
Oneida[b]			
NPTOC = 25.0 (TOTL-UV) + 0.40	0.86	0.0001	29
TTHMFP = 1578.2 (TOTL-UV) + 42.4	0.94	0.0001	31

[a] *Range in Data:* TOTL-UV, 0.016–0.666 cm^{-1}; NPTOC, 0.80–14.7 mg/L; TTHMFP, 69–1108 µg/L.
[b] *Range in Data:* TOTL-UV, 0–0.309 cm^{-1}; NPTOC, 0.21–10.6 mg/L; TTHMFP, 50–597 µg/L.

to measurement of UV to normalize the data to a standard pH value. Floc or turbidity in samples may interfere with the light absorbance measurements, requiring filtering of samples. Strong oxidant addition in water plants can be a problem because of a reduction in UV absorbance due to an alteration of the structure of the organic molecules without any real reduction in TOC or THM precursors. This may require selecting sampling locations ahead of points of strong oxidant addition, otherwise poor correlations may result.

8.3.5.2 Predicting Inst TTHMs

UV absorbance as a surrogate for THM precursors as developed above does not tell you directly whether you meet the maximum contaminant level (MCL) for THMs in drinking water. The MCL is based not on the concentration of THM precursors remaining, but on the concentration of THMs formed and existing in the treated water at the time of sampling—that is, the Inst TTHM concentration. Development of a general predictive equation for the Inst TTHM concentration of water leaving a treatment plant is difficult. Theory dictates that THM formation through a treatment plant shall depend on raw water quality variables, water plant operational variables, and plant design variables. Raw water quality variables include THM precursor concentration, temperature, and for some waters, the bromide concentration. Plant operating variables include pH during treatment, chlorination practice (locations and dosages), and the efficiency of the removal of THM precursors. Plant design variables are hydraulic detention time through the plant, the processes used to accomplish removals of THM precursors, and where removal takes place in the plant relative to chlorine application.

Correlations were developed on an individual basis for the Canton and Oneida plants. Therefore, plant design variables are fixed, and implicitly part of the developed predictive equations. Linear correlations between the Inst TTHMs of the finished water (water leaving the clearwell) and single independent variables produced the following results. The Inst TTHM concentration was highly correlated with total chlorine consumption for the Canton water plant and prechlorine consumption for Oneida (19). Total chlorine consumption is the sum of all chlorine dosages less the free chlorine residual of the finished water leaving the clearwell. For Oneida prechlorine consumption is the difference between the total chlorine dose applied to both the rapid mix tank and to the filters less the free chlorine residual following filtration.

Linear multiple regression analysis was used to obtain the predictive equations presented in Table 8.10. The correlations are good, significant, and satisfy theoretical criteria. Inst TTHMs through the plant depend on raw water quality (THM precursor concentration) and plant operation (pH and chlorine consumption). The precursor concentration is measured by a surrogate parameter, UV (254 nm). Raw water temperature is included in the Canton predictive equation, but is not needed for Oneida. Finished water pH

TABLE 8.10
Multiple Regression Predictive Equations for Inst TTHMs in Finished Drinking Waters

Equation[a]	r^2	α	n
Canton			
Inst TTHM $= -278.0 + 11.1(A) + 47.1(B) + 12.8(C) + 2.71(D)$	0.88	0.001	16
Oneida			
Inst TTHM $= -111.4 + 11.8(B) + 163.6(C) + 30.1(E)$	0.96	0.0001	16

[a] A = total chlorine consumption; B = finished water pH; C = raw water UV (254 nm); D = raw water temperature; E = prechlorination consumption.

was used to account for pH effects on THM formation rather than other sampling points because a large fraction of the hydraulic detention time for both plants occurs in the clearwell. These predictive equations are good for Canton and Oneida only, because the plant design and operational variables (removals of precursors, points of chlorination, and detention times) for these plants are implicitly included in their development. They do illustrate, however, that predictive relationships can be obtained for water treatment plants. The framework used here could be used by water utilities to develop their own predictive equations. Predictive equations could provide water plant personnel with an estimate of Inst TTHMs in their finished water.

8.4 DIRECT FILTRATION

It was pointed out earlier that direct filtration facilities for potable water supplies in North America have been designed primarily on the basis of turbidity removal. There is little published information on the treatment of colored waters by direct filtration, particularly with cationic polyelectrolytes as the sole coagulants (23–25,27,28). This section addresses direct filtration with regard to traditional measures of performance (turbidity, color, and head loss development) and also to removals of NPTOC, THM precursors (TTHMFPs), and UV (254 nm) absorbance using cationic polymers and alum as the primary coagulants. Other points covered are the selection of cationic polymer dosages, the performance of small-scale versus large-scale pilot plants, in-line direct filtration versus direct filtration with flocculation, and the use of UV absorbance as a surrogate parameter for monitoring direct filtration performance.

8.4.1 Pilot Plant Studies

The two model water supplies described in Section 8.3.1 were used in the direct filtration studies—the Grasse River, a highly colored supply, and the Glenmore Reservoir, a low turbidity, protected upland water source.

Small-scale studies were conducted using Plexiglass* columns of 4.45 cm i.d. Dual media filters were used with an anthracite depth of 38.1 cm (ES (effective size) of 1.0–1.2 mm) over 15.2 cm of sand (ES of 0.45–0.55 mm). The large-scale studies employed a Waterboy* pilot plant with a filter area of 0.372 m² (4 ft²), and was operated at flows of 0.5–1.51 L/s (8–24 gpm). A dual media filter was used with 50.8 cm of anthracite (ES of 1.0–1.2 mm) over 25.4 cm of sand (ES of 0.45–0.55 mm). A schematic of the large-scale pilot plant is presented in Figure 8.11.

The pilot plants were operated, as needed, over a 2-year period to evaluate seasonal changes in raw water quality on performance. THM precursor removal was calculated as the difference between the 7-day TTHMFPs (pH 7.5, 20°C) of the raw and filtered waters. The pilot plant studies were also designed to evaluate the effects of chemical treatment variables (pH, various cationic polymers as sole coagulants, and alum) and physical variables (filtration rate, in-line direct filtration versus direct filtration with flocculation, and water temperature) on filtration performance.

8.4.2 Cationic Polymer Dosage Selection

Jar tests can be used to select cationic polymer dosages for direct filtration (23,29). Testing of jar test data against direct filtration data is shown by the

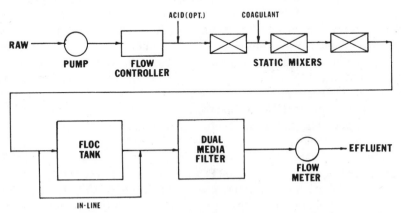

Figure 8.11. Schematic of direct filtration pilot plant.

* Mention of trade names or commercial products in this chapter does not constitute endorsement or recommendation for use.

results in Figure 8.12. The jar test results plotted in the top figure show a sharp drop in true color at cationic polymer dosages of 2–4 mg/L; above 4 mg/L, little change occurs in the true color readings. These results indicate that the required polymer dose for coagulation is 4 mg/L. The bottom figure shows direct filtration results where the polymer dose was varied during the course of the experiment. Best results were obtained at 4 mg/L, as the jar tests predicted. Underdosing and overdosing occur in direct filtration and can be predicted from the jar test data.

For colored waters such as the Grasse River, the cationic polymer dose required for coagulation depends on the raw water concentration of organic matter. This is illustrated by the data in Figure 8.13 which shows a correlation between the polymer dose (Magnifloc 573C) and total UV (254 nm). UV absorbance is used here as a surrogate for TOC. These results are highly significant and practical. The predictive equation developed in this study is specific to the Grasse River, but the principle is generally applicable. In practice, a water utility could make on-line spectrophotometric UV measurements of the raw water quality. From a predictive equation relating polymer dose to raw water UV (254 nm), the utility could then estimate the polymer dosage needed for coagulation.

8.4.3 Performance with Cationic Polymers

8.4.3.1 Grasse River

Figure 8.14 shows results for two pilot plant experiments conducted in the summer when the color and TOC concentration are high. Consequently, polymer coagulant doses are high. The experiments compare in-line direct filtration with direct filtration preceeded by flocculation at a filtration rate of 4.88 m/h and without pH adjustment of the raw water. Excellent filter performance is obtained using only a cationic polymer to treat a highly colored water. Furthermore, there is little difference in the apparent color and turbidity of the filtered water when treated by in-line direct filtration or direct filtration with flocculation. A major difference does exist, however, in the head loss development, with much greater head loss in the in-line direct filtration mode. This latter observation will be discussed later.

Cationic polymers can be used as sole coagulants in the treatment of low turbidity, humic waters by direct filtration. Raw water quality with apparent color values ranging from 50 to 200 Pt-Co color units and turbidity of 1–6 ntu was treated effectively with filtered water quality of 5–10 Pt-Co color units and turbidity of 0.2 ntu. Filtration rate (varied from 4.88 to 14.6 m/h), type of filtration (in-line versus direct filtration with flocculation), and water temperature had little effect on filter performance based on removals of turbidity, color, and organic matter. The effect of pH on removals of organic matter is discussed later.

Cationic polymer dosages for direct filtration vary seasonally with

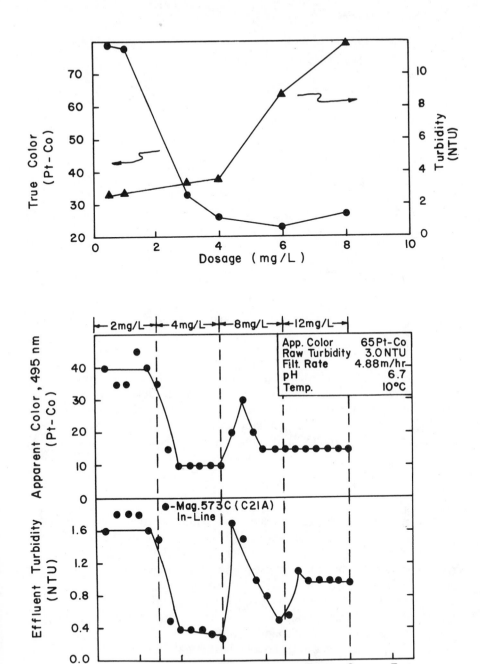

Figure 8.12. Testing of jar test predictions with direct filtration pilot plant data for the Grasse River. (Top): Jar test data using a cationic polymer. (Bottom): Direct filtration results for varying dosages of cationic polymer during the filter run. Note polymer feed pump off briefly after 3 hours.

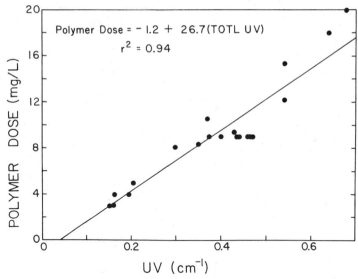

Figure 8.13. Correlation between cationic polymer dose and raw water total UV (254 nm) for the Grasse River.

changes in the concentrations of organic matter in the Grasse River. Polymer dose requirements are relatively low (3–4 mg/L) during the winter and spring when the raw water quality is good (see Fig. 8.6). In the summer when the TOC and color are high, polymer dosages of 9–12 mg/L are needed. Natural waters of lower TOC and color than the Grasse River would require lower polymer doses.

8.4.3.2   Glenmore Reservoir

Cationic polymers can also be used successfully to treat a low turbidity supply like the Glenmore Reservoir. This is illustrated by the results presented in Figure 8.15 in which a polymer dose of only 2.5 mg/L is needed. Over the course of the study, filtered water turbidities of less than 0.3 ntu and color of 5-15 Pt-Co units were achieved on a regular basis.

Polymer dosages vary with seasonal changes in the concentration of organic matter and particulates in the raw water. However, the variation in dosage requirements is less extreme than that for the Grasse River; it varied from 2 to 6 mg/L with an estimated average annual concentration of 4 mg/L.

8.4.3.3   In-Line versus Flocculation

It was stated earlier that the removals of organic matter, filtered water turbidity, and color did not differ between in-line direct filtration and direct filtration with a flocculation period provided by a flocculation basin. For the

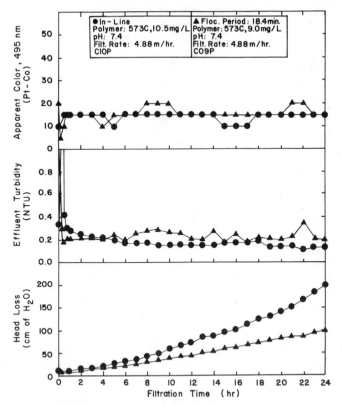

Figure 8.14. Direct filtration of the Grasse River with a cationic polymer: In-line treatment versus direct filtration with a flocculation period prior to filtration. Raw water quality: 2.5 ntu turbidity, 108–115 Pt-Co units apparent color, 0.37–0.40 cm^{-1} UV (254 nm), 7.3–8 mg/L NPTOC, and 657–674 µg/L TTHMFP.

Grasse River in particular, there was significantly less head loss development when treating the same water by direct filtration with flocculation compared to in-line direct filtration (see Fig. 8.14). These results are significant and have important design ramifications.

In the treatment of a water like the Grasse River, direct filtration with flocculation will give longer filter runs. This is explained by considering that water with large amounts of humic matter contains high concentrations of submicron particles (the humic particles). By providing a flocculation period through the use of a flocculation basin or otherwise, the humic particles after coagulant (cationic polymer) addition can aggregate to larger sizes. This causes a reduction in the number and surface area of the floc particles retained in the filter compared to in-line filtration. The larger particles through flocculation produce less head loss development and longer filter runs.

A flocculation period is recommended in direct filtration since it will

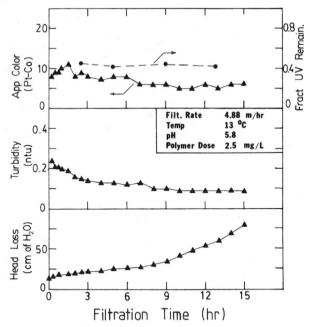

Figure 8.15. In-line direct filtration of the Glenmore Reservoir with a cationic polymer at pH 5.8. Raw water quality: 0.74 ntu turbidity, 38 Pt-Co units apparent color, 0.105 cm^{-1} UV (254 nm), 3.2 mg/L NPTOC, and 175 μg/L TTHMFP.

lengthen filter runs, especially for waters containing submicron particles (humics, viruses, asbestos, etc.). The flocculation tank also provides the water plant operator with time to adjust chemical (coagulant) dosages for direct filtration when changes in raw water quality occur.

8.4.4 Alum

Alum is frequently used in direct filtration practice either with or without a polymeric filter aid. Its use is generally limited to high-quality raw waters of low turbidity and low TOC to avoid high alum dosages.

Alum can be quite effective in treating a colored water by direct filtration, as illustrated by the results in Figure 8.16 for the Grasse River. Excellent performance is noted in terms of filtered water turbidity and color. However, the alum dose is relatively high at 14 mg/L, even at pH 5.5, causing high head loss development and a short filter run. This filter experiment was conducted during the winter when the TOC concentration of the Grasse River was relatively low, 4.1 mg/L.

In general, alum dosages are high for the Grasse River due to the high color and high TOC. Under summer water quality conditions, alum dosages at pH 5.5 can exceed 20 mg/L. Alum works well in treating the Glenmore Reservoir by direct filtration (20). To produce water of low turbidity with

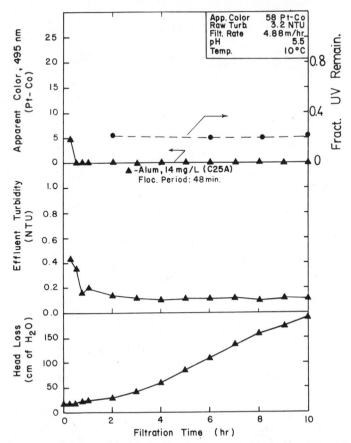

Figure 8.16. Direct filtration with flocculation of the Grasse River using alum at pH 5.5. Raw water quality: 3.2 ntu turbidity, 58 Pt-Co units apparent color, 0.168 cm^{-1} UV (254 nm), 4.1 mg/L NPTOC, and 252 µg/L TTHMFP.

low concentrations of TOC and THM precursors, alum dosages of about 10 mg/L at pH 5.5–6 are necessary. Even under well-controlled conditions of pH and applied alum dosages, direct filtration is difficult since precipitation of aluminum hydroxide can occur. Parallel reactions described in Section 8.2.3.3 and Figure 8.4 illustrate the problem. Treatment above pH 6 or treating waters with TOC above 4–5 mg/L probably results in the precipitation of aluminum hydroxide. The additional suspended solids applied to the filters would cause short filter runs.

8.4.5 Removals of NPTOC and THM Precursors

8.4.5.1 General

Average percent removals for UV (254 nm), NPTOC, and THM precursors (TTHMFP) achieved by direct filtration, using either a cationic polymer as

TABLE 8.11
Average Percent Removals by Direct Filtration

Percent Removals	Cationic Polymer		Alum	
	Grasse River ($n = 18$)	Glenmore Reservoir ($n = 8$)	Grasse River ($n = 7$)	Glenmore Reservoir ($n = 9$)
Total UV	58 (6.2)	51 (8.7)[a]	76 (4.3)	71 (7.8)
NPTOC	37 (6.0)	39 (12.2)	57 (9.7)	56 (17.6)
TTHMFP[b]	42 (8.8)	41 (5.1)	59 (8.2)	59 (9.1)

[a] Standard deviation.
[b] 7-day, pH 7.5, 20°C.

the sole coagulant or alum are summarized in Table 8.11. These data are for pH conditions of 5.2–7.5, filtration rates of 4.88–14.6 m/h, and for in-line direct filtration versus direct filtration with flocculation. Filtration rate, in-line versus flocculation, and water temperature had little effect on performance. A slight improvement in removal efficiencies can be achieved with cationic polymers by operating at lower pH conditions as illustrated by the data in Figure 8.17.

The data in Table 8.11 demonstrate that cationic polymers remove approximately 40% of the THM precursors. Alum in direct filtration yields higher removals with THM precursor reductions of approximately 60%. It is interesting that the removal efficiencies of each coagulant do not depend on the water source. The Grasse River and Glenmore Reservoir are different type waters, yet the removal efficiencies are approximately the same. A discussion of the applicability of the two coagulants in conventional water treatment and direct filtration is presented in Section 8.5.

8.4.5.2   UV as a Surrogate Parameter

Edzwald et al. (19) have shown that UV (254 nm) absorbance is an excellent surrogate parameter for monitoring direct filtration pilot plants. For pilot plant studies using Grasse River water, UV explained 90% of the variation of the TTHMFP data and 86% of the variation of NPTOC. For Glenmore Reservoir pilot plant data, it explained 89% of the variation of TTHMFP and 72% of the variation of NPTOC. An advantage of using UV to monitor NPTOC and TTHMFP in pilot plant studies is that it can provide an instantaneous assessment of performance due to any raw water quality changes or changes in pilot plant treatment or operation.

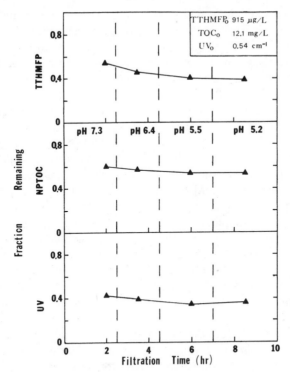

Figure 8.17. Direct filtration of the Grasse River with a cationic polymer: Effect of pH on removals of TTHMFP, NPTOC, and UV (254 nm).

8.4.5.3 Performance of Small-Scale and Large-Scale Pilot Plants

Pilot plant studies of direct filtration were conducted on both a small-scale (filter columns of 4.45 cm i.d.) and a large-scale (filter bed surface of 0.372 m^2). An important finding was that the small-scale pilot plant gave similar results to the large-scale plant in evaluating various coagulants and coagulant dosages, filtered water turbidity and color, and removals of organic matter and THM precursors. This was true for both the Grasse River and the Glenmore Reservoir. This is important since small-scale pilot plants are cheaper to construct and operate. Large-scale pilot plants are necessary, however, to evaluate length of filter runs, head loss development, and backwashing.

8.5 SUMMARY

Alum coagulation is an effective treatment process for the removal of TOC and THM precursors from water supplies containing naturally occurring organic matter, especially humic materials. The results of field studies on two water supplies support the view that optimum pH for alum coagulation occurs at pH 5–6. Two coagulation mechanisms are described in the chapter. The mechanism depends on pH, alum dose, and raw water TOC concentration. When alum coagulation is practiced at pH 5–6 on waters containing 4–5 mg/L TOC or less, the mechanism involves precipitation of aluminum-humate particles. At higher TOC concentrations or higher pH conditions, the removal of dissolved organic matter occurs by adsorption on aluminum hydroxide.

The field studies reported in this chapter show that conventional-type water treatment facilities are effective in removing TOC and THM precursors and in controlling THM formation. The two water supplies examined achieved similar removals (approximately 70% for TOC and THM precursors); yet, the two supplies are contrasting systems. The Grasse River is a highly colored water supply in which the TOC and 7-day TTHMFP reached 14.7 mg/L and 1108 μg/L, respectively. The Glenmore Reservoir, on the other hand, is a protected upland reservoir with much lower concentrations of organic matter and THM precursors. Both plants use alum coagulation near optimum pH conditions. It is noted that both water supplies are low in alkalinity and hardness.

THM formation through a treatment plant depends on raw water quality variables, water plant operating variables, and plant design variables. Both plants are effective in controlling THM formation because of the effectiveness of the coagulation process, pH conditions during treatment, and their chlorination practice. Where prechlorination is practiced, low doses are used. Seasonal problems with Fe and Mn in the Glenmore Reservoir require use of a preoxidant. Substitution of potassium permanganate for chlorine significantly reduced THM formation.

Direct filtration with cationic polymers as the sole coagulant is a feasible treatment process. Waters containing high color can be treated yielding excellent filtered water quality in terms of turbidity and color. Approximately 40% of the TOC and THM precursors are removed by direct filtration with cationic polymers. This process is not as effective as direct filtration with alum or conventional water treatment employing alum in the removal of TOC and THM precursors. In practice, the performance of direct filtration must be evaluated through pilot plant studies. The following guidelines are proposed, however, regarding direct filtration with cationic polymers as the sole coagulant. Direct filtration is a potentially feasible method of treatment for waters of low to moderate TOC and TTHMFP levels, say mean annual TOC of 4–5 mg/L and TTHMFP of 350–400 μg/L. For waters with higher concentrations, conventional water treatment is preferred for removals of TOC and THM precursors and control of THM formation.

UV (254 nm) absorbance is a good surrogate parameter for TOC and THM precursors. It can be used to monitor their removal in laboratory studies, pilot plants, and full-scale treatment plants.

ACKNOWLEDGMENTS

The author gratefully acknowledges the research funding of Allied Chemical Company (Grant No. 328233) and the U.S. Environmental Protection Agency (Cooperative Agreement CR 807034). The cooperation and assistance of Dr. R.E. Highsmith of Allied and Dr. Gary S. Logsdon of the EPA are greatly appreciated. In the preparation of this chapter, funding was provided by the Office of Research and Development of the EPA under Grant R810492 (Donald F. Carey, Project Officer). The content and conclusions are the author's views and do not necessarily reflect the views and policies of Allied or the EPA. Mention of trade names or commercial names does not constitute endorsement or recommendation for use. The author thanks W.C. Becker, B. Gong, R.E. Hubel, A.J. Laffin, C.A. McGowan, S.J. Tambini, and K.K. Wattier for their contributions to this work.

REFERENCES

1. Rook, J.J., *Water Treatment Examination, 23,* 234, 1974.
2. Bellar, T.A. Lichtenburg, J.J., and Kroner, R.C., *J. Am. Water Works Assoc., 66,* 703, 1974.
3. Symons, J.M., Bellar, T.A., Carswell, J.K., DeMarco, J., Kropp, K.L., Robeck, G.G., Segar, D.R., Slocum, C.J., Smith, B.L., and Stevens, A.A., *J. Am. Water Works Assoc., 67,* 634, 1975.
4. Federal Register, National interim primary drinking water regulations; control of trihalomethanes in drinking water; final rule, *44,* (No. 231), 68624, Nov. 29, 1979.
5. Federal Register, National interim primary drinking water regulations; trihalomethanes, *47* (No. 44), 9796, March 5, 1982.
6. Black, A.P., and Willems, D.G., *J. Am. Water Works Assoc., 53,* 589, 1961.
7. Black, A.P., and Christman, R.F., *J. Am. Water Works Assoc., 55,* 897, 1963.
8. Hall, E.S., and Packham, R.F., *J. Am. Water Works Assoc., 57,* 1149, 1965.
9. Mangravite, F.J., Jr., Buzzell, T.D., Matijevic, E., and Saxton, G.B., *J. Am. Water Works Assoc., 67,* 88, 1975.
10. Babcock, D.B., and Singer, P.C., *J. Am. Water Works Assoc., 71,* 149, 1979.
11. Kavanaugh, M.C., *J. Am. Water Works Assoc., 70,* 613, 1978.
12. Semmens, M.J., and Field, T.K., *J. Am. Water Works Assoc., 72,* 476, 1980.
13. Randtke, S.J., and Jepsen, C.P., *J. Am. Water Works Assoc., 73,* 411, 1981.
14. Chadick, P.A., and Amy, G.L., *J. Am. Water Works Assoc., 75,* 532, 1983.
15. Dempsey, B.A., Ganho, R.M., and O'Melia, C.R., *J. Am. Water Works Assoc., 76,* 141, 1984.
16. Fleischacker, S.J., Johnson, D.E., and Randtke, S.J., *J. Am. Water Works Assoc., 75,* 132, 1983.
17. Reckhow, D.A., and Singer, P.C., *J. Am. Water Works Assoc., 76,* 151, 1984.
18. Symons, J.M., Stevens, A.A., Clark, R.M., Geldreich, E.D., Love, O.T. Jr., and DeMarco, J., *Treatment Techniques for Controlling Trihalomethanes in Drinking Water,* EPA-600/2-81-156, Cincinnati, Ohio, 1981.
19. Edzwald, J.K., Becker, W.C., and Wattier, K.L., *J. Am. Water Works Assoc., 77* (No. 4), 122, 1985.
20. Edzwald, J.K., Removal of trihalomethane precursors by direct filtration and conventional treatment, EPA Research Report EPA-600/2-84-068, Cincinnati, 1984.
21. Snodgrass, W.J., Clark, M.M., and O'Melia, C.R., *Water Res., 18,* 479, 1984.

22. Davis, J.A., and Gloor, R., *Environ. Sci. Technol., 15,* 1223, 1981.
23. Glaser, H.T., and Edzwald, J.K., *Environ. Sci. Technol., 13,* 299, 1979.
24. Scheuch, L.E., and Edzwald, J.K., *J. Am. Water Works Assoc., 73,* 497, 1980.
25. Edzwald, J.K., Becker, W.C., and Tambini, S.J., Aspects of direct filtration in treatment of low turbidity, humic waters, in Weiler, R., and Janseens, J.G. (eds.), *Proceedings of the International Symposium, Water Filtration.* Published by the Flemish Chapter of the Filtration Society, Antwerp, Belgium, 1982.
26. Edzwald, J.K., and Lawler, D.F., Mechanisms of particle destabilization for polymers in water treatment, in Proceedings of Amer. Water Works Assoc. Seminar, *Use of Organic Polyelectrolytes in Water Treatment,* Am. Water Works Assoc., Denver, CO, 1983, pp. 17–35.
27. Rebhun, M., Fuhrer, Z., and Adin, A., Contact flocculation-filtration of organics colloids, in Weiler, R., and Janseens, J.G. (eds.), *Proceedings of the International Symposium, Water Filtration,* Flemish Chapter of the Filtration Society, Antwerp, Belgium, 1982.
28. Rebhun, M., Fuhrer, Z., and Adin, A., *Water Res., 18,* 963, 1984.
29. Habibian, M.T., and O'Melia, C.R., *J. Environ. Engr. Div., ASCE, 101,* 567, 1975.

CHAPTER 9

Removal of Organic Contaminants in Drinking Water by Adsorption

Francis A. DiGiano

Department of Environmental Sciences and Engineering
School of Public Health
University of North Carolina at Chapel Hill
Chapel Hill, North Carolina

9.1 TREATMENT OBJECTIVES

Activated carbon is generally recognized as an effective and economical adsorbent for use in water treatment. Its application to date, however, has been limited in large measure to the removal of taste and odor. While providing a very effective solution to this problem, activated carbon is certain to be used on a broader scale for removal of a wide spectrum of organic contaminants. The organic chemicals of concern include pesticides from agricultural runoff, synthetic organic chemicals (SOCs) from industry, and naturally occurring precursors to chlorinated compounds generated in water treatment itself.

The treatment objectives of adsorption may be expressed either in terms of the desire to remove a specific organic contaminant for which a maximum contaminant level (MCL) has been set or a group of contaminants represented by a surrogate parameter. MCLs already exist for four pesticides—endrin, lindane, toxaphene, and methoxychlor—and two herbicides—2,4-D and 2,4,5-TP (Silvex)—and for THMs (collectively representing the chloro- and bromomethane compounds). Other MCLs may follow for synthetic organic chemicals of industrial origin. The removal of surrogate parameters, for example, trihalomethane formation potential (THMFP), total organic carbon (TOC), total organic chlorine (TOCl), and UV absorbance, may provide equally valid ways of expressing treatment objectives. Moreover, several treatment objectives, as defined by specific chemicals and surrogate parameters, may have to be met simultaneously.

9.2   IMPORTANT ADSORPTION CONCEPTS

9.2.1   Service Time

The adsorption process involves transport of contaminants to adsorption sites within the macro- and micropore structure of the activated carbon. Both the rate of transport and the final position of equilibrium between the solid and liquid phase concentrations must be considered in process design. These act together to determine the service time of the adsorption bed, that is, the maximum time of operation before the product water exceeds some specified limit of contaminant concentration.

Figure 9.1 illustrates the importance of service time as a process design criterion. In this hypothetical example, three different contaminant control objectives—removal of taste and odor, TOC, and specific SOCs—are presented. Each of these gives a different definition of service time for the same empty bed contact time (EBCT). Generally speaking, if taste and odor control is the treatment objective, service time is very long (Fig. 9.1a) because the components producing taste and odor are strongly adsorbed and are present in very low concentrations. On the other hand, TOC as a surrogate parameter is not well adsorbed and service time is therefore shorter (Fig. 9.1b). The definition of service time is more difficult to establish when deal-

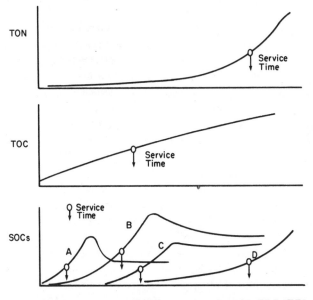

TIME OF BED OPERATION OR VOLUME TREATED

Figure 9.1. Examples of three criteria used to define the service time of a GAC bed: threshold odor number (TON); total organic carbon (TOC); and synthetic organic chemicals (SOCs), as illustrated by simultaneous presence of components A, B, C, and D.

ing with specific SOCs (Fig. 9.1c). In this example, the effects of competitive adsorption and changing feed composition are seen in the complex pattern of effluent concentrations. If component A has the most restrictive MCL, service time is relatively short. On the other hand if components A, B, and C are not of concern from a health standpoint but component D is, then service time is extended and better use is made of the adsorptive capacity of the bed.

9.2.2 Adsorption of THM Precursors versus SOCs

A sharp distinction exists between adsorption of THM precursors and SOCs. THM precursors are ill-defined, macromolecular structures. Their large size causes slow adsorption and may even restrict entry into some of the pores. Moreover, because of their rather polar nature, these structures have relatively low adsorption affinity. In contrast, most of the SOCs of concern are of much lower molecular weight and easily transported into the pore structures. While some of the more volatile organic compounds may not be well adsorbed, many other SOCs are strongly adsorbed, particularly if they are aromatic in nature. THM precursors also vary in their adsorbability depending on the source and type of pretreatment provided (chlorination, coagulation, or ozonation). The adsorption isotherms of various SOCs and THM precursors (expressed as TOC) are shown in Figures 9.2a and b, respectively. Well-adsorbed SOCs (Fig. 9.2a) reach surface concentrations of 10^{-3} mol/g, which for a typical molecular weight of 100 g/mol means an adsorption of 0.1 g/g and higher. In contrast, the adsorption of TOC is less than 0.05 g/g (Fig. 9.2b).

THM precursors and SOCs also differ markedly in concentrations found in surface waters and this affects the breakthrough pattern observed in granular activated carbon (GAC) beds. The THM precursors, typically measured as TOC, may be present in concentrations ranging from 1 to about 10 mg/L (and even higher in special cases such as swampy waters). SOCs are in the range of 1–100 μg/L. For this reason the SOCs are often referred to as micropollutants. The high concentration of THM precursors coupled with their low adsorbability cause the sorptive capacity of GAC beds to be exhausted relatively quickly. In contrast, the strong adsorption of most SOCs and their low concentrations means that sorptive capacity is saturated relatively slowly.

9.2.3 Competitive Adsorption

Although adsorption principles have evolved from study of single-component systems, multicomponent systems are the rule in water treatment. The term "competitive adsorption" is often used to describe how rate and equilibrium are modified by the presence of other components. Three categories of competitive adsorption are identifiable: (1) among SOCs; (2) between SOCs and TOC; (3) among the molecular weight fractions comprising TOC.

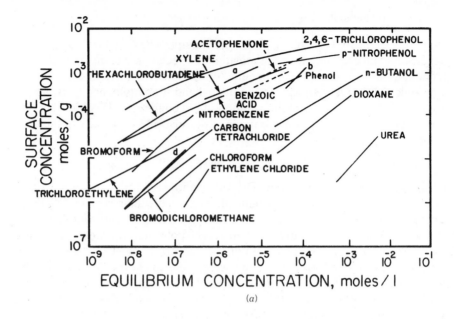

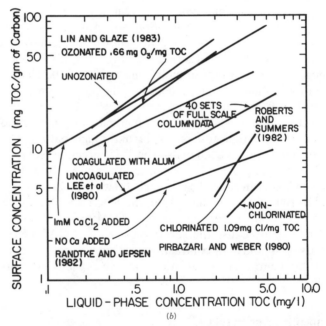

Figure 9.2. (*a*) Adsorption isotherms for SOCs; (a) (di)-*n*-butyl phthalate; (b) bis (2-chloroethyl) ether; (c) dimethyl phthalate; (d) dibromochloromethane (1). [Reprinted from Journal AWWA, Vol. 72, No. 1 (January 1980), by permission. Copyright © 1980, The American Water Works Association.] (*b*) Summary of adsorption isotherms found in the literature for humic substances before and after typical chemical treatments (2). [Reprinted from AWWA Seminar Proceedings *Strategies for the Control of Trihalomethanes* 1983, by permission. Copyright © 1983, The American Water Works Association.]

240

The most well-studied of these is competitive adsorption among SOCs. The actual breakthrough curves for as many as six components adsorbing simultaneously have been predicted with sophisticated mathematical modeling (3). However, the description of simultaneous adsorption of SOCs and TOC is much more difficult to predict. This is due to the large differences in concentrations and molecular structure that complicate the description of competitive equilibria and diffusional mass transport.

Humic substances, which comprise most of the TOC, may have limited access to adsorption sites. High-molecular-weight fractions of these naturally occurring organic substances are not able to enter the smaller pores, that is, the micropores. Thus activated carbons that differ in pore size distribution will also differ in ability to adsorb THM precursors. Figure 9.3 correlates the Freundlich adsorption capacity term* K of various types of humic substances with total pore volume contained within pores of less than the stated radius for several different brands of carbon. Adsorption of the lower-molecular-weight fraction of peat fulvic acid (<1000 MW) correlates with the volume of pores associated with pores of radius less than 70 nm while that of the higher-molecular-weight fraction (>50,000) correlates with the volume of pores associated with pores of a much larger maximum radius (400 nm). This suggests that the macromolecular structure of humic substances can limit access to micropores. It may also explain why competitive adsorption is not always observed between humic substances and trihalomethanes (THMs) (5).

Competitive adsorption may occur among various fractions of TOC even in the absence of SOCs. It may even be possible for some fraction to compete rather favorably with SOCs for adsorption sites, although there is little evidence of this in the literature. An empirical approach has been developed to classify adsorbability of TOC into as few as three fractions—nonadsorbable, weakly adsorbable, and strongly adsorbable (6); in more sophisticated models, the adsorbable fractions are further subdivided. An example of successful modeling of a TOC breakthrough curve for humic substances by this procedure is presented in Figure 9.4. Pretreatment by chemical coagulation or ozonation can alter the molecular weight distribution of humic substances or cause a chemical change that increases the adsorption affinity. It may also increase the competition with SOCs, although there is little experimental evidence of this effect.

9.2.4 Variability in Feed Composition

The variability in composition of the feed water to the adsorption unit with time complicates process performance predictions. Of practical concern is

---

* The adsorption capacity, K, refers to the amount adsorbed (q) when the equilibrium fluid phase concentration (C) is unity in any dimensions used to express the relationship where $q = KC^{1/n}$.

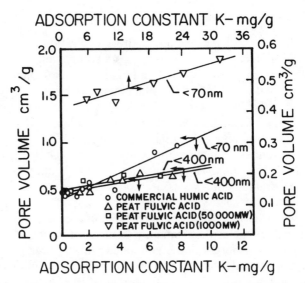

Figure 9.3. Adsorption constant of humic substances as a function of pore volume in. pores smaller than a certain radius (4). [Reprinted from Journal AWWA, Vol. 71, No. 9 (September 1979), by permission. Copyright 1979, The American Water Works Association.]

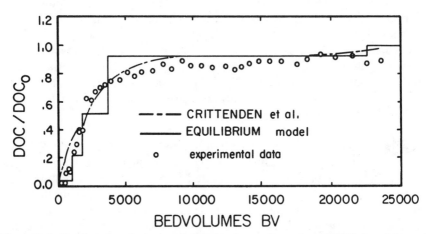

Figure 9.4. Predicted and observed dissolved organic carbon (DOC) breakthrough behavior (DOC_0 = influent concentration and DOC = effluent concentration) obtained in pilot plant column tests of GAC adsorption; source of TOC is concentrated humic substance added to tap water in feed to column (7). The Crittenden et al. model (8) includes the influence of adsorption kinetics by mass transfer resistance of the film and intraparticle diffusion on competitive adsorption; this is termed the homogeneous surface diffusion model (HSDM). The equilibrium model includes competitive equilibrium among five adsorbable components in the mixture. The simplified competitive adsorption model (SCAM) was used in both approaches for equilibrium predictions (6). [Reprinted from the *Proceedings of the ASCE National Specialty Conference on Environmental Engineering* 1984, by permission of the American Society of Civil Engineers.]

the resulting effluent concentration profile of an SOC that suddenly decreases or disappears from the influent after having been adsorbed by the bed for some time. Desorption of the SOC should be expected because the carbon particles must reequilibrate with the new solution concentration. An illustration of this effect for chloroform is presented in Figure 9.5. The unloading of the carbon particles may take a very long time. During this time, the effluent concentration can be higher than the influent concentration.

A change in composition may also mean that the degree of competition among SOCs is altered. This causes unloading of weakly adsorbed components if, for example, the concentration of a more strongly adsorbed component appears in the influent. An example of this effect is presented in Figure 9.6. The appearance of dichlorophenol (DCP) in the feed causes displacement of dimethylphenol (DMP) from sorbed phase and thus raises its concentration in the effluent. This figure also shows that good agreement is obtained between experimental data and model predictions. While encouraging, much more complex mixtures are encountered in actual plant operations and these are not as easily described by mathematical models.

The results of a pilot plant study of GAC at Rotterdam, The Netherlands (11), illustrate the principle of desorption in practice. In this plant, chlorination of the raw water supply is interrupted in winter operation. This sud-

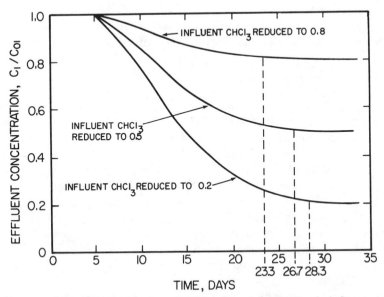

Figure 9.5. Desorption of chloroform ($CHCl_3$) due to a reduction in influent concentration where C_{01} = original influent concentration and C_1 = effluent concentration (9). [Reprinted from Journal AWWA, Vol. 75, No. 1 (March 1983) by permission. Copyright © 1983, The American Water Works Association.]

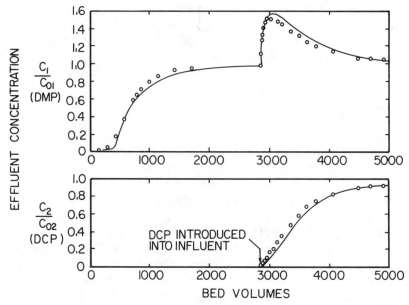

Figure 9.6. Breakthrough curves for sequential feed of dimethylphenol (DMP) and dichlorophenol (DCP) where C_{01} and C_{02} are the influent concentrations (10). [Reprinted from Journal WPCF, Vol. 56, No. 3 (March 1984), by permission. Copyright © The Water Pollution Control Federation.]

denly lowers the influent chloroform concentration to the GAC beds and causes desorption of previously sorbed chloroform as shown in Figure 9.7.

9.2.5 Microbial Activity

There is considerable debate concerning the importance of microbial activity in GAC beds (12,13). It is clear, however, that the surface of activated carbon provides for attachment of bacteria. Some argue that service time is extended due to biodegradation in the biofilm that develops around the activated carbon particle. This is evidenced by attainment of some residual TOC removal (on the order of 30%) in GAC beds after long periods of operation (12). The amount of removal will increase with increasing EBCT (although less than proportionally) and may vary seasonally (more removal in warmer months). The additional removal of TOC obtained through biodegradation may not be considered important unless it is sufficient to keep the THM precursor concentration below that which will violate the MCL for THMs.

TOC removal is the most obvious evidence of biodegradation because it is easily measured. However, much less is known about the removal of trace concentrations of SOCs in the biofilm and about interaction between the processes of adsorption and biodegradation. There are laboratory data

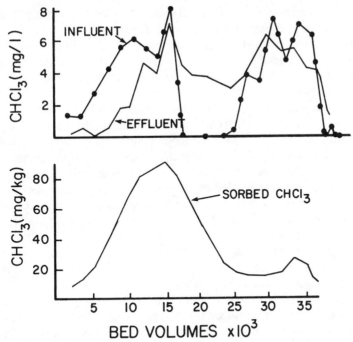

Figure 9.7. Results of pilot plant testing of GAC column at Rotterdam Waterworks in The Netherlands (11). Adsorption of chloroform (CHCl₃) was followed by desorption during period when influent chloroform decreased to below detection limit due to temporary elimination of prechlorination.

showing that microbial activity causes desorption of SOCs from internal sorption sites and biodegradation in the biofilm (13); this is termed "bioregeneration." Bioregeneration is not well understood and remains difficult to quantify in actual plant operations. It promises a way of lengthening service time by providing additional sorptive capacity for SOCs.

9.2.6 Regeneration

For GAC beds to be economical when service time is relatively short (on the order of a year or less), on-site regeneration must be included. This is a significant departure from many traditional applications of GAC for control of taste and odor where service time is long enough (on the order of 2 or 3 years) to justify use on a throwaway basis. It also represents a significant increase in complexity of plant operations.

The regeneration process consists of: drying and volatilization of sorbed organic contaminants; charring of the carbon; and activation. The objective is to minimize the losses of GAC through the process while producing an adsorbent that can perform as well as the virgin material. Some of the more important properties of GAC to control are its hardness, particle size, pore size distribution, and surface oxides (other functional groups on the surface may also be important). These properties may be related. For example, hardness can be reduced due to loss of micropores and an increase of macropores that weakens the skeletal structure of the carbon; similarly a large increase in surface oxides may also weaken the material. Pore size distribution and surface oxides may also affect adsorption properties. Moreover, surface oxides and other surface functional groups may have some role in catalytic surface reactions that produce unwanted by-products from sorbed organic contaminants (14). The reactivation process requires that the proper time and temperature (or temperature profile) be selected to give the desired properties, and tests are needed to determine if these properties have been achieved. The regenerated carbon must be able to adsorb a range of contaminants of interest and further, unwanted by-products of surface reactions should be minimized.

Although experience is somewhat limited in water utilities, regeneration has been shown to be effective. Figure 9.8 gives one practical measure of recovery of adsorptive capacity in an actual plant operation. This is the amount of UV absorbing contaminants that can be removed in batch tests as function of the adsorbent dosage. The results indicate that after five regenerations, the activated carbon is as good if not better an adsorbent than the virgin material.

9.2.7 Use of Powdered Activated Carbon versus Granular Activated Carbon

Powdered activated carbon (PAC) differs from GAC adsorption in two important respects: (1) kinetics of adsorption and (2) process performance characteristics. The kinetics of adsorption are controlled by mass transfer from the solution to adsorption space. Because of its much smaller particle diameter and much larger external surface area, mass transfer is much faster for PAC than GAC. When added to the rapid mix tank, therefore, PAC rapidly adsorbs contaminants. Nevertheless, it may not be reasonable to assume that adsorption equilibrium will be attained in the time allotted for flocculation and sedimentation (16); testing of adsorption rate is recommended.

Despite the considerably slower rate of adsorption on GAC than PAC, the fixed bed configuration provides a very large mass of GAC compared to the volume of water in the adjacent void spaces. This produces a much greater driving force for adsorption than in a PAC system. In addition, a definable mass transfer zone is created in a GAC bed. Ahead of this zone is fresh adsorbent and behind it is exhausted adsorbent. Therefore, the concentration of adsorbable contaminants in the effluent is close to zero during the early stages of operation because the mass transfer zone is contained within the bed.

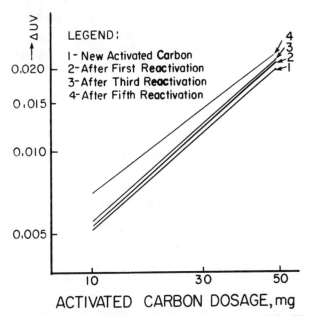

Figure 9.8. Adsorption isotherms after repeated regeneration (15). The measure of adsorption is the decrease in UV absorbance in the water after equilibrium with the activated carbon.

The process performance characteristics of PAC and GAC systems differ significantly. Dosing PAC to the rapid mix tank cannot produce a consistent effluent quality if the mix of contaminants is changing in the influent. Under this condition, the residual concentrations will be constantly changing if the PAC dosage remains fixed and the process is controlled by equilibrium. In contrast, GAC can insure low effluent concentrations regardless of the variation in influent concentration so long as the mass transfer zone is contained within the bed.

PAC does offer one significant advantage over GAC in addition to being far less costly. That is, effluent concentration will never exceed influent concentration as is possible in a GAC bed. This is because fresh PAC is fed continuously such that equilibrium adsorption occurs only with the local solution in which the particles are suspended as they move through the plant. GAC particles, however, are fixed in bed and so experience all solution compositions; as the composition changes, re-equilibration must occur. The result is displacement of previously sorbed components when a more strongly adsorbed component appears.

PAC has been used more extensively than GAC because it can be easily dosed to the rapid mix tank. In contrast, GAC requires a separate unit operation and thus capital expenditures. However, GAC provides more positive control over process performance and enables more complete removal of contaminants than PAC over long periods of operation. The drawback to GAC is economic rather than technical. Nevertheless, GAC offers water utilities a very reasonable method for control of organic chemicals.

9.3 DESIGN CONSIDERATIONS

9.3.1 Steps in Process Design

The following steps are needed in process design:

- Select the contaminants to be removed
- Estimate the service life and, thus, the regeneration frequency of each adsorption unit
- Select the arrangement of the adsorption units
- Select the method of regeneration

Selection of the contaminants to be removed is difficult because the composition of the feed stream is largely unknown and, in fact, quite variable. Nonetheless, a decision may be made on the basis of some specific MCL that must be met or on the basis of the desire to provide generally good removal of a broad spectrum of organic contaminants. In the latter case, a surrogate parameter such as UV absorbance, TOC, or TOX may be useful.

The arrangement of beds is usually parallel although a combination of parallel and series may be better. Series operation becomes advantageous when the mass transfer zone is long, that is, for slow adsorption kinetics as illustrated in Figure 9.9. In this case, at the point of exhaustion the lead bed is regenerated and replaced with column 2, column 2 is replaced by column 3, and column 3 by fresh GAC to maximize the utilization of sorptive capacity.

The most common methods of regeneration are the multiple hearth fur-

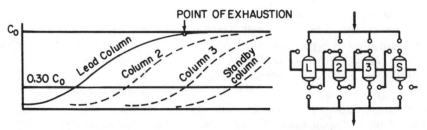

Figure 9.9. Hypothetical breakthrough curves for three beds operating in series with service time defined by the effluent concentration reaching 0.30 of the influent concentration (C_0). When the service time is reached, Column 2 is switched to the Lead Column position, Column 3 to the Column 2 position, and the Standby Column to the Column 3 position; the Lead Column is regenerated having reached the point of exhaustion of sorptive capacity.

nace and the fluidized bed furnace (17). The fluidized bed is a relatively recent innovation. It offers greater flexibility in process control by allowing the time and temperature (or temperature gradient) of reactivation to be more easily adjusted. It has been used successfully in the Federal Republic of Germany and has been pilot plant tested extensively in the United States. However, there is little guidance available concerning the best combination of operating conditions to use. This is because experience has been limited and the criteria for effective regeneration are not well established. For example, carbon manufacturers use the molasses number and the iodine number as measures of adsorptive capacity. These are thought to represent the extremes in molecular weight and type of sorbing components. However, they are not a substitute for use of adsorption data for specific components of interest in each specific situation. Minimizing the loss of carbon per cycle and preventing the formation of unwanted by-products of surface reactions involving sorbed species are equally important concerns of regeneration. The time, temperature, and combination of gases and steam used in regeneration may control these properties.

9.3.2 THM versus THM Precursor Removal

In contrast to moving the point of chlorination or using an alternative disinfectant, adsorption is a capital intensive alternative. It must therefore be considered very carefully. THMs are very poorly adsorbed as are their precursors, the humic substances. However, adsorption is more easily justified for removal of THM precursors (THMFP) than for THMs. In either case, service life is far shorter than if many other synthetic organic contaminants were selected for control. It is also clear that the sorptive capacity of GAC is being underutilized.

PAC may be selected over GAC in the special case where high THMFP values are only experienced seasonally and the percent removal required of the process is not large. Figure 9.10 is an adsorption isotherm for THMFP obtained from a New Orleans, LA study of PAC (18). The dosage W of PAC required to achieve an equilibrium THMFP concentration C_f is obtained by:

$$W = \frac{C_0 - C_f}{q} \tag{9.1}$$

where C_0 is the initial THMFP concentration and q, the sorbed phase concentration in equilibrium with C_f. The PAC dosage can be calculated using Figure 9.10 for the hypothetical case where all the THMFP is due $CHCl_3$. In this case, the PAC dosage to reduce the THMFP concentration from 130 $\mu g/L$ to 100 $\mu g/L$ (the MCL) is:

$$W = \frac{(1.08 - 0.83) \ \mu mol/L}{0.011 \ \mu mol/mg}$$

$$W = 23 \ mg/L$$

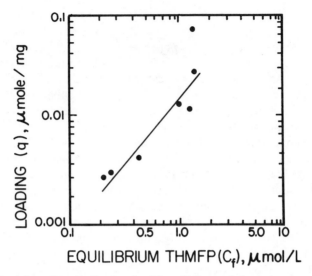

Figure 9.10. Adsorption isotherms for New Orleans, LA study sponsored by the U.S. EPA using PAC to remove THM precursors from Mississippi River water (18). THMFP (THM formation potential) conditions: pH 10; 29°C (85°F); storage time 5 days. [Reprinted with permission of authors of Ref. 18. (Note: the abscissa scale is incorrect in the Ref. 18 (Fig. 83).]

However, higher influent THMFP would require PAC dosages that may be uneconomical; for example, if the influent THMFP were 200 μg/L (as $CHCl_3$), the same calculation procedure gives a dosage of 76 mg/L to achieve 100 μg/L (as $CHCl_3$).

The best procedure to establish the PAC dosage is to add PAC in the standard coagulation jar test and not to rely solely on separate determination of adsorption isotherms for either the raw water or water after chemical coagulation as is illustrated above. In this way, the combined effect of adsorption and chemical coagulation on precursor removal is determined. Equally important, the conditions of chemical coagulation that will provide for effective capture of PAC in the chemical sludge can be checked.

The design considerations for removal of THMFP in a GAC bed are more complex than for PAC. Of most concern is the establishment of service time from the breakthrough curve. This is determined in large part by the feed concentration of the component being adsorbed, EBCT, and the length of the mass transfer zone. There is a minimum EBCT necessary to contain the mass transfer zone. For TOC or THMFP, diffusional transport through the porous carbon structure is very slow and thus the mass transfer zone is long. The shape of the mass transfer zone is reflected in the shape of the breakthrough curve. An example of a breakthrough curve for THMFP is shown in Figure 9.11. Service time is about 100 days when based on reaching an effluent concentration of 100 μg/L of terminal TTHM if the EBCT is 23 min.

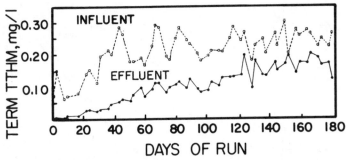

Figure 9.11. The removal of terminal TTHM (total THM) by a postfilter GAC adsorber (EBCT = 24 min) at Jefferson Parish, LA, study sponsored by the U.S. EPA (18). The terminal TTHM is a measure of THM precursors. [Reprinted with the permission of the authors of Ref. 18.]

9.3.3 Effect of Prechlorination

All indications from research are that chlorination should not precede GAC adsorption. Aside from the obvious detrimental effect of THM formation (THMs are weakly adsorbed), prechlorination can lead to unwanted reactions with synthetic organic chemicals. The compounds formed by chlorination of the solution may be less adsorbable than the original contaminants. As important is the reaction of chlorine with sorbed organic compounds (14) which then produces new and possibly more complex compounds that may be of significant health concern. The identification and fate of these newly formed compounds remain significant issues to be addressed.

9.3.4 Effect of EBCT and Application Rate

Figure 9.12 is an example of the relationship between EBCT and service time obtained from pilot plant studies at the Andijk Waterworks in The Netherlands (16). In this case the service time criterion is 50% removal of TOC. In general, a linear relationship is expected between service time and EBCT. However, there is also a minimum EBCT that is required to establish the so-called "constant pattern" of the mass transfer zone which defines its length and shape. Figure 9.12 indicates that the increase in service time is more than proportional to EBCT. This may be attributable to additional removal of TOC through biodegradation at longer EBCT.

A different relationship will be obtained between service time and EBCT for each treatment objective. That is, the relationship depends on the components or surrogate parameters being used to judge service time and on the acceptable effluent concentration for any of these.

Application rate is typically between 2 and 5 gpm/ft^2 (or 4.8 and 12 m/h respectively, when expressed as a filtration velocity in metric units). Within this practical range, application rate does not usually have much influence

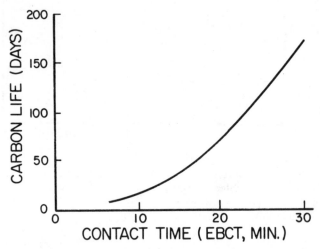

Figure 9.12. Service time of GAC bed as a function of EBCT at the Andijk Water-works of the PWN in The Netherlands (12). The criterion for service time is 50 percent reduction in TOC. [Reprinted from *Activated Carbon in Drinking Water Technology,* by permission. Copyright © 1983, American Water Works Association Research Foundation.]

on the shape of the breakthrough curve from which the service time is obtained. Obviously, the greater the application rate, the deeper the bed to achieve the same EBCT. The choice of application rate could, therefore, be influenced by the plant hydraulics and pumping costs. Although no definitive criteria exist for selecting the application rate, a more thorough analysis is deemed prudent through mathematical modeling of breakthrough behavior.

9.3.5 Backwashing Frequency

There are no specific operating rules governing the frequency of backwashing. GAC beds are usually preceded by filtration and therefore head loss accumulation is unlikely to control the backwashing policy. Plants in The Netherlands backwash as infrequently as possible to avoid disturbing the bed. Bed disturbance can cause rearrangement of particles such that unwanted shifts occur in the shape and position of the mass transfer zone. Moreover, backwashing causes particle loss and thus increased sludge handling. On the other hand, backwashing has been found necessary in some plants to control sloughing of attached bacteria into the product water (16). A typical, but arbitrary, backwashing frequency is once per week.

9.3.6 Examples of Design Criteria

Design criteria have evolved as a result of research and practical experience. Table 9.1 is an example from the Federal Republic of Germany (19). The

TABLE 9.1
Design Criteria for GAC Filters in Germany

Treatment Method	Filtration Velocity (m/h)	Bed Height (m)	Empty Bed Retention Time (min)	Throughput Ratio Before Regeneration[a] (m³/m³)
Dechlorination	25–35	2	2–4	1,000,000
Taste and odor removal	20–30	2–3	8–10	100,000
Organics removal	10–15	2–3	8–15	25,000
Biological acti- vated carbon	8–12	2–4	15–25	100,000

Source: Ref. 19.
[a] m³ of water processed per m³ of activated carbon.

design criteria are: application rate (filtration velocity), bed height, empty bed contact time, and throughput ratio before regeneration (i.e., service time). Four specific objectives of treatment are indicated: dechlorination, taste and odor control, organics removal, and biological activated carbon. Taste and odor control requires a fairly short EBCT and gives a relatively long service time. This is expected because taste and odor causing compounds are present in very low concentrations and are very strongly adsorbed. In contrast, the use of GAC beds for organics removal infers some service life criterion, such as removal of UV absorbing substances. This results in a much shorter service time. It is difficult to assert that Table 9.1 actually represents the translation of research findings into practice. For example, there is still considerable debate over the practical implications of microbial activity on GAC (biological activated carbon) and whether this process can be designed and controlled to extend service time. Table 9.1 is based more on practice than on an engineering analysis from process fundamentals. Therefore, it is a guideline for design but certainly not a substitute for process analysis.

Table 9.2 summarizes the design criteria used in various waterworks in The Netherlands (16). Included are the type of filter (either pressure or open), the EBCT, regeneration frequency, regeneration criteria, and regeneration losses. There is considerable variation in EBCT and regeneration frequency. EBCT ranges from 12 to 37 min and regeneration frequency from 10 to 24 months. The regeneration criteria are also different; several are expressed as either a maximum loading of TOC or trichloroethylene on GAC. The assumed or measured regeneration losses range from 5 to 10%; these are typical of U.S. experience as well. A significant fraction of this loss may occur in transport of GAC to and from the reactivation facility.

TABLE 9.2
GAC Design Criteria in The Netherlands

Plant	Type of Filters	EBCT (min)	Regeneration Frequency	Bed Volumes Treated	Regeneration Criteria	Cost as Percent of Total Cost	Regeneration Loss (%)
Kralingen	Pressure	12	15 mos.	55,000	Not given	8.6	5–10
Andijk	Open	35	11–12 mos.	14,000	70 µg/L effluent THM	27.0	10
Zevenbergen	Open	37	12–24 mos.	15,000	85 g TOC/kg	26	10
Ouddorp	Open	11–55	Nonregenerable		80 g TOC/kg		
Zeist	Pressure	12	10 mos.	21,000–38,000	Not given <1.0 µg/L effluent TCE 12–12 g TCE/kg	7	5%
Hilversum	Pressure	17–20	12	31,000	<1 µg/L effluent TCE 43 g TCE/kg	Not given	10%
Zwolle	Pressure	12	24[a]	50,000[b]	bis-chlorisopropylether		8–10%
Dordrecht	Pressure	9	Not built yet				

Source: Ref. 16.
[a] Anticipated from pilot plant data.
[b] From pilot plant data.

The influence of preozonation on process design criteria is not yet well understood. Ozonation is capable of cleaving the macromolecular structures comprising humic substances. It may also produce more polar by-products; this is true for the trace organic contaminants in the mixture as well as for the humic substances. The most often observed effect is an increase in microbial activity as measured by greater, steady-state removal of TOC across the adsorption bed after long periods of operation (i.e., after several months); however, the increase in biodegradability of TOC is not always significant (20). Associated with more active microbial degradation is the problem of aftergrowth of bacteria in the distribution system. In this sense, the GAC system must necessarily take on the characteristics of a biological process in order to remove the biodegradable organic matter and limit aftergrowth.

Although not well documented, ozonation may also change the adsorbability of both humic substances and trace organic contaminants. The expected result is that adsorption will decrease because of the increase in polarity; however, cleavage of macromolecules may also increase access to smaller pores in the carbon structure.

Table 9.3 is a summary of findings from a pilot study of ozonation performed at the Philadelphia water works (21). The results of parallel operation of a GAC bed with and without preozonation are used to develop design criteria of bed life, that is, service time, based on removal of TOC and volatile halogenated organics (VHO). The data show that if TOC removal is the design criterion, preozonation produces a significant extension of bed life. However, there is little or no advantage if VHO removal is the design criterion. Moreover, even if preozonation does produce a longer bed life, its costs may outweigh the cost savings accompanying less frequent regeneration. Such pilot plant studies can provide useful design information. However, without also conducting controlled laboratory studies, there will always be questions raised about the generality of pilot plant findings. This is very much the case for preozonation: fundamental understanding of the effects on biodegradability and adsorbability is lacking and thus the implications for process design remain unclear.

9.3.7   Use of Adsorption in Treatment of Ground Water

The treatment objectives are quite different for ground water than surface water. Surface waters may need to be treated to remove a wide variety of SOCs and perhaps TOC, whereas treatment of ground waters may only be necessary because of the presence of one or two SOCs (most often these SOCs are volatile cleaning solvents). Moreover, it is more likely that the quality of ground water may not vary as much as surface water. This not only simplifies analysis of process design but also limits consideration of such effects as desorption and displacement.

Although air stripping is an economical method of removing VOCs from

TABLE 9.3
GAC Bed Life and Use Rate for Different Design Criteria
With and Without Preozonation

	GAC			Ozone-GAC		
	Bed Life	Use Rate		Bed Life	Use Rate	
Criterion	Days	g/m^3	lb/mil gal	Days	g/m^3	lb/mil gal
<0.5 mg TOC/L	38	110	916	57	73	611
<1.0 mg TOC/L	69	61	505	148	28	235
Chloroform						
<1 µg/L	35	119	995	32	131	1088
<10 µg/L	51	82	683	44	95	791
<100 µg/L	78	54	446	80[a]	52	435
Dichlorophenol						
<1 µg/L	109	38	319	113	37	308
<5 µg/L	203	21	172	272	15	128

Source: Ref. 21.
[a] Highest effluent level recorded on day 80 when chloroform concentration was 97.6 µg/L.

ground water, some situations may require that air stripping be followed by adsorption to achieve acceptable product water. This is because economic air stripping may be limited in practical terms to about 90% removal, regardless of the concentration entering the unit. Thus if the feed concentration is 500 µg/L, the residual concentration will be 50 µg/L after air stripping, which may still violate the MCL.

9.4   DATE ACQUISITION FOR PROCESS DESIGN

9.4.1   Alternatives to Pilot Plant Studies

Pilot plant studies have been encouraged by concern over the uniqueness of the mix of contaminants to be removed at each site. Moreover, long-term analysis of the process is considered important in order to measure the response to seasonal variability in the quality of the feed water. Data acquisition in such pilot plant studies can be very expensive. While perhaps necessary in many situations, there are less expensive sources of data that can, if nothing else, help design a better pilot plant study.

9.4.1.1   Batch Equilibrium and Rate Tests

Before considering dynamic column testing, one should take advantage of the simplicity of batch equilibrium and rate tests to gather as much information as possible in a short time. At the very least, these data can help in the design of the column testing program by providing some estimate of service time and mass transfer resistance. At the outset of any study, however, decisions have to be made concerning which contaminants are of most interest to remove in what is usually a complex mixture and how much resolution of the mixture composition is needed, that is, how many individual components or group parameters must be measured. These decisions determine the importance of describing competitive adsorption and influence the complexity of the process design approach.

The construction of the adsorption isotherm from batch equilibrium tests is fairly straightforward. This is often referred to as the bottle point method because data are collected using small bottles (100–500 mL) to contact the adsorbent with the contaminant in solution. There is a rather large published compendium of isotherms for both specific contaminants and background natural organic matter usually expressed by TOC. As important, the test procedure itself has been refined such that true equilibrium can be approached (22).

Existing isotherm data and the guidelines for collection of new data are the logical beginning points for assessing process design. At the very least, these data can be used to make a rough calculation of service time by ignoring the kinetic effects that influence the shape of the breakthrough curve. This is even true for complex mixtures where competitive adsorption effects must be considered.

Batch rate data can be used to calculate surface diffusivities relatively easily (23). The governing mass transport model has been put into a form that allows for convenient nomagraphic solution rather than use of a computer. The surface diffusivity is an important measure of mass transfer resistance in operation of fixed beds and it is an essential piece of information for computer simulation of bed service time. Table 9.4 indicates the variation

TABLE 9.4
Typical Surface Diffusivities (cm^2/S)

TOC	10^{-10}–10^{-11}
Carbon tetrachloride	3×10^{-10}
Chloroform	1×10^{-10}
Chlorophenol	1.3×10^{-9}
Dieldrin	4.4×10^{-9}
Dichlorophenol	1.5×10^{-8}
Dichlorobenzene	8×10^{-8}

Source: Ref. 2.

observed in this design parameter. As is expected from molecular size considerations alone, the surface diffusivity of humic substances, that is, TOC, is much smaller than that of simple aromatic compounds such as DCP. The slow diffusion of humic substances explains the shape of the breakthrough curve observed so often for TOC. Rather than being S-shaped, as is the case for trace organic contaminants, the TOC breakthrough curve rises steadily (see Fig. 9.4). Thus a process design based on TOC removal gives a different effluent concentration profile than does that based on a trace organic contaminant.

9.4.1.2 Dynamic Column Tests

Dynamic column tests require very careful planning. The relationship between EBCT and service time is not well enough appreciated in practice. Once the EBCT is selected in pilot plant studies to duplicate the intended design EBCT, then service time will also be the same as design. This means that regardless of the actual size of the pilot scale adsorber units (i.e., length and diameter of the columns), the time of breakthrough for many contaminants that are present in low concentration and fairly well adsorbed could be on the order of many months, if not years.

Well-known chemical engineering principles of scale-up show how a shorter EBCT can be used in pilot plant studies to lessen the cost of data collection. This is the basis of the so-called "minicolumn" test, an example of which is given in Figure 9.13. In this test, only a few grams of GAC are packed into a small column having a diameter of about 2.5 cm. Even without sophisticated mathematical modeling, minicolumn tests can yield valuable information. The procedure allows for rapid collection of breakthrough data. In this example, just 20 days were needed to obtain complete breakthrough when influent concentration of carbon tetrachloride was about 200 μg/L. A larger-scale, column experiment, that is, with an EBCT approaching that of process design, would have required many months to obtain this same profile. This figure also shows how such column tests are used to measure the effects of background humic substances on adsorption of a trace contaminant.

If careful attention is paid to scale-up principles, it should be possible to get an estimate of service time directly from mini- or small-scale laboratory columns without use of a sophisticated mathematical model. This requires that the EBCT be decreased and the filtration velocity (application rate) increased from the full-scale design specification. A recent study showed that the bed volumes of feed (BVF) to breakthrough and the shape of the breakthrough curve obtained from a properly designed, small-scale column will be the same as that in the full-scale system (25). This permits direct calculation of service time. The scaling equations for EBCT and application

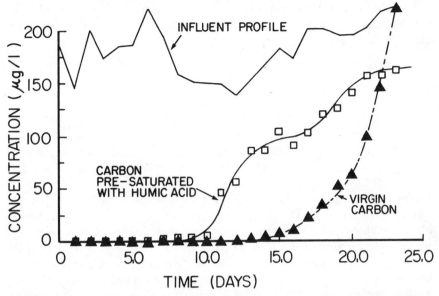

Figure 9.13. Experimental breakthrough profiles for carbon tetrachloride (CCl₄) in minicolumn adsorbers (24); effect of presaturation with humic acid on the adsorption capacity (flow rate = 5 mL/min, 2 g of 50/60 US sieve size carbon). [Reprinted from Journal AWWA, Vol. 74, No. 4 (April 1982), by permission. Copyright © 1982, The American Water Works Association.]

rate V_s are determined by particle radius in the small-scale (R_{SS}) and large-scale (R_{LS}) systems are shown below:

$$\frac{\text{EBCT}_{SS}}{\text{EBCT}_{LS}} = \left[\frac{R_{SS}}{R_{LS}}\right]^2$$

$$\frac{(V_S)_{SS}}{(V_S)_{LS}} = \frac{R_{LS}}{R_{SS}}$$

An example of the comparison between large-scale and small-scale conditions using these scaling equations follows:

	Large Scale	Small Scale
R	12 × 40 mesh	100 × 120 mesh
V_S	5 m/h	48.9 m/h
EBCT	5 min	3.14 s
BVF	66,100	66,100
Run time	231 days	2.41 days

Thus, with proper scaling the time to observe breakthrough was reduced from 231 days to 2.41 days. There are limits, however, to the amount of scaling that is practically possible in laboratory columns. For example, decreasing the GAC particle size and increasing the application rate can cause head loss to be unacceptable.

9.5   PROCESS COSTS

There are no full-scale adsorption facilities yet in operation in the United States that include on-site regeneration. Thus costs can only be projected. The U.S. EPA provides a detailed analysis of projected costs that is very useful (26). In addition, some cost data are available from existing European facilities (16).

The EPA cost analysis shows quite clearly the importance of many design parameters (26). The most general picture of total production costs is given in Figure 9.14. Total production costs include amortized capital and operation and maintenance (O&M) costs. All of the designs presented include on-site regeneration using the multiple hearth furnace. The two factors shown in Figure 9.14 that influence costs are plant capacity and reactivation frequency. Plants of less than 10 MGD capacity are very expensive, regardless of the reactivation frequency. These costs could be reduced significantly if on-site regeneration were not included. This is easily seen in Table 9.5 in which the costs associated with reactivation, the contactor, and the activated carbon are listed separately. In all size plants, regeneration contributes most to the costs. However, for the smaller plants the cost of regeneration is far too high to be practical. One solution to this problem may be to use regional regeneration facilities where the distances between plants are not

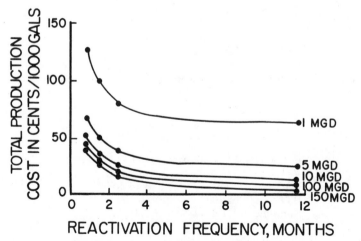

Figure 9.14. Total production cost versus reactivation frequency for postfilter adsorption (26). [Reprinted from Journal Env. E. Div., ASCE, Vol. 110, No. 4 (August 1984), by permission. Copyright © 1984, the American Society of Civil Engineers.]

TABLE 9.5
Amortized Capital and O&M Costs for GAC in Various Size Plants with On-Site Multiple Hearth Reactivation (¢/1000 Gal)

| Description | Plant Capacity—MGD (m³/day) | | | | | |
| | 1 (3800) | | 5 (19,000) | | 10 (38,000) | |
	Capital	O&M	Capital	O&M	Capital	O&M
Reactivation	50.3	17.5	14.5	8.5	9.8	7.4
Contactor	6.6	4.9	5.3	2.2	5.2	1.7
Activated carbon	3.1	0.0	2.7	0.0	2.6	0.0
Total	60.0	22.4	22.5	10.7	17.6	9.1

Source: Ref. 26.

prohibitively long. The total cost of treatment by GAC for a 38,000 m³/day (10 MGD) plant is 6.9 cents/m³ or 26.7 cents per thousand gallons.

Figure 9.14 shows that total production costs are reduced for any size plant as the reactivation frequency increases, that is, as the time between reactivations increases. This is logical because as reactivation frequency increases, the required capacity of the regeneration furnace decreases. The curvilinear nature of the relationship in Figure 9.14, however, suggests that the cost benefits are diminishing as the reactivation frequency is extended. Nevertheless, the cost savings are still significant throughout the range given. The reactivation frequency is extended for the same EBCT by making the assumption that service time is longer. That is to say, the better adsorbed the target component or the lower its concentration in the feed, the longer the service time and the less expensive is adsorption. This is why the treatment objective for the adsorption process needs to be carefully examined. The objective should not, for example, be control of THMs; these components are relatively poorly adsorbed thus producing a service time that is too short. The earlier discussion of Figure 9.1c illustrated this same point.

Given the plant capacity and the reactivation frequency, the designer may choose different configurations of the contactor volume. For example, Table 9.6 lists two options for design of a 10 MGD facility having an EBCT of 18 min. The choice is between using six contactors or 12 contactors in parallel; each option provides the same volume of activated carbon. The table indicates that the capital costs for six contactors are less than for 12. The general conclusion is that fewer contactors of larger volume are always less expensive.

The choice of EBCT also affects production costs. Increasing the EBCT to increase the service time may not decrease costs. For example, the EPA

TABLE 9.6
Influence of Size of Contactors on Their Annual Capital Cost for a 10 MGD (38,000 m³/day) Plant

Number of Contactors	Volume per Contactor in cubic feet (m³)	Total Contactor Volume[a] in cubic feet (m³)	Annual Capital Cost (Dollars/ year)
12	1,500 (42.5)	18,000 (510.0)	194,000
6	3,000 (84.9)	18,000 (510.0)	174,000

Source: Ref. 26.
[a] This corresponds to an EBCT of 18 min.

analysis shows that the production costs are only equivalent if an increase in EBCT from 18 to 22 min results in an increase of service time from 2.4 to 4.2 months. Such a disproportionate increase in service time with EBCT cannot be expected unless microbial activity accounts for additional removal of contaminants.

9.6 SUMMARY

The use of activated carbon in water treatment is sure to increase as more attention is given to control of specific organic contaminants or surrogate parameters. The beginning point for design is to decide which contaminants are targeted for control. The removal of THMFP will result in a different design and costs than, for example, the removal of a specific SOC. Moreover, it is well recognized that the mix of contaminants is unique in every plant and that seasonal, and sometimes daily, variations are important. Thus the removal objectives are site specific. While this prevents generalizing a set of design guidelines, there are still ways of predicting performance and of estimating costs.

Computer algorithms are available to predict the breakthrough behavior of fixed beds for a variety of conditions that result from competitive adsorption and variability in feed composition. These are still admittedly simplistic because they do not yet accommodate the full measure of complexity encountered under plant conditions. Nonetheless, they are helpful in understanding the general principles governing dynamic performance. Mathematical modeling, together with a fairly simple, laboratory data acquisition program, can reduce the costs of long-term, pilot plant studies. This does not necessarily mean that such pilot plant studies can be avoided. Even though expensive, they can be justified if the results produce a more cost-effective final design.

There are many factors influencing the cost of GAC treatment. Among the most important are service time, configuration of beds, and EBCT. All of these depend on accurate knowledge of adsorption equilibrium and kinetics for each application. The standard case put forth by the EPA in their cost analysis is for a relatively short service time of 2.4 months and an EBCT of 18 min. Total treatment cost (capital plus operation and maintenance) amounts to about 6.9 cents/m^3 or 26.7 cents per thousand gallons for a 38,000 cu^3/day (10 MGD) plant. This cost needs to be expressed as a percentage of the total costs for treatment. According to experience in The Netherlands, for example, GAC treatment contributes from 9 to 27% of the total for water treatment (16). A much larger percentage contribution may be expected in the United States because conventional treatment is often less complex than in European water works. Less expensive GAC facilities can be expected if service times greater than 2.4 months are realistic. However, service time is determined by the contaminant removal objectives and these must be well understood before performing a cost analysis.

REFERENCES

1. Suffet, I.H., *J. Am. Water Works Assoc., 72,* 41, 1980.
2. Crittenden, J.C., and Hand, D.W., Design considerations for GAC treatment of synthetic organic chemicals and TOC, AWWA Preconference Seminar on Strategies for the Control of Trihalomethanes, Las Vegas, NV, 1983.
3. Crittenden, J.C., Luft, P., Hand, D.W., and Friedman, G., Proceedings of the 1984 ASCE National Specialty Conference on Environmental Engineering, Los Angeles, CA, p. 370, 1984.
4. Lee, M.C., Snoeyink, V.L., and Crittenden, J.C., *J. Am. Water Works Assoc., 73,* 440, 1981.
5. Chudyk, W.A., Snoeyink, V.L., Beckmann, D., and Temperly, T.J., *J. Am. Water Works Assoc., 71,* 529, 1979.
6. Frick, B.R., and Sontheimer, H., Adsorption equilibria in multi-solute mixtures of known and unknown composition, in Treatment of Water by Granular Activated Carbon, McGuire, M.J., and Suffet, I.H. (eds.), *Adv. in Chemistry Series,* 101 ACS, Washington, DC, 1983.
7. Huebele, C., and Sontheimer, H., Proceedings of the 1984 ASCE National Specialty Conference on Environmental Engineering, Los Angeles, CA, p. 376, 1984.
8. Crittenden, J.C., Wong, B.W.C., Thacker, W.E., Snoeyink, V.L., and Hinrichs, R.L., *J. Water Poll. Control Fed., 52,* 2780, 1980.
9. Thacker, W.E., Snoeyink, V.L., and Crittenden, J.C., *J. Am. Water Works Assoc., 75,* 144, 1983.
10. Thacker, W.E., Crittenden, J.C., and Snoeyink, V.L., *J. Water Poll. Control Fed., 56,* 243, 1984.
11. Rook, J.J., Comparison of the removal of halogenated and other organic compounds by six types of carbon in pilot filters, in *Treatment of Water by Granular Activated Carbon,* McGuire, M.J., and Suffet, I.H. (eds.), *Adv. in Chemistry Series,* 202 ACS, Washington, DC, 1983.
12. Committee Report, *J. Am. Water Works Assoc., 73,* 447, 1981.

13. Chudyk, W.A., and Snoeyink, V.L., *Environ. Sci. Technol., 18,* 1, 1984.
14. McCreary, J.J., Snoeyink, V.L., and Larsen, R.A., *Environ. Sci. Technol., 16,* 339, 1982.
15. Strack, B., Operation, problems and economy of activated carbon regeneration, in EPA-600/9-76-030, translation of Report on Special Problems of Water Technology, Sontheimer, H. (ed.), U. S. EPA, 1975.
16. Activated Carbon in Drinking Water Technology, Cooperative Research Report with Keuringsinstituut voor Waterleidingartikclen, AWWA Research Foundation, Denver, CO, 1983.
17. Weber, W.J., Jr. and Bernadin, F.E., Jr., Removal of organic substances from municipal wastewaters by physicochemical processes, in *Control of Organic Substances in Water and Wastewater,* Berger, B.B. (ed.), EPA-600/8-83-011, U.S. EPA, 1983.
18. Symons, J.M., Stevens, A.A., Clark, R.M., Geldreich, E.E., Love, O.T., Jr. and De-Marco, J., *Treatment Techniques for Controlling Trihalomethanes in Drinking Water,* EPA-600/2-81-156, 1981.
19. Sontheimer, H., *J. Am. Water Works Assoc., 71,* 618, 1979.
20. Glaze, W.H., and Wallace, J.L., *J. Am. Water Works Assoc., 76,* 2, 68, 1984.
21. Neukrug, H.M., Smith, M.G., Maloney, S.W., and Suffet, I.H., *J. Am. Water Works Assoc., 76,* 4, 158, 1984.
22. Randtke, S.J., and Snoeyink, V.L., *J. Am. Water Works Assoc., 75,* 406, 1983.
23. Hand, D.H., Crittenden, J.C., and Thacker, W.E., *J. Env. E. Div. ASCE, 109,* 82, 1982.
24. Weber, W.J., and Pirbazari, M., *J. Am. Water Works Assoc., 74,* 203, 1982.
25. Berrigan, J.K., Crittenden, J.C., Friedman, G., and Hand, D.W., Scale-up of rapid small-scale adsorption tests to field-scale adsorbers: Theoretical and experimental basis, Presented at the Annual WPCF Meeting Research Symposium on Adsorption, New Orleans, LA, Oct., 1984.
26. Clark, R.M., Eilers, R.G., and Lykins, B.W., Jr., *J. Env. E. Div. ASCE, 110,* 737, 1984.

CHAPTER 10

Alternative Disinfection Processes

Alan A. Stevens

Drinking Water Research Division
Water Engineering Research Laboratory
U.S. Environmental Protection Agency
Cincinnati, Ohio

James M. Symons

Environmental Engineering Program
University of Houston-University Park
Houston, Texas

10.1 INTRODUCTION

Trihalomethanes (THMs) are formed during drinking water treatment when the free chlorine used as a disinfectant combines with THM precursors present in the water (1,2). The four THMs most commonly found in finished drinking water are chloroform, bromodichloromethane, dibromochloromethane, and bromoform (3). A fifth, dichloroiodomethane is found much less frequently (4). A source of organic precursors that reacts with chlorine is aquatic humic materials found in virtually all drinking water sources, especially surface supplies (1,5). These humic materials include both humic and fulvic acids that, when present in high concentration, account for much of the color in those waters. Natural bromide and iodide are a source of the nonchlorine halogens (6).

The concentrations of THMs observed in water treatment and distribution are heavily dependent on reaction time, humic material concentration, pH, temperature, and to a smaller extent, chlorine dose and residual (5,7). Formation is strongly dependent on chlorine dose when the reaction is chlorine limited, but little change is usually seen in chloroform concentration once the demand requirement has been met and a reasonable residual maintained (4,8).

Bromide ion, even at very low concentrations relative to the applied chlorine, strongly influences the mixture of THMs formed. This occurs as a result of rapid oxidation of Br^- to HOBr followed by reaction of this active species with the organic precursor. The reaction of bromine with humic substances is much more rapid than that of chlorine, resulting in both the disproportionate share of bromine-substituted species and approaching a final THM concentration much earlier (usually minutes rather than hours or days) (4).

One approach (4) to controlling THM concentrations is the use of a disinfectant other than free chlorine that does not participate in these reactions. Several agents are possible alternatives to free chlorine for disinfection: chloramines (combined chlorine), chlorine dioxide, ozone, potassium permanganate, hydrogen peroxide, bromine chloride, bromine, iodine, ferrate ion, hydroxide ion (high pH), silver, and UV radiation. Of these, chloramines, chlorine dioxide, and ozone are the most commonly used as disinfectants in drinking water treatment practice today and have been studied in detail. The others are less significant in use for disinfection. Discussion in this chapter is weighted according to this use pattern.

10.2   GENERAL CONSIDERATIONS: BIOCIDAL ACTIVITY

The primary reason for the use of disinfectants in drinking water is to ensure the destruction of pathogenic microorganisms during the treatment process, thereby preventing the transmission of waterborne disease. Secondarily, the presence of a disinfectant in the water distribution system helps to maintain the quality of water by preventing the growth of microorganisms.

10.2.1   Kinetics and Comparative Efficiencies (Free Chlorine, Chloramines, Chlorine Dioxide, and Ozone)

Biocidal activity by chemical disinfectants has frequently been considered a kinetic process similar to a chemical reaction, the microorganism being considered as one of the substances involved in the reaction. The effectiveness or efficiency of biocidal agents is determined by the rate at which the reaction or inactivation of the microorganism population proceeds.

For example, data from the results of a number of experiments conducted using different disinfectants at various concentrations can be used to construct plots of the type shown in Figure 10.1. As indicated, these results show the exposure times and concentrations of several disinfectants needed to produce a given level of inactivation of a given microorganism. Figure 10.1 is a composite of results obtained in one laboratory over a period of years using consistent experimental methods and microorganisms (8,9). The results show that chlorine dioxide at pH 7 and HOCl at pH 6 produce similar rates of inactivation of *Escherichia coli*. Hypochlorite ion was less effective, and monochloramine at pH 9 and dichloramine at pH 4.5 were even less so.

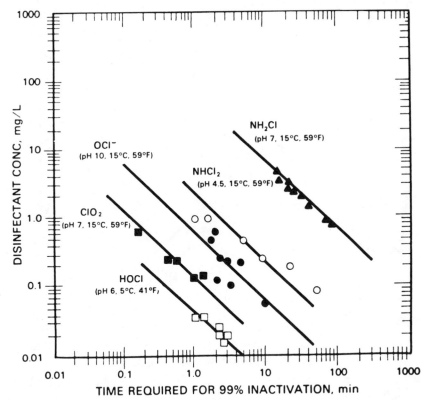

Figure 10.1. Inactivation of *E. coli* (ATCC 11229) by free and combined chlorine species and chlorine dioxide (9,10).

From the data shown in Figure 10.1, the degree of difference in efficiency between the disinfectants can be calculated and expressed quantitatively. For example, HOCl at pH 6 is 35 times as effective as OCl⁻ at pH 10. A similar plot showing virucidal efficiency of these disinfectants for poliovirus 1 is shown in Figure 10.2 (9,10). Note that higher disinfectant concentrations and longer contact times, in general, are needed for inactivation of poliovirus 1 than for *E. coli*, although the differences are less than 1 to 2 orders of magnitude, depending on the disinfectant used. Also, the difference in efficiency between HOCl at pH 6 and OCl⁻ at pH 10 is only about fourfold, and the efficiency order of the two types of combined chlorine is reversed as compared to the *E. coli* data.

Constructing similar curves when ozone is used as the disinfectant is experimentally more difficult than with the disinfectants shown, primarily because ozone is such a powerful and unstable disinfectant causing limitations on sampling times and ozone measurements. In spite of this experimental difficulty, ozone is known to inactivate microorganisms rapidly.

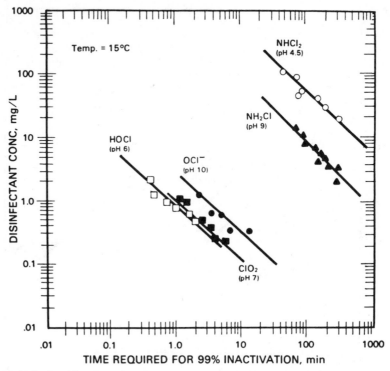

Figure 10.2. Inactivation of poliovirus 1 (Mahoney) by free and combined chlorine species and chlorine dioxide (9,10).

10.2.1.1   Microorganism Effects on Comparative Efficiencies

As shown in Figures 10.1 and 10.2, neither the order of efficiency nor the degree of difference between the disinfectants is the same for *E. coli* as for poliovirus 1. Further evidence of such differences is shown in Table 10.1. This study examined the inactivation rates of six different enteroviruses by HOCl at pH 6 and by OCl⁻ at pH 10 (11). The results indicate that the degree of difference in disinfection efficiency of HOCl at pH 6 and of OCl⁻ at pH 10 ranged from fivefold for Coxsackie A9 virus to 192-fold for ECHO 1 virus. Also note the occurrence of differences of 10-fold and 53-fold in the rates of inactivation of all other viruses by HOCl at pH 6 and OCl⁻ at pH 10.

10.2.1.2   Disinfectant Chemistry Effects on Comparative Efficiencies

Assessing the efficiencies of different free and combined chlorine species also is complicated by the nature of the chemical reactions that determine the chemical species present and the chemical equilibria established under various pH conditions. For instance, in the reaction

$$HOCl \rightleftarrows H^+ + OCl^-$$

(10.1)

TABLE 10.1
Virus Inactivation by Free Residual Chlorine

| Virus Strain | Minutes Required for 99% Inactivation at 5.0 ± 0.2°C (41 ± 0.4°F) | | |
	pH 6.0	pH 10.0	Ratio[a]
Coxsackie A9 (Griggs)	0.3	1.5	5
ECHO 1 (Farouk)	0.5	96.0	192
Polio 2 (Lansing)	1.2	64.0	53
ECHO 5 (Noyce)	1.3	27.0	21
Polio 1 (Mahoney)	2.1	21.0	10
Coxsackie B5 (Faulkner)	3.4	66.0	19

Source: Ref. 11.
[a] Time required at pH 10.0/time required at pH 6.0.

a rapidly achieved equilibrium exists that is drastically influenced by pH. Even at pH 10, however, approximately 0.5% of the free residual chlorine is still present as HOCl, and because it is a much more powerful biocide than OCl^-, its presence could substantially influence the biocidal activity observed.

Similarly, Eq. (10.2)

$$HOCl + NH_3 \rightleftarrows NH_2Cl + H_2O \qquad (10.2)$$

is reversible, and a solution of 2 mg/L NH_2Cl is estimated to be 0.58% hydrolyzed (0.58% HOCl) at pH 7 and 25°C (77°F) (12). Because of the much higher biocidal efficiency of HOCl, its influence on the disinfection rate observed could be substantial and could explain the influence of pH on the biocidal efficiency of monochloramine.

Furthermore, the equation:

$$H^+ + 2NH_2Cl \rightleftarrows NH_4^+ + NHCl_2 \qquad (10.3)$$

indicates that although mostly monochloramine is formed when excess ammonia is present at high pH (>8), addition of hydrogen ion (lowering pH) will cause formation of dichloramine, with the position of this equilibrium being determined by the pH of the treated water. Thus, with chlorine and chloramines, pure species are never present, and pH determines their identities. The influence of pH therefore cannot be experimentally separated from species effectiveness for disinfection.

Nevertheless, in the case of chlorine, disinfection efficiency does decline rapidly as the pH is increased from 7 to 9. The efficiency of chlorine dioxide also changes substantially over this pH range; but in contrast to chlorine, the effectiveness increases as the pH increases (9). In this case, the change appears to be in microorganism sensitivity rather than in disinfectant species present, because unlike chlorine, chlorine dioxide does not quickly dissociate or disproportionate into different chemical species within this pH range. In earlier studies, a similar effect was shown with *E. coli* (i.e., more rapid inactivation at pH 8.5 than at lower pH by equivalent concentrations of chlorine dioxide) (13).

The pH of the water also affects ozone chemistry. At high pH values, ozone decay is accelerated, proceeding through hydroxyl radical intermediates; therefore, the pH of the water being treated may also influence ozone effectiveness.

Thus, the influence of many factors on disinfection make precise rankings of these three alternative disinfectants ozone, chlorine dioxide, and chloramines difficult. Nevertheless, chloramines are generally ranked as disinfectants that are weaker than free chlorine at all pH values. The National Research Council (NRC) Safe Drinking Water Committee rated chloramines moderate to low in relative disinfection ability and unsuitable as a primary drinking water disinfectant (14). In general, ozone and chlorine dioxide are ranked as strong disinfectants that are nearly equal to or better than free chlorine for destruction of pathogens.

Despite the generally weaker biocidal efficiency of chloramines, the chlorine-ammonia treatment process has been used successfully for primary disinfection for years by a number of utilities. Chloramine formation as accomplished in these treatment plants differs significantly, however, from the procedures used in preparing chloramine for use in the laboratory chloramine disinfection studies described above. In experimental work, the chloramines are generally preformed and the microorganisms are added subsequently. In chlorine-ammonia treatment for primary disinfection as practiced in the field, ammonia and chlorine are added to the water either simultaneously or in close succession.

Depending on pH, temperature, mixing efficiency, and the chlorine/ammonia ratio present, free chlorine may persist in practice for several minutes and result in rapid inactivation of microorganisms during that time. Subsequently, chloramines will persist for long periods of time, continuing its disinfection action in waters that exhibit a strong chlorine demand that may rapidly deplete a free residual. Given sufficient contact time, the weaker disinfectants still possess an adequate germicidal ability.

Often a free chlorine residual is maintained for some period of time intentionally before ammonia is added, taking advantage of a rapid organism inactivation, followed by a stable chloramine residual in the distribution system. This usefulness of chloramines as an effective secondary disinfectant is recognized in the NRC report (14).

10.2.2 Comparative Efficiencies of Other Alternatives to Free Chlorine

Because of the lack of widespread use of the remaining disinfectant chemicals listed in Section 10.1 of this chapter, detailed discussion of experimental results is beyond the scope of this chapter. The NRC Report (14) attempted to summarize the relative efficiencies shown in Table 10.2 for demand free systems. Free chlorine is shown for comparison purposes. Thus, some alternatives listed in Table 10.2 have disinfectant properties that are as useful as free chlorine whereas others are clearly not as good or have unknown comparative worth. Considerations other than efficacy in laboratory situations, such as engineering or cost considerations, ultimately determine their reliable usefulness. Bromine chloride was not included in Table 10.2 but is essentially an alternate method for introducing bromine as the disinfection agent (14).

10.2.3 Extrapolation from Laboratory Studies to Field Situations

Although information derived from laboratory studies is very useful in assessing the relative biocidal efficiency of disinfectants, other factors are important in the application of this information to actual drinking water treatment in the field. In water treatment, pure cultures of organisms are not present as clean suspensions in a medium free of extraneous materials that might react with the disinfectant used, thereby destroying or altering its biocidal capability. Rather, in the field, a variety of microorganisms are present in their natural state, suspended in a medium containing a variety of other solid and dissolved materials, some of which can have pronounced effects on disinfectant concentration and activity. Because of these factors, disinfection in the field does not behave as a constant rate process as it does

TABLE 10.2
Summary of Efficacies of Less Common Alternate Disinfectants
Biocidal Activity by Organism Class

Agent	Bacteria	Virus	Cysts
Free chlorine[a](HOCl)	Excellent	Excellent	Moderate
Permanganate	Questionable	—[b]	—
Hydrogen peroxide	Questionable	Questionable	—
Bromine	Excellent	Excellent	Good
Iodine (as I_2)	Excellent	Good	Good
Iodine (as HOI)	Excellent	Excellent	Low
Ferrate	Moderate	Good	—
Elevated pH (>12)	Good	Good	—
Silver	Low	—	Low
UV radiation	Good	Good	—

Source: Ref. 14.
[a] For comparison to others.
[b] Unknown.

in laboratory studies. These effects change the shape of the decay curves and perhaps even the order of overall effectiveness observed. Additionally, particulate matter (turbidity) is thought to shield both viruses (15) and bacteria (16) from the disinfection process. Indeed, supposed protection of microorganisms by their association with particulate matter has been the major consideration in establishing a primary turbidity limit for drinking water. Thus, site-specific water quality considerations are very important in determining the most appropriate disinfectant to accomplish the microbiological quality goals.

10.3   FORMATION OF THMs BY ALTERNATIVES TO FREE CHLORINE

One of several alternatives for preventing formation of THMs from chlorination of drinking water is to use a disinfectant that does not produce this undesirable side reaction product. Below, the most commonly considered alternatives are discussed with respect to their tendency to form THMs.

10.3.1   Chloramines

The reaction of free chlorine with ammonia to form mono- and dichloramine have been discussed above (Section 10.2.1.2). In addition to Eqs. 10.2 and 10.3, other reactions are important in the "breakpoint" chlorination process. Sequential reactions occur to form monochloramine, dichloramine, and nitrogen trichloride. Oxidation occurs in the process when sufficient excess chlorine is present to oxidize the ammonia to nitrogen gas, the chlorine being reduced to chloride (Eq. 10.4).

$$2NHCl_2 + H_2O \rightarrow N_2 + HOCl + 3H^+ + 3Cl^- \qquad (10.4)$$

Eqs. 10.2–4 present a somewhat simplified picture, and the end result is heavily dependent on pH. The general process is graphically depicted in Figure 10.3. The breakpoint curve has been divided into three regions that are important to study if the influence of ammonia/chlorine on the formation of THMs, when "chloramines" are used as a disinfectant, is to be understood.

10.3.1.1   Region A

In Region A (Fig. 10.3) the reactions of chlorine and ammonia to form combined chlorine are very fast relative to THM formation; very little free chlorine is present, and when chloramines are preformed, or ammonia is present in stoichiometric excess, THMs do not form to a significant degree (Fig. 10.4) (7). These data show little development of chloroform during the 70 h of exposure when preformed combined chlorine was the disinfectant. In

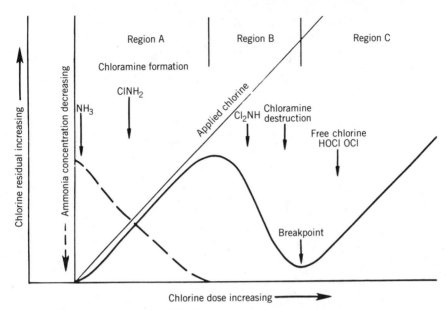

Figure 10.3. Chlorine-ammonia breakpoint curve.

contrast, much higher concentrations of chloroform were formed in the presence of a free chlorine residual.

Symons et al. (4) document 12 instances when chloramines were used or when ammonia was added after a period of free chlorination to form chloramines at treatment plants. These data show lower resulting THM concentrations when compared with situations in which free chlorine was the disinfectant. Example utilities included St. Louis County, MO, Beaver Falls, PA, Jefferson Parish, LA, Huron, SD, Casitas, CA, and Stuttgardt, West Germany, indicating that over a wide geographic area, the conclusions drawn from Figure 10.4 appear valid. More recently, many utilities have changed to combined chlorine to meet the requirements of the THM regulation.

10.3.1.2 Region B

In Region B (Fig. 10.3) oxidation reactions, such as that of Eq. 10.4, are occurring in the presence of excess chlorine. The oxidation reactions are much slower than the initial chloramine formation reactions, and, thus, free chlorine can persist for many minutes, although probably less than an hour (17). This will allow significant THM formation during that time after adding the chlorine. This phenomenon has been observed at at least one utility where breakpoint chlorination has been historically necessary to remove ammonia from the source water (18). Thus, THM control is not complete when chloramine treatment is attempted in Region B.

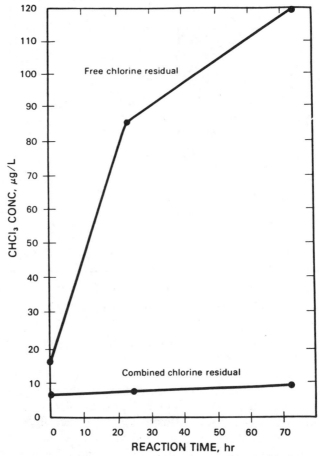

Figure 10.4. Chloroform formation by free and combined chlorine residual (4).

10.3.1.3   Region C

After breakpoint (Fig. 10.3) mainly free chlorine exists and no THM control is expected. Chlorination into Region C is usually a strategy to remove ammonia by oxidation, counter to the intent of THM control by chloramine treatment.

10.3.1.4   Bromide Problems with Chloramine/Breakpoint

Lange and Kawczynski, in their efforts to control THM concentrations at the Contra Costa County Water District, experimented with the use of chloramines (19). They conducted jar tests arranged to resemble treatment at the water plant with source water chlorination, ammonia being added to the chlorinated water at a weight ratio of $3:1$ (NH_3/Cl_2). The data (Table

TABLE 10.3
Results of Chloramine Studies at Contra Costa, CA, September 1977

Cl_2 Contact Time before Adding NH_3 (h)	Trihalomethanes ($\mu g/L$)					Total $\mu g/L$
	pH	$CHCl_3$	$CHBrCl_2$	CHB_2Cl	$CHBr_3$	
0	7.0	3	2	1	<1	6
0.5	7.0	15	16	39	50	120
1.0	7.0	7	18	45	55	125
1.5	7.0	8	20	51	60	139
4.0	7.0	9	26	58	50	143
Control treatment sample (excess Cl_2)	8.2	5	18	84	189	296

Source: Ref. 19.

10.3) show that the addition of ammonia did arrest the formation of THMs. Because the high bromide concentration caused a rapid formation of bromine-substituted THMs, however, very little time could be allowed to elapse between the addition of chlorine and ammonia if significant reductions in THM concentrations were to be achieved. The California State Department of Health required a free chlorine residual be maintained for a minimum of 10 min before the addition of ammonia. Other Primacy States may have similar requirements. Thus, utilities with source waters having a high bromide content within these Primacy jurisdictions may experience difficulty controlling THMs with chlorine-ammonia treatment. Recent studies have shown that the free chlorine concentration and contact time may be reduced considerably with adequate disinfection, minimizing THM formation if mixing intensity is sufficiently great (20).

Thus, the presence of bromide in source waters lessens the effectiveness of THM control by chloramine treatment even in Region A of the breakpoint curve.

10.3.2 Chlorine Dioxide

In 1976 Miltner (21) completed work that demonstrated that pure chlorine dioxide would not cause formation of THMs. A study was conducted in which a humic acid solution (5 mg/L) was dosed with 8 mg/L chlorine dioxide. After 48 h of contact time, 1.7 $\mu g/L$ of chloroform was formed (Fig. 10.5), but no other THM species occurred. For comparison, a similar humic acid solution was dosed with 8 mg/L of free chlorine. In the same time period, 108 $\mu g/L$ of chloroform (Fig. 10.5) and 1.5 $\mu g/L$ of bromodichloromethane were formed—about 110 $\mu g/L$ total THM (TTHM). This study indicates that chlorine dioxide does not produce THMs from precursor ma-

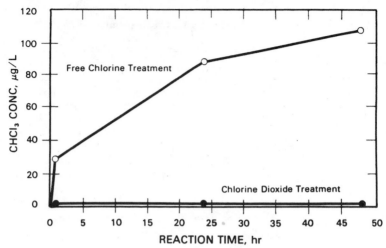

Figure 10.5. Chloroform formation in water containing 5 mg/L humic acid dosed with chlorine-free chlorine dioxide or free chlorine (4).

terials that will react with free chlorine to produce THMs. Miltner's work with Ohio River water confirmed that this general conclusion applies to the natural aqueous medium.

During normal continuous flow operation, chlorine dioxide is usually generated under conditions that result in mixtures of free chlorine and chlorine dioxide (see Section 10.5.4). Although the presence of chlorine suggests that THMs would be formed under these circumstances, for reasons not yet understood, chlorine dioxide alters certain THM precursors so that the yield of THMs is decreased when free chlorine reacts with them. THM concentrations tend to be inversely related to the chlorine dioxide/free chlorine ratio (4). Concentrations of THMs, therefore, can be kept low if the excess free chlorine used to produce chlorine dioxide is carefully controlled.

Symons et al. (4) document several cases where chlorine dioxide has been used successfully, either alone or in conjunction with chlorine to successfully lower THM concentrations at both the pilot and full scale.

10.3.3   Ozone

THM formation during ozonation under practical drinking water treatment conditions, especially with waters low in bromide content, does not appear to occur (4). The issue of formation of HOBr from bromide and, thus, subsequent bromoform formation is controversial. The subject was recently reviewed by Cooper (22). This review indicates that competing reactions involving ozone, precursors, bromide, bromate, and HOBr/OBr⁻, depend on the concentrations of each and on pH. At normal water treatment pH values, ozone doses between those necessary to oxidize humics (low, to meet demand) and those necessary to oxidize $HOBr/OBr^-$ to BrO_3^- (high) will result in bromoform production in waters relatively high in bromide.

10.3.4 Bromine, Bromine Chloride

Hypobromous acid is the expected active species upon hydrolysis of either applied Br_2 or $BrCl$; in both cases, bromoform would be expected to form. Symons' (4) review describes an EPA study that verified these expectations. In this study, when free chlorine was used as a disinfectant, the primary THM was chloroform, but when bromine chloride was used as a disinfectant, nearly all of the THM content appeared as bromoform (Table 10.4).

Furthermore, when summed, the TTHM concentration would be much higher for the BrCl-treated water, indicating that use of these agents as substitutes for chlorine for THM control would be counterproductive.

10.3.5 Iodine

The formation of THMs during iodination was studied by Rickabaugh and Kinman (23). Their findings showed that none of the regulated THMs (24) were formed when iodine was used as the disinfectant. Some iodoform may have been formed, but analysis did not include this compound.

Indeed, iodination to form iodoform is the basis for the classic haloform test for methyl ketones. Iodomethanes form during chlorination of waters containing iodide ion, presumably through an iodine intermediate, in a manner analogous to bromomethane species formation from chlorination of waters containing bromide (6).

10.3.6 THM Formation from Other Alternative Disinfectants

No evidence could be found by the authors to indicate that permanganate, hydrogen peroxide, ferrate, silver, high pH, or UV radiation would themselves cause formation of THMs in natural waters.

TABLE 10.4
THM Formation in Treated Water Disinfected with Chlorine and Bromine Chloride

Reaction Time (h)	THMs Formed with Cl_2 ($\mu g/L$)				THM Formed with BrCl ($\mu g/L$)			
	$CHCl_3$	$CHBrCl_2$	$CHBr_2Cl$	$CHBr_3$	$CHCl_3$	$CHBrCl_2$	$CHBr_2Cl$	$CHBr_3$
6	44	16	3.4	0.2	0.3	0.1	1.7	149
24	85	23	4.5	1.3	0.4	0.1	2.0	177
48	106	28	5.2	0.5	0.5	0.1	2.7	194
72	116	30	5.8	0.6	0.6	0.2	3.2	209
96	118	41	5.9	0.5	0.5	0.1	3.4	209

Source: Ref. 4.

10.4 FORMATION OF BY-PRODUCTS OTHER THAN THM

When any disinfectants are used in water treatment, by-products other than THMs may be found. Most is known about the by-products of the reactions of chlorine, chlorine dioxide, ozone, and the chloramines because these commonly used oxidants have been subjected to the most intense study. Little can be found in the literature about by-products from the other disinfectants addressed in this chapter; thus, the organic by-products of reactions of only chlorine, chlorine dioxide, ozone, and chloramines are reviewed in a relatively detailed way, and separately, the inorganic end products from all the disinfectants discussed in this chapter are summarized.

10.4.1 Organic By-Products

10.4.1.1 Specific Compounds from Chlorine

Nonpolar compounds other than THMs that were either not detectable in the source water or were present in lower concentrations have been detected in finished water at ng/L to µg/L concentrations. The sources of most of these are poorly understood. At least 19 non-THM, halogenated, volatile compounds were found by Rook (25) in the Rotterdam Storage Reservoir. Stieglitz et al. found additional compounds formed at low concentrations in a Rhine River bank filtrate sample upon chlorination (26). Rook speculated on a possible pathway to explain the formation of some of the observed by-products as related to his proposed mechanism for THM formation from *m*-dihydroxyphenyl moieties. Stieglitz suggested no mechanism. Coleman et al. reported the co-presence of chloropicrin, chlorobenzene, a chlorotoluene isomer, and a chloroxylene isomer as well as their respective logical precursors (nitromethane, benzene, toluene, and *m*-xylene) in finished chlorinated tap water (27). With the exception of benzene, all of the above precursors were shown to react with free chlorine to form the expected products.

In other studies reported by Symons et al. (4), haloacetonitriles were observed in a finished tap water. Concentrations of acetonitrile in the mg/L range could not be made to react with free chlorine under realistic reaction conditions to form detectable chlorinated derivatives. Trehy, however, showed that dichloro-, bromochloro-, and dibromoacetonitrile were formed upon low pH chlorination of a south Florida drinking water source (28). At high pH, such as in lime-softening systems, these by-products are less of a problem. Suffet et al. found 1,1,1-trichloroacetone in two tap waters, but not in the respective source waters (29). Later work at Manchester, NH, has shown the formation of 1,1,1-trichloroacetone upon chlorination (4). Dichloro- and trichloroacetic acids are now thought to be major products of chlorination of humic materials, under some conditions exceeding the concentrations of THMs (30). At this writing, literally hundreds of discrete minor by-products resulting from treatment of humic materials with chlorine have been detected, mostly aromatic and aliphatic acids, some containing halogen (31–33). The majority of these, determined to be unique by their mass spectra, have not been fully characterized or identified.

10.4.1.2   Specific Compounds from Chlorine Dioxide

Although chlorine dioxide does not react to produce THMs, chlorine dioxide does react with organic material during water treatment and, like chlorine, is therefore likely to produce other organic by-products. Chlorine dioxide reacts with natural humic acids, lowering the color in drinking water supplies (34). Similarly, large quantities of chlorine dioxide are used for bleaching operations in the pulp and paper industry. Much work on the by-products of this process has been accomplished and formed the basis for a review of the subject in a way pertinent to drinking water applications by Stevens (35). Briefly, the review described chlorinated and nonchlorinated derivatives (including acids, epoxides, quinones, aldehydes, disulfides, and sulfonic acids) that are products of reactions carried out under conditions somewhat different from those experienced at water treatment plants.

As a follow-up to this review, some studies have been conducted to explore the formation of by-products under conditions more closely related to drinking water treatment conditions. Semiquantitative results of one study have been described where C_2 through C_8 aldehydes were noted to increase in concentration after treatment of a natural water with chlorine dioxide (35). In that work, no other dramatic differences were observed between treated and untreated samples with regard to compounds amenable to the type of gas chromatographic analysis used. Other studies have since shown that many of the same aromatic and aliphatic acidic compounds seen upon chlorine treatment are also seen after chlorine dioxide treatment of natural humic materials (36).

10.4.1.3   Specific Compounds from Chloramines

The potential for formation of organic by-products as a result of disinfection with chloramines is not as obvious as with chlorine dioxide. Chloramines are weaker disinfectants (less reactive with cells) compared to chlorine and chlorine dioxide, and waters generally exhibit a much lower disinfectant demand when chloramines are used. Because chloramines do hydrolyze to form traces of free chlorine, some reaction products of this oxidant might be expected, but at much lower concentrations in a given time than when free chlorination is practiced. Except for chlorine exchange reactions with primary and secondary amines present in treated waters, information regarding specific by-product formation from chloramines under drinking water treatment conditions is virtually absent from the literature (14).

10.4.1.4 Specific Compounds from Ozone

Ozone is a highly reactive oxidant that might be expected to produce oxidation products of organic materials found in water supplies. Unlike the oxidants chlorine, chlorine dioxide, and chloramines, however, ozone would not be expected to produce chlorinated by-products unless indirectly by way of oxidation of halides present.

Although much is known about ozone reactions in other media, surprisingly little information exists about the action of ozone as an oxidant of organic compounds in aqueous solution. This lack of data exists even though ozone has been in widespread use for decades as a water and waste water disinfectant. The reviewed data suggest that oxygenated products such as ketones, aldehydes, and acids are most likely formed from alcohols and olefinic double-bond and aromatic ring cleavage (14).

Of the few studies performed in connection with drinking water treatment, a study by Schalekamp was revealing concerning by-product formation (37). Schalekamp analyzed water before and after an ozone treatment step at various ozone doses. He found that the concentration of total C_6–C_{14} aldehydes rose by a factor of more than 10 as the ozone dose increased from 0 to 5 mg/L. Absolute concentrations were low, however, usually a few hundred nanograms per liter after ozonation and always less than 1 μg/L, depending on the aldehyde.

Sievers et al. also found the same aldehydes and reported some apparent hydrocarbon formation upon ozonation of the effluent from a secondary waste treatment plant in Estes Park, CO (38).

10.4.2 Inorganic End Products

The above discussion of by-products ignores the possible problems surrounding the inorganic end products of chemical reduction of the oxidant disinfectant. A brief summary of these products is presented in Table 10.5. Most of the end products are natural constituents of the water matrix, and at the low concentrations normally applied, they blend into insignificance. Two major exceptions to this are the end products from chlorine dioxide and iodine. The U.S. EPA has issued a recommendation limiting total oxidant residuals from chlorine dioxide (ClO_2, ClO_2^-, ClO_3^-) to 1 mg/L (39), and the possibility of adverse health effects of increased intake of iodine or iodide to susceptible individuals has been raised (14). Ferrate and permanganate produce end products that are usually easily removed in the treatment process. Silver ion residual is limited by the National Interim Primary Drinking Water Regulations to 0.05 mg/L (40).

10.4.3 Organic Halogen

One approach to characterizing and quantifying organic by-product formation from the disinfectants that contain halogen is to measure total organic

TABLE 10.5
Summary of Inorganic End Products from Use of Disinfectants

Agents	End Product	End Product Significance
Cl_2, $HOCl/OCl^-$	Cl^-	Insignificant
Chloramines	Persistent residual	Effects under study
ClO_2	ClO_2^-, ClO_3^-, Cl^-	Recommend 1 mg/L limit on $ClO_2 + ClO_2^- + ClO_3^-$ on health effect basis
O_3	O_2, OH^-, H_2O	Insignificant
$KMnO_4$	MnO_2	Precipitates in treatment
H_2O_2	O_2, OH^-, H_2O	Insignificant
Br_2, $BrCl$, $HOBr/OBr^-$	Br^-	Insignificant
I_2, HOI/OI^-	I^-	Large iodine or iodide intake discouraged
FeO_4^{2-}	Ferric oxides/hydroxides	Precipitate in treatment
Elevated pH	H_2O	None
Ag/Ag^+	Residual/solids	Residual maximum contaminant level is 0.05 mg/L
UV	None	Not applicable

halogen (TOX) by use of an activated carbon adsorption-pyrolysis technique (41). This analytical technology has been applied to solutions of a soil humic acid that was treated separately with each of the three disinfectants: free chlorine, combined chlorine (chloramines), and chlorine dioxide. Time, temperature, pH, disinfectant dose, and organic substrate concentrations were varied. Selected examples of the results obtained are presented in Figure 10.6 and 10.7 (42).

The bar graphs in Figures 10.6 and 10.7 represent nonpurgeable organic halogen (NPOX) above the zero line and THMs below the zero line. The THMs were exclusively chloroform in these experiments and are expressed as μg Cl^-/L. Expressed in this way the THMs are a purgeable organic halogen (POX) measurement and thus can be compared with common units to NPOX. The full length of a bar, therefore, represents TOX. Arranged in this manner, the concentration of TOX in solution at 20°C (68°F) can be seen to be mostly composed of NPOX at pH values of 5 and 7, while at pH 11 the amount of THMs present is the greater (Fig. 10.6). The ratio of NPOX to THMs is shown to diminish with increasing pH, which occurs because increasing pH favors an increase in THM formation while at the same time promoting a drastic decrease in NPOX concentration.

The effect on organic halide formation of increasing the dose of chlorine from 8.1 mg/L to 20 mg/L at 20°C (68°F) is also demonstrated in Figure 10.6. Compared with the low dose, TOX is increased by 29, 36, and 27% at pH

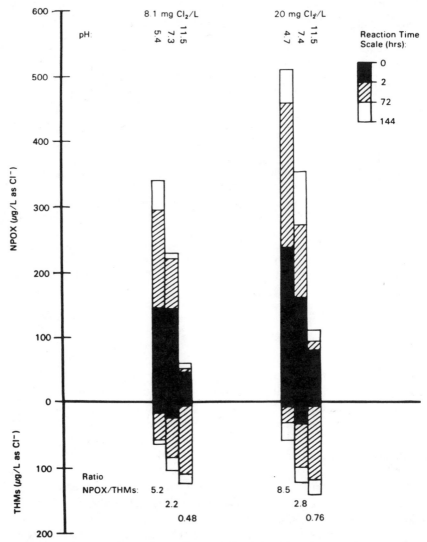

Figure 10.6. A comparison of the effects of pH and oxidant dose on the formation of NPOX and THMs (CHCl₃) at 20°C (68°F) in distilled water solutions of 5 mg humic acid/L.

values near 5, 7, and 11, respectively, at the higher dose. This is caused almost entirely by an increase in NPOX formation.

Figure 10.6 also reveals a relatively rapid formation of NPOX within the first 2 h after dosing with chlorine when compared with THM formation. All of these results of chlorination are typical of those reported for humic materials from other sources by Fleischacker and Randke (43).

Figure 10.7 compares the NPOX and THM formation at 20°C (68°F) at

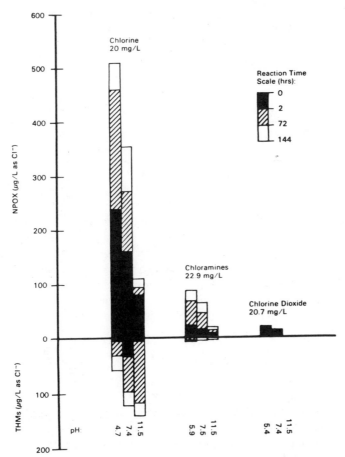

Figure 10.7. A comparison of the formation of NPOX and THMs (CHCl₃) at 20°C (68°F) in distilled water solutions of 5 mg humic acid/L dosed with various disinfectants.

three different pH levels for each of the three disinfectants, chlorine, chloramines, and chlorine dioxide. NPOX formation is reduced by 85% when comparing the use of chloramine as the disinfectant to chlorine and was even much lower when chlorine dioxide is used. THM formation is reduced by greater than 95% when chloramines were used, and no THMs were detected when chlorine dioxide was used. Unlike with chlorine, which increases the formation of THMs with increasing pH, the use of chloramines and chlorine dioxide as disinfectants results in a decrease in the formation of all organic halides (including THMs) with increasing pH. NPOX formation with chlorine dioxide was rapid with no difference observed between 2-h and 144-h sample concentrations.

TABLE 10.6
Formation of Total Organic Halogen (TOX) in Ohio River Water
Treated with Various Disinfectants (μg/L as Cl$^-$)

	Free Chlorine	Chloramines	Chlorine Dioxide	Ozone	Control (No Disin- fectant)
Test 1	194 (2.5 mg/L)[a]	101 (2.0 mg/L)	61 (3.0 mg/L)	9 (3.0 mg/L)	17
Test 2	53 (3.3 mg/L)	26 (0.8 mg/L)	17 (2.4 mg/L)	11 (1.0 mg/L)	13

Source: Ref. 4.
[a] Disinfectant dose is indicated in parentheses.

Similarly, two studies described by Symons et al. (4) compared the formation of organic halogen when four different disinfectants were used to treat a natural surface water. In these experiments, Ohio River water that had been coagulated, settled, and filtered in a pilot plant was disinfected with free chlorine, chloramines, chlorine dioxide, and ozone. For this experiment, these data show that organic halogen is formed by the action of these disinfectants in the following order of yield (Table 10.6):

Free chlorine > Chloramines > Chlorine dioxide > Ozone

These tests confirm that the trends seen in the humic acid studies are applicable to natural waters as well.

Data from eight utilities demonstrating the variability that may be caused by a number of the factors described above are presented in Table 10.7. The health significance of the organic halogen measurement as an indicator of water quality has not been determined; thus, the usefulness of this approach to estimating the significance of overall organic by-product formation is not yet known.

10.5   APPLICATION OF DISINFECTANTS

Engineering and other practical considerations of applying disinfectants is a subject area of rather large scope and has been the topic of a number of reviews (12,14,44). For this chapter, discussion is limited to a brief overview of several disinfectants.

10.5.1   Chlorine

Since the beginning of this century, both liquid chlorine and dry forms (hypochlorites) have been used for drinking water chlorination in this country. During this time, the off-site manufacture and on-site feeding of these chemicals has become commonplace technology. Although good safety

TABLE 10.7
Organic Halogen (OX) in Finished Waters

Utility	NPOX Concentration (μg/L as Cl$^-$)	POX (mostly TTHM) (μg/L as Cl$^-$)	NPOX/POX Ratio
A	17	9.8	1.7
B	NF[a]	NF	—
C	52	64	0.8
D	36	31	1.2
E	165	180	0.9
F	136	114	1.2
G	66	133	0.5
H	98	27	3.6

Source: Ref. 4.
[a] None found.

practices are required to handle these chemicals, chlorine is a good disinfectant, and a disinfectant residual can be maintained.

10.5.2 Chloramines

Chloramines can be formed either by chlorination of ammonia present naturally in the water or ammonia that is added artificially. If ammonia is present in the source water, simply adding chlorine, as noted previously, will produce chloramines. If no ammonia is present, feeders are available that will add any one of a variety of ammonia containing chemicals to the water. The ammonia can be added either before or after the chlorine, depending on the optimum arrangement at a given site. Although chloramines must be produced on-site, they do form a disinfectant residual and are widely used.

10.5.3 Ozone

Ozone must be produced on site. This is accomplished by use of an ozonator, a device within which a stream of clean dry air or oxygen passes through an electrical discharge. On-site power requirements are a major consideration, about 13–22 kWh/kg of ozone generated from air and approximately half that when oxygen is used. Compressors and dryers may increase overall power requirements by 20–50%. Other factors affecting efficiency of ozone production are the rate of gas flow, applied voltage, and the temperature of the gas. Approximately twice the percent of ozone by weight is obtained if oxygen, rather than air, is used as the feed stream. Heat produced during the process is a major problem and cooling must be provided.

The gas stream containing ozone is fed into the water to effect the transfer of ozone from the gas phase into the water. The usual methods are to inject the ozone gas stream through an orifice at the bottom of a co- or countercurrent contact chamber or to aspirate the gas into a contact chamber where it is mixed with the water mechanically.

The chief disadvantages to the use of ozone are the relatively complex generating equipment that must be operated on site, its poor solubility in water, its short half life, and, thus, the lack of residual. In practice, however, these problems have been successfully overcome, and ozone is widely used in Europe.

10.5.4 Chlorine Dioxide

The preparation and distribution of chlorine dioxide in bulk is not practical. Therefore, it is generated and used on site. Sodium chlorate or sodium chlorite may be used to generate chlorine dioxide, the method of production depending upon the amount of chlorine dioxide that is required. Compared to some industrial applications, use at a drinking water treatment plant is small, and sodium chlorite is the chemical of choice.

Chlorine dioxide can be prepared from the reaction of chlorine with sodium chlorite through the following:

$$Cl_2 + H_2O \rightarrow HOCl + HCl \tag{10.5}$$

$$HOCl + HCl + 2NaClO_2 \rightarrow 2ClO_2 + 2NaCl + H_2O \tag{10.6}$$

The theoretical weight ratio of sodium chlorite to chlorine is $1:0.39$. With commercial grade sodium chlorite (80%), the weight ratio is $1:0.30$. In practice, a chlorite to chlorine ratio of $1:1$ is recommended. The excess chlorine lowers the pH, thereby increasing the reaction rate and optimizing the yield of chlorine dioxide.

Chlorine dioxide also may be prepared from sodium hypochlorite and sodium chlorite. The sodium hypochlorite is acidified to yield hypochlorous acid and the chlorine dioxide is generated according to Eq. 10.6. Each of the methods produces a solution containing both chlorine and chlorine dioxide. Chlorine-free chlorine dioxide may be prepared by the addition of a strong acid, such as sulfuric acid or hydrochloric acid, to sodium chlorite.

Although chlorine dioxide must be generated on site, the equipment is relatively simple, and a disinfectant residual is produced. Care must be taken in handling and storage of the sodium chlorite because it is unstable. Chlorine dioxide is widely used in Europe.

10.5.5 Other Disinfectants

The more common method of adding iodine to a municipal water supply produces the required concentration of iodine by passing water through a bed of crystalline iodine called a saturator. This has been used in many small, semipublic and private home water systems. Because the maximum concentration is limited by solubility to 200–300 mg/L at the ambient temperatures that are expected for drinking water, some physical complications would accompany the introduction of this method into the large waterworks system in view of the large saturation bed required. Another method is to employ nonhazardous solvents and solubilizing agents, such as ethyl alcohol or aqueous potassium iodide, to overcome the low concentration of aqueous iodine stock for solution feeders. A disinfectant residual is produced.

Some difficulties that are encountered when handling liquid bromine and its corrosive nature have encouraged the use of bromine chloride as a technique for introducing bromine into water. Like chlorine, bromine chloride is produced off-site and shipped as the dry liquid in steel containers. Liquid bromine chloride is removed from cylinders under moderate pressure, then vaporized, and the gas is metered in equipment that is similar to that used for chlorine. Some special materials must be used in the gas feeders because bromine chloride is more reactive than chlorine with polyvinylchloride plastics. A disinfectant residual is produced.

High pH values in water are usually obtained by addition of either calcium hydroxide or sodium hydroxide. Calcium hydroxide is most often prepared by reaction of calcium oxide with water at the plant. All three chemicals may be easily obtained and none are particularly dangerous to store, handle, or feed.

Off-site commercially produced hydrogen peroxide is available in aqueous solution, usually ranging from 30 to 90%. A concentrated solution would be diluted and then applied with a chemical metering pump. Little experience exists with this chemical in water plant operations partly because hydrogen peroxide is not presently considered to be a disinfectant in the water industry. Fire or explosion hazards exist with the concentrated solutions, requiring care in handling.

Potassium permanganate is readily available in crystalline form. In water treatment, a dilute solution of a few percent is usually prepared and applied with a chemical metering pump. Permanganate may also be added by using conventional dry-feed equipment. Although not normally used as a disinfectant, potassium permanganate is widely used in water treatment practice for controlling taste and odor.

A wide variety of metallic ferrate salts have been prepared. However, only a few of the preparations yield ferrates of sufficient purity and stability for use in the treatment of water. In contrast to permanganate, ferrates typically are not presently available in large bulk quantities.

Silver has been applied principally by electrolytic dissolution of the metal, principally in swimming pool applications. Incorporating silver in a filter medium, such as activated carbon, is an alternative approach. Silver is incorporated into granular activated carbon to prevent bacterial growth in filters used in home applications.

UV radiation can be applied to water in two distinctly different ways, each having advantages and disadvantages. Tubular reactors have the advantages of being sealed and under pressure, but they suffer the disadvantage of lamp fouling and the need for cleaning. The alternative approach is to place the lamps above the water surface, between the surface and a reflecting material. This approach is efficient, does not have a lamp fouling problem, but must operate under atmospheric pressure within an open vessel that can permit contamination. Problems that are common to both applications are adsorption of radiation by particulates as natural color, hydroxy radical scavengers, and the lack of a residual.

10.5.6 Summary

All of the above mentioned approaches may be used to disinfect drinking water. The systems vary in cost, complexity, the hazard of the chemicals used, and the production of a disinfectant residual, but careful design and operation can produce satisfactory disinfection.

REFERENCES

1. Rook, J.J., *Water Treatment and Examination, 23,* 234, Part 2, 1974.
2. Bellar, T.A., Lichtenberg, J.J., and Kroner, R.C., *J. Am. Water Works Assoc., 66,* 703, 1974.
3. Symons, J.M., Bellar, T.A., Carswell, J.K., DeMarco, J., Kropp, K.L., Robeck, G.G., Seeger, D.R., Slocum, C.J., Smith, B.L., and Stevens, A.A., *J. Am. Water Works Assoc., 67,* 634, 1975.
4. Symons, J.M., Stevens, A.A., Clark, R.M., Geldreich, E.E., Love, O.T., Jr., DeMarco, J., *Treatment Techniques for Controlling Trihalomethanes in Drinking Water,* American Water Works Association, Denver, 1982, 289 pp.
5. Rook, J.J., *J. Am. Water Works Assoc., 68,* 168, 1976.
6. Bunn, W.W., Haas, B.B., Deane, E.R., and Kleopfer, R.D., *Environ. Lett., 10,* 205, 1975.
7. Stevens, A.A., Slocum, C.J., Seeger, D.R., and Robeck, G.G., *J. Am. Water Works Assoc., 68,* 615, 1976.
8. Kajino, M., and Yagi, M., Formation of trihalomethanes during chlorination and determination of halogenated hydrocarbons in drinking water, in Afghan, B.K., and Mackay, D. (eds.), *Hydrocarbons and Halogenated Hydrocarbons in the Aquatic Environment,* Plenum, New York, 1980, p. 491.
9. Scarpino, P.V., Cronier, S., Zink, M.L., Brigano, F.A.O., and Hoff, J.C., Effect of particulates on disinfection of enteroviruses and coliform bacteria in water by chlorine dioxide, in Proceedings, Fifth Water Quality Technology Conference, Kansas City, MO, December 4–7, 1977, Paper 2B-3 American Water Works Association, Denver, CO, 1978, 11 pp.
10. Esposito, M.P., The Inactivation of Viruses in Water by Dichloramine, M.S. Thesis, University of Cincinnati, Cincinnati, OH, 1974.
11. Engelbrecht, R.S., Weber, M.J., Salter, B.B., and Schmidt, C.A., *Appl. Environ. Microbiol., 40,* 249, 1980.
12. White, G.C., *Handbook of Chlorination,* Van Nostrand Reinhold, New York, 1972.
13. Benarde, M.A., Israel, B.M., Olivieri, V.O., and Granstrom, M.L., *Appl. Microbiol., 13,* 776, 1965.

14. Safe Drinking Water Committee, *Drinking Water and Health,* Vol. 2, National Academy Press, Washington, DC, 1980.

15. Hoff, J.C., The relationship of turbidity to disinfection of potable water, in Hendricks, C.H. (ed.), *Evaluation of the Microbiology Standards for Drinking Water,* EPA-570/9-78-002, Washington, DC, 1978, NTIS Accession No. PB 297119.

16. Hejkal, T.W., Wellings, F.M., LaRock, P.A., and Lewis, A.L., *Appl. Environ. Microbiol., 38,* 114, 1979.

17. Palin, A.T., *Chemistry and Control of Modern Chlorination,* 2nd ed., Lamotte Chemical Products, Chestertown, MD, 1983, p. 19.

18. Arber, R.P., Speed, M.A., Scully, F., Significant findings related to the formation of chlorinated organic compounds in the presence of chloramines, in Jolley, R.L., et al. (eds.) *Water Chlorination: Chemistry, Environmental Impact and Health Effects,* Vol. 5, Lewis Publisher, Inc., Chelsea, MI, 1985, pp. 951–963.

19. Lange, A.A., and Kawczynski, E., *J. Am. Water Works Assoc., 70,* 653, 1978.

20. Sorber, C.A., Williams, R.F., Moore, B.E., and Longley, K.E., *Alternative Water Disinfection Schemes for Reduced Trihalomethane Formation: Volume 1. Prototype Studies,* EPA-600/S2-82-037, 1982. NTIS Accession No. P.B.82 227471.

21. Miltner, R., The Effect of Chlorine Dioxide on Trihalomethanes in Drinking Water, M.S. Thesis, University of Cincinnati, Cincinnati, OH, 1976.

22. Copper, W.J., Zika, R.G., and Steinhauer, M.S., The effect of bromide in water treatment II. A literature review of ozone and bromide interactions and the formation of organic bromide, *Ozone Sci. Eng., 7*(4), 313–325, 1985.

23. Rickabaugh, J., and Kinman, R.N., Trihalomethane formation from iodine and chlorine disinfection of Ohio River water, in Jolley, R.L., Gorchev, H., and Hamilton, R.D., Jr. (eds.), *Water Chlorination: Environmental Impact and Health Effects,* Vol. 2. Ann Arbor Science Publishers, Inc., Ann Arbor, MI, 1978, pp. 583–591.

24. *Federal Register, 44,* No. 231, 68624–68707, November 29, 1979; *45,* 15542–15547, March 11, 1980.

25. Rook, J.J., *Environ. Sci. Technol., 11,* 478, 1978.

26. Stieglitz, L., Roth, W., Kühn, W., and Leger, W., *Vom Wasser, 47,* 347, 1976.

27. Coleman, W.E., Lingg, R.D., Melton, R.G., and Kopfler, F.C., The occurrence of volatile organics in five drinking water supplies using gas chromatography/mass spectrometry, in Keith, L.H. (ed.), *Identification and Analysis of Organic Pollutants in Water,* Ann Arbor Science Publishers, Inc., Ann Arbor, MI, 1976, pp. 305–327.

28. Trehy, M.L., and Bieber, T.I., Dihaloacetonitriles in chlorinated natural waters, in Jolley, R.L., et al. (eds.), *Water Chlorination: Environmental Impact and Health Effects,* Vol. 4, Ann Arbor Science, Ann Arbor, MI, 1983, pp. 85–96.

29. Suffet, I.H., Brenner, L., and Silver, B., *Environ. Sci. Technol., 10,* 1273, 1976.

30. Christman, R.F., and Norwood, D.L., Millington, D.S., Johnson, J.D., Stevens, A.A., *Environ. Sci. Technol., 17,* 625, 1983.

31. Johnson, J.D., Christman, R.F., Norwood, D.L., and Millington, D.S., *Environ. Health Perspect., 46,* 63, 1982.

32. Norwood, D.L., Johnson, J.D., and Chrsistman, R.F., Chlorinated products from aquatic humic material at neutral pH in Jolley, R.L., et al. (eds.), *Water Chlorination: Environmental Impact and Health Effects,* Vol. 4, Ann Arbor Science, Ann Arbor, MI, 1983, pp. 191–200.

33. Seeger, D.R., Moore, L.A., and Stevens, A.A., The formation of acidic trace organic byproducts from the chlorination of humic acids, in Jolley, R.L., et al., (eds.), *Water Chlorination: Chemistry, Environmental Impact and Health Effects,* Vol. 5, Lewis Publishers, Inc., Chelsea, MI, 1985, pp. 859–873.

34. Black, A.P., and Christman, R.F., *J. Am. Water Works Assoc., 55,* 897, 1963.

35. Stevens, A.A., *Environ. Health Perspect., 46,* 101, 1982.

36. Colclough, C.A., Johnson, J.D., Christman, R.F., and Millington, D.S., Organic reaction

products of chlorine dioxide and natural aquatic fulvic acid, in Jolley, R.L., et al. (eds.), *Water Chlorination: Environmental Impact and Health Effects,* Vol. 4, Ann Arbor Science, Ann Arbor, MI, 1983, pp. 219–229.

37. Schalekamp, M., Experience in Switzerland with ozone, particularly in connection with the neutralization of hygenically undesirable elements present in water, in *Proceedings—1977 Annual Conference American Water Works Association,* Anaheim, CA, May 8–13, 1977, American Water Works Association, Denver, CO, 1978, Paper 17-4, 22 pp.

38. Sievers, R.E., Barkley, R.M., Eiceman, G.A., Shapiro, R.H., Walton, H.F., Kolonko, K.J., and Field, L.R., *J. Chromatogr., 142,* 745, 1977.

39. U.S. EPA, Office of Drinking Water, *Trihalomethanes in Drinking Water: Sampling, Analysis, Monitoring and Compliance,* EPA 570/9-83-002, 1983, p. 31.

40. *Federal Register, 40,* No. 248, 59566–59588, December 24, 1975.

41. Dressman, R.C., Najar, B.A., and Redzikowski, R., The analyses of organohalides (OX) in water as a group parameter, in *Proceedings 7th Annual AWWA Water Quality Technology Conference,* American Water Works Assoc., Denver, CO, 1980, pp. 69–92.

42. Stevens, A.A., Moore, L., Dressman, R.C., and Seeger, D.R., Disinfectant chemistry in drinking water—Overview of impacts on drinking water quality, in Rice, R.C. (ed.), *Safe Drinking Water—The Impact of Chemicals on a Limited Resource,* Drinking Water Research Foundation, Alexandria, VA, 1985, pp. 87–108.

43. S.J., Fleischacker, and Randke, S.J., *J. Am. Water Works Assoc., 75,* 132, 1983.

44. Johnson, J.D. (ed.), *Disinfection—Water and Wastewater,* Ann Arbor Science, Ann Arbor, MI, 1975, 425 pp.

PART FOUR

Assessing the Risks Associated with Organic Contaminants in Drinking Water

CHAPTER 11

Toxicologic Assessment of Organic Carcinogens in Drinking Water

Edward J. Calabrese
Andrew T. Canada

Division of Public Health
University of Massachusetts
Amherst, Massachusetts

11.1 INTRODUCTION

This chapter is designed to provide the reader with an overview of the process by which a chemical is determined to be a carcinogen. This process is not specifically designed for carcinogens in drinking water. It is assumed that any substance that is found to be a carcinogen in animals following oral feedings would also be a carcinogen if ingested from drinking water.

The judgment made regarding the potential for a xenobiotic to produce cancer in humans is, more often than not, based on less than adequate data. The most relevant studies from which predictions of the relative potential for carcinogenicity of a chemical in humans are epidemiologic investigations carried out following inadvertent workplace or environmental exposures. This will be addressed specifically in the next chapter. More than 50 agents have been identified as probable carcinogens in humans through epidemiologic studies (Table 1). Epidemiologic studies, however relevant, do have specific limitations: (1) due to ethical considerations, humans cannot be deliberately exposed to known or potential carcinogens, as is the case in a controlled experiment; (2) the study group is often not representative of the population as a whole which has been or may be exposed to the possible carcinogen; (3) exposure may be so widespread as to make the identification of unexposed controls and therefore the carrying out of an epidemiologic study very difficult; (4) doses or exposure levels are difficult to document as they often involve retrospective unverifiable histories which make dose-

TABLE 11.1
List of Specific Chemicals, Substances, and Processes Demonstrably
Involved in the Causation of Human Cancer

Part 1—Specific chemicals

β-Naphthylamine	Arsenicals
Benzidine	Bis-choloroethyl sulfide (mustard gas)
4-Aminobiphenyl	Benzene
4-Nitrobiphenyl	Chromates
Chlornaphazine	Radium
Diethylstilbestrol	Thorotrast (thorium dioxide)
Bis-(chloromethyl)ether	Auramine[a]
Vinyl chloride	Magenta[a]
Aflatoxin	Isopropyl oil[a]

Part 2—Mixtures, substances, and processes

Asbestos	Petroleum oils (high boiling)
Estrogenic compounds	Effluents (coke ovens)
Tobacco	Combustion products
Tobacco smoke	Betel nuts
Soots	Uranium ores (radon, radon daughters)
Tars	
Pitches	Other radioactive material
Asphalts	Immunosuppressants—cytotoxic drugs
Cutting oils	
Shale oils	Wood dust[a]
Creosote oils	Nickel refining[a]

[a] Manufacturing exposures identified and showing evidence of carcinogenic effects in exposed people.

response relationships difficult to establish; (5) other risk factors that can confound the results, such as cigarette smoking, are widespread in the population.

For these reasons, a number of methods using both whole animals (in vivo) and isolated cell systems (in vitro) have been developed in an attempt to predict the probability that the exposure of a human to a particular chemical will result in the development of cancer.*

11.1.1 Carcinogenicity Testing Programs

Prior to the 1978 establishment of the National Toxicology Program (NTP) within the U.S. Department of Health, Education and Welfare (now HHS),

---

* On November 23, 1984, the U.S. EPA published proposed guidelines for carcinogen risk assessment in the *Federal Register*. It represents the first proposed revision of the 1976 Interim Procedures and Guidelines for Health Risk Assessment of Suspected Carcinogens, *Federal Register, 41*, 21402–21405, 1976.

carcinogenicity and toxicity testing were carried out by a diverse group of government agencies (particularly by the Carcinogenesis Studies Branch of the National Cancer Institute), private contract testing laboratories, chemical and pharmaceutical companies, and various academic institutions. Due to a wide diversity of methods and standards, such as different animal species and strains, numbers of animals tested, and pathological review standards, the results of these studies were often difficult to reproduce as well as to then correlate with human evidence. Presently over 65,000 individual chemicals are used in the United States and 700–1000 more are synthesized each year (1). It is clear that toxicity, mutagenicity, and carcinogenicity data that can be qualitatively and quantitatively extrapolated to humans are needed.

The magnitude of the problem is overwhelming, since for a large majority of chemicals, there is either no toxicity or minimal toxicity information available from which to perform an adequate human health hazard assessment (Fig. 11.1). Proportionally more animal testing has occurred on pesticides and their inert ingredients and on drugs and excipients used in drug dosage formulations than in any other chemical category. The virtual absence of any actual human exposure/toxicity information on most chemicals impedes the development of standards that will protect humans against the adverse effects of chemicals. This lack of human data is in part due to limited voluntary reporting of human workplace exposure data and adverse effects. The NTP program was created to coordinate efforts of federal agencies involved in toxicity testing and to establish common standards and procedures for the various toxicity tests. The ultimate goal was to provide useful, interpretable, and reproducible data on the toxicity of a chemical no matter which federal agency was the source of the information. The NTP program has as one of its goals the identification of the general toxicity to various organs of the compounds it chooses to study. However, the primary mission of the NTP program is, using cell and whole animal bioassays, to identify chemicals that are potential carcinogens and mutagens and to develop short-term in vivo and in vitro tests for detecting chemicals that may cause cancer, birth defects, and genetic mutations.

11.1.2 Choice of Chemicals Studied

The selection of a compound for study from the tens of thousands for which there is little or no toxicological information begins with the nomination of a chemical for the testing program. This nomination can be made by private individuals, academics, industrial sources, and members of government agencies.

The individual or group that nominates a particular chemical is requested to give their reasons for the nomination and to supply all information on the chemical they are aware of. This information includes the overall toxicity, various uses, amount produced per year, existing or potential hu-

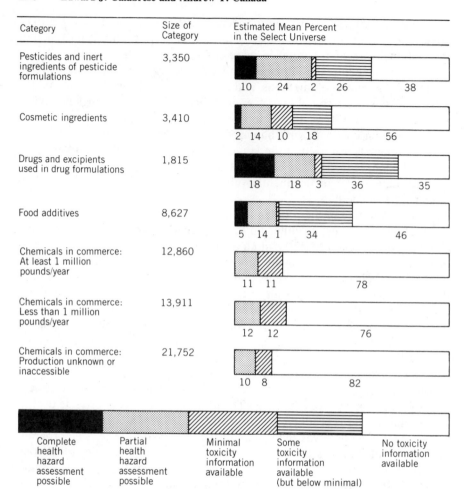

Figure 11.1. Ability to conduct health-hazard assessment of substances in seven categories of select universe. (1)

man exposure data, and animal and human metabolism studies. This nomination process is shown in Table 11.2. Each nomination, regardless of the amount and completeness of the information provided, is then reviewed by the NTP Chemical Evaluation Committee. The decision whether or not to include a chemical in the program is based on the potential for human exposure and the adequacy of the existing toxicity data. The potential for human exposure can be further broken down to:

1. Number of people potentially exposed.
2. Per capita exposure in that population.
3. Frequency of exposure.
4. Probability of a toxic response occurring at that exposure.

TABLE 11.2
Outline of Procedure for Decision-making in Evaluating Adequacy of Toxicity Information on Specific Substance[a]

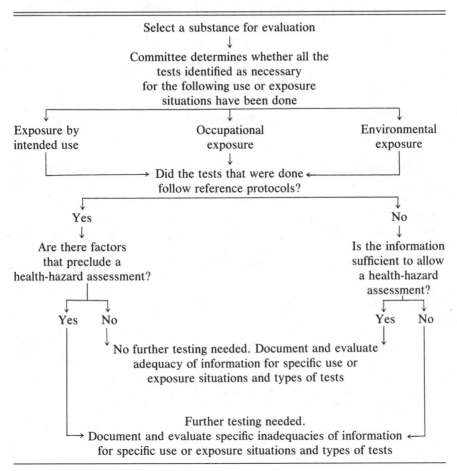

[a] *Source:* Ref. 1.

5. Severity of a toxic response at that exposure.
6. Potential cost to society if adverse health effects result from an exposure.

A number of widely varying methods designed to bring a degree of objectivity to the NTP selection process have been suggested for determining the chemicals that will undergo intensive screening. Among these are:

1. A scoring system such as that proposed by Astill et al. (2), in which a numerical value is placed on various toxicity test results and the

chemical(s) with the highest score then becomes the one with the highest program priority.

2. Modeling-based systems that use overall toxicity data or specific elements from various toxicology studies combining them into an index that represents the overall health hazard (3).

3. Screening procedures in which the answers to predesigned questions of relative toxicity determine the importance category into which the chemical is placed [the decision-tree method of Cramer et al. (4) and Brusick (5)].

4. The combined judgment of experts representing various toxicologic disciplines. This is probably the most accurate assessment method, but is also the most expensive.

Before agents will be tested for their carcinogenicity in animal models, a case must be made to justify the time and considerable expense (i.e., up to a million dollars). To make the case for carcinogenicity testing in NTP bioassays, review panels consider the chemical-physical properties, including possible structure-activity relationships and the capacity of the agent in question to cause genetic alterations (i.e., mutations) in a variety of short-term genotoxic assays and the result of acute and subchronic assays.

11.1.3   Structure-Activity Relation Correlates

A method that is gaining acceptance as a method for determining that a chemical may be a potential carcinogen is that of structure-activity relationships (SAR). Dipple et al. (6) were the first to suggest that the ability of a chemical to react with cellular nuclear elements was a property common to a large number of the organic chemicals that were known carcinogens. The degree of chemical reactivity of a parent compound or a compound requiring metabolic activation (procarcinogen) can frequently be predicted from the chemical structure. For the majority of carcinogens, the body metabolizes the procarcinogen into highly reactive intermediate metabolites that are electrophilic. These electrophilic compounds are attracted to the electron-rich sites found in cellular nucleic acids and proteins (Fig. 11.2).

Evidence seems to point to a requirement that a chemical carcinogen must interact with DNA for the carcinogenic-mutagenic process to be initiated. Much of the data for this observation has been produced as a result of the study of the metabolic activation of benzo[a]pyrene (6a,7). Of interest is the failure of the Salmonella assay (Ames test) to detect the carcinogen dimethylbenzanthracene (DMBA). This is due to the failure of DMBA to bind to the Salmonella DNA, thus giving rise to a false-negative assay report (8). This group also found that although mouse skin activated DMBA to form adducts with DNA, this phenomenon did not appear to occur in mouse liver microsomes.

Binding to DNA in and of itself is not enough to produce the DNA

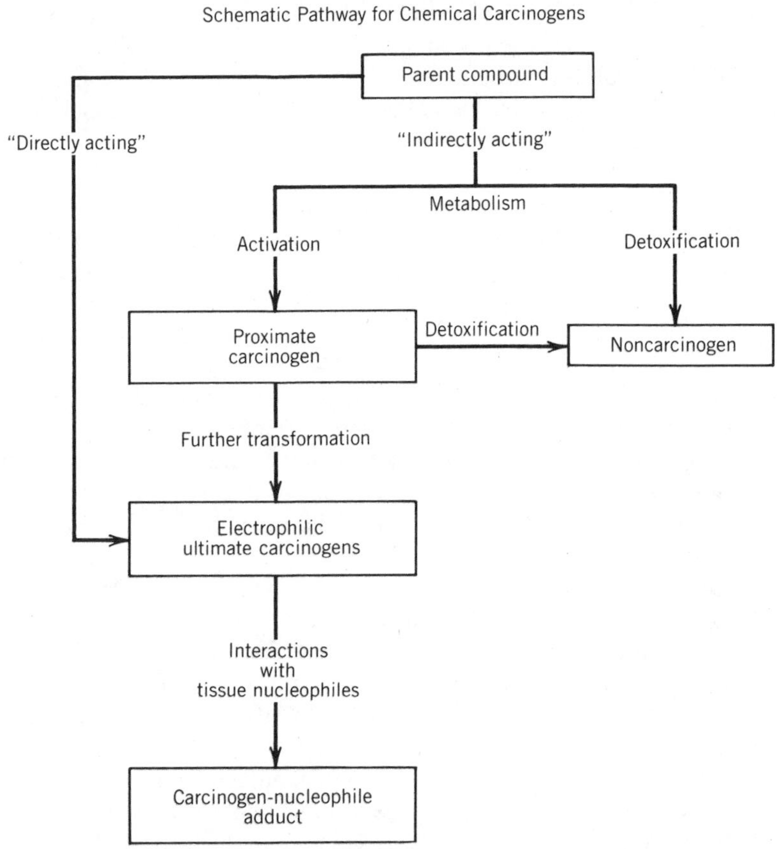

Schematic Pathway for Chemical Carcinogens

Figure 11.2. Examples of electrophilic activation of major classes of chemical carcinogens.

changes associated with mutagenicity. An example of this is methyl methane sulfate which binds rapidly to DNA but is not a carcinogen, while dimethylnitrosamine (DMN) which binds as quickly and to the same degree is a carcinogen (9,10). Apparently methyl nitrosamine binds the DNA nucleotide guanine in the O^6 and O^4 position, which is a critical step to effecting DNA replication, in contrast to methyl methane sulfate, which binds to the N^7 position (11).

One problem in utilizing SAR as a predictive factor between species is the difference, both qualitative and quantitative, in the way various carcinogens or procarcinogens are metabolized. An example of this is the *l*-isomer of 2-naphthylamine, which is *N*-hydroxylated to a powerful carcinogen in both rodents and humans. However, it has not been found to be a carcinogen in guinea pigs because they apparently lack the ability to hydroxylate the chemical to its active carcinogenic metabolite (12).

The SAR models under current investigation utilize three specific molecular properties: (1) the probability of transport from outside the cell to the ultimate site of the anticipated interaction with DNA, a function of lipophilicity; (2) the specificity and reversibility of the xenobiotic-DNA interaction; and (3) the degree of chemical reactivity of the xenobiotic or metabolite with DNA.

Once a chemical has been chosen for the NTP testing program, the chemical then moves to both acute (short-term) and chronic (long-term) test procedures.

11.1.4 Short-term Testing

The majority of short-term carcinogenicity tests are designed to evaluate the potential for a chemical to produce either a quantifiable mutation in a test system or an intracellular genetic change that would then predispose to a later mutation (genetic toxicity tests). The use of these mutagenicity studies for identifying potential carcinogens is based on the observation that there is a strong relationship between the results of the short-term mutagenicity tests in vitro and long-term in vivo carcinogenicity studies (13). However, the exceptions to this observed relationship are numerous both for apparent mutagens that do not seem to be human carcinogens and known human carcinogens that are not mutagenic in any current short-term tests, for example, asbestos. It is assumed that the initiation of the carcinogenesis process involves an alteration of genetic material in cells. This has provided the primary rationale for the use of the in vitro mutagen assays (genotoxicity studies) as screens or as evidence of the need for moving on to the next step, that of carcinogenicity testing in whole animals (14).

In vitro mutagenic test methods may include metabolic activation by liver microsomes and covalent bond formation with DNA. However, they do not take into consideration other important cellular processes such as transport across various membranes, concentration effects in specific organ systems, or removal from the cell, all of which may be important to whether a tumor ultimately develops or not.

11.1.5 Genotoxicity Tests

The Organization for Economic Cooperation and Development (OECD) has adopted guidelines for the 10 genetic toxicity tests that they considered to represent the most well-established and reproducible (Fig. 11.2). These are: (1) Ames *Salmonella*/liver microsome reverse-mutation assay; (2) *Escherichia coli* reverse-mutation assay; (3) rodent micronucleus assay; (4) in vitro chromosomal-aberration assay in mammalian cells; (5) sex-linked recessive lethal assay in *Drosophilia melanogaster;* (6) forward gene-mutation assay in mouse lymphoma L5178Y (tk +/−) cells; (7) forward gene mutation assay in Chinese hamster ovary (CHO) (HGPRT) cells; (8) forward gene-mutation assay in Chinese hamster V79 (HGPRT) cells; (9) in vivo chromosomal-aberration analysis in rodent bone marrow; and (10) the rodent dominant-lethal assay (1).

The frequent starting point and primary test for genetic toxicology testing is an assay for the ability of a chemical to induce a gene mutation (Ames *Salmonella* or *E. coli*. reverse mutation assays) coupled with an assay for chromosomal damage (rodent micronucleus assay or in vitro chromosome aberration assay) (Fig. 11.3). The reason for starting with short-term in vitro tests is that the cost of performing a whole-animal, two-species (usually mice and rats) carcinogen test is approximately $500,000 (15), compared with $800–$7000 for the majority of the standard in vitro tests (Table 11.2) (1).

A number of investigators (16,17) have suggested that the use of a specific combination of short-term tests could reduce the number of the false positives and false negatives that occur when only one test is employed. A tiered system (Table 11.3) employing three to four short-term tests in the decision-making process resulted in a predictive accuracy of the in vitro tests to whole animal carcinogenicity studies of 81.6–89.7% (18). However, no combination of tests as of yet can avoid false positives or false negatives. Yet, when a false-positive Ames *Salmonella* test results, it is more likely

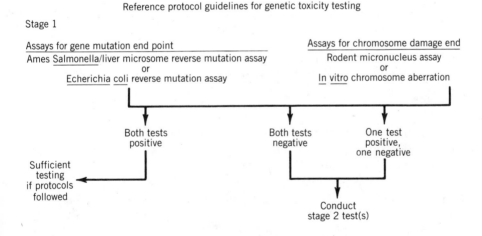

Reference protocol guidelines for genetic toxicity testing

Stage 1

Assays for gene mutation end point
Ames Salmonella/liver microsome reverse mutation assay
or
Echerichia coli reverse mutation assay

Assays for chromosome damage end
Rodent micronucleus assay
or
In vitro chromosome aberration

Both tests positive

Both tests negative

One test positive, one negative

Sufficient testing if protocols followed

Conduct stage 2 test(s)

Stage 2

Assays for gene mutation end point
Drosophila melanogaster sex-linked recessive lethal assay
Mouse lymphoma L5178Y (TK+/−) forward gene mutation assay
CHO (HGPRT) forward gene mutation assay
V79 (HGPRT) forward gene mutation assay

Assays for chromosome damage end point
In vivo chromosome aberration analysis in rodent bone marrow
Rodent dominant lethal assay

Figure 11.3. End points and tests defining them in the two-stage assessment of genetic toxicity. (53)

TABLE 11.3
Predictive Accuracy of Several Short-Term Tier Testing Regiments[a]

Tier Number	Tests Included in Tier	Number of Chemicals on Which Results Are Based	Predictive Accuracy
1	S. typhimurium/plate Differential killing, B. subtilis M45 Rec⁻ Cell transformation, BHK-21 cells	38	84.2
2	S. typhimurium/plate Cell transformation, BHK-21 cells Forward mutation, S. pombe	29	86.2
3	S. typhimurium/plate Forward mutation, S. pombe Cell transformation, BHK-21 cells Unscheduled DNA synthesis, HeLa cells	29	89.7
4	S. typhimurium/plate Unscheduled DNA synthesis, HeLa cells Cell transformation, BHK-21 cells	38	81.6
5	S. typhimurium/plate Cell transformation, BHK-21 cells Rabin's test (degranulation) rat liver cells	30	83.3
6	S. typhimurium/plate Unscheduled DNA synthesis, HeLa cells Cell transformation, BHK-21 cells	38	81.6

[a] Source: Ref. 18.

than not that the animal bioassay test was carried out improperly, resulting in a negative result, than that the short-term tests were wrong in the mutagen/carcinogen prediction (1).

Brusick (5) has proposed an interesting model system for evaluating the results of mutagen testing for a particular chemical. This system gives a predetermined weight to the positive and negative results from each of the various tests. The scoring system takes into account whether the agent is given with or without the bioactivating S-9 fraction as well as the extrapolative relevance of each genotoxicity assay. The final score then determines the likelihood that the test substance is a mutagen and a potential carcinogen. This type of system is attractive since it offers a potentially effective way to integrate all types of genotoxicity data and arrives at a collective genotoxicity score and a relative ranking of the mutagenic potential of chemical agents. This type of methodology is invaluable to decision makers who have to decide on which specific agents will be selected for the chronic cancer bioassays.

 The remainder of this genotoxicity section will focus on a description and analysis of the Ames *Salmonella* genotoxicity assay because of its widespread use and applicability to the field of environmental carcinogenesis.

11.1.5.1   The Salmonella Mutation Assay

The most widely used of all the short-term in vitro tests is the *Salmonella* reverse-mutation assay developed by Ames et al. (19). This assay tests for the ability of a chemical to induce histidine-requiring strain of *Salmonella* to revert (backward mutate) to a nonhistidine-requiring wild strain (which is normally found). To simulate in vivo metabolic activation, microsomes containing the mixed function oxidase system from rat liver are added to the media. (This is also known as the S9 fraction.)

The Ames test yields positive reversion results with many chemicals which have been identified as carcinogens in rodents (20,21). The test appears to be very good in predicting carcinogenicity for certain classes of compounds such as the aromatic amines, nitrosamines, and polyaromatic hydrocarbons (PAHs) but poorly predicts the carcinogenicity of some other classes (22). More specifically, the test does not appear to be useful for metal carcinogens because of the large amounts of various chelating metals including magnesium found in the culture media (23). In general, inorganic carcinogens do not appear to be mutagenic except in certain select systems. These inorganics appear to alter DNA replication through their effect on the fidelity of DNA synthesis (24). Other known carcinogens such as asbestos, DDT, and the class known as tumor promoters (discussed later) also give negative test results (21). Despite the widespread use of the Ames *Salmonella* test, a recent study found significant discrepancies between a group of collaborating laboratories performing the Ames assay on apparently identical compounds (25).

 A major concern is the ability of the Ames test to detect chemicals that will be carcinogens in vivo. One of the factors generating this concern is due to the loss of cellular and tissue organization of the metabolizing enzymes and cofactors between the intact cell in vivo and the microsomal fraction in vitro. The cellular disorganization is the result of grinding up the liver and centrifuging the resultant homogenized tissue, removing all but the microsomal (S9) fraction (26). Another consideration is the now well-recognized difference between various tissues in the same animal to metabolize the same substrate (27). The use of liver microsomes that might be lacking in a specific microsomal isoenzyme needed to bioactivate a carcinogen that may be lung specific is an example of this potential problem. This appears to be the case for safrole which, although a carcinogen in the in vivo bioassay test, is not converted in the Ames test to its active carcinogenic metabolite (23). Other carcinogens may be activated by enzymes found in the cystolic fraction of the cell (which is discarded) rather than the microsomal fraction. An example of this is cycasin which is activated by intestinal β-glucosidases. The lack of this enzyme in the Ames test produces negative results (23). Another one of the other factors that may lead to false negatives for the Ames test is the inability to use an appropriate dose of a test chemical due to its inherent toxicity to the *Salmonella* test organism (23).

False positives may occur if a bacterial tester strain contains enzymes capable of converting a chemical to an active mutagen and if these enzymes are deficient or absent in the mammals in which the bioassay is conducted. A specific example of this problem is seen with the nitro carcinogens. This chemical class is reduced by bacterial nitroreductase to its active carcinogen (nitroso) forms. Although mammalian liver contains similar enzymes as do organisms, the "in situ activation" by the enzymes within the bacteria results in a higher activation rate. This results in an increase in the apparent mutagenic potential of the nitro chemical compared with what would be expected in the intact animal.

Another potential reason for a false positive is that the bacterial strain used in the Ames assay contains a mutation that inhibits the ability of the bacteria to carry out DNA excision repair. This increases the chance that any DNA damage that might occur will not be repaired. This again would not be the case in the whole mammal where excision repair remains intact.

The presence of slight impurities in the test compound can also give false positives if the impurity is a very strong mutagen (27a). As mutagens can differ by a factor of 10^6 in their relative potency (Table 11.4), the ability of a 1% impurity that is a strong mutagen to produce a positive Ames test for the test compound could be predicted in advance. Either 99.9% + pure compounds must be used or each impurity should also be tested individually for its individual respective mutagenic potency before a decision is made as to whether a chemical is a mutagen in the Ames test. One negative aspect of using highly purified compounds in this test is that in the real world of human exposures, exposures to the impure or technical grade chemical is generally the rule, while exposure to a 99.9% pure compound is rare. This false negative potential is pointed out by Cragg et al. (27b), where a negative Ames test was found for a complex mixture of hydrocarbons from a petroleum distillate mixture as well as from a number of distillate fractions from the same mixtures. However, in a mouse dermal carcinogenesis bioassay, both the mixture and many of the individual fractions were found to be positive.

11.1.5.2 In Vivo Bioassay

When the National Cancer Institute (NCI) initiated the rodent bioassay system to detect possible human carcinogens, the issue was simply to identify the agent as carcinogenic or not. However, over the ensuing years the data from such studies have become the object of biostatistical evaluation in the form of regression analysis by which the incidence of cancer was estimated at doses many times lower than those actually used in the studies. This

TABLE 11.4
Range of Mutagenic Potency in the *Salmonella*/Microsome Test[a]

Chemical	Rev/nmol	Ratio
1,2-Epoxybutane	0.006	1
Benzyl chloride	0.02	3
Methyl Methanesulfonate	0.63	105
2-Naphthylamine	8.5	1,400
2-Acetylaminofluorene	108	18,000
Aflatoxin B_1	7,057	1,200,000
Furylfuramide	20,800	3,500,000

[a] *Source:* Ref. 50.

extrapolation process, which will be discussed in Chapter 17, literally forced the traditional cancer bioassay to provide information on not only which chemicals were carcinogens but how carcinogenic they were.

Due to the costs of labor and laboratory resources, laboratory studies are usually limited to no more than 50–100 animals of both sexes of two species, usually mice and rats, at each dose selected for study (Table 11.5). To make up for the logistic and financial limit on the number of animals that can be tested practically, one of the doses of the chemical is chosen to represent the highest dose that can be given and still not affect the long-term health or anticipated survival of the animal test population. This dose is known as the maximum tolerated dose (MTD).

The most commonly cited definition of the MTD is that of Sontag et al.

TABLE 11.5
The U.S. National Cancer Bioassay Protocol for Chemical Carcinogens[a]

Treatment Group		Species I	Species II
Level A	Male	50[b]	50
	Female	50	50
Level B	Male	50	50
	Female	50	50
Control[c]	Male	50	50
	Female	50	50

[a] *Source:* Ref. 28.
[b] Number of animals required.
[c] Some reduction in the number of animals is possible if controls are run jointly for a group of test compounds.

(28) in the NCI Guidelines for Carcinogen Bioassay in Small Rodents: "The MTD is defined as the highest dose of the test agent during chronic study that can be predicted not to alter the animal's normal longevity from effects other than carcinogenicity." According to Haseman (29) this definition provided the basis for dose selection in the NCI carcinogenicity studies and still strongly influences those performed by the NTP.

In practical reality, the MTD has been estimated from the concentration used in the subchronic (i.e., 90 day) study that "causes no more than a 10% weight decrement, as compared to the appropriate control groups, and does not produce mortality, clinical signs of toxicity, or pathologic lesions (other than those that may be related to a neoplastic response) that would be predicted [in the chronic study] to shorten an animal's natural life span." Since that 1976 publication, Haseman (29) asserts that weight gain is no longer considered the primary variable in dose selection but that pathology and pharmacokinetic data are given greater weight.

With respect to the pharmacokinetic perspective, a number of investigators feel that a dose is unacceptable if it "overlooks" or "saturates" the metabolic pathways. It is felt that such excessive doses may overwhelm the enzymatic processes for activation of the chemical to the toxic form or deactivation of the toxic form to a less harmful metabolite (30). To address this possible issue, the NTP now conducts a series of chemical disposition experiments for many agents that are studied in 2-year cancer bioassays. These pharmacokinetic studies are designed principally for those agents for which chemical disposition data are lacking. These studies develop information on the rates and routes of excretion and discern a dose range in which chemical absorption, metabolism, and disposition are linear. If needed, a more extensive study is then conducted to assess kinetic parameters such as degree of absorption, half-lives (specific tissue and whole body), time to reach the target site, and tissue concentration at the target site. Other objectives of such a study are assessments of chemical concentrations in all major tissues and possible target tissues, as well as determination of parent compound and metabolic ratios at six to nine times. With such information, the NTP hopes to improve its dose-selection process for the MTD. More specifically, the NTP along with IARC (31) operates on the premise that "the metabolism and pharmacokinetics of the test compound not be the primary criteria for selecting the high dose, but that pharmacokinetic data . . . should be taken into consideration in selecting lower doses for study."

In addition to the metabolic loading objection that has been used in selection of the MTD is the concern for secondary carcinogenesis. Secondary carcinogenesis occurs when excessive doses produce nonspecific tissue damage that is considered to be a secondary mechanism leading to carcinogenic effects. Such secondary effects may arise from a variety of causes, including the creation of nutritional imbalances, immune suppression, chronic physical injury, and others (32). An example of such a phenomenon is the use of high doses of hepatotoxins which may lead to initiation of liver tumors (33). Presumably, if the liver were not injured the tumor would not have developed. It has been argued that such recurrent liver injury plays a determining role in the induction of hepatic tumors by perchloroethylene. Chloroform-induced carcinogenicity is also thought to be acting via a secondary mechanism or a nongenetic (i.e., epigenetic) process (34,35).

According to Haseman (29), metabolic overload and secondary carcinogenesis are "legitimate issues that should be considered in the design and interpretation of carcinogenicity studies." However, he argued that they should not be the justification for routinely explaining away all tumor increases seen only at high dose levels.

The major argument to support the use of the MTD in cancer testing is statistical in nature. Since current testing protocols are known to be relatively insensitive for detecting carcinogenic effects (see 29,36,37); due especially to financial limitations concerning the size of the study, it is deemed necessary to use a high dose to evoke any possible carcinogenic effect.

Page (38) proposed a list of major recommendations (Table 11.6) for a carcinogen bioassay in small rodents. In addition to issues such as the minimum number of animals and the doses chosen, he felt that the route of administration should, if possible, duplicate that anticipated in humans. He also placed a great deal of emphasis on animal husbandry, such as the use of filtered air together with sterilized bedding and feed. This is not only to avoid unnecessary animal mortality but also to reduce the chance of an intercurrent tumorigenic viral infection that could confound the results by producing false positives.

11.1.5.3 Problems with In Vivo Carcinogenicity Testing

There are a number of agents that can modify the carcinogeneic process in vivo serving either to increase or decrease the incidence of observed tumors. These include: enzyme inducers, specific enzyme cofactors, metabolizing enzyme poisons, antioxidants, hormones, and any agent that affects the overall nutritional status of the test animal (26).

Enzyme inducers can either increase or decrease the effect of a carcinogen on the animal in vivo. The administration of the enzyme inducer 3-methylcholanthrene (3-MC) resulted in a failure of the hepatocarcinogen methyl-dimethyl aminoazobenzene to produce liver tumors in rats. The reason was that the 3-MC increased the metabolism of the carcinogen, resulting in a much lower effective dose to the liver (39). Other agents that are inducers of metabolizing enzymes and thus reduce the carcinogenicity of chemicals not requiring activation are: polycyclic aromatic hydrocarbons; quinones; chlorinated hydrocarbons, such as DDT and PCBs; certain antioxidants, such as BHT; phenobarbital; and some steroids (40). For the enzyme-inducing agents to be effective in reducing tumor incidence, they must be administered prior to exposure to the carcinogens.

TABLE 11.6
Major Recommendations for Carcinogen Bioassay in Small Rodents[a]

Animals	At least two species, usually rat, mouse, or hamster; both sexes; weanlings started on treatment
Animal care	Improved facilities and care; filtered enclosed cages; sterilized bedding and feed
Chemistry	Analysis for purity, stability, and proper preparation; strict safety requirements
Route of administration	Same as usual human exposure, where possible
Doses and dose levels	At least two doses; highest—that tolerated
Treatment period	Major portion of natural lifespan—24 months on study
Sample size	Fifty (50) per dose/sex, including control groups
Pathology examination	Careful and complete gross and microscopic examination of all animals in study
Data reporting	All aspects including individual animal clinical effects, tumor and nontumor pathology results

[a] Source: Ref. 38.

Carcinogenicity can also be affected if a substance is present that can inhibit metabolism and reduce the subsequent inactivation of a carcinogen. This is true in the case of the enzyme epoxide-hydrolase, which is the primary enzyme responsible for converting the strong epoxides formed by many compounds to their less toxic dihydrodiols. The epoxide hydrolase inhibitor 1,1,1-trichloro-2-propeneoxide will increase the tumor incidence for epoxide-forming chemicals such as benzo[a]pyrene and 3-MC on the skin (41).

Some investigators have felt that carcinogenicity testing should involve administering the test agent to the parents of the test animals during gestation and during the neonatal period as well as throughout adult life (42). This is based on the belief that for many chemicals exposure begins in the fetal stage as a result of ingestion of the chemical by the mother.

The ideal animal system in which to study whether or not an in vitro mutagen is a potential carcinogen in humans would be one where the chemical under study was absorbed and metabolized by the same processes and at the same rate as in the human. Primates are highly desirable in this respect due to their similarities to humans (43). However, both their cost and their life span, which is too long to permit timely decisions on which to base human risk assessment decisions, prevent their use in standard bioassay programs. Additionally, the long gestational period in primates would substantially delay mutagenicity results long past that which is politically acceptable.

Rodents have become the standard animal used in bioassay systems despite substantial differences in metabolism of many xenobiotics between rodents and humans. Additional differences between species make rodent-human extrapolations even more difficult. They are: the transport of the mutagenic metabolite to specific target organs; the efficiency of DNA repair and misrepair mechanisms; and differences in tissue levels of intracellular protectant systems (44). An additional problem in carcinogenicity testing is that the choice of animal strain selection is controversial because all the current methods utilize highly inbred strains. The advantage in the use of inbred strains is the greater uniformity of response. However, a predictable incidence of spontaneous tumors may result in a failure to detect tumorigenic potential due to a more narrow range of animal sensitivity to the potential carcinogen than that which exists in random-bred (outbred) animals (38).

The significant difference among rodent strains in the development of organ-specific spontaneous tumors due to unknown factors further complicates extrapolation of the results to humans. Without any explanation for the strain difference in spontaneous tumors, it is difficult to know what strain best reflects humans. This is illustrated in the fact that the incidence of liver and bladder tumors differs greatly among five different strains of rats following administration of the potent carcinogen N-2-fluorenylacetamide (Table 11.7).

Both the number of animals and the incidence of spontaneous tumors in the controls (animals not receiving the test chemical) are important in determining how many tumors must be detected to reach the 5% level of statistical significance (Table 11.8). As an example, if a mouse strain had a spontaneous tumor evidence of 20% for an organ-specific tumor (controls), the animals given the chemical would have to have a tumor incidence of 40% to be statistically significant if 50 animals are employed in each group! *Any* degree of spontaneous tumor incidence in controls diminishes the chance of detecting a weak carcinogen producing a low number of tumors (38). It should also be kept in mind that all of the various carcinogen tests assume a "null hypothesis," that is, there is no difference in tumor incidence between

TABLE 11.7
Strain Differences in Cancer Induction by N-2-Fluorenylacetamide in Rats[a]

Strain	Average Survival (weeks)	Liver Tumors (%)	Bladder Tumors (%)
A × C	54	50	10
August	43	70	0
Copenhagen	78	20	40
Fischer	45	80	0
Marshall	31	80	0

[a] Animals started on test from 10–15 weeks of age. N-2-Fluorenylacetamine was administered in the diet at a level of 0.05% for 1 year. Data from Ref. 51.

TABLE 11.8
Incidence of Tumors in Treated Groups Required for Significance
($p = 0.05$) Depending on Group Size and Spontaneous Tumors[a]

Percent Incidence of Tumors in Controls	Number of Animals per Group[b]				
	10	25	50	75	100
0	50%	20%	12%	8%	6%
10	70	40	28	24	21
20	80	52	40	36	34
30	90	64	52	47	45
40	100	72	62	58	55

[a] Calculations based upon tabulations of Mainland and Murray (52).
[b] Controls and treated groups of same size.

treated and control groups. If the level of statistical significance approaches but does not reach the widely accepted 0.05 level, the test chemical is then considered a noncarcinogen, whereas change from tumor-free to tumor-present in but one out of 100 test animals would have resulted in the chemical being called a carcinogen.

The many difficulties experienced in the current bioassay program have caused Salsburg (45) to feel that the current program lacks acceptable specificity and validity and that a drastically different design is required to address those concerns.

11.1.5.4   Tumor Promoters

A group of compounds are recognized which, although they do not produce cancer themselves, appear to increase the qualitative and quantitative potential for a known carcinogen (initiator) to induce a malignant change. The promoter is therefore not carcinogenic alone; it only affects the carcinogenic potential of a known initiator if given after the initiator. Other differences from initiators are seen in Table 11.9 (46). The failure of any currently available in vitro mutagenicity test to detect promoters has led to the American Association for the Advancement of Science (AAAS) Panel on Carcinogenicity to state, "The relevance of [promoters] in human carcinogens cannot be determined, and for the moment, such bioassays (*for promoters* can only be used as ancillary evidence in assessing data for potential human hazard" (47).

11.1.5.5   Interpretation of Testing Results

Evidence of genotoxicity or mutagenic potential of a chemical should not in and of itself be considered to be any more than a warning that the material

TABLE 11.9
A Comparison of Biologic Properties of Initiating Agents
and Promoting Agents[a]

Initiating Agents	Promoting Agents
1. Carcinogenic by themselves "solitary carcinogens"	1. Not carcinogenic alone
2. Must be given before promoting agent	2. Must be given after the initiating agent
3. Single exposure is sufficient	3. Require prolonged exposure
4. Action is "irreversible" and additive	4. Action is reversible (at early stage) and not additive.
5. No apparent threshold	5. Probable threshold
6. Yield electrophiles: bind covalently to cell macromolecules	6. No evidence of covalent binding
7. Mutagenic	7. Not mutagenic

[a] *Source:* Ref. 46.

may be a carcinogen in test animals. However, no direct assessment of human risk can be made from any of the in vitro tests now in use. The absence of detectable genotoxicity in vitro and positive results in an animal bioassay system will require a careful analysis for an explanation of the observed discrepancy.

When the carcinogen results in the treated animals are compared with control animals for the purpose of human extrapolation, the following factors need to be considered (12):

1. The percentage of animals with a given tumor in a specific organ in relation to tumors in the same organ in controls.
2. The multiplicity of such tumors in certain tissues like lung, breast, and intestine.
3. The latent period leading to a particular neoplasm.

These data taken together give a rough indication of the relative potency of a particular carcinogen. One of the gray areas in interpolating in vivo animal bioassay results to humans is where there is a low (10–20%) yield of tumors with very high doses of a test chemical. This type of bioassay data does not provide a firm foundation to: (1) state that the substance is a proven carcinogen; and (2) attempt to make an assessment of risk to humans. One thing is clear, it is only after a positive bioassay result that a substance is labeled a carcinogen. Positive in vitro tests can only raise a suspicion, not result in a conclusion.

According to the 1984 proposed carcinogen guidelines of the EPA, the "question of how likely an agent is to be a human carcinogen should be answered in the framework of a weight-of-evidence judgment. Judgments about the weight of evidence involve considerations of the quality and adequacy of the data and the kinds of responses induced by a suspect carcinogen. There are three major steps to characterizing the weight of evidence for carcinogenicity: (1) Characterization of the evidence from human studies and from animal studies individually, (2) Combination of the characterizations of these two types of data into a final indication of the overall weight of evidence for human carcinogenicity, and (3) Evaluation of all supportive information to determine if the overall weight of evidence should be modified.''

A system for evaluating the weight of evidence has been proposed by the EPA. Their approach represents a modification of the classification scheme used by the International Agency for Research on Cancer (IARC) (48). The IARC approach asserted that the evidence that an agent causes cancer in man is considered at one of three levels of scientific strength: sufficient, limited, and inadequate. The EPA (49) has recently offered an adaptation to the IARC approach for classifying the weight of evidence for animal and human data. The EPA classification system for the characterization of the overall weight of evidence for carcinogenicity includes:

Group A—Carcinogenic to Humans. The category is employed only when there is sufficient* evidence from epidemiologic studies to establish a causal association between exposure to agent(s) and cancer.

Group B—Probably Carcinogenic to Humans. This category involves agents for which the evidence of human carcinogenicity from epidemiologic investigations changes from almost ''sufficient'' to ''inadequate.'' Because of this variation in quality of the data, Group B is divided into two subgroups, B_1, for which there is at least limited evidence of carcinogenicity to humans from epidemiologic studies, and B_2, for which there is inadequate human data but adequate (sufficient) animal study evidence.

Group C—Possibly Carcinogenic to Humans. This category includes agents with limited† evidence of carcinogenicity in animals in the absence of human data. It includes a wide variety of evidence such as definitive malignant responses in a single well-conducted experiment, marginal tumor response in studies having inadequate design or reporting, benign but not malignant tumors with an agent showing no response for mutagenicity, and marginal responses in a tissue known to have a high and variable background rate.

---

* Sufficient evidence is achieved when there is an increased incidence of malignant tumors or combined malignant and benign tumors. If the benign tumors are not judged to have the potential to progress to malignancies of the same morphologic type, they will not be combined with the malignant neoplasms.

† Limited evidence of carcinogenicity is defined by having data reflecting a single species, strain, or experiment or that the experiments are restricted by inadequate dosage levels, inadequate duration of exposure, inadequate follow-up period, poor survival, too few animals, or only an increase in benign tumors.

Group D—Not Classified. This category is reserved for substances for which there is inadequate evidence of carcinogenicity.

Group E—No Evidence of Carcinogenicity for Humans. This final grouping is employed for substances for which there is no evidence for carcinogenicity based on both animal and human epidemiologic evidence.

11.2 CONCLUSION

This chapter is intended to provide the reader with an overview concerning how toxicologists assess the carcinogenic potential of chemical agents. The methodologies employed do not differ between suspected carcinogenic agents whether they are found in water, food or air. The only deviation may be via the route of administration. Thus, the procedures discussed have applicability beyond the scope of this book.

REFERENCES

1. National Research Council, *Toxicity Testing: Strategies to Determine Needs and Priorities,* National Academy Press, Washington, DC, 1984, p. 276.
2. Astill, B.D., Lockhart, H.B., Jr., Moses, J.B., Nast, A.N.M., Raleigh, R.L., and Terhaar, C.J., Sequential testing for chemical risk assessment, in Conway, R. (ed.), *Environmental Risk Analysis of Chemicals,* Second International Congress of Toxicology, Brussels. Van Nostrand-Reinhold, New York, 1981.
3. Brown, S.L., Cofer, R.L., Eger, T., Liu, D.H.W., Mabey, W.R., Suttinger, K., and Tuse, D., A Ranking Algorithm for EEC Water Pollutants, CRESS Report No. 136, SRI International, Menlo Park, California, 1980. 8 pp.
4. Cramer, G.M., Ford, R.A., and Hall, R.L., Estimation of toxic hazard—A decision tree approach, *Food Cosmet. Toxicol., 16,* 255–276, 1978.
5. Brusick, D., *Principles of Genetic Toxicology,* Plenum Press, New York, 1980, p. 95.
6. Dipple, A., Lawley, P.D., and Brookes, P., Theory of tumor initiation by chemical carcinogens: Dependence of activity on structure of ultimate carcinogen. *Eur. J. Cancer, 4,* 493–506, 1968.
6a. Baird, W.M., Dipple, A., Grover, P.L., Sims, P., and Brookes, P. Hydrocarbon-deoxyribonucleoside products formed by the binding of derivatives of 7-methyl-benz(A)anthracene to DNA, *Br. J. Cancer, 28,* 84–85, 1973.
7. Borgen, A., Darvey, H., Castagnoli, N., Crocker, T.T., Rasmussen, R.E., and Wang, I.Y., Metabolic conversion of benzo(a)pyrene by Syrian hamster liver microsomes and binding of metabolites to deoxyribonucleic acid, *J. Med. Chem., 16,* 502–506, 1973.
8. Dipple, A., Baird, W.M., Bigger, C.A.H., Moschel, R.C., and Andrews, A.W., DNA-carcinogen interactions, in Symposium on Structural Correlates of Carcinogenesis and Mutagenesis from *Proceedings of the Second FDA Office of Science Summer Symposium,* August 31, 1977, Office of Science, FDA, 1977, pp. 4–7.
9. O'Connor, P.J., Capps, M.J., Craig, A.W., Lawley, P.S., and Shah, S.A., Differences in the patterns of methylation in rat liver ribosomal RNA after reaction *in vivo* with methyl methanesulphorate and N-N-dimethylnitrosamine, *Biochem. J., 129,* 519–528, 1972.
10. O'Connor, P.J., Capps, M.J., and Craig, A.W., Comparative studies of the heptatocarcinogen N,N-dimethylnitrosamine in vivo: Reaction sites in rat liver DNA and the significance of their relative stabilities, *Br. J. Cancer, 27,* 153, 1973.

11. Gershman, L.L., and Ludlum, D.B., The properties of O⁶-methylguanine in templates for RNA polymerase, *Proc. Am. Assn. Canc. Res., 14,* 13, 1973.

12. Weisburger, J.H., and Williams, G.M., Decision point approach to carcinogen testing, in Structural Correlates of Carcinogenesis and Mutagenesis from *Proceedings of the Second FDA Office of Science Summer Symposium,* August 31, 1977, Office of Science, FDA, 1977, pp. 45–52.

13. ICPEMC, Mutagenesis testing as an approach to carcinogenesis, *Mutat. Res., 99,* 73–91, 1982.

14. Squire, R.A., Carcinogenicity testing and safety assessment. *Fund. Appl. Toxicol., 4,* 5326–5334, 1984.

15. Weinstein, M.C., Cost-effective priorities for cancer prevention, *Science, 221,* 19–23, 1983.

16. Bridges, B.A., Use of a three-tier protocol for evaluation of long-term hazards, particularly mutagenicity and carcinogenicity, in Montesano, R., Bartsch, H., and Tomatis, L. (eds.), *Screening Tests in Chemical Carcinogenesis,* IARC Scientific Publications No. 12, Lyon, France, International Agency for Research on Cancer, 1976.

17. Weisburger, J.H., and Williams, G.M., Carcinogen testing: Current problems and new approaches, *Science, 214,* 401–407, 1981.

18. Lave, L.B., Omenn, G.S., Heffernan, K.D., and Dranoff, G., Analysis of the cost-effectiveness of tier-testing for potential carcinogens as reported in *Toxicity Testing: Strategies to Determine Needs and Priorities,* National Research Council, National Academy Press, Washington, DC, 1982, p. 227.

19. Ames, B.N., McCann, J., and Yamasaki, E., Methods for detecting carcinogens and mutagens with the Salmonella/mammalian microsome mutagenicity test, *Mutat. Res., 31,* 347–364, 1975.

20. Ames, B.N., Identifying environmental chemicals causing mutations and cancer, *Science, 204,* 587–593, 1979.

21. Purchase, I.F.H., An appraisal of predictive tests for carcinogenicity, *Mutat. Res., 99,* 53–71, 1982.

22. Rinkus, S.J., and Legator, M.S., Chemical characterization of 465 known or suspected carcinogens and their correlation with mutagenic activity on the Salmonella typhimurium system. *Cancer Res., 39,* 3289–3318, 1979.

23. McCann, J., and Ames, B.N., The Salmonella/microsome mutagenicity test: Predictive value for animal carcinogenicity, in *Mutagenesis,* Flamm, W.G., and Mehlman, M.A. (eds.), Hemisphere Publishing, Washington, 1978, pp. 87–108.

24. Sirover, M.A., and Loeb, L.A., Metal-induced infidelity during DNA synthesis, *Proc. Nat. Acad. Sci., 73,* 2331, 1976.

25. de Serres, F.J., and Ashby, J. (eds.), Evaluation of short-term tests for carcinogenesis. Report of the International Collaborative Program. *Progress in Mutation Research,* Vol. 1, Elsevier-North Holland, New York, 1981, p. 828.

26. Clayson, D.B., Nutrition and experimental carcinogenesis: A review, *Cancer Res., 75,* 3292–3300, 1975.

27. Gram, T., Xenobiotic metabolism on mammalial lung, in *Extrahepatic Metabolism of Drugs and Other Foreign Compounds,* Gram, T. (ed.), SP Medical and Scientific Books, New York, 1980, pp. 159–209.

27a. Donahue, E.V., McCann, J., and Ames, B.N., Detection of mutagenic impurities in carcinogens and non-carcinogens by high-pressure liquid chromatography and the Salmonella/microsome test, *Cancer Res., 38,* 431–438, 1978.

27b. Cragg, S.T., Conaway, C.C., and MacGregor, J.A., Lack of concordance of the mouse dermal carcinogenesis bioassay for complex petroleum hydrocarbon mixtures, *Fund. Appl. Toxicol., 5,* 382–390, 1985.

28. Sontag, J.M., Page, N.P., and Saffiotti, U., Guidelines for carcinogen bioassay in small rodents, NCI-CG-R-1, National Cancer Institute Carcinogenesis Technical Report Series No. 1, U.S. Department of Health, Education and Welfare, Washington, DC, 1976.

29. Haseman, J.K., Issues in carcinogenicity testing: Dose selection, *Fund. Appl. Toxicol., 5,* 66–78, 1985.

30. Gehring, P.J., Watanabe, P.G., and Park, C.N., Resolution of dose response toxicity data for chemicals requiring metabolic activation: Example vinyl chloride, *Toxicol. Appl. Pharmacol., 44,* 581–591, 1978.

31. International Agency For Research on Cancer (IARC), Long-term and short-term screening assays for carcinogens: A critical appraisal, *IARC Monogr. Eval. Carcinog. Risk Chem. Man.,* Suppl. 2, pp. 21–83, 1980.

32. International Life Sciences Institute (ILSI), The selection of doses in chronic toxicity/carcinogenicity studies, In *Current Issues in Toxicology,* Grice, H.C. (ed.), Springer-Verlag, New York, 1984, pp. 9–49.

33. Shank, R.C., and Barrows, L.R., Toxicity-dependent DNA methylation: Significance to risk assessment, In *Health Risk Analysis,* Richmond, C.R., Walsh, P.J., and Copenhaver, E.D. (eds.), Franklin Institute Press, Philadelphia, 1981, pp. 225–233.

34. Reitz, R.H., Quast, J.F., Stott, W.T., Watanabe, P.G., and Gehring, P.J., Pharmacokinetics and macromolecular effects of chloroform in rats and mice: Implications for carcinogenic risk estimation, In *Water Chlorination: Environmental Impact and Health Effects,* Jolley, R.L., Brungs, W.A., Cumming, R.B., and Jacobs, V.A. (eds.), Vol. 3, Ann Arbor Science Pub., Ann Arbor, MI., 1980, pp. 983–993.

35. Stott, W.T., Reitz, R.H., Schumann, A.M., and Watanabe, P.G., Genetic and nongenetic events in neoplasia, *Food Cosmet. Toxicol., 19,* 567–576, 1981.

36. Food Safety Council (FSC), Chronic toxicity testing: Proposed system for food safety assessment, *Food Cosmet. Toxicol., 16* (Suppl. 2), 97–108, 1978.

37. Occupational Safety and Health Administration (OSHA), Identification, classification and regulation of potential occupational carcinogens, *Fed. Register, 45*(15), 5001–5296, 1980.

38. Page, N.P., Chronic toxicity and carcinogenicity guidelines, in Animal Toxicity and Carcinogenesis from *Proceedings of the Conference on the Status of Predictive Tools in Application to Safety Evaluation: Present and Future,* Pathotox, Illinois, 1977.

39. Raha, C.R., Gallagher, C.H., Shubik, P., and Peratt, S., Covalent binding to protein of the K-region oxide of benzo[a]pyrene formed by microsome incubation, *J. Natl. Canc. Inst., 57,* 33–38, 1976.

40. Clayson, D.B., The importance of metabolic considerations, in Symposium on Structural Correlates of Carcinogenesis and Mutagenesis from *Proceedings of the Second FDA Office of Science Summer Symposium,* August 31, 1977, Office of Science, FDA, 1977a, pp. 38–44; Clayson, D.B., Relationships between laboratory and human studies, *J. Environ. Pathol. Toxicol., 1,* 31–90, 1977b.

41. Burki, K., Stoming, T.A., and Bresniche, E., Effects on an epoxide hydrase inhibitor on *in vitro* binding of polycyclic hydrocarbons to DNA and on skin carcinogenesis, *J. Natl. Canc. Inst., 52,* 785–788, 1974.

42. Friedman, L., *Carcinogenesis Testing of Chemicals,* Goldberg, L. (ed.), CRC Press, Cleveland, 1973.

43. Calabrese, E.J., *Principles of Animal Extrapolation,* John Wiley & Sons, New York, 1983.

44. Kornbrust, D.J., and Mavis, R.D., Relative susceptibility of microsomes from lung, heart, liver, kidney, brain and testes to lipid peroxidation: Correlation with vitamin E content, *Lipids, 15,* 315–322, 1980.

45. Salsburg, D., The lifetime feeding study in mice and rats—An examination of its validity as a bioassay for human carcinogenesis, *Fund. Appl. Toxicol., 3,* 63–67, 1983.

46. Weinstein, I.B., Evaluating substances for promotion, co-factor effects and synergy in the carcinogenic process, *J. Environ. Pathol. Toxicol., 3,* 89–101, 1980.

47. AAAS, Interdisciplinary Panel on Carcinogenicity, Criteria for evidence of chemical carcinogenicity, *Science, 225,* 682–687, 1984.

48. International Agency for Research on Cancer (IARC), IARC Monographs on the Evalua-

tion of the Carcinogenic Risk of Chemicals to Humans, Supplement 4, Lyon, France: International Agency for Research on Cancer, 1982.

49. U.S. Environmental Protection Agency (EPA), Proposed guidelines for carcinogen risk assessment, *Fed. Register, 49:* 46294–46301, 1984.

50. McCann, J., Choi, E., Yamasaki, E., and Ames, B.N., Detection of carcinogens as mutagens in the Salmonella/microsome test: Assay of 300 chemicals, *Proc. Natl. Acad. Sci., 72,* 5135–5137, 1975.

51. Dunning, W.F., Curtis, M.R., and Madsen, M.E., Induction of neoplasms in five rats with acetylaminofluorene, *Cancer Res., 7,* 134–140, 1947.

52. Mainland, D., and Murray, I.M., Tables for use in fourfold contingency tests, *Science, 116,* 591–594, 1952.

53. Organization for Economic Cooperation and Development, Short-term and Long-term Toxicology Groups, Final Report, Paris, 1979, 185 pp.

54. Organization for Economic Cooperation and Development, *Guidelines for the Testing of Chemicals. Organization for Economic Cooperation and Development,* Paris, 1981, 700 pp.

55. Watanabe, P.G., Young, J.D., and Gehring, P.J., The importance of non-linear (dose-dependent) pharmacokinetics in hazard assessment. *J. Environ. Pathol. Toxicol., 1,* 147–159, 1977.

CHAPTER 12

Epidemiologic Approaches to the Assessment of Carcinogens in Drinking Water

Kenneth P. Cantor

Environmental Epidemiology Branch
National Cancer Institute
National Institutes of Health
Bethesda, Maryland

Marty S. Kanarek
Theresa B. Young

Department of Preventive Medicine
Medical School
University of Wisconsin
Madison, Wisconsin

12.1 INTRODUCTION

Epidemiology as an instrument of public health research and policy has origins in the water-borne cholera epidemics of mid-nineteenth century London. John Snow, a physician, collected information in a door-to-door survey and determined that cholera was much more common in households that used water piped from a reach of the Thames River heavily contaminated by sewage than in those served by cleaner Thames water (1). The needed preventive measures were taken, although the mechanism of disease causation was poorly understood, it being decades before the microbial basis of infectious disease was firmly established. In the industrialized world, the past 100 years have witnessed important shifts in the major causes of human illness and mortality. With the understanding and control of infectious diseases such as cholera, typhoid fever, and tuberculosis, slowly developing chronic conditions such as arteriosclerotic heart disease, chronic obstructive lung disease, and cancer have become the major debilitating conditions and causes of death. This change in disease patterns has led to the modification of epidemiologic techniques and the introduction of many new approaches.

In recent years, there is a resurgent interest in water-borne contaminants, but the current focus is on low levels of organic chemicals that may be human carcinogens. Among these are solvents, industrial process intermediates, and other industrial chemicals that are carcinogenic in laboratory feeding experiments (2). Contamination of ground water with industrial chemicals, especially from hazardous waste disposal sites, has received special attention (3–7). Chlorination by-products, including chloroform, other trihalomethanes (THMs), and higher-molecular-weight halogenated organics (8), are also of concern (9,10). Among chlorination by-products, chloroform is a carcinogen (11), others are mutagenic in the Ames *Salmonella* assay (12,13), and concentrated mixtures of higher-molecular-weight organic fractions transform mammalian cells in tissue culture (14) and induce skin-painted tumors in rats (15). The presence and toxicity of chlorination by-products has raised the possibility that the time-tested benefits of chlorine disinfection may be in part offset by an increase in the chronic disease burden.

How are these potential risks best evaluated? Laboratory experimentation can provide important information regarding the exposures to chemicals in drinking water that should be of major concern (2). However, interpretation of positive laboratory findings is "less straightforward than one would wish, particularly if attempts have been made to extrapolate quantitatively" (16). In addition, laboratory approaches are generally not well suited to evaluate risks from exposure to complex mixtures, especially those found in drinking water that exhibit day-to-day variation in chemical makeup and concentration. Epidemiologic observations continue to be needed. The use of epidemiologic findings as a basis for preventive action does not require extrapolation from very high to low doses, or across species. In this chapter, we address the major epidemiologic methods for evaluating chronic disease risk from drinking water contaminants, with emphasis on strengths and limitations of the different approaches.

12.1.1   Factors That Influence Epidemiologic Assessment

Epidemiologic assessment of potential risk from low levels of drinking water contaminants is complicated by the special characteristics of both the possible health effects and the exposure. Rather than the hours or days that separate exposure to an infectious agent and subsequent illness, cancer and other chronic diseases develop over a period of several decades. Although many factors, such as age, nutritional status, and presence of other diseases may influence the occurrence and outcome of typhoid fever or cholera, the presence of the infectious agent is a necessary precondition for development of the disease. For most chronic diseases, in contrast, any of several environmental, genetic, metabolic or other factors, acting alone or in concert, can result in similar disease outcomes in different individuals. Some types of cancer are strongly linked to very specific exposures (e.g., mesothelioma and asbestos, angiosarcoma of the liver and vinyl chloride), but these are exceptions to the general rule of multiple, interacting risk factors (17).

The long latency for cancer requires knowledge of exposures that occurred many years in the past, up to 5 or 6 decades before manifestation of the disease. Methods for quantifying contaminants in municipal water supplies are only recently available and no direct data exist for water supplies of previous decades. Fortunately, estimation of past contaminant levels in drinking water is facilitated by several factors, among them relative long-term stability of water treatment practices associated with large fixed investments, and the availability of historical records from larger water utilities detailing water source and treatment practices. These data permit characterization of past water quality for many large populations, especially those with low levels of in-migration. In epidemiologic studies that collect information from individuals, personal residential histories may be gathered, and a link made to historical water utility records to create a lifetime exposure profile.

With the availability of gas chromatography/mass spectroscopy analytical methods, there is increasing emphasis on routine surveillance of many organic water pollutants, and quantitative data on many water supplies is accumulating. These data are useful for epidemiologic studies in two respects. Current measurements can be used directly as estimates of past levels for contaminants that show little temporal variation. For other compounds, measures of current levels can be used in conjunction with routinely collected water utility data to identify characteristics that are useful in estimating contamination levels. Statistical models that incorporate these characteristics can then be applied to estimate past exposures of study subjects in epidemiologic investigations.

Additional features of water contaminant exposures that affect epidemiologic approaches are their relatively low levels and their usual occurrence as mixtures. Levels of carcinogens in the general environment are generally much lower than occur in occupational settings, where epidemiologic studies have been successful in identifying and quantifying health effects. However, exposure to general environmental factors often commences at an early age or prenatally, when the person may be more sensitive, and exposures are often continuous rather than limited to a 40-hour work week. It is likely that low-level environmental exposures result in only small risk increases, since adverse effects of most known carcinogens are dose-related. John Snow's success in linking cholera to polluted water (1), and the more recent discovery of smoking as the major cause of lung cancer were facilitated by the profound impact of the exposure on the probability of disease. In these cases, exposed persons are 10 or more times as likely to become ill as the nonexposed. In contrast, long-term exposure to low levels of water contaminants is not expected to increase risk by much more than 50 or 100% above that of the unexposed. This places severe requirements on the care with which epidemiologic investigations must proceed. Among other considerations, it is important that the statistical power of studies to detect small increases in risk be maximized through the use of accurate exposure estimates and large sample sizes.

Epidemiologic studies of the dilute organic mixtures that occur in drinking water are not expected to provide the level of specificity possible in occupational studies, where exposures are often of high intensity and limited variety. Positive associations observed in studies of drinking water contaminants can only lead to estimates of risk related to mixtures. For the purpose of prevention, this may not greatly matter, since chemicals in mixtures, such as chlorination by-products, often share a similar origin, and most engineering practices to control them are not specific to single chemicals, but decrease them all.

12.1.2 Epidemiologic Measures of Risk

Risk in a population is defined as the probability of disease, usually on a unit time and population basis, such as the number of lung cancer cases occurring per 100,000 persons per year. The expression of risk is often further confined to particular subpopulations, as defined by age, sex, or race; for example, number of lung cancer cases among white females of age 60–69 per 100,000 (white females of age 60–69) per year.

The outcome of analytic epidemiologic studies, including historical cohort and case-control designs, is usually expressed as a relative risk. The relative risk is an estimate of the disease rate in an exposed population relative to a comparable, but unexposed group. The relative risk for lung cancer among smokers, for example, has been estimated as about 10. Another measure of risk, sometimes more meaningful from a public health standpoint, is the attributable risk. The attributable risk estimates the number of cases that result from a given type of exposure factor, or alternatively, the number of cases that would not have occurred had the exposure not been present. The public health impact of an exposure is not necessarily indicated by the magnitude of the relative risk. The relative risk for coronary heart disease due to smoking is only 1.5 or 1.8, yet the number of smoking-related coronary heart disease deaths is greater than smoking-related lung cancer deaths, due to the higher underlying rates of heart disease among nonsmokers (18). If they are etiologic factors, drinking water contaminants are thought to have a relatively small effect on cancer rates, resulting in risk increases of no greater than 50 or 100% (relative risks no greater than 1.5–2.0). However, because of the large size of exposed populations, and underlying rates of cancer, modest increases in risk can imply important public health consequences.

12.2 INDIRECT STUDIES: DEVELOPING THE HYPOTHESIS

12.2.1 Definition

Indirect studies often comprise the first steps in epidemiologic assessment of cancer etiology. They seek to compare geographic or temporal patterns of disease occurrence with the distribution of environmental factors, occupations, socio-demographic characteristics, or other features of populations and their surroundings. The indirect study is used primarily to generate and evaluate the feasibility of hypotheses. It was the first approach used to assess the potential relationship of organics in drinking water and human disease occurrence. This type of study design has been called ecologic, aggregate, correlational, descriptive, or geographic. The compared groups are usually defined by geopolitical boundaries. Comparisons utilizing U.S. county cancer mortality data (19–24) and studies of cancer incidence and water quality in towns (25–27) are notable examples. Studies of smaller geographic areas, such as census tracts (28,29) or small towns, take advantage of data from regional tumor registries. Results can be more meaningful because populations from smaller places are likely to be more homogeneous than county populations with respect to disease rates, environmental exposures, and other demographic and socio-economic factors. The theoretical bases of indirect studies are fully developed (30,31).

12.2.2 Mortality and Morbidity Records

There has been national record keeping of deaths in the United States since 1902. The National Death Registration Area started with 10 states and several additional cities, and increased in geographical scope until the entire country was included by 1933. Legal registration of death is now required in all states. The death certificate in most states is a variant of a standard form developed by the National Center for Health Statistics. The death certificate, completed by a physician, funeral director, or coroner, records name, place of residence, age, date and place of birth, and usual occupation and industry of the deceased. There are spaces to enter multiple causes of death—the underlying cause and several contributing causes. Causes of death are routinely coded using a numeric system developed by the World Health Organization. The International Classification of Diseases (ICD) is revised every 10 years or so and is now in its 9th revision (32). Inaccuracies in the death certificate entry for underlying cause, for example, from misdiagnosis or lack of pathologic data, or errors in the proper numeric coding of cause of death, can lead to erroneous cause-specific mortality rates. However, for most cancers there is very good correlation between the underlying cause entered on the death certificate and hospital diagnosis (33).

Population-based tumor registries are central data collection systems that record all newly diagnosed cancers (other than nonmelanomic skin cancer) that occur within a defined geographic area. Information is gathered from physicians, clinics, and hospitals. At least 30 states require reporting, and 26 of these have population-based tumor registries (34). Ten additional states also have population-based tumor registries. Cancer incidence rates for geographic areas are calculated by combining tumor registry information with census data from the same population. Cancer rates from different places must be age-adjusted before they can be compared with one another. Cancers are primarily diseases of old age, and older populations can be expected to have higher cancer rates than younger groups, even though age-specific rates may be similar and risk factors are equally distributed. Direct comparison of rates in two or more populations may therefore lead to erroneous conclusions simply due to differences in their age structures. To correct for this possibility, rates are usually age-adjusted before such comparisons are made. A standard population is selected and the rate is adjusted by calculating the hypothetical rate that would be observed if the age distribution of the study population were the same as that of the standard population, given the age-specific rates in the study population. Populations with similar age-specific rates, but with very different balances of age groups, would reveal similar age-adjusted rates, and could be properly compared. All indirect epidemiologic studies use age-adjusted mortality or morbidity rates for their comparisons.

12.2.3   Mathematical Model

Indirect studies typically use a variation of the standard multiple linear regression model

$$E(y_i) = B_0 + B_e X_{ei} + B_1 X_{1i} + B_2 X_{2i} + \cdots\cdots + B_n X_{ni}$$

where y_i = age-adjusted sex-, site-, and race-specific geopolitical unit cancer mortality or incidence rate

X_{ei} = "exposure" variable

X_{1i} to X_{ni} = a series of variables directly or indirectly in causal sequence influencing cancer rates, such as population density, percentage foreign born, dummy variables denoting the region of the geopolitical unit's location, percentage of geopolitical unit's population employed in each of several manufacturing industries, etc.

i = 1, 2, . . . n is a subscript denoting the geopolitical unit.

This model assumes that the disease (or mortality) rate can be represented by a linear combination of geopolitical unit-level factors, including environmental exposures. The linear model assumes additivity of risk, whereas the interaction for several known etiologic factors could be multiplicative. Logarithmic or other transformation of the dependent variable (cancer rate) may be appropriate in situations where multiplicative interactions are suspected.

Results from indirect studies are reported as correlation coefficients or regression coefficients calculated from the regression model. The correlation coefficient expresses the strength of association between risk and an exposure or other factor of interest, usually after adjusting for the effects of other factors included in the regression model. The correlation coefficient does not reveal the magnitude of risk variation that is associated with changes in the level of the exposure. This is expressed by the regression coefficient, a measure of the level that the risk is increased (or decreased) by a unit increase in the level of the exposure, after adjusting for other factors in the model.

12.2.4 Regression Considerations

The precision of cancer mortality rates from geographic units of different size, as estimated by the inverse variance, is proportional to population size (35). Most ecologic studies of water or air quality weight the rates in the regression model directly by population size or its square root (proportional to the standard deviation). Some ecologic studies report the effects of different weighting procedures on study results. Hogan et al. (22) reported important and variable differences in patterns of association in weighted versus unweighted models. There is no single weighting method appropriate for all applications, but it may not matter greatly, so long as the more stable rates from places with large populations are accorded more weight in regressions than rates from smaller places. A strong and consistent association usually retains its statistical significance under several different weighting schemes.

The regression model should ideally include as many "independent" variables as are reasonably related to the disease and to exposure, but no more. Inclusion of irrelevant variables in the model (overspecification) can decrease the precision of correlations and, to the extent that the variables are statistically associated with both exposure and effect, can also diminish the strength of associations (confounding). Under some conditions, bias may be introduced if the outcome (dependent) variable is age-adjusted and the predictor variables are not (36).

Associations may result from statistical links with causal factors that are incidentally associated with the exposure variable, but whose levels are unknown and therefore cannot be included in the regression model. Failure to include such confounding factors can lead to spurious associations. Among the known or suspected causal factors often not considered in ecologic studies of water quality are cigarette smoking and dietary patterns. These deficits in the information base may compromise the specificity of results, especially in ecologic studies that implicate smoking-related diseases, such as cancers of the bladder or respiratory sites. If county smoking rates are statistically related to county measures of water exposures, the causal associations with smoking could mistakenly be attributed to ambient environmental factors.

12.2.5   Estimates of Exposure

Accurate estimation of historical exposures poses special problems for ecologic studies, because of the opportunity for exposure misclassification due to migration, in addition to the difficulty of estimating area-wide exposures many years ago. Between 1955 and 1960, 17% of the U.S. population moved between counties, ranging from 10 to 36% among the states (37). Migration diminishes the strength of ecologic associations, because new migrants to study areas are, on the average, randomly distributed with regard to past exposures (38).

Past exposures to air or water pollutants are derived in ecologic studies from industrial surveys, census information, or are reconstructed from knowledge of past engineering or water treatment practices. Many indirect studies of cancer mortality and water quality rely on data from a 1963 Inventory of Municipal Water Supplies (39) to calculate the percentage of county populations served by surface or ground sources, by chlorinated or nonchlorinated sources, or by sources treated by prechlorination. Thus, Salg (20) examined the association between water quality factors and cancer mortality in the 346 counties of the Ohio River drainage basin, using as exposure variables the percentage of each county's population served by surface water and served by prechlorinated water. Page et al. (24) used the percentage population of Louisiana parishes drinking water from the Mississippi River. Kuzma et al. (21), studying associations of water quality and cancer rates in Ohio counties, used a dichotomous exposure variable indicating the type of source (surface or ground) used by the majority of a county's population. Simplifying the exposure variable to this level may unnecessarily obscure exposure gradients. Other studies used trihalomethane levels derived from U.S. Environmental Protection Agency surveys or special state surveys as exposure estimates (22,23,40).

Most ecologic studies use mortality rates from 1950 to 1969, so exposures that may be implicated took place prior to 1945. Although some changes in water treatment practices have occurred since, relative patterns of chlorination among different source types have been stable. In addition, the use of recent THM data to represent past exposures is supported by observations of a strong year-to-year correlation in THM levels among water supplies (22,41). Migration probably contributes more to errors in estimating exposures than unrecorded changes in water source or treatment.

12.2.6   Strengths and Weaknesses of Indirect Studies

Indirect explorations of the geographic differences in disease occurrence are integral to advancement of epidemiologic knowledge because of their usefulness in formulating and evaluating the plausibility of hypotheses. Results are used to determine if more detailed study is warranted, and if so, to help frame the appropriate questions. Indirect studies are usually performed rapidly and at relatively low cost because they rely on previously collected data.

Inferences from indirect studies are limited because information on potentially important factors such as migration and historical exposure patterns is often partially or completely missing. The measure of effect is a regional cancer incidence or mortality rate, a group characteristic, and not the particular experience of individual group members. Individuals who are diagnosed with or die from cancer may not, by dint of exposure, be at highest risk in the population. Information is often lacking on other factors related to disease risk, such as cigarette smoking and dietary practices, and observed associations may be due to the confounding influence of such uncontrolled variables. Positive associations must therefore be interpreted with caution since they may not reflect a cause-effect relationship (42).

Quantitative estimates of risk from indirect studies are usually not appropriate, although many observers have been tempted to go beyond hypothesis elaboration and apply results of indirect studies in risk assessments. This may be appropriate if the risk assessment is qualitative and if statements of possible association are framed in the context of limitations of the indirect study design.

12.3   ANALYTIC STUDIES

In contrast to indirect studies, the unit of measure in analytic studies is the individual, not the group. Links are sought between exposures to individuals and their subsequent disease status. The use of data on an individual level makes the analytic study a more powerful instrument than the indirect study. Measures of association between exposure and disease in analytic studies are commonly expressed as relative risks, the probability (risk) of disease among exposed persons relative to the unexposed (see Section 12.1.2). Analytic studies thus directly furnish quantitative estimates of risk. The historical cohort and the case-control study are the approaches most commonly used.

12.3.1   Historical Cohort Studies

Although the cohort design is not the most efficient to evaluate the elevation in risk of rare diseases due to low level exposures, it deserves mention. Entry of subjects into an historical cohort study is based on past exposures, such as to occupational or environmental factors, therapeutic treatments (X-rays, specific pharmaceuticals), or shared host characteristics. Cohort members are followed over time, and the frequency of disease occurrence is compared with the frequency of like conditions in comparable but nonexposed populations. The cohort study is most useful to assess risks associated with exposures that are relatively rare in the general population but common in the study population, such as certain occupational exposures. The cohort method is generally inefficient in studying low-level chronic exposures and risk of rare conditions (such as most cancers), because of the need to gather exposure information for many thousands of individuals to evaluate risk among the relatively few. For this reason, it is infrequently used to evaluate exposures to ambient environmental contaminants. A cohort study of cancer risk as related to water quality was reported from a Maryland county where a population-based registry of exposures and cancer cases had been created for other purposes many years before water quality exposures were evaluated (43). It was not necessary in this case to gather additional information about individual exposures to evaluate the hypothesis.

12.3.2   Case-Control Studies Based on Mortality Records

Entry into a case-control study is governed by disease status, and exposures are determined retrospectively. The level of association of a disease with a factor is based on a high or low frequency of that factor among diseased persons (cases) relative to its frequency among healthy individuals (controls). When a relatively rare disease such as site-specific cancer is the object of study, the case-control approach has many advantages. The case-control study is a principal tool of modern cancer epidemiology, and the methodology has been studied in detail (44–46). Case-control studies of water contaminants have taken two forms: those based on mortality records and those that use data directly from patients (or next of kin) and a healthy comparison group.

The first case-control studies of drinking water quality and cancer risk selected deceased cases and controls from computerized listings of state vital statistics bureaus (47). These studies share some weaknesses with indirect studies. Among them are the potential for exposure misclassification due to migration and an inability to account for many other exposures or risk factors that can confound relationships with water contaminants, or interact with them. A major strength resides in the large numbers of cases and controls that can be readily drawn from state mortality records. As mentioned above, the relatively small excess risks expected from low-level exposures necessitates large numbers of study subjects to ensure statistical stability of risk estimates.

Cases are those who died of the disease of interest. Controls are decedents selected randomly from among other causes of death (in some cases, only noncancer causes), and matched to cases on demographic characteristics such as sex and age. Measures of past exposure that have been used include categorical descriptions of source type (for example, surface or ground water) and treatment (for example, chlorinated or non chlorinated), sometimes modified by information on upstream contamination (for surface sources), by chlorination levels, or by other treatment practices. An exposure category for each decedent is obtained by linking the residence address from the death certificate with local water utility information. This method assumes that study subjects have had the same address—and water source—for a period at least as long as the induction period of the cancer under study. The assumption is not always valid, and the potential for misclassifying exposure increases with the level of (recent) in-migration into the study area (38). To decrease the possibility of assigning erroneous exposures to subjects in case-control mortality studies, investigators have: (1) restricted studies to places with low rates of in-migration (48), (2) stratified by the number of years that the decedent was served by his or her last water source, as documented by water company records (49), and (3) stratified and independently computed relative risks for study subjects (a) born in the same county as listed on the death certificate, (b) born in the same state but a different county, and (c) born in a different state (50). The last approach assumes that decedents with the same residence county at the beginning and end of life are more likely to have had longer exposures to their last water source than those with different counties. Some studies have been restricted to females, because of the higher probability that they spent a larger proportion of their time at home than males (48).

12.3.2.1   Strengths and Weaknesses of Case-Control Mortality Studies

Several aspects of the death certificate case-control study design are of particular importance when investigating environmental exposures that vary geographically. Mortality records are complete, easily accessed, and represent state-wide exposure variability. Larger samples of case and control decedents can be assembled than are normally possible when consenting patients and control subjects make up the sample. Individual death record data can often be linked, through the "usual place of residence," to other pre-existing data, such as water supply characteristics.

Because of the long latent period between most carcinogenic exposures and onset of disease, the validity of this technique rests on the assumption that the decedent lived in the same place for at least 20 years before death. Also, the lack of ability to directly gather personal or life-style information on various confounding factors can limit the confidence with which findings are interpreted. Death records are used as surrogates for incidence information, with the possibility of error in cause-of-death classification, and underascertainment for malignancies with good long-term survival.

12.3.3   Confounding, Bias, and Effect Modification

12.3.3.1   Confounding

Positive associations in indirect or case-control studies may not signal causal relationships but rather statistical associations of the factor under study with a real causal factor (the confounder). The error arises from attributing causality to a factor that is not itself a cause of the disease, but is asociated with a causal factor. When the link is unknown, the disease can mistakenly be associated with the unrelated factor. If, for example, cigarette smoking rates were elevated in geographic areas with poor drinking water quality, but the researcher studying bladder cancer had no information on smoking levels, elevated bladder cancer due to cigarette smoking might falsely be attributed to drinking water contaminants. Cigarette smoking would be said to "confound" the link and cigarettes would be the confounder. When no information on the suspected confounder is available, as with smoking levels in the completed indirect or case-control mortality studies, there is no simple way to distinguish the influence of confounders from other putative risk factors. In some cases, confounding goes unnoticed because the confounder is not recognized as a causal factor and is therefore not assessed nor used in the analysis. In addition to creating spurious links, confounding can also mask real ones. If the hypothetical association between smoking and water quality were inverted (that is, if smoking were more prevalent in places with clean water), and water quality were a real risk factor, elevated risk from water contaminants could be obscured by the confounder.

Careful control for confounding is especially important in studies of low-dose exposures, where low elevations of risk are under study. Studies of incident cases and controls, where much more information on potential confounders is gathered by interview or mail questionnaire directly from the study subjects, provide a much greater opportunity than indirect or case-control mortality studies to control for confounding due to diet, smoking, occupation, and other factors. When such data are available, confounding can be controlled in the analysis by stratifying on the level of the potential confounder. Recent developments in logistic regression methods facilitate analyses that adjust for confounding by two or more factors simultaneously (46).

12.3.3.2   Bias

Bias in case-control mortality studies can arise if controls died of causes related to the exposure of interest. Bias can also occur if digitized records of death are grouped by town or by village, and potential controls are not randomized before selection into the study. When controls are selected by the method of next best "match" (on age, race, sex) in the record sequence, the probability is enhanced that the control resided in the same place as the case, and thereby overmatched on drinking water source. Overmatching obscures real risk differences, because it reduces the basis for distinguishing exposure differences.

Selection bias can occur when persons who routinely seek medical care are more likely to be diagnosed with certain conditions than those who use health services infrequently. If personal characteristics or exposure factors are related to this trait, cases may more (or less) frequently report exposure than noncases. This type of bias may arise where certain causes of death appear more frequently among certain occupations than others. It has been suggested, for example, that the observation of elevated risk of dying from brain cancer observed among some white collar professions is the result of bias associated with a relatively higher level of medical care in these groups that results in more accurate death certificate diagnoses (51).

12.3.3.3   Age, Sex, Race, and Effect Modification

Age, sex, and race are not causal in the same sense as external exposures, but they can influence disease probability, often strongly. In most case-control studies, the influence of age, sex, and race on disease probability is controlled by matching controls to cases on these characteristics. In indirect studies, these factors are controlled by age-adjusting rates to a standard population and restricting analysis to specific sex/race groups. Age, sex, and race can be effect modifiers because they may also alter the interaction of etiologic factors with the organism.

12.3.4   Studies of Incident Cases and Controls

Case-control studies of newly diagnosed (incident) cases are conducted to test the refined hypotheses developed from other studies. Information on a large number of host and environmental factors is gathered by questionnaire or interview directly from individuals with the disease of interest (or next-of-kin) and from a matched series of randomly selected controls without the condition. Case-control incidence studies of environmental exposures are no different, in principle, than case-control studies of other types of exposures. However, the types of exposures and the relatively small excess risks expected demand that special care be taken in study design and execution to minimize bias and confounding and to maximize the possibility of discovering risk elevations, if present. Among the more important design considerations are:

1. Size of the study population.
2. Hospital-based versus population-based study.
3. The choice of study population(s).
4. Questionnaire design and/or interview procedures.
5. Exposure assessment.

These issues will be discussed, using as examples features of a large collaborative bladder cancer study conducted by the National Cancer Institute (NCI) (52,53) and a case-control study of colon cancer conducted in Wisconsin.

12.3.4.1   Size of the Study Population

The magnitude of the odds ratio that can be detected in a case-control study depends on the sample size, the number of controls per case, and the proportion of exposed individuals. Due to the small excess risk expected, studies of ambient air and water contaminants require large numbers of respondents. Detection of an odds ratio of 1.5 in a study with a 1:1 control/case matching ratio and an exposure rate among controls of 20% requires at least 419 cases, if the desired statistical significance (*p* value) is 0.05, with an 80% probability of detecting an association (54). Detection of larger odds ratios (relative risks) could be achieved with fewer cases.

In general, the largest study population that is logistically and economically feasible should be planned. It is desirable that major study subgroups (by age, sex, cigarette smoking status, etc.) be large enough in themselves to permit relatively stable risk estimates, since consistency of risk patterns across subgroups bolsters confidence in overall estimates. With adequate numbers of respondents at different levels of exposure, dose-response gradients can also be determined with some confidence. Cases in the NCI bladder cancer study included all residents of 10 U.S. locations diagnosed with bladder cancer in 1978. Approximately 3000 cases and 6000 controls were interviewed. The Wisconsin study included approximately 400 colon cancer cases, 800 incident cases with diagnoses of other cancers who served as one control series, and 800 general population controls.

12.3.4.2   Hospital-Based versus Population-Based Studies

Cases in hospital-based studies are patients newly diagnosed during a defined period from one or more institutions. Controls are selected after matching to cases on hospital of diagnosis, as well as demographic characteristics (sex, age, race). Controls are randomly selected from a range of diagnoses other than the disease of interest or related conditions. Population-based studies draw cases from the general population of one or more well-defined geographic locations, with the intention of including all patients newly diagnosed with the study condition, or a randomly selected fraction. Several methods for selecting controls are available. Both the NCI bladder cancer study and the Wisconsin colon cancer study were population-based, using existing population-based cancer reporting systems to identify eligible cases. The NCI study included all bladder cancer cases, aged 25–84, newly diagnosed in 1978 and resident in 10 places of the U.S. (5 states and 5 metropolitan areas). Controls less than 65 years of age were selected by a random digit dialing method (55) and those 65 and over from a random 1% listing of the over-65 population from the U.S. Health Care Financing Administration (53). The Wisconsin study included colon cancer cases, aged 35–89, who had been reported to Wisconsin's population-based cancer reporting system from January, 1980, through July, 1982. For each of the 400 cases, 2 controls with cancer of a different site were selected (frequency matched on age, sex, and race). The cancer control group was chosen to control for factors governing selection for being reported to the system and to control for recall bias in recently diagnosed cancer cases. In addition, 800 general population controls were selected using records of drivers license holders in the state.

Population-based studies have several advantages over hospital-based designs, especially for studies in which low elevations of risk are expected. The source (denominator) population in the population-based study is unambiguous, the total population of the study area. In hospital-based studies, the source population is never so clearly defined. Extrapolation of results to the general population is therefore on less secure grounds than with a population-based study. The population-based design also enhances the ability to control for all suspected sources of bias and confounding.

While hospital-based studies have logistical advantages, such as ease of case ascertainment and access, they can be compromised by biases inherent in the types of available controls. A hospital-based study of bladder cancer observed similar levels of saccharin use in cases and controls (56). This finding was questioned because many conditions leading to hospitalization are linked to obesity, and it is likely that the obese consume greater-than-average levels of saccharin. Use of this control group might have biased the result and obscured a real association, if present, between saccharin ingestion and bladder cancer (57,58). Many such biases are possible when using hospital-based control series.

Hospital patients are usually drawn from a restricted geographical area with limited variation in levels of air and water contaminants. If controls are matched to cases by hospital, as is customary, overmatching on pollutant exposure is very likely, decreasing the possibility of detecting an effect. Because population-based studies seek to include all cases in a region, estimates of absolute risk are facilitated, as are estimates of exposure frequency in the general population, as inferred from the experience of control groups. Given these drawbacks, the hospital-based study is not generally a good choice to study environmental exposures.

12.3.4.3   Choice of the Study Population

Geographic areas appropriate for population-based case-control studies are those where a significant number of persons are exposed at each of several exposure levels. It is also helpful if the region has had limited in-migration, because characterization of past exposures usually requires knowledge of environmental conditions in places of former residence. Study populations and historic (and predicted) rates of the disease under study must be large enough to assure availability of the desired number of newly diagnosed cases within the study time period.

Having been designed primarily for other purposes, the NCI study did not use all of the above criteria in selection of the source populations. Nevertheless, at least three of the five states that were included (Iowa, New Jersey, and Connecticut), with more than half the total study population, satisfied the requirements for numbers of expected cases, exposure variety, and limited in-migration.

12.3.4.4 Questionnaire Design and Interviews

The data collection instrument should seek information on known and suspected risk factors, effect modifiers, and confounders to the greatest extent possible, while keeping the completion time to a reasonable length. When an in-person interview is used, interviewers must be carefully selected and trained to minimize bias due to different approaches to cases and controls that might involve collecting more information from one group or the other, or by asking leading questions. In the ideal study, the interviewer would be blinded to the respondent's status as case or control, but this is often not possible. A form of recall bias may arise when cases, in seeking reasons for their disease, remember past events differently than controls. This type of recall bias can be addressed by selecting as controls persons diagnosed with diseases not thought to be related to the exposure of interest, as was done for one control group in the Wisconsin colon cancer study.

In the NCI bladder cancer study, information was obtained on demographic background, detailed smoking and occupational histories, information on relevant medical conditions, and a history of artificial sweetener use. The interview also covered coffee drinking habits, use of hair dyes, and fluid ingestion patterns. A lifetime residential history was recorded, and for each residence, the respondent was asked if the primary drinking water source was a community supply, a private well, bottled water, or another source. The residential/water source history was subsequently linked to historical water supply information. The Wisconsin colon cancer study utilized a self-administered questionnaire, which took about 1 hour to complete. Complete residential histories were collected and information was sought for three age periods (childhood, early adult, and over age 35) on medical and occupational histories, bathing frequency, diet, smoking, and other factors potentially relevant to colon cancer etiology, such as laxative use.

12.3.4.5 Exposure Assessment

The ultimate success of low-level exposure studies rests in large part on the accuracy of estimates of past exposures. Because direct environmental measurements are usually not available for the exposure periods of greatest interest, we must rely on other types of information sources that permit modeling of past environmental conditions.

In air quality studies, the exposure modeling might integrate several types of information: wind patterns and air mixing zones, inventories of past industrial patterns and fuel consumption, and locations of the residence and workplace of respondents relative to known past air pollution sources.

The quality of drinking water in the past may be modeled with more confidence. Water source and treatment practices are often stable over the years, because treatment and distribution systems represent large fixed investments. Past changes in sources, treatments, and distribution are well documented. Thus, it is possible to estimate accurately many aspects of drinking water quality delivered to the residence of study respondents in the past.

In our experience, most subjects know if their primary water source at each residence was the local community supply or a private well or spring. Thus, it is feasible to gather individual residence histories, and to link these data with historical information from the water utilities likely to have served most of the study population. The linkage is aided by the fact that distribution zones for most water utilities follow the geopolitical boundaries that also define towns or cities.

In Wisconsin, past exposure levels of study subjects were estimated by applying a statistical model developed from current data on THM levels and water utility records, to historic water works data and then merging with residential histories. Eighty Wisconsin soures were sampled seasonally and analyzed for levels of volatile organic chemicals. THM concentrations were then modeled as a function of recent data routinely collected and maintained by waterworks, such as pH, temperature, and chlorine dose. The resulting equations were applied to similar past data to estimate past THM exposures for each year and Wisconsin town or city. A quantitative estimate of each respondent's yearly THM dose was derived by merging these data with each residential history and information on drinking water consumption.

In the NCI bladder study, past residence, and water supply records were linked to construct a year-by-year lifetime record of water source and treatment for each study participant, going back to 1900 or the year of birth, whichever was the more recent. Using this method, water supply source and treatment for 75% of all the years lived by study respondents was successfully defined (59).

12.3.4.6 Summary of Case-Control Interview Studies

Case-control interview studies can be efficient in evaluating effects of long-term low-dose exposures, so long as the relative risk is at least 1.4 or 1.5. Even if the study population is large enough so risk ratios of this magnitude are statistically significant, results might be questionned because low relative risks can arise from confounding or bias in control selection, interview procedures, or selective memory of respondents. Carefully designed population-based case-control studies decrease the possibility of most types of bias, and are preferable to hospital-based studies. To enhance study sensitivity, it is important that the study setting be chosen to maximize the exposure differential between persons on the high and low ends of the exposure scale.

A retrospective study, by itself, cannot establish or disprove a causal association between exposure and disease, because the possibility of confounding, however remote, always exists. Case-control studies, however, in conjunction with other epidemiologic, clinical, and laboratory investigations, can provide strong and convincing evidence for or against causal patterns, and thus establish the basis for effective preventive measures. If adverse effects can, in fact, be measured, the case-control study is one of the best tools to obtain risk estimates directly from human experience.

12.4 GUIDELINES FOR EVALUATION OF RESULTS

Many factors that influence the interpretation of results from epidemiologic studies have already been discussed in the context of strengths and weaknesses of particular study designs and descriptions of confounding and bias. Here we will outline several additional factors that have been suggested as guides to decisions on causal inference from epidemiologic findings (60,61).

12.4.1 Temporal Considerations of Disease Rates

If causality is to be ascribed to an exposure, changes in disease rates should follow changes in exposure levels or frequency of exposure in the overall population. Thus, when it was first observed that lung cancer rates were rapidly increasing in the 1930s and 1940s, changes in environmental conditions and personal habits several years earlier were considered as possible causes. Among them were increases in automobile use (exhaust), paved roads (bitumin), and cigarette smoking. The impact of temporal changes in risk factors that confer small increases in risk may be masked by changes in other, more influential, risk factors. This implies that analysis of time trends in cancer rates following changes in exposure to water contaminants may not be a fruitful avenue of investigation, because other risk factors, also changing in the affected populations, may have been more important in effecting disease outcomes.

The latency period, the time between first exposure to an agent and expression of disease, is considered for most cancers to be at least 2 decades. Latency associated with low-level exposures such as drinking water contaminants may be much greater. Analytic studies must consider this by examining exposures among diseased and nondiseased persons that occurred several decades in the past.

12.4.2 Strength of Associations

One suggested criterion for assessing causality is the strength of the associa-tion between exposure and effect, as reflected by the magnitude of relative risk estimates. Relative risks of less than 2.0 may indicate an unperceived bias or confounding factor, whereas those over 5.0 are unlikely to do so. The finding of a strong association can be quite helpful in establishing causality, but when a relatively common tumor is the subject of study, the additional risk imposed by weak or low level carcinogens may be relatively small (16). This criterion is important in studies of water contaminants in the sense of suggesting caution in the causal interpretation of results, because relative risks of less than 2.0 are the expected outcome. Causal interpretation of these studies must therefore rely more strongly on other criteria relating to the consistency of findings and observation of dose-response relationships.

12.4.3 Consistency of Findings and Dose-Response Relationships

12.4.3.1 Consistency among Studies

Similar observations from independently conducted studies in different pop-ulations and geographical locations by independent investigators can strengthen the plausibility of hypotheses generated by indirect studies, and can tighten the argument for a causal interpretation of results from analytic studies. The larger the number of parallel independent observations, the lower the probability that the associations are due to chance alone or to a common confounding factor, although these possibilities can never be to-tally excluded. Thus, the suggestion of a link between low levels of carcino-gens in drinking water and elevated risk of bladder, colon, and rectal cancers was strengthened by independent observations from indirect studies in sev-eral settings (62–65).

12.4.3.2 Consistency within Studies

Similar findings among various subgroups within a study enhance the proba-bility that findings relate to disease etiology, and do not result from con-founding or bias. If subgroups are of sufficient size, it is helpful to evaluate the effects of exposure within groups defined by the presence or absence of other risk factors (e.g., among smokers and nonsmokers, among the occupa-tionally exposed and nonexposed) and by demographic characteristics (e.g., by sex, age).

12.4.3.3 Dose-Response Relationships

Where possible, information should be gathered on degree of exposure, both with respect to duration and intensity. In studies of drinking water quality, this may first involve collection of historical data from water utilities as well as information from study subjects on residence, water source, and past levels of relevant fluid ingestion, and then combining them to derive meaningful exposure indices. A causal interpretation is enhanced by showing that risk increases with the level of duration and/or intensity of exposure, since it is unlikely that levels of a confounder would vary in a similar manner.

12.4.4   Biologic Plausibility

The criterion of biologic plausibility requires that the effect is a reasonable outcome of the exposure, based on observations of the same or related effects observed under tightly controlled laboratory conditions. This is not usually a very stringent test since many different biological mechanisms can be invoked to explain an adverse health impact. On the other hand, the lack of such correlative evidence does not imply that epidemiologic observations are in error (61). With respect to water contaminants, the observation of genotoxic and carcinogenic effects of selected compounds and mixtures (12–15) adds weight to the plausibility of positive epidemiologic associations.

REFERENCES

1. Snow, J., *Snow on Cholera*, Hafner, New York, 1965.
2. National Research Council, Safe Drinking Water Committee, *Drinking Water and Health*, National Academy of Sciences, Washington, DC, 1977.
3. U.S. Congress, Office of Technology Assessment, *Protecting the Nation's Groundwater from Contamination*, OTA-0-233, Washington, D.C., 1984.
4. Council on Environmental Quality, *Contamination of Ground Water by Toxic Chemicals*, Washington, DC, 1981.
5. Wilkens, J.R., III, and Reiches, N.A., Epidemiologic approaches to chemical hazard assessment, in *Hazard Assessment of Chemicals: Current Developments*, Vol. 2, Academic Press, 1983, pp. 133.
6. Heath, C.W., Jr., *Environ. Health Perspect., 48*, 3, 1983.
7. Landrigan, P.J., *Environ. Health Perspect., 48*, 93, 1983.
8. Glaze, W. H., Seleh, F.Y., and Kinstley, W., Characterization of non-volatile halogenated compounds formed during water chlorination, in Jolley, R.L., Brungs, W.A., Cumming, R.B. (eds.), *Water Chlorination: Environmental Impact and Health Effects*, Vol. 3, Ann Arbor Science Publishers, Inc., Ann Arbor, MI, 1980, pp. 99.
9. Bellar, T.A., Lichtenberg, J.J., and Kroner, R.C., *J. Am. Water Works Assoc., 66*, 703, 1974.
10. Rook, J.J., *J. Soc. Water Treat. Exam., 23(2)*, 234, 1974.
11. Page, N.P., and Saffiotti, U., *Report on Carcinogenesis Bioassay of Chloroform*, National Cancer Institute, Division of Cancer Cause and Prevention, Bethesda, MD, 1976.
12. Loper, J.C., *Mutat. Res., Rev. Genet. Toxicol., 76*, 241, 1980.
13. Simmon, V.F., and Tardiff, R.G., The mutagenic activity of halogenated compounds found in chlorinated drinking water, in Jolley, R.L., Gorchev, H., Hamilton, D.H., Jr. (eds.), *Water Chlorination: Environmental Impact and Health Effects*, Vol. 2, Ann Arbor Science Publishers, Inc., Ann Arbor, MI, 1979, pp. 417.
14. Land, D.R., Kurzepa, H., Cole, M.S., and Loper, J.C., *J. Environ. Pathol. Toxicol., 4*, 41, 1980.

15. Robinson, M., Glass, J.W., Cmehil, D., Bull, R.J., and Orthoefer, J.G., Initiating and promoting activity of chemicals isolated from drinking waters in the SENCAR Mouse—a five city survey, in Waters, M.D., et al. (eds.), *Short-Term Bioassays in the Analysis of Complex Environmental Mixtures, II. Environmental Science Research,* Vol. 22, Plenum Press, New York, 1980, pp. 177.

16. Doll, R., *Int. J. Epidemiol., 14,* 22, 1985.

17. Rothman, K.J., *Am. J. Epidemiol., 104,* 587, 1976.

18. Cantor, K.P., Human Case-Control Studies in Risk Assessment, in Richmond, C.R., Walsh, P.J., Copenhaven, E.D. (eds.), *Health Risk Analysis,* Franklin Institute Press, Philadelphia, PA, pp. 109–120.

19. Mason, T.J., and McKay, F.W., *U.S. Cancer Mortality by County:* 1950–1969, USDHEW Publ. No. NIH 74-615, National Cancer Institute, Bethesda, MD, 1974.

20. Salg, J., Cancer Mortality and Drinking Water in 346 Counties of the Ohio River Valley Basin, Ph.D. Thesis, Department of Epidemiology, University of North Carolina, Chapel Hill, NC, 1977.

21. Kuzma, R.J., Kuzma, C.M., and Buncher, C.R., *Am. J. Public Health, 67,* 725, 1977.

22. Hogan, M.D., Chi, P.Y., Hoel, D.G., and Mitchel, T.J., *J. Environ. Pathol. Toxicol., 2,* 873, 1979.

23. Cantor, K.P., Hoover, R., Mason, T.H., and McCabe, L.J., *J. Natl. Cancer Inst., 61,* 979, 1978.

24. Page, T., Harris, R.H., and Epstein, S.S., *Science, 193,* 55, 1976.

25. Bean, J.A., Isacson, P., Hausler, W.J., Jr., and Kohler, J., *Am. J. Epidemiol., 116,* 912, 1982.

26. Bean, J.A., Isacson, P., Hahne, R.M.A., and Kohler, J., *Am. J. Epidemiol., 116,* 924, 1982.

27. Isacson, P., Bean, J.A., and Lynch, C., Relationship of cancer incidence rates in Iowa municipalities to chlorination status of drinking water, in Jolley, R.L., et al. (eds.), *Water Chlorination: Environmental Impact and Health Effects,* Vol. 4, Ann Arbor Science Publishers, Inc., Ann Arbor, MI, 1982, pp. 1353.

28. Kanarek, M.S., Conforti, P.M., Jackson, L.A., Cooper, R.C., and Murchio, J.C., *Am. J. Epidemiol., 112,* 54, 1980.

29. Conforti, P.M., Kanarek, M.S., Jackson, L.A., Cooper, R.C., and Murchio, J.C., *J. Chronic Dis., 34,* 211, 1981.

30. Morgenstern, H., *Am. J. Public Health, 72,* 1336, 1982.

31. Langbein, L.I., and Lichtman, A.J., *Ecological Inference,* Series on quantitative applications in the social sciences, No. 07-010, Sage, Beverly Hills, CA, 1978.

32. World Health Organization, *Manual of the International Classification of Disease, Injuries, and Causes of Death, 9th Revision,* Geneva, 1977.

33. Percy, C., Stanek, E., and Gloeckler, L., *Am. J. Public Health, 71,* 242, 1981.

34. Enterline, J.P., Kammer, A., Gold, E.B., Lenhard, R., and Powell, G.C., *Am. J. Public Health, 74,* 449, 1984.

35. Kleinbaum, D.G., and Kupper, L.L., *Applied Regression Analysis and Other Multivariable Methods.* Duxbury Press, North Scituate, MA, 1978.

36. Rosenbaum, P.R., and Rubin, D.B., *Biometrics, 40,* 437, 1984.

37. U.S. Bureau of the Censes, *United States Census of Population, 1960, Characteristics of the Population,* United States Printing Office, Washington, DC, 1963.

38. Polissar, L., *Am. J. Epidemiol., 111,* 175, 1980.

39. U.S. Public Health Service, Division of Water Supply and Pollution Control, Basic Data Branch: *1963 Inventory of Municipal Water Facilities,* Report No. 775, Washington, DC, 1964.

40. Tuthill, R.W., and Moore, G.S., *J. Am. Water Works Assoc., 72,* 570, 1980.

41. Cantor, K.P., in Part V, Public Health Aspects: Regulatory Programs on Environmental Carcinogens, panel discussion, *Ann. NY Acad. Sci., 298,* 576–580, 1977.

42. Yerushalmy, J., On inferring causality from observed association, in Ingelfinger, F.J., Relman, A.S., and Finland, M. (eds.), *Controversy in Internal Medicine,* W.B. Saunders Co., Philadelphia, PA, 1966, pp. 659.

43. Wilkins, J.R., III, and Comstock, G.W., *Am. J. Epidemiol., 114,* 178, 1981.
44. Cole, P., *J. Chronic Dis., 32,* 15, 1979.
45. Schlesselman, J.J., *Case-control Studies: Design, Conduct, Analysis,* Oxford University Press, New York, 1982.
46. Breslow, N.E., and Day, N.E., *Statistical Methods in Cancer Research,* Vol. 1, The Analysis of Case-Control Studies, IARC, Sci. Publ. Issue 32, Lyon, 1980.
47. Crump, K.S., and Guess, H.A., *Annu. Rev. Public Health, 3,* 339, 1982.
48. Young, T.B., Kanarek, M.S., and Tsiatis, A.A., *J. Natl. Cancer Inst., 67,* 1191, 1981.
49. Gottlieb, M.S., Carr, J.K., and Morris, D.T., *Int. J. Epidemiol., 10,* 117, 1981.
50. Struba, R.J., *Cancer and Drinking Water Quality,* Ph.D. Dissertation, University of North Carolina, Chapel Hill, NC, 1979.
51. Thomas, T.L., and Waxweiler, R.J., *Scand. J. Work Environ. Health, 12,* 1, 1986.
52. Hoover, R.N., Strasser, P.H., Child, M., et al., *Lancet, 1,* 837, 1980.
53. Hartge, P., Cahill, J.I., West, P., Hauck, M., Austin, D., Silverman, D., and Hoover, R., *Am. J. Public Health, 74,* 52, 1984.
54. Rothman, K.J., and Boice, J.D., Jr., *Epidemiologic Analysis with a Programmable Calculator,* NIH Publication No. 79-1649, U.S. Department of Health, Education, and Welfare, Washington, DC, 1979.
55. Waksberg, J., *J. Am. Stat. Assoc., 73,* 40, 1978.
56. Kessler, I.I., and Clark, J.P., *J. Am. Med. Assoc., 240,* 349, 1978.
57. Silverman, D.T., Hoover, R.N., and Swanson, G.M., *Am. J. Epidemiol., 117,* 326, 1983.
58. Goldsmith, D.F., *Environ. Res., 27,* 298, 1982.
59. Cantor, K.P., Hoover, R., Hartge, P., Mason, T.J., Silverman, D.T., and Levin, L.I., Drinking water source and risk of bladder cancer: A case-control study, in Jolley, R.L., Bull, R.J., Davis, W.P., Katz, S., Roberts, M.H., Jr., and Jacobs, V.A. (eds.), *Water Chlorination: Environmental Impact and Health Effects,* Lewis Publishers, Chelsea, MI, 1985, pp. 143.
60. Lilienfeld, A.M., *Foundations of Epidemiology,* Oxford University Press, New York, 1976.
61. Cole, P., Introduction, in Breslow, N.E., and Day, N.E. (eds.), *Statistical Methods in Cancer Research: Vol. I. The Analysis of Case-Control Studies,* IARC, Sci. Publ. Issue 32, Lyon, 1980.
62. National Research Council, Safe Drinking Water Committee, *Drinking Water and Health,* Vol. 3, National Academy of Sciences. Washington, DC, 1980.
63. Wilkins, J.R., III, Reiches, N.A., and Kruse, C.W., *Am. J. Epidemiol., 110,* 420, 1979.
64. Cantor, K.P., and McCabe, L.J., *Am. Water Works Assoc.,* Proceedings, 1978 Annual Conference 32-5:5, 1979.
65. Shy, C.M., and Struba, R.F., Air and water pollution, in Shottenfeld, D., and Fraumeni, J.F., Jr. (eds.), *Cancer Epidemiology and Prevention,* W.B. Saunders Co., Philadelphia, PA, 1981, pp. 336.

CHAPTER 13

Principles of Quantitative Risk Assessment and Their Application to Drinking Water

David W. Hosmer

Division of Public Health
University of Massachusetts
Amherst, Massachusetts

13.1 INTRODUCTION

The Wednesday, October 26, 1983, issue of the *Boston Globe* carried a story headlined, "State tells four cities of drinking water risk." The article reported that tests done by the Massachusetts Department of Environmental Quality Engineering yielded water samples from a variety of Massachusetts communities that contained trihalomethanes (THMs) in quantities greater than 100 ppb, the maximum allowable level under Federal law. The article stated that Federal cancer risk calculations estimate that if a group of 10,000 people consumed 2 L of water containing 100 ppb of THMs for 70 years there would be three to four additional instances of cancer.

The problem addressed in this article is a typical example of the problem faced when attempting to assess quantitatively the risks associated with exposure to a particular compound which may be present in seemingly low concentrations. The particular issues that are most relevant to quantitative risk assessment in the above example of increased levels of THMs include: (1) What is the basis of the Federal standard of 100 ppb? (2) If the Federal standard was developed to provide a certain level of protection, how much additional risk can one expect when the Federal standard is exceeded, say, by 50 ppb? Answers to these questions are often sought through the application of statistical modeling techniques. The goal of this chapter is to present: (1) the historical background of these methods, (2) the methods and models that are currently being used to assess risk, (3) a discussion of the properties of the models and their use in assessing risk, and (4) a discussion of the

339

statistical problems in extrapolating from high to low doses. The presentation will be nontechnical and will provide an overview of the issues, avoiding unnecessary computational and mathematical details. References will be provided for those wishing a more detailed or technical treatment of the subject.

13.2   HISTORICAL BACKGROUND ASSOCIATED WITH THE STATISTICAL METHODS USED IN QUANTITATIVE RISK ASSESSMENT

The earliest attempts at quantitative risk assessment (QRA) were derived from statistical methods used in quantal response bioassays. In a typical quantal response bioassay, a number of animals (plants, etc.) are exposed to a substance at a number of levels or doses. The response of each animal is then recorded as being present or absent. The basic study used for the animal model in a QRA consists of animals exposed to varying levels of a potential carcinogen over their lifetimes. At the conclusion of the study the presence or absence of tumors in each animal is noted. These studies are designed to assess risk over a lifetime of chronic exposure to a substance. A second type of exposure—acute—may occur, but very little work has been done to develop models for QRA for this type of exposure, thus it will not be discussed in this chapter. The statistical models that were used first to analyze data of these type hypothesize a dose-response relationship that results from the concept of a tolerance distribution. In this model each animal is assumed to have its own individual tolerance to the substance. If the dose is less than this tolerance, then the response will always be absent, and if the dose is greater than or equal to the tolerance, then the response will always be present. A particular dose-response model arises from assumptions about the distribution of tolerances in the "population" of all animals. Models used in bioassay have assumed that the log of the dose follows a normal distribution or logistic distribution. The normal distribution gives rise to the well-known probit model. This class of models, called tolerance distribution models, will be referred to as TDMs throughout the remainder of this chapter. Each of the TDMs used in QRA will be presented in the next section.

Interest increased in QRA with the advent of the Delaney Clause and the development of instruments that could detect the presence of a substance at very minute levels in food or drinking water. Investigators felt that while a TDM might be a reasonable model for estimating the LD_{50} of a pesticide to kill boll weevils, it is too simplistic an approach to assessing cancer risk in animals or humans. The TDMs did not use any of the then developing knowledge of the biology of cancer. This spurred development of another class of models, which are called stochastic models. This class of models considers the occurrence of a tumor to be a random event at the individual level rather than at the population level. To date the models of this type that have been used in QRA are either the "hit" or "stage" type models, and these will also be presented in the next section.

It should be noted that there is far from unanimous opinion on which is the best model or even type of model to use in QRA. The issue of which model to use is further clouded by the one problem which differentiates QRA from traditional bioassay, extrapolation. In a typical risk assessment based on an animal model, animals are exposed to doses that are usually much more concentrated than levels occurring in the environment. The reason for this is practical, namely cost. The animals must be exposed to high enough doses so that a dose-related relationship, if one exists, will be observed with high probability and with a manageable number of animals. A typical risk assessment study will be conducted at three doses: zero, a "maximum level," and one-half of the "maximum level." The problem addressed in QRA is how to use these data to extrapolate downward to best estimate that dose which produces a minimal excess risk or probability of developing cancer. This dose is called the virtual safe dose (VSD) and is usually associated with an excess risk of the order 10^{-6}, or one additional cancer among 1 million people per lifetime of exposure. The source of the controversy and confusion is that while a number of statistical models will fit the animal data equally well, they will yield widely divergent estimates of a VSD. The methods used to estimate a VSD will be discussed in subsequent sections. The length and depth of the issues surrounding the estimation of a VSD are largely due to the fact that there is really no way actually to verify the predicted result. Its appropriateness is an exclusive function of the biologic plausibility of the method used to derive the estimate.

A second problem makes QRA even more difficult. The final goal of a QRA is to obtain a VSD for human exposures. As noted above, the exposure studies are done on animals, hence not only must extrapolation be performed out of the observed response range for animals, it must be further extrapolated to a human model from an animal model. Some work has been done on this problem; but most of the research represents a first step or attempt at more biologically plausible models.

Biologic plausibility is the goal of current research in QRA. This research is directed at incorporating new knowledge of the biology of cancer, making existing models more sensitive to the recognized extrapolation problems, and developing new models and new approaches to the fundamental goal of estimating a biologically plausible VSD.

The literature of QRA is now quite extensive and continues to expand at a rapid rate. The reader interested in QRA must review not only the statistical literature, but the relevant subject matter journals as well, where a number of important new developments have first appeared. The bibliography compiled by Krewski and Brown (8) is an excellent guide to the literature.

13.3 STATISTICAL MODELS USED IN QRA

As noted in the preceding section, the typical QRA will have as its basis an experiment where the potential carcinogen has been exposed to animals at a number of levels or doses with the presence or absence of a response recorded for each animal exposed. Let $P(d)$ denote the probability that an animal exposed at dose d has the response present at the conclusion of the experiment. In this section the models currently being used to relate d to $P(d)$ will be presented.

13.3.1 Tolerance Distribution Models

A tolerance distribution model assumes that each animal has its own tolerance and that these follow some specified statistical distribution in the population of all animals. Hence, if one assumes that the animals under study are a random sample from this population, then the probability of observing a response corresponds to the chance of selecting an animal whose tolerance is less than the dose the animal is exposed to. If we assume that the cumulative distribution function of the tolerance distribution is $F(d)$, then $P(d) = F(d)$. The most frequently used tolerance distribution models are the probit, logistic, and Weibull distributions.

The probit TDM is obtained by assuming that the log of the tolerance follows a normal distribution. It should be noted that the archaic procedure of adding 5 to a normal deviation to obtain a probit is not used in applications of the log-normal tolerance distributions in QRA. In this case the mathematical form of the model is given by the equation

$$P(d) = \Phi(\alpha + \beta \log d) \qquad (13.1)$$

where $\Phi(\cdot)$ represents the cumulative normal distribution and is defined by the equation $\Phi(x) = 1/\sqrt{2\pi} \int_{-\infty}^{x} \exp(-u^2/2) du$. The text by Finney (5) is the classic reference on the use of the probit model in quantal response bioassays.

The logistic TDM has a dose-response curve that is very similar in shape to the probit. It assumes that the log of the tolerances follows the logistic distribution. The response probability is given by the equation

$$P(d) = \{1 - \exp - (\alpha + \beta \log d)\}^{-1}. \qquad (13.2)$$

The third among the commonly employed TDMs assumes that the log of the tolerances follows the Weibull distribution. This assumption generates a response probability that is given by the equation

$$P(d) = 1 - \exp[-\exp(\alpha + \beta \log d)]. \qquad (13.3)$$

Because of their long history of use in bioassay, TDMs were the first models to be employed to quantitatively assess risk of exposure to a carcinogen. The TDMs are deterministic on the individual level and are thus now thought not to model correctly the current theory that carcinogenesis is a stochastic process.

13.3.2 Stochastic Models

There are two stochastic models that are currently being used in QRA, the multihit and multistage models. The multihit model was the first to be proposed. The simplest case is the one-hit model, which is based on the probability of a target site being hit by a biologically effective dose. The biologic details of the model are given in Iverson and Arley (7) and the necessary mathematical details are presented in Whittemore (13) and Whittemore and Keller (14). The multihit model is obtained by assuming that the number of hits required to transform a cell into a cancer cell follows a homogeneous Poisson process. If we let k represent the required number of hits, then for a given dose d, the probability of response is given by the equation

$$P(d) = 1 - \exp(-\beta d) \sum_{j=0}^{k-1} (\beta d)^j / j! . \tag{13.4}$$

This function may be expressed in an equivalent form as

$$P(d) = \Gamma(k)^{-1} \int_0^d (\beta^k u^{k-1}) \exp(-\beta u) du, \tag{13.5}$$

where $\Gamma(\cdot)$ denotes the gamma function. If k in the above equation is also thought to be an unknown parameter, then $P(d)$ in Eq. 13.5 is a TDM where the tolerances are assumed to follow a gamma probability distribution. Thus, this model is often referred to as the gamma-multihit model, reflecting the fact that the number of hits may be unknown as well as the potency.

The second stochastic model arises from the assumption that a cell will generate into a tumor only after k random biologic stages of development. It is assumed that the occurrence of these events or stages is independently and exponentially distributed. This assumption after some simplification gives the probability of response as

$$P(d) = 1 - \exp\left(- \sum_{j=1}^{k} \beta_j d^j\right) . \tag{13.6}$$

By comparing Eqs. 13.4 and 13.6 it can be seen that the one-hit and one-stage model are identical, but that for $k \geq 2$ the two models have different mathematical forms. It is generally believed that the multistage model is the more biologically plausible among the two stochastic models. One problem with the multistage model is that k, the number of stages, is rarely known and must be determined empirically. The consequences of misspecification of the number of stages will be discussed in the next section.

The investigator who wishes to perform a QRA on some substance must

TABLE 13.1
**A Comparison of the Predicted Responses for the Five Most
Commonly Employed QRA Models**

	Number (%)		Predicted Response Rate (%) for					
Dose	Tested	Responding	Probit	Logit	Weibull	Gamma Multihit	Multi-stage	One-Hit/stage
0	50	4	0.0	0.0	0.0	0.0	0.0	0.0
90	50	20	20.0	20.0	20.0	19.9	20.0	21.6
180	50	40	40.0	40.0	40.0	40.2	40.0	38.6

first decide which model to use. From a statistical point of view, this choice is difficult, since each of the models given above may fit the observed animal data equally well. Table 13.1 presents the results of fitting each of these models to a hypothetical animal experiment.

It is clear from the fact the predicted response rates are nearly identical that it would be impossible to select one model over another on the pure statistical grounds of goodness of fit. However, choice of the model will have a critical effect on the estimate of VSD.

13.4   STATISTICAL METHODS USED TO ESTIMATE RESPONSE AT LOW DOSES

The problem of extrapolation of fitted dose-response models to low doses continues to be an area of considerable interest, research, and controversy. Simply stated, the problem is to estimate the dose, d^*, associated with a particular response probability, say p^*, where p^* is usually in the range 10^{-4} to 10^{-8} with 10^{-6} being the most commonly employed value. The mathematics are simple, requiring the solution to the equation $P(d^*) = p^*$ where $P()$ is any one of the five models described in the previous section. The reason this problem is controversial is that the five models may yield estimated values of d^* which differ by as much as 1000-fold. Table 13.2 below provides the estimates of VSD(d^*) for different values of p^* for each of the five models used in Table 13.1.

The values in Table 13.2 point out quite clearly the problem faced when animal data from a dose-response experiment conducted at high doses are the basis of the QRA. At a response (risk) of 10^{-6} the predicted VSD range from a high of 9×10^{-1} for the probit to a low of 3.7×10^{-4} for the one-hit stage. This difference is typical of the disparity one finds in the estimated values of VSD from the five models.

The difference in the estimated VSD arises because each of the models

TABLE 13.2
Estimates of VSD Using the Five Fitted Models of Table 13.1

Response (p^*)	Probit	Logit	Weibull	Multihit	Multi-stage	One-Hit/stage
10^{-4}	3.0	3.5×10^{-1}	1.4×10^{-1}	2.3×10^{-1}	4.7×10^{-2}	3.7×10^{-2}
10^{-6}	9.0×10^{-1}	1.4×10^{-2}	3.0×10^{-3}	6.6×10^{-3}	4.7×10^{-4}	3.7×10^{-4}
10^{-8}	3.3×10^{-1}	5.3×10^{-4}	6.4×10^{-5}	1.9×10^{-4}	4.7×10^{-6}	3.7×10^{-6}

generates a curve that approaches zero as d goes to zero at a distinctly different rate. In virtually every animal experiment the slope of the dose-response curves has been convex upwards or sublinear. (There are some exceptions, most notable being data on vinyl chloride exposure in mice; but even these data generated a sublinear response function when the dose was corrected to dose at site rather than observed dose.) When the curve is sublinear the predicted VSD will usually be in the order one-hit stage < multistage < Weibull < logistic, multihit < probit. The one-hit stage model becomes for small values $P(d) = \beta d$ or linear. This line lies above the fitted response curves of all the other models, thus giving the lowest estimate of VSD for a given value of p^*.

Several modifications of the estimation of VSD from the solution of the equation $P(d^*) = p^*$ have been proposed. One is to use the lower 95% confidence limit of the VSD, d_1^*, rather than the value of d^*. Several different methods have been proposed for determining d_1^*, but all rely on the application of the statistical theory of the distribution of the estimated parameters of the dose-response function. One method is based on considering d^* as a function of the estimated parameters and derives its distribution and thus confidence interval from this view. A second method extrapolates linearly downward from the fitted model at the 1% or 10% fitted response rate. The rationale for the latter method is to incorporate the known conservative properties of the linear extrapolation approach with a presumably more realistic or flexible model than the one-hit stage. Mantel and Bryan (10) were the first to use this type of approach. They suggested the use of the probit model with a slope 1. They chose a slope of 1 because this was a shallower slope than they had ever seen in fitted dose-response models. It is now generally regarded that the Mantel-Bryan procedure is too conservative.

In practice the most frequently used approach to estimation of VSD is to use the multistage model and the lower 95% confidence limit as the estimate of VSD. In most applications, the number of stages will not be known. The multistage model used to estimate VSD is the one that best fits the observed data. Portier and Hoel (12) show via Monte Carlo simulation that for the three-dose, 50 animals/dose design, if the fitted model is not the correct model, then the estimated VSD may be too small or large by a factor of as much as 4 depending on the type of misspecification (e.g., fitting a two-stage when it should be one, etc.). In addition, they show that estimates of VSD based on the distribution of estimated parameters for large samples may be grossly inaccurate. Their work demonstrates that as many as 100,000 animals per dose may be needed for estimators based on this theory to be accurate. It is unlikely that these particular problems with the multistage model also apply to other models as they arise from the fact that each power of d in the multistage model has its own parameter. This is not the case in the gamma multihit or the TDMs.

The importance of accurate estimates of VSD and the still unresolved controversy over which method to use has generated research into alternative approaches to the problem that attempt to acknowledge directly the uncertainty in extrapolating downward as well as the problem of converting from an animal model to human exposures. The animal-human extrapolation issue has generated its own rather large literature. Crouch et al. (3) use a model based on an extension of the one-hit stage model to estimate the risks associated with organic contaminants of drinking water. Their approach is based on an assumption of constant relative potency. If β_1 and β_2 denote the potency in two species, then the constant relative potency model states that $\beta_1/\beta_2 = K_{12}$ overall substances. They demonstrate the plausibility of this assumption for different species of rodents and to a lesser extent human and rodents. The human data are just not very good or as extensive as animal data and it is difficult to conclude one way or the other if the constant relative potency model applies to animal-human extrapolation. The model proposed by Crouch et al. (3) states that the estimated risk (response probability) for humans at a dose d is given by an equation of the form

$$R = \beta \times K \times E \times d \qquad (13.7)$$

where β is the potency from the animal data, K is the relative potency constant for human to animals, E is an extrapolation dose factor. They assume that β, K, and E are each log normally distributed and thus so is R. They suggest use of upper 98th percentiles of this distribution as the estimator or risk. The problem of reversing this process to find an estimated VSD is technically a fairly straightforward extension of estimating risk. Crouch, et al. (3) apply this methodology to a variety of situations where conventional model-based QRA have been performed and note a number of similar estimates of risk, but also indicate some important differences. Which is the correct or better method remains to be determined. The Crouch et al. (3) approach is intuitively appealing but its actual application is limited to those compounds where sufficient data exist to determine values for β, K, E, and variance estimates of the respective log normal distributions.

Park and Snee (11) have suggested an approach that is similar in form to that proposed by Crouch et al. (3) but does not depend on the one-hit stage model. They propose that the virtual safe dose be estimated via an equation

$$\text{VSD} = D \times F \times (A_1 \times A_2 \times \ldots \times A_1) \times (M_1 \times M_2 \times \ldots \times M_j)$$

$$(13.8)$$

where D is a model based estimate of VSD, F is an extrapolation factor, the A's are factors associated with risk estimation such as animal to human conversion, and the M's are risk management factors. This approach is proposed within a content of a model that moves from risk management, of which QRA is a portion, to the ultimate regulatory decision on risk management. Park and Snee propose their model to stimulate discussion and further research toward developing a more integrated and realistic approach to risk assessment and management. To date this model has not been applied to actual data.

It is unfortunate that there is no clear solution to the problem of low-dose extrapolation; but the current situation reflects the complexity of the problem and the uncertainty of the process being dealt with. As more knowledge is gained about the process of carcinogenesis one can expect parallel developments in models and extrapolatory procedures.

13.5 INCORPORATION OF BACKGROUND RESPONSE

It is well known that humans and animals will develop tumors in the absence of exposure to a substance, and this is especially true of certain rodent strains that are commonly employed in dose-response experiments. In mathematical terms this implies that $P(0) > 0$. Two methods have been suggested for incorporating the chance of background response and these are called additive and independent background. The independent assumption generates a modified dose response model $P'(d)$ where $P'(d) = \gamma + (1 - \gamma)P(d)$. The interpretation of this model is that a fraction γ of the population will develop the tumor in absence of exposure, and, among the proportion $(1 - \gamma)$ who do not develop the tumor spontaneously, a fraction $P(d)$ will develop the tumor due to exposure to a dose d. The additive background model states that the spontaneous response rate is equivalent to a dose δ and the modified model is $P'(d) = P(\delta + d)$. It is usually the case when models incorporating background are used to estimate risk that a VSD is reported that corrects for the background response rate through the use of $P(d) - P(0)$ or $P(d) - P(0)/1 - P(0)$. The choice of which of these corrections to use does not seem to matter much as each will yield approximately the same estimates of risk or VSD. However, it does make a difference which type of background one chooses to incorporate into the model. Crump et al. (4) have shown that the use of the additive background model will induce low-dose linearity. Thus, estimates of VSD based on additive background will always be higher than those based on the independent background model. The actual difference between the two will vary from model

TABLE 13.3
Estimates of VSD for Various Models Incorporating Additive (A) and Independent (I) Background Response

Response (p^*)	Probit		Logit		Weibull		Multihit		Multistage[a]	One-hit Stage[a]
	A	I	A	I	A	I	A	I		
10^{-4}	8.7×10^{-2}	4.8	8.5×10^{-2}	7.6×10^{-1}	7.8×10^{-2}	3.8×10^{-1}	7.7×10^{-2}	6.2×10^{-1}	7.2×10^{-2}	4.2×10^{-2}
10^{-6}	8.7×10^{-4}	1.5	8.5×10^{-4}	4.1×10^{-2}	7.8×10^{-4}	1.3×10^{-2}	7.7×10^{-4}	3.1×10^{-2}	7.2×10^{-4}	4.2×10^{-4}
10^{-8}	8.6×10^{-6}	6.3×10^{-1}	8.5×10^{-6}	2.3×10^{-3}	7.8×10^{-6}	4.5×10^{-4}	7.7×10^{-6}	1.5×10^{-3}	7.2×10^{-6}	4.2×10^{-6}

[a] The use of independent and additive yield equivalent estimates for these models.

to model. Table 13.3 presents estimates of VSD for the data of Table 13.1 incorporating a background response using a definition of excess risk of $P(d) - P(0)$.

The estimates of VSD in Table 13.3 demonstrate the low-dose linearity that results from the assumptions of additivity as well as the differences between the independent additive models. It should be noted that the models are much less disparate in their estimates of VSD under the additive model. In this case the TDMs and stochastic models yield nearly equivalent estimates of VSD.

13.6 SUMMARY

The previous sections have been written with the goal to provide the reader with an overview of the current methods, models, issues, and controversies in QRA. For related articles discussing many of the topics considered here as well as other topics, see Brown and Koziol (1) and Park and Snee (11), and for a more mathematical treatment see Krewski and Van Ryzin (9). A very brief overview of some of the details regarding parameter estimation and computer software is given in the appendix.

Current research in QRA is directed at developing models that are biologically more plausible than those in use. One direction this research is proceeding in is to combine the best of the properties of the TDMs and stochastic models. The TDMs account for such well-known properties as differential susceptibility but fail to recognize the stochastic nature of carcinogenesis. The opposite is true for the stochastic models, which assume each person or animal is equally susceptible to a given dose. Another important area of research is looking into the implications of nonlinear kinetics (see Ref. 6). Each of the models discussed has the implicit assumption that the dose at the site of action is proportional to the administered dose. As more becomes known about the process of carcinogenesis, considerable progress should be made in sharpening the models' ability to account for this knowledge. The fundamental goal of each of these efforts is to make QRA a more useful tool for assessing risk and subsequent standard setting. As the technology to measure the presence of substances advances, it will become increasingly important to use QRA as part of a more global view of risk assessment.

APPENDIX: ESTIMATION OF THE UNKNOWN PARAMETERS IN A QRA MODEL

In any QRA one is faced with the practical problem of using the observed data to estimate the unknown parameters in the particular model chosen. The method of choice is either maximum likelihood or constrained maximum likelihood. The more sophisticated reader should see Krewski and Van Ryzin (9) for discussion of the technical details. This section will only provide an overview.

Let K denote the number of doses that are used in a particular animal experiment, and let d_k, r_k, and n_k denote the dose, the number of animals responding, and the total number of animals exposed at the kth level $k = 1$, $2, \ldots, K$. Under the assumption that the animals respond independently of one another, the likelihood of the experiment for a model say $P'(d)$, where $P'(d)$ is a generic representation of one of the models discussed earlier, is

$$L = \prod_{k=1}^{K} \binom{n_k}{r_k} \{P'(d_k)\}^{r_k} \{1 - P'(d_k)\}^{n_k - r_k} ,$$

The function L above depends only on the unknown parameters in the model P'. The best-fitting model is chosen to be that model whose parameter values maximize the value of L. These estimated parameters are called the maximum likelihood estimates and are usually found as the values that maximize $\log(L)$ rather than L because the computations involved in maximizing $\log(L)$ are simpler than for L.

Suppose as an example we consider fitting the one-hit model with independent background to data in Table 13.1. In this example

$$P'(d) = \gamma + (1 - \gamma)(1 - e^{-\beta d}),$$

the unknown parameters are γ and β. Maximization of the function L in this case yields estimated values of $\hat{\gamma} = 0.039$ and $\hat{\beta} = 0.00242$. These values may then be used to predict a response for a given dose or to solve for a dose given a specified excess risk over background, the VSD. The VSD for an excess risk over background of 10^{-6} would be that dose d^* that solves equation

$$10^{-6} = \hat{P}'(d) - \hat{P}'(0)$$
$$= (1 - \hat{\gamma}) \cdot (1 - e^{-\hat{\beta} d^*}).$$

The solution in this case is $d^* = 4.2 \times 10^{-4}$. To estimate the excess risk over background for a given environmental dose requires that $\hat{P}$ be evaluated at the environmental dose. The excess risk over background for an environmental dose of 2 mg/kg/day would be

$$\hat{R} = \hat{P}'(2) - \hat{P}'(0)$$
$$= (1 - \hat{\gamma})(1 - e^{-\hat{\beta} 2})$$
$$= 1.88 \times 10^{-4}$$

or about 2 per 10,000. Methods for confidence interval estimation may be employed to provide lower limits for the estimated VSD or upper limits for the estimated risk $\hat{R}$.

Computer Software

Computer programs are available that will perform the tasks of parameter estimation and confidence interval estimation for each of the models discussed. A program for the TDMs may be obtained from Dr. Daniel Krewski, Environmental Health Center, Tunney's Pasture, Ottawa, Ontario, Canada K1A 0L2. A computer program for the multistage model may be obtained from Dr. Kenny Crump at Science Research Systems, 120 Gaines St., Ruston, LA 71270. There is a charge for each of these programs, and the exact cost as well as details on how to obtain copies of the programs may be obtained from Drs. Krewski and Crump. The computations presented in this chapter were performed on versions of these programs converted to run on the University of Massachusetts CDC CYBER computer.

REFERENCES

1. Brown, C., and Koziol, J., Statistical aspects of estimation of human risk from suspected environmental carcinogens, *SIAM Rev., 25,* 151–181, 1983.
2. Crouch, E.A.C., and Wilson, R., Regulation of carcinogens, *Risk Analysis, 1,* 47–57, 1981.
3. Crouch, E.A.C., Wilson, R., and Zeise, L., The risks of drinking water, *Water Resources Res., 19,* 1359–1375, 1983.
4. Crump, K.S., Hail, D.G., Langley, C.H., and Peto, R., Fundamental carcinogenic processes and their implications for low dose risk assessment, *Cancer Res., 36,* 2973–2979, 1976.
5. Finney, D.J., *Probit Analysis,* 3rd ed., Cambridge University Press, London, 1971.
6. Hoel, D., Kaplan, N.L., Anderson, M.W., Implication of nonlinear kinetics on risk estimation in carcinogenesis, *Science, 219,* 1032–1037, 1983.
7. Iverson, S., and Arley, N., On the mechanism of experimental carcinogenesis, *Acta Pathol. Microbiol. Scand., 27,* 773–803, 1950.
8. Krewski, D., and Brown, C., Carcinogenic risk assessment: A guide to the literature, *Biometrics, 37,* 353–366, 1981.
9. Krewski, D., and Van Ryzin, J., Dose response models for quantal response toxicity data, in Csorgo, M., Dawson, D., Rao, J.N.K., Saleh, A. (eds.), *Statistics and Related Topics,* North Holland Publishing Co., New York, 1981.
10. Mantel, N., and Bryan, W.R., Safety testing of carcinogenic agents, *J. Natl. Cancer Inst., 27,* 455–470, 1961.
11. Park, C. and Snee, R., Quantitative risk assessment: State of the art for carcinogenesis, *Fund. Appl. Toxicol., 3,* 320–333, 1983.
12. Portier C., and Hoel, D., Low dose extrapolation using the multistage model, *Biometrics, 39,* 897–906, 1983.
13. Whittemore, A., Quantitative theories of oncogenesis, *Adv. Cancer Res., 27,* 55–88, 1978.
14. Whittemore, A., and Keller, J., Quantitative theories of carcinogenesis, *SIAM Rev., 20,* 1–30, 1978.

CHAPTER 14

Carcinogenic Hazards Associated with the Chlorination of Drinking Water

R. J. Bull

College of Pharmacy
Washington State University
Pullman, Washington

14.1 INTRODUCTION

14.1.1 General Background

In 1974 Rook (1) in the Netherlands and Bellar et al. (2) in the United States reported the formation of trihalomethanes (THMs) as by-products of the chlorination of drinking water. This simple observation was the first hint that there may be a reason for seriously examining the safety of one of the most commonly employed and effective public health measures. In the past decade, research efforts have concentrated heavily on studying the mechanisms of chlorine reactions and the nature of the products that result and their toxicological properties. Impetus was added to these efforts by the fact that chlorination by-products are usually the major organic chemicals present in finished drinking water (3).

It must be made clear that the study of hazards associated with chlorination is not aimed at doing away with drinking water disinfection. This process has served as a very important barrier to the spread of water-borne infectious disease, still the largest practical problem in the provision of safe drinking water on a global scale. Rather, research examining potential hazards of the by-products of disinfection is directed at providing the information that is necessary to determine how to produce drinking water in a way that minimizes all types of risk. In this regard, chlorination must be considered against a variety of alternative measures that could be employed or certain pretreatment options that could reduce by-product formation by reducing precursor concentrations. The consideration of the health risks associated with such alternatives is in its infancy and will require a much more sophisticated approach to risk assessment than has been employed in the past (4). The present paper deals only with the chemical hazards that arise from the chlorination of drinking water. However, the correction of any of the potential hazards mentioned should be undertaken only after first considering their impact on the microbiological quality of the finished drinking water.

14.1.2   Major Identified By-Products of Chlorination

Rather diverse chemical structures have been shown to arise through the reaction of chlorine with organic material that is present in sources of drinking water or with chemicals felt to be reasonable models of this material. For a comprehensive review of the types of products that have been identified, the reader is referred to the comprehensive series edited by Jolley et al. (5). Table 14.1 lists the products that have been identified following the chlorination of humic acid (6) to provide some indication of the types of chemicals that can be formed. Inclusion of Br^- in the reaction mixture leads to the formation of corresponding brominated and/or mixed bromine and chlorine-substituted products. These have been omitted from Table 14.1 for the sake of simplicity. The toxicological properties of some of these chemicals will be considered below.

Although all the chemicals in Table 14.1 are halogenated, it should not be concluded that only halogenated by-products are formed with chlorination. The two major forms of chlorine present in aqueous solution are $HOCl$ and OCl^-, depending on pH (pKa = 7.5). Both forms are capable of acting as oxidants as well as forming halogenated by-products. The degree to which nonchlorinated oxidized products are formed is not clear since they are much more difficult to detect in complex mixtures than the chlorine-addition products. However, the extent to which the halogenated by-products also include oxidized functional groups such as aldehydes, ketones, and carboxylic acids is evidence of chlorines activity as an oxidant.

The formation of oxidation products with chlorine is not simply of academic interest. Many of the disinfectants being considered as alternatives to chlorine (e.g., chlorine dioxide, ozone, and hydrogen peroxide) act primarily as oxidants and do not participate, at least directly, in halogen addition reactions. Consequently, a part of the decision concerning potential health effects of alternate means of disinfection will eventually have to focus on the relative hazards of oxidized versus halogenated by-products. This is a complex problem that is beyond the scope of the present chapter. However, the interested reader is referred to a recent overview of this problem prepared by Bull and McCabe (4).

14.1.3   Evidence for Other Products of Potential Concern

A variety of studies have demonstrated formation of by-products of chlorination that possess mutagenic activity (7–11). It is clear that most of the

TABLE 14.1
Tentative Identification of Constituents of Chlorinated Humic Acid[a]

Trihalomethanes
  Choroform

Acids
  Dichloroacetic acid
  Trichloroacetic acid

Aldehydes
  Dichloroacetaldehyde
  Trichloroacetaldehyde
  Dichloropropanal
  Trichloropropanal
  2-Chloropropenal
  2,3-Dichloropropenal
  3,3-Dichloropropenal
  2,3,3-Trichloropropenal
  Trichlorobutanal
  Dichlorobutenal

Ketones
  1-Chloro-2-propanone
  1,1-Dichloro-2-propanone
  1,3-Dichloro-2-propanone
  1,1,1-Trichloro-2-propanone
  1,1,3-Trichloro-2-propanone
  1,1,1,3-Tetrachloro-2-propanone
  1,1,3,3-Tetrachloro-2-propanone
  Pentachloropropanone
  3-Chloro-2-butanone
  1,1,1,-Trichloro-2-butanone

Ketones (cont.)
  1,1,3-Trichloro-2-butanone
  1,3-Dichloro-2-butanone
  1,1,-Dichloro-2-butanone
  3,3-Dichloro-2-butanone
  1-Chloro-3-buten-2-one
  3-Chloro-3-buten-2-one
  Dichloro-3-buten-2-one
  Trichloro-3-buten-2-one
  Tetrachloro-3-buten-2-one
  Pentachloro-3-buten-2-one
  Trichlorocyclopentenedione

Nitriles
  Chloroacetonitrile
  Dichloroacetonitrile
  Trichloroacetonitrile
  Dichloropropanenitrile
  Dichloropropenenitrile
  Trichloropropenenitrile

Aromatics
  2,4,6-Trichlorophenol
  Trichlorodihydroxy benzene
  Tetrachlorothiophene

Alkanes and Alkenes
  Hexachloroethane
  Pentachloropropene
  Tetrachlorocyclopropene
  Hexachlorocyclopentadiene

[a] *Source:* Ref. 6.

chemicals responsible for the activity generated by chlorine in complex mixtures have yet to be identified. On the basis of studies of humic acid as a model substrate, the formation and decay of these products are very dependent upon pH and they appear to be much less volatile and more polar than identified products of chlorination such as the THMs (12,13).

Because of the lack of appropriate toxicological information for large numbers of the by-products of chlorination, it must be recognized that any assessment of the hazards that is simply based upon products for which such information exists may be underestimates of the actual risks. Simple evidence of mutagenic activity in bacteria, in many cases the only data that exist for most of the identified by-products of chlorination, is not a particularly accurate indication of the degree of carcinogenic potency (14), despite the usefulness of these systems in helping to identify those by-products that might possess carcinogenic properties. As a result, current estimates of the carcinogenic risk associated with the chlorination of drinking water on the basis of toxicological studies must remain qualitative. The present paper will review data that has recently become available that bear on this issue.

14.2   TRIHALOMETHANES

The THMs are the most widely recognized by-products of the chlorination of drinking water and are found ubiquitously. The most common forms found include trichloromethane (chloroform), dichlorobromomethane, dibromochloromethane, and tribromomethane (bromoform). Occasionally, iodinated derivatives are also observed. Of this class of chemicals, the toxicological effects of chloroform have been the best characterized. In the near future the results of carcinogenesis bioassays on the other three major THMs should be available from the National Toxicology Program (NTP). In the absence of this information, data on chloroform carcinogenicity will have to suffice for the purposes of this review. Chloroform's carcinogenic properties have been recently reviewed by IARC (15) and by Davidson et al. (16). Consequently, the present section will focus primarily on data that has become available since these earlier reviews.

14.2.1   Reassessment of Chloroform Carcinogenesis

The principal evidence that chloroform possessed carcinogenic activity arose from a study sponsored by the National Cancer Institute (NCI) (17), although there was prior evidence of its carcinogenic properties available in the literature (18). In the NCI study B6C3F1 mice of both sexes developed a high incidence of hepatocellular carcinomas and male Osborne-Mendel rats developed renal tumors following chronic exposure to chloroform. These studies employed high concentrations of chloroform dissolved in corn oil that were administered by stomach tube five times weekly for 78 weeks. A reanalysis (19) of the pathological slides obtained from this study indicated that the chloroform-treated mice had a greater incidence of malignant lymphoma than controls whereas similarly treated rats had developed thyroid tumors, cholangiofibromas, and cholangiocarcinomas (females only).

For a variety of reasons, it was felt that the results of the NCI study were inadequate for the purpose of estimating the cancer risk to humans consuming very low concentrations of chloroform in their drinking water. A major consideration was the early observation by Eschenbrenner and Miller (18) that hepatoma development following chloroform administration to strain A mice was confined to those doses that produced frank liver necrosis. To address this and other related issues, a larger study was cosponsored by the U.S. Environmental Protection Agency (EPA) and the NCI (20). This study differed from the previous NCI study in that chloroform was administered in the drinking water and the treatments were extended to lower doses with substantially expanded group sizes. This design was to avoid problems that may have been associated with gavage dosing and the corn oil vehicle. The expanded group sizes at low doses also provide a better basis for the extrapolation of the data to the lower concentrations encountered in drinking water. In addition, biochemical measurements were made periodically in both serum and liver in animals that had been subjected to treatments that paralleled those in the lifetime study. Because of the aversive reaction the animals had to chloroform in the drinking water, it was necessary to include control groups that were restricted in their water intake to that consumed by the highest dose group for each species. To avoid multiplying the costs of this large study, only female B6C3F1 mice and male Osborne-Mendel rats were included in the experiment.

The Jorgenson et al. study (20) substantially supported the previous results obtained in male Osborne-Mendel rats. It was necessary to adjust the tumor incidence in this experiment for survival since animals that received the highest doses of chloroform survived significantly longer than the control animals that were allowed to drink water ad libitum. It was reasonably clear that this increased survival was not attributable to the chloroform treatment because control animals that were restricted in their water consumption also lived longer. A more likely explanation for the longer survival in these groups was that animals that consumed less water were also much leaner; a situation that is widely known to contribute to longer survival. In Figure 14.1 the adjusted tumor incidence rate (renal tubular adenomas + adenocarcinomas) in the Jorgenson et al. study (20) is directly compared with the rates observed in the prior NCI study (17). It should be noted that the effective doses of chloroform in the NCI study are somewhat smaller than indicated because they were administered for only 5 days a week. Aside from the differences observed in the control incidence, the dose-response curves are very similar. The relatively small control group used in the NCI study (20 animals) is probably responsible for the difference in the control incidence of renal tumors.

In sharp contrast with the results with the rat, the results obtained from the Jorgenson et al. study (20) in female B6C3F1 mice were substantially different from those observed in the original NCI study (Table 14.2). In the former case, there was no evidence of a dose-related increase in the incidence of hepatocellular carcinomas. The two highest doses utilized in this study bracket the low dose in the earlier NCI study where an 80% incidence of hepatocellular carcinoma was observed. The Jorgenson et al. study maintained treatments for 7 days a week for 104 weeks, whereas treatments were administered for only 5 days a week for 78 weeks in the NCI experiments; thus the differences in responses cannot be accounted for by a smaller total dose in the former study.

Table 14.3 summarizes the available data dealing with the carcinogenic

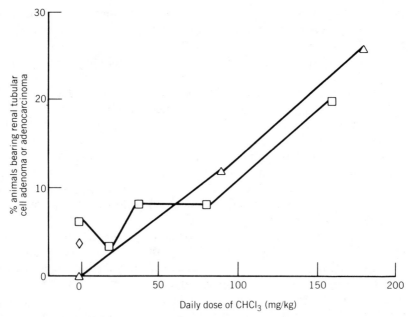

Figure 14.1. Comparison of the carcinogenic response in the kidney of the male Osborne-Mendel rat on the NCI study ($\triangle$) and the Jorgenson et al. (20) study ($\square$). Control animals matched with the high dose group for water consumption ($\diamond$). Indicated doses were administered for 5 days/week for 78 weeks in the NCI study (17) and for 7 days a week for 104 weeks in the Jorgenson et al. (20) study.

effects of chloroform in experimental animals. It is notable that those studies that were unable to demonstrate any carcinogenic activity with chloroform treatment uniformly involved dosage levels below those used in either the Jorgenson et al. (20) or the NCI (17) studies. The only positive results not previously mentioned were the finding of increased renal tumors in ICI mice when chloroform was administered in a toothpaste base at a maximum dose of 60 mg/kg per day (21). Renal tumors were not observed in C57BL, CBA, or CF/1 mice treated with the same dose in the same study.

Because of the doses of chloroform that have been required to produce tumors approximate those that produce overt toxicity, it has been argued that the carcinogenic response observed is secondary to tissue necrosis. The growth of already transformed cells would be stimulated by the regenerative hyperplasia that follows such tissue damage. Support for this hypothesis was obtained by Reitz et al. (22) with their demonstration of increased incorporation of [3H]thymidine into DNA of both the liver and the kidney of B6C3F1 mice. Moore et al. (23) also found that doses of chloroform approximating 60 mg/kg administered in corn oil resulted in damage to the kidney of mice that was also accompanied by evidence of tubular regeneration as measured by [3H]thymidine uptake. Similar doses in a toothpaste base were without ef-

TABLE 14.2
**Comparison of Liver Tumor Incidence in Female B6C3F1 Mice
in the NCI[a] and Jorgenson et al.[b] Studies**

NCI		Jorgenson et al.	
Dose (mg/kg)	Incidence of Hepatocellular Carcinoma (%)[c]	Dose (mg/kg)	Incidence of Hepatocellular Adenoma + Carcinoma (%)[d]
0	6	0	5
		Matched 0	0
		34	4
		65	6
		130	0
238	80		
		263	2
477	95		

[a] Data obtained from National Cancer Institute Study (17).
[b] Data obtained from Jorgenson et al. (20).
[c] Data is not corrected for longevity.
[d] Data is corrected for intercurrent mortality.

fect. At 240 mg/kg, similar effects were observed with both vehicles. Such responses were not, however, strongly apparent in either the liver or kidney of Osborne-Mendel rats.

In the Jorgenson et al. study (20), there was little evidence of overt hepatic or renal pathology in the female B6C3F1 mice treated with chloroform. However, there was some evidence that the liver of B6C3F1 mice was unable to adequately handle triglyceride. In Table 14.4 the percent liver fat observed in B6C3F1 mice is compared with the same parameter in Osborne-Mendel rats after 90 or 180 days of chloroform treatment via the drinking water. The mice displayed lipid accumulation in the liver at much lower doses than the rats and the severity of the accumulation was greater at the high doses for the mice than the rats (24). In themselves, these changes in hepatic fat are relatively mild. The question is, do they become more severe if the treatment is complicated by the high lipid intake that results from administering the chloroform in corn oil? In the rat these changes in liver fat were preceded by a substantial depression of serum triglycerides that became particularly apparent in the aging animal (24). Data on serum triglycerides of the mice under these conditions were not available. At this time one can only speculate that the use of corn oil may have contributed to the chloroform-induced development of hepatocellular carcinomas in the first NCI study by imposing a large lipid load on a liver whose function has been compromised by chloroform.

The argument that chloroform only induced tumors secondary to overt

TABLE 14.3
Evidence for Carcinogenicity of Chloroform in Experimental Animals

Species and Strain	Vehicle	Highest Dose Tested	Route	Tumor	References
A mice	Olive oil	2350 mg/kg	Oral	Liver	(18)
C57BL mice	Toothpaste	60 mg/kg/day	Oral	None	(21)
CBA mice	Toothpaste	60 mg/kg/day	Oral	None	(21)
CF/1	Toothpaste	60 mg/kg/day	Oral	None	(21)
ICI mice	Toothpaste	60 mg/kg/day	Oral	Kidney	(21)
Sprague-Dawley rat	Toothpaste	60 mg/kg/day	Oral	None	(21)
Beagle dog	Toothpaste	20 mg/kg/day	Oral	None	(21)
Beagle dog	Toothpaste	20 mg/kg/day	Oral	Total Neoplasms	(19)
Osborne-Mendel rat	Corn oil	180 mg/kg/day	Oral	Kidney	(17)
Osborne-Mendel rat	Corn oil	180 mg/kg/day	Oral	Thyroid	(19)
Osborne-Mendel rat	Corn oil	180 mg/kg/day	Oral	Cholangiofibromas	(19)
Osborne-Mendel rat	Corn oil	180 mg/kg/day	Oral	Cholangiocarcinomas	(19)
B6C3F1 mice	Corn oil	477 mg/kg/day	Oral	Liver	(17)
B6C3F1 mice	Corn oil	477 mg/kg/day	Oral	Lymphoma	(19)
Osborne-Mendel rat	Drinking water	130 mg/kg/day	Oral	Kidney	(20)
B6C3F1 mice	Drinking water	400 mg/kg/day	Oral	None	(20)

TABLE 14.4
Liver Fat Content in Mice and Rats Treated with Chloroform in Drinking Water

| | Percent Liver Fat[a] | | | |
| | Female B6C3F1 Mice | | Male Osborne-Mendel Rats | |
Chloroform (mg/L)	90 days	180 days	90 days	180 days
0	3.33	5.82	3.32	4.49
200	3.45	7.93[b]	3.31	4.50
400	3.89[c]	6.77	3.20	4.59
900	4.51[c]	7.11[b]	3.58	4.77
1800	6.36[c]	10.40[c]	3.46	5.13[b]

[a] Average value for 6–10 mice or 10–20 rats at each dose and time.
[b] Significantly different from corresponding control at $p < 0.05$ by ANOVA and t test.
[c] Significantly different from corresponding control at $p < 0.01$ by ANOVA and t test.

tissue necrosis is indirectly supported by the difficulties encountered in demonstrating genotoxic properties of the chemical (15). More recent data (Table 14.5) has indicated a very low association of chloroform or its metabolites with DNA in the liver or kidney of mice or rats (22,25,26). However, it has been shown that chloroform possesses the capability of inducing chromosome breakage and increased sister-chromatid exchange (SCE) frequencies in human lymphocytes in vitro and increased SCE in the bone marrow of mice treated with chloroform in vivo (27). Chloroform has also been shown to enhance the transformation of Syrian hamster embryo cells by S7 adenovirus (28). It is notable that the concentrations of chloroform tested in these positive experiments tend to be quite high and certainly exceed those tested in similar experiments (29) that gave rise to negative results (Table 14.5). The difficulties in interpetation that are raised by the lack of strong evidence of genotoxic effects by chloroform are further complicated by the failure to demonstrate clearly the existence of an alternative mechanism for chloroform-induced cancer. For example, Pereira et al. (26) were unable to demonstrate clearly either tumor-initiating or tumor-promoting activity of chloroform in the rat liver.

In estimating the carcinogenic risk to man from chloroform in drinking water, all of the above data must be considered. At present, it is clear that chloroform is capable of producing renal tumors in one strain of mice and one strain of rats. Under circumstances that involved gavage with a vehicle that contributed substantially to the lipid intake of the animal, chloroform also induced liver tumors in one strain of mice. In a separate experiment where the influence of corn oil gavage was removed, tumors did not develop

TABLE 14.5
Evidence for Genotoxic Activity of Chloroform

Test System	Special Conditions	Highest Dose Tested	Result	Reference
S. typhimurium	± Mouse and Rat Liver and Kidney S-9	10 mg/plate	Neg.	(49)
TA1535				
TA1537				
TA1538				
TA98				
TA100				
Syrian hamster embryo cells—enhanced viral transformation	Applied in sealed chamber	0.5 ml/chamber	Pos.	(28)
V79 cells—8-azaguanine locus	Applied in flow through system, then sealed	3% in air	Neg.	(51)
Hyman lymphocyte—sister chromatid exchange induction	± Rat liver S-9	400 µg/plate	Neg.	(29)
Hyman lymphocyte chromosome damage	± Rat liver S-9	400 µg/plate	Neg.	(29)
E. coli WP2p and WP2uvr A⁻ p	± Rat liver S-9	10 mg/plate	Neg.	(29)
Human lymphocyte—sister chromatid exchange, in vitro		6 mg/ml	Pos.	(27)
Mouse bone marrow—sister chromatid exchange, in vivo		4 × 200 mg/kg/L	Pos.	(27)
Rat Liver, DNA—binding in vivo	Liver	N.A.	Neg.	(25)
Sprague Dawley rat—DNA binding, in vivo	Kidney	N.A.	Neg.	(26)
B6C3F1 mice—DNA binding, in vivo	Liver	N.A.	Neg.	(26)
B6C3F1 mice—DNA binding, in vivo	Liver	N.A.	Neg.	(22)

at similar dose levels. These data are sufficient to conclude that chloroform is probably carcinogenic in humans. However, the application of these results to a quantitative estimate of the risks that chloroform in drinking water presents to humans depends very much on the interpretation of how the results apply to the human situation. Americans, at least, consume considerably more fat in their diet than that present in the usual laboratory rodent diet. Consequently, one might conclude that experiments conducted with corn oil gavage best represent the human situation and that the mouse liver tumors must be considered significant. On the other hand, administration of chloroform in drinking water might be considered most relevant because it avoids bolus doses and corresponding high peaks in systemic chloroform concentrations. In this case the renal tumors observed in male Osborne-Mendel rats and ICI mice might be the most appropriate for estimating the carcinogenic risk to humans. Finally, the question of whether chloroform can be treated as a chemical capable of initiating the carcinogenic process or simply one that can accelerate the development of already initiated cells into overt tumors must be considered. The possibility that chloroform can act only in the latter way in no way decreases its importance as a drinking water contaminant. However, it might significantly affect the model or the parameters used in a model to extrapolate its effects to the low doses encountered in drinking water.

14.3 HALOACETONITRILES

To date, the haloacetonitriles that have been identified in chlorinated drinking water have been the dihalogenated chlorine and bromine derivatives (30,31). The occurrence of the haloacetonitriles in water at the tap is at least partially dependent upon the pH of the finished drinking water, since all members of the class normally found are alkaline labile (32).

Investigations into the carcinogenic properties of this class of chemicals have been confined to three members of the chlorinated series chloroacetonitrile (CAN), dichloroacetonitrile (DCAN), and trichloroacetonitrile (TCAN); bromochloroacetonitrile (BCAN); and dibromochloroacetonitrile (DBAN). The first experimental evidence of the potential of these derivatives as carcinogens was the observation that DCAN was mutagenic in *Salmonella typhimurium* (33). More recently, Bull et al. (34) confirmed the mutagenic effects of DCAN in *Salmonella* and demonstrated similar, but more potent, mutagenic effects were associated with BCAN. CAN, TCAN, and DBAN were found to be inactive in the presence or absence of a 9000 × *g* supernatant fraction (S-9) of a liver homogenate taken from rats previously treated with Arochlor 1254. On the other hand, all five haloacetonitriles significantly increased the frequency of SCE in Chinese hamster ovary cells, in vitro, in the absence of rat liver S-9 fraction. However, none of these same five haloacetonitriles could be shown to possess mutagenic activity in the mouse micronucleus assay (34) or to increase the frequency of mouse spermhead abnormalities (35) when administered in vivo.

Direct evidence of carcinogenic activity of these chemicals in experimental animals is limited to studies of their tumor-initiating activity in the skin of Sencar mice (34) and their ability to increase lung adenoma incidence in strain A mice (36). High doses applied topically (1200, 2400, and 4800 mg/kg body weight split between six applications over a 2-week period) followed by a 20-week promotion schedule involving topical application of 12-O-tetradecanoyl-phorbol-13-acetate (TPA) yielded significantly higher tumor incidences with CAN, BCAN, and DBAN. DCAN and TCAN produced small increases in cumulative tumor yields, but the changes were not significantly different from the incidence of tumors observed in animals that had only received TPA. The relative activity of the haloacetonitriles in mutagenesis and carcinogenesis assays is compared in Table 14.6.

In a separate group of experiments, the haloacetonitriles were administered orally to strain A/J mice at 10 mg/kg three times weekly for 8 weeks. At 9 months of age, small but significant increases in the incidence of lung adenomas were observed in animals treated with CAN, TCAN, and BCAN. The tumor incidence was increased to 32, 28, and 31%, respectively, whereas the incidence in the control group was 10% in this experiment. The historically high and variable spontaneous rate of lung adenoma development in this strain makes this result difficult to interpret in the context of the haloacetonitriles representing a hazard to humans at the low concentrations found in drinking water.

The available evidence that would suggest that haloacetonitriles in drinking water represent a serious carcinogenic hazard to humans is weak at present. The relatively low potency the class exhibits in virtually all test

TABLE 14.6
Activity of Selected Haloacetonitriles in In Vitro Tests for Mutagenic Effects and as Tumor Initiators in the Mouse Skin

Compound	TA100[a]	SCE in CHO[b]	ED$_{35}$ Sencar[c]
Chloroacetonitrile	0	50	2.9
Dichloroacetonitrile	71	103	—
Trichloroacetonitrile	0	230	—
Bromochloroacetonitrile	860	396	2.2
Dibromoacetonitrile	0	469	1.6

[a] Results expressed in net revertants/μmol. Negative results are indicated by a 0.
[b] Results expressed as induced SCEs/mM concentration. Experiments conducted in Chinese hamster ovary cells.
[c] The ED$_{35}$ is the total dose in g/kg that produced a 35% incidence in Sencar mice treated topically with the indicated compound. Chemicals that did not produce a statistically significant response are indicated by —.

systems employed would suggest that they pose a minimal risk. There is no evidence available that clearly demonstrates the ability of these chemicals to initiate cancer by a systemic route of administration. Nevertheless, these chemicals are important because they are the first group of by-products of chlorination other than the THMs that have been shown to possess both carcinogenic and mutagenic properties (albeit weak). In addition, these compounds can be derived from reaction between chlorine and various amino acids and proteins (32) and their formation in vivo has been demonstrated following intubation of rats with aqueous solutions of chlorine (37). Because they are felt to be fairly widespread in their occurrence in drinking water and because of the fact that drinking water often contains a significant residual of unreacted chlorine, it is important that the relative hazards of these chemicals be determined in lifetime carcinogenesis bioassays.

14.4   CHLORINATED PHENOLS

Chlorination of drinking water often gives rise to small quantities of chlorinated phenol derivatives and they are observed as by-products of the chlorination of humic acid (6). The concentrations in drinking water could be considerably higher, but the taste and odor problems that arise from their presence usually acts as a strong incentive to minimizing their formation. The principal products formed include 4-chlorophenol, 2,4-dichlorophenol, and 2,4,6-trichlorophenol (38).

There is substantive evidence of carcinogenic effects only with 2,4,6-trichlorophenol. The NCI conducted a feeding study (39) of the effects of 5,000 and 10,000 ppm of 2,4,6-trichlorophenol in the diet of Fischer 344 rats and 5214 and 10,428 ppm in the diet of B6C3F1 mice. The incidence of lymphomas and leukemias was increased in male rats in a dose-related manner. Hepatocellular carcinomas were increased in both male and female mice and the incidence was also dependent on the dose of 2,4,6-tirchlorophenol. These data would suggest that 2,4,6-trichlorophenol could pose a carcinogenic risk to man if it were to be produced at high enough concentrations in the finished drinking water.

While neither 4-chlorophenol or 2,4-dichlorophenol have been shown capable of producing cancer independent of other factors, they do seem capable of acting as cocarcinogens in producing nervous system tumors when animals are treated with ethylurea and nitrite in utero (40). These two chemicals also are capable of acting as tumor promoters in the mouse skin (41). Consequently, their occurrence in finished drinking water cannot be neglected. However, it is not clear what means should be utilized to determine the acceptable levels of these chemicals in drinking water partly because of limited data but also because of their indirect activity.

14.5 HALOGENATED ALDEHYDES AND KETONES

Chlorinated aldehyde and ketone derivatives are prominent by-products of chlorination reactions with a variety of organic substrates. In Table 14.7 a list of those products for which mutagenicity data exists is displayed along with a comparison of their relative mutagenic activity in *Salmonella typhimurium* strain TA100. The range in mutagenic potency is quite remarkable; from 4.8 net revertants/μmol for 1,1-dichloropropanone to 220,000 for 2,3,3-trichloropropenal. In all cases the activity associated with these chemicals is observed without the addition of an exogenous means for metabolic activation (e.g., S-9), suggesting that they act directly.

Very little information exists demonstrating carcinogenic responses to these chemicals in intact animals. Preliminary information would indicate that 1,1-dichloroacetone and 1,1,1-trichloroacetone are incapable of initiating tumors in the skin of Sencar mice (36). On the other hand, 1,3-dichloroacetone and 2-chloropropenal seem to be quite active in this regard (R.J. Bull and M. Robinson, unpublished data). Although these data are suggestive, the extent to which these compounds contribute to an overall carcinogenic hazard attributable to chlorination by-products cannot be estimated on the basis of the available information.

14.6 OTHER BY-PRODUCTS OF POTENTIAL CONCERN

Chlorination of drinking water (7–11), aqueous solutions of humic acids (12,42), or pulpmill effluents (43) results in the formation of mutagenic activ-

TABLE 14.7
Mutagenic Activity of Chlorinated Acetone and Aldehyde Derivatives in *Salmonella typhimurium* Strain TA100

Chemical	Net Revertants/μmol[a]
1,1-Dichloropropanone	4.8[b]
1,3-Dichloropropanone	14,500
1,1,1-Trichloropropanone	120[b]
1,1,3-Trichloropropanone	3,970[c]
1,1,3,3-Tetrachloropropanone	1,530[b]
Pentachloropropanone	860[b]
2-Chloropropenal	114,000[d]
3,3-Dichloropropenal	730[b]
2,3,3-Trichloropropenal	408,000[b]

[a] The net revertants/μmol were calculated from the linear portion of the dose-response curves.
[b] Based on the data of Meier et al. (44).
[c] Based on the data of Douglas et al. (43).
[d] Based on the data of Rosen et al. (50).

ity that cannot be accounted for by summing the activity of individual chemicals than can be identified and measured in the reaction mixture. In the case of chlorinated humic acid mixtures, it was estimated that approximately 7% of the mutagenic activity in these mixtures could be accounted for by the sum of the measured concentrations of DCAN, 1,1-dichloropropanone, 1,3-dichloropropanone, 1,1,1-trichloropropanone, 1,1,3-trichloropropanone, 1,1,3,3-tetrachloropropanone, pentachloropropanone, 3,3-dichloropropenal, and 2,3,3-trichloropropenal (44). The remainder of the activity had to be associated with chemicals that (1) have not been identified, (2) have not been quantified, and/or (3) have not been assayed for mutagenic activity in *Salmonella*. The progression of research to remedy all three of these deficiencies depends heavily upon obtaining pure samples of the individual components. Since many of the by-products of chlorination are not available commercially, appropriate accounting for the mutagenic activity requires the synthesis of some rather unique chemicals based on preliminary identifications by mass spectrometric analysis. An additional difficulty arises from the fact that most of the compounds produced appear to be polar chemicals that do not lend themselves easily to coupled gas chromatographic/mass spectrometric analysis.

Introduction of Br^- into reaction mixtures of chlorine and organic material further complicates the picture. As indicated earlier the presence of Br^- results in the formation of a greater variety of chemicals with varying degrees of bromine substitution (6). Rather small additions of Br^- also sharply increase the amount of mutagenic activity that is observed as well (11,44).

The fact that the mutagenic activity in reaction mixtures of chlorine and organic material similar to that found in many drinking water sources cannot be accounted for does not necessarily mean that a large cancer risk exists. Rather, this rather large gap in our knowledge simply makes it very difficult to estimate the overall dimensions of any carcinogenic hazard that might be associated with the use of chlorine in drinking water disinfection. Consequently, at present there is little reason to abandon the use of chlorine for the much higher risks of water-borne infectious disease. Neither would it be wise to switch to alternate forms of drinking water disinfection about which we know even less except in those rare instances where the concentrations of particular undesirable by-products exceed acceptable levels. Even in this latter case, decisions should be made deliberately to be certain that the microbiological quality of the finished product is not compromised. Because of the widespread use of chlorine in drinking water disinfection, it is obvious that further research is needed to resolve these uncertainties.

14.7 POTENTIAL CARCINOGENIC RISK FROM INGESTION OF CHLORINE

Periodically, some fears arise as to the possible carcinogenic properties of chlorine itself. Both common species of chlorine that exist in aqueous solution (HOCl and OCl⁻) are very reactive chemicals and it has been generally considered unlikely that they would ever be systemically absorbed as such. However, recent papers (37,45) point out that the possibility that chlorine forms potentially hazardous by-products by reaction with the rich organic substrate present in the gastrointestinal tract. These observations suggest that there may be an indirect hazard.

Sodium hypochlorite (OCl⁻) has been shown to be mutagenic in *Salmonella* (46) and to inhibit preferentially growth of DNA polymerase-deficient bacteria (47). Again because of the extreme reactivity of the chemical it is not possible to determine whether these results were directly due to hypochlorite or some reaction product produced with the organic components of the media.

More recently, attempts have been made to detect mutagenic activity of chlorine in vivo in the mouse micronucleus assay, looking for alterations in bone marrow cytogenetics in mice and for the production of spermhead abnormalities in mice (35). The maximum doses utilized in these studies were 8 mg/kg body weight because of the limited aqueous concentrations of chlorine that were possible. Data in the micronucleus assay and bone marrow cytogenetics studies gave no indication of a mutagenic effect by either HOCl or OCl⁻. However, increases in the percent of abnormal sperm were observed with doses of OCl⁻ of 4 mg/kg body weight and above 3 weeks following the last dose (five daily doses were administered). This effect was not observed with HOCl. The extent of the effect was small and appeared to be self-limited in that higher doses did not further increase the response. Consequently, it is unlikely that an effect on reproductive function could be demonstrated (although this is presently being investigated). However, these data do suggest the in vivo formation of a mutagenic by-product that has sufficient stability to reach the testis.

Again it is not possible to relate these results directly to a carcinogenic hazard. Although many carcinogens are known to produce spermhead abnormalities in mice (48), such effects cannot be used to predict qualitatively a carcinogenic hazard or to estimate reliably its magnitude. Consequently, we are faced with another unknown in attempting to estimate the carcinogenic hazard that may be associated with the use of chlorine as a disinfectant of drinking water.

14.8   SUMMARY AND CONCLUSIONS

In the decade that has elapsed since THMs were first shown to be formed during the chlorination of drinking water, the perception of the problem has become much more sophisticated and to some extent more realistic. (1) The microbiological quality of drinking water is still the primary concern of public water systems and in many instances appropriate disinfection is essential for this purpose. (2) As long as reactive chemicals such as chlorine or any of its suggested alternatives are used for disinfection of drinking water, there will be some alteration of the chemical composition in the finished water. (3) The reaction of these chemicals will result in the formation of a very complex mixture of by-products that will be very difficult to deal with from an analytical point of view. (4) In most instances the concentrations of these by-products will be low, but not necessarily negligible. (5) There are some aspects of disinfectant use whose hazards are relatively well understood (e.g., the potential carcinogenic hazard associated with chloroform as a by-product of chlorination). (6) There is considerably more about the hazards of disinfection that we do not understand than we do understand. (7) We have no evidence to indicate that we have a very large and widespread hazard associated with current disinfection practice at present (this does not exclude the possibility of a serious problem in an individual water supply). (8) Even in the absence of overt evidence of damage in human populations exposed to chlorine and its by-products, it is important to understand the nature of any potential hazards that arise from the use of chlorine or its alternatives because of the size of the populations exposed (i.e., the relatively small effect in a large population affects a large number of people). (9) Changing disinfection practice in response to only a partial understanding of the hazards associated with the alternative procedures is as likely to increase the risks as it is to decrease them. Finally, cancer represents only one of the potential risks involved in alternative methods of disinfecting drinking water and it is not necessarily the most important, as was pointed out by a number of authors at the Fifth Water Chlorination Conference held in 1984 (5).

In terms more specific to the carcinogenic hazards associated with chlorination, it is clear that the greatest problem is associated with the 80–90% of the organic material that is poorly characterized in finished drinking water. The presence of chemicals in this fraction that possess mutagenic activity is sufficient cause to encourage both analytical chemical and toxicological work in this area for the foreseeable future. For the toxicologist this will raise a number of problems that are somewhat unique. Not only does the problem involve an extremely complex mixture of, in all probability, several thousand chemical compounds; but it involves these chemicals at very low concentrations. Consequently, not only do the problems of whether the effects of a single compound can be extrapolated over several orders of magnitude exist, but also whether the interactions between these chemicals can be extrapolated to much lower doses than have been studied. Can the hazard be considered the sum of each individual chemical's activity or is there the possibility of substantial potentiation of effects as might be expected from the coexistence of tumor initiators and promoters? Can a bioassay of a specified synthetic mixture of the chemicals found in drinking water or a complex mixture of these chemicals be considered representative of the type of interactions that will go on at much lower doses? What would the impact be if one or more components were removed from or added to the mixture? It is clear that resolution of these issues will take time, and the resolution of many of them may have to await substantive developments in the sciences of toxicology and analytical chemistry as a whole.

REFERENCES

1. Rook, J.J., *Water Treat. Examin., 23,* 234, 1974.
2. Bellar, T.A., Lichtenberg, J.J., and Kroner, R.C., *J. Am. Water Works Assoc., 66,* 703, 1974.
3. Symons, J.M., Bellar, T.A., Carswell, J.K., DeMarco, J., Kropp, K.L., Robeck, G.G., Seeger, D.R., Slocum, C.J., Smith, B.L., and Stevens, A.A., *J. Am. Water Works Assoc., 67,* 708, 1975.
4. Bull, R.J., and McCabe, L.J., Risk assessment issues in evaluating the health effects of alternate means of drinking water disinfection, in Jolley, R.L., et al. (eds.), *Water Chlorination: Environmental Impact and Health Effects,* vol. 5, Butterworth, pp. 111–130, 1985.
5. Jolley, R.L., et al. (eds.), *Water Chlorination: Environmental Impact and Health Effects,* vol. 5, Butterworth, pp. 111–130, 1985.
6. Coleman, W.E., Munch, J.W., Kaylor, W.H., Streicher, R.P., Ringhand, R.P., and Meier, J.E., *Environ. Sci. Technol., 18,* 674, 1984.
7. Cheh, A.M., Skochdopole, J., Koski, P., and Cole, L., *Science, 207,* 90, 1980.
8. Zoeteman, B.C.J., Hrubec, J., de Greef, E., and Kool, H.J., *Environ. Health Perspect., 46,* 197, 1982.
9. Kool, H.J., van Kreijl, C.F., de Greef, E., and van Kranan, H.J., *Environ. Health Perspect., 46,* 207, 1982.
10. Kool, H.J., and van Kreijl, C.F., *Water Res., 18,* 1011, 1984.
11. Meier, J.R., and Bull, R.J., Mutagenic properties of drinking water disinfectants and by-products, in Jolley, R.L., et al. (eds.), *Water Chlorination: Environmental Impact and Health Effects,* vol. 5, Butterworth, pp. 207–220, 1985.
12. Meier, J.R., Lingg, R.D., and Bull, R.J., *Mutat. Res., 118,* 25, 1983.
13. Kopfler, F.C., Ringhand, H.P., Coleman, W.E., and Meier, J.R., Reactions of chlorine in drinking water, with humic acids and *in vivo,* in Jolley, R.L., et al. (eds.), *Water Chlorination: Environmental Impact and Health Effects,* vol. 5, Butterworth, pp. 161–173, 1985.
14. Ashby, J., Styles, J.A., and Paton, D., *Br. J. Cancer, 38,* 34, 1978.
15. *IARC Monographs on the Evaluation of Carcinogenic Risk of Chemicals to Man.* Chloroform, *20,* 401, 1979.
16. Davidson, I.W.F., Sumner, D.D., and Parker, J.C., *Drug and Chemical Toxicol., 5,* 1, 1982.
17. *National Cancer Institute Carcinogenesis Bioassay of Chloroform,* NTIS No. PB264018/ AS, 1976.
18. Eschenbrenner, A.B., and Miller, E., *J. Natl. Cancer Inst., 4,* 385, 1943.
19. Reuber, M.D., *Environ. Health Perspect., 31,* 171, 1979.
20. Jorgenson, T.A., Meierhenry, E.F., Rushbrook, C.J., Bull, R.J., Robinson, M., and Whitmire, C.E., *Fund. Appl. Toxicol., 5,* 760–769, 1985.
21. Roe, F.J.C., Palmer, A.K., Worden, A.N., and Van Abbe, N.J., *J. Environ. Pathol. Toxicol., 2,* 799, 1979.
22. Reitz, R.H., Fox, T.R., and Quast, J.F., *Environ. Health Perspect., 46,* 163, 1982.
23. Moore, D.H., Chasseaud, L.F., Majeed, S.K., Prentice, D.E., Roe, F.J.C., and Van Abbe, N.J., *Fd. Chem. Toxicol., 20,* 951, 1982.
24. Jorgenson, T.A., Rushbrook, C.J., and Jones, D.C.L., *Environ. Health Perspect., 46,* 141, 1982.
25. Diaz-Gomez, M.I., and Castro, J.A., *Cancer Lett., 9,* 213, 1980.
26. Pereira, M.A., Lin, L.-H.C., Lippitt, H.M., and Herren, S.L., *Environ. Health Perspect., 46,* 151, 1982.

27. Morimoto, K., and Koizimu, A., *Environ. Res., 32,* 72, 1983.
28. Hatch, G.G., Mamay, P.D., Ayer, M.L., Casto, B.C., and Nesnow, S., *Cancer Res., 43,* 1945, 1983.
29. Kirkland, D.J., Smith, K.L., and Van Abbe, N.J., *Fd. Cosmet. Toxicol., 19,* 651, 1981.
30. Trehy, M.L., and Bieber, T.I., Detection, identification and quantitative analysis of di-haloacetonitriles in chlorinated natural waters, in Keith, L.H. (ed.), *Advances in the Identification and Analysis of Organic Pollutants in Water,* vol. 2, Ann Arbor Science Publ., Ann Arbor, MI, p. 941, 1981.
31. Oliver, B.G., *Environ. Sci. Technol., 17,* 80, 1983.
32. Bieber, T.I., and Trehy, M.L., Dihaloacetonitriles in chlorinated natural waters, in Jolley, R.L., et al. (eds.), *Water Chlorination: Environmental Impact and Health Effects,* vol. 4, Ann Arbor Science, Ann Arbor, MI, pp. 85–96, 1983.
33. Simmon, V.F., Kauhanen, K., and Tardiff, R.G., *Prog. Genet. Toxicol.,* 249, 1977.
34. Bull, R.J., Meier, J.R., Robinson, M., Ringhand, H.P., Laurie, R.D., and Stober, J.A., *Fundam. Appl. Toxicol. 5,* 1065–1074, 1985.
35. Meier, J.R., Bull, R.J., Stober, J.A., and Cimino, M.C., *Environ. Mutagen., 7,* 201–211, 1985.
36. Bull, R.J., and Robinson, M., Carcinogenic activity of haloacetonitriles and haloacetone derivatives in the mouse skin and lung, in Jolley, R.L., et al. (eds.), *Water Chlorination: Environmental Impact and Health Effects,* vol. 5, Butterworth, pp. 221–227, 1985.
37. Mink, F.L., Coleman, W.E., Munch, J.W., Kaylor, W.H., and Ringhand, H.P., *Bull. Environ. Contam. Toxicol., 30,* 394, 1983.
38. Rockwell, A.L., and Larson, R.A., Aqueous chlorination of some phenolic acids, in Jolley, R.L., et al. (eds.), *Water Chlorination: Environmental Impact and Health Effects,* vol. 2, Ann Arbor Science, Ann Arbor, MI, p. 67, 1978.
39. *National Cancer Institute Bioassay of 2,4,6-Trichlorophenol for Possible Carcinogenicity,* DHEW Publ. No. (NIH) 79-1711.
40. Exon, J., and Koller, L., Co-carcinogenic and reproductive effects of chlorinated phenols, in Jolley, R.L., et al. (eds.), *Water Chlorination: Environmental Impact and Health Effects,* vol. 5, Butterworth, pp. 307–330, 1985.
41. Boutwell, R.K., and Bosch, D.K., *Cancer Res., 19,* 413, 1959.
42. Kringstad, K.P., Ljungquist, P.O., de Sousa, F., and Stromberg, L.M., *Environ. Sci. Technol., 17,* 553, 1983.
43. Douglas, G.R., Nestmann, E.R., McKague, A.B., San, R.H.C., Lee, E.G.-H., Lie-Lee, V.W., and Kowbel, D.J., Determination of potential hazard from pulp and paper mills: Mutagenicity and chemical analysis, in Stich, H.F. (ed.), *Carcinogens and Mutagens in the Environment,* vol. 4, *The Workplace,* CRC Press, Inc., 1984. (In press.)
44. Meier, J.R., Ringhand, H.P., Coleman, W.E., Munch, J.W., Streicher, R.P., Kaylor, W.H., and Schenck, K.M., *Mutation Res., 157,* 111–122, 1985.
45. Vogt, C.R., Liao, J.C., Sun, G.Y., and Sun, A.V., *Proceedings 13th Annual Conference on Trace Substances in Environmental Health,* University of Missouri, Columbia, MO, p. 453, July 4–7, 1979.
46. Wlodkowski, T.J., and Rosenkranz, H.S., *Mutat. Res., 31,* 39, 1975.
47. Rosenkranz, H.S., *Mutat. Res., 21,* 171, 1973.
48. Wyrobek, A.J., Gordon, A., Burkhart, J.G., Francis, M.W., Kopp, R.W., Letz, G., Malling, H.V., Topham, J.C., and Whorton, M.D., *Mutat. Res., 115,* 1, 1983.
49. Van Abbe, N.J., Green, T.J., Jones, E., Richold, M., and Roe, F.J.C., *Fd. Chem. Toxicol., 20,* 557, 1982.
50. Rosen, J.D., Segall, Y., and Casida, J.E., *Mutat. Res., 78,* 113, 1980.
51. Sturrock, J., *Br. J. Anaesth., 49,* 207, 1977.

CHAPTER 15

Epidemiologic Assessment of Health Risks Associated with Organic Micropollutants in Drinking Water

Shirley A. A. Beresford

Department of Epidemiology
School of Public Health
The University of North Carolina at Chapel Hill
Chapel Hill, North Carolina

15.1 METHODOLOGICAL CONTEXT

The approaches open to the epidemiologist when the health outcome under study is rare, and the supposed risk is geographically defined, are more restricted than in other types of epidemiologic investigation. Three major classes will be discussed here, the first of which has considerable weight in studies of rare outcomes in spite of the traditional criticisms.

Aggregate population studies, sometimes called ecologic studies, are concerned with the experience of geographically defined population groups and are the least expensive studies. The health data used are routinely collected, and may relate to deaths (mortality data) or to episodes of sickness (morbidity data). There is usually no information about drinking water quality or socioeconomic characteristics that can be attached to an individual death or episode or to an individual person in the population. It is necessary to examine the relationships between average mortality measures, average water quality measures, and average socioeconomic measures. Consequently, much power and precision is lost in such aggregate population studies, although these can be improved by the use of special statistical techniques to remove differential sampling error and allow for geographical clustering. The problem of possible confounding is particularly great in these studies, but there are some techniques for minimizing its effect. Ways to adjust for different kinds of potential confounding to improve power and precision are outlined in Table 15.1. Examples of the application of these considerations will be provided in Section 15.4.

More expensive are the case-control studies, which are retrospective studies using information on individuals. These may use death certificates, when the risk to be assessed is specific to a cause of death, or they may use newly diagnosed or incident cases. The quality of retrospective information obtained is likely to be higher for incident cases than for persons identified from death certificates. In general, case-control studies are more powerful than aggregate population studies, although they are liable to selection bias and recall bias. This undermines their value in the view of some investigators.

The most expensive studies are the longitudinal or cohort studies. Birth cohort studies, with a short follow-up, are feasible if the health risks of interest include infant mortality or incidence of congenital malformations. Cohorts defined in middle-age and followed for 5–10 years are appropriate for assessing risks of diseases that are common in later life, such as cancer or heart disease. Some relevant historical information may have to be gathered retrospectively at the start of the study. This type of study has undoubtedly the greatest power of those discussed, but may not be feasible because of expense if the health risks to be assessed are small.

It is the art of good study design to reconcile the ideal and the possible in such a way as to maximize the acquisition of useful data (1), and clearly the appropriate epidemiological method to use depends on the particular health outcomes under investigation. The health effects that may be associated with organic micropollutants in drinking water include cancer and infant mortality. Studies of cancer risks have other methodological considerations in addition to those already discussed. Because of the very long lag time associated with most cancers between exposure to a causative agent and detection of cancer in an individual, it is particularly important to obtain historical measurements of exposure to different water qualities. Site-specific cancers, even of the more common sites, such as lung or stomach, occur infrequently (less than 2/1000 per year), and the number of cases that might be attributable to a particular environmental agent are smaller still. In such circumstances, studies need to use very large populations with very different exposure levels to the environmental agent of concern in order to detect a cancer risk. From this it follows that, for acute contamination of ground water or spills which do not affect a large population, epidemiological methods are not likely to provide very precise estimates of risk.

Aggregate population studies have traditionally been viewed as hypothesis-generating studies. Interesting associations uncovered by such studies are usually intepreted with caution, and await confirmation by a more powerful study, closer to the experimental mode. However, since one of the criteria suggested for inferring causality from association (2) is replication of findings by different investigators in different places at different times, it seems reasonable, that once a hypothesis is formed from one study, aggregate population studies can play a role in hypothesis testing.

TABLE 15.1
Key Methodological Considerations in Aggregate Studies

1. Age-standardized health indices
2. Historical estimates of water quality
3. Adjustment for known risk factors
    Socioeconomic characteristics
    Genetic factors
    Climatic factors
4. Acknowledgment of other possible confounding variables (including migration)
5. Unequal sized areas—some form of weighted analysis should be used
6. Geographical clustering—model should be adapted to allow for correlated
    residuals

Source: Reprinted from *First Atlantic Workshop Proceedings*, by permission. Copyright © 1982, The
    American Water Works Association.

15.2 HISTORICAL PERSPECTIVE:
HYPOTHESIS-GENERATING STUDIES

The early studies examined source of water supply as one of a number of
factors possibly associated with local or regional differences in cancer mortality. Stocks (3), in assessing variation between London boroughs in death
rates for 1921 and 1930, found stomach cancer standardized mortality ratios
(at ages 45 and over) to be lower in London boroughs served with water from
the river Thames by the Western Division of the Metropolitan Water Board
(as it was then) than in boroughs served by well water, and these differences
could not be explained by differences in social conditions. Boroughs served
by water from the river Lee had the highest rates, but these boroughs also
had the poorest social conditions, as measured by number of persons per
room in a household. Stocks did not carry out any multivariate statistical
analyses, and made no attempt to interpret these results in terms of pollution
of the water supplies. In the Netherlands, Diehl and Tromp (4), studying
cancer death rates for the period 1900–1930 in the municipalities of the 12
provinces, found that municipalities without water systems (i.e., public water systems) had higher cancer death rates. They also showed that the highest cancer death rate was found in areas served by river water compared
with well water. The mortality figures were not broken down by site, and the
main interest of the authors was in associations with soil type. No adjustment was made for socioeconomic factors.
    Another investigation that was predominantly interested in soil distribu-
tions, was that carried out using cancer deaths from 1943 to 1952 in Anglesey, a rural county in north Wales (5). They too found an excess of cancer
of the stomach, as a proportion of all cancer deaths, in those people not
served by public mains. The public mains, although mostly supplied with
surface water, were treated and monitored, whereas other sources (wells,

springs, stored rain water) were not monitored but still might have been subject to occasional pollution. This excess in stomach cancer deaths was not significant after adjusting for social class and soil type. The study suffered from the problem of small numbers. Another study of parishes in west Devon, also a largely rural area, investigated cancer deaths in the period 1939 to 1958 with respect to rock geology (6). Allen-Price found that the parishes served by well and spring water had either the highest cancer death rates in general or the lowest, with upland surface supplies from moorlands having intermediate rates. He speculated that the radioactivity of the water might be a factor of importance. Cook and Watson (7) examined the geographical location of residents of Missouri with multiple cancers, and noticed an excess in counties downstream from Kansas City, a source of high pollution. However, they concluded that such an excess could hardly by attributed to river water since most of the population obtained their drinking water from deep wells. This was another descriptive study, with no analysis of socioeconomic factors.

Finally, Stocks (8) examined cancer mortality in the county boroughs of the United Kingdom from 1958 to 1967 using cancer deaths as a proportion of all deaths. He found variations with respect to water source in rates for cancer of the esophagus, stomach, intestine and rectum, breast, prostate, and bladder. These included a possible excess of cancer of the stomach in areas served by upland surface water. This paper was criticized by Clayton (9), who recommended the use of the age-specific death rate for 45–64 years rather than the proportionate mortality rate, which is dominated by deaths in old age. The findings of these early studies do not form a coherent pattern, possibly because they were undertaken more as exploratory studies, rather than with a clear hypothesis, concerning source of drinking water, in mind.

Following a report on the presence of chemicals in New Orleans water, performed by the Environmental Protection Agency (EPA) in 1972, a number of studies were undertaken both in the United States and in the United Kingdom. The EPA and the Environmental Defense Fund (EDF) initiated chemical and epidemiological studies in July, 1974, which were published in early November (10,11). The chemical study found a number of organic chemicals, including suspect carcinogens, in the New Orleans supply, and the epidemiological study found an association between certain cancers and the consumption of Mississippi River water. As a result, the House of Representatives voted by an overwhelming majority for the passing of the Safe Drinking Water Act, which enabled the EPA to set limits for chemical contaminants (12).

15.3   THE MISSISSIPPI RIVER STUDY

The EDF study (11) was one of the first to use the cancer death rates for the years 1950–1969 collated for the whole United States by Mason and McKay (13). Classifying the parishes of Louisiana according to the proportion of their drinking water supply that came from the Mississippi River, they found a significant association with total cancer mortality in white males. This association was interpreted as showing a 15% excess cancer mortality in white males drinking water from other sources (11). Some association persisted when median income of the parish, degree of urbanization, and type of industrial employment were taken into account.

The study was severely criticized with respect to its methodology (14), because of poor definition of the water variable (percent water from Mississippi River), because no account was taken of the differing sizes of the parishes, and because no result had been found with respect to liver cancer, although the chemicals identified in the water as suspect carcinogens induced cancer of the liver in animals (15). There was no discussion of other race-sex groups in the early report, or of alternative hypotheses that could explain the results. This was a particular problem because parishes served by the Mississippi River were all clustered together (16). The authors of the EDF report published further analyses of their work, showing a relationship between parishes drinking Mississippi water and mortality from cancers of the urinary tract (in white males and nonwhite females) and from cancers of the gastrointestinal tract (in white and nonwhite males and females). The results were similar when New Orleans was removed from the analysis, and also when the analysis was restricted to the 29 southern parishes (17). DeRouen and Diem (18) again referred to the problem that toxicological studies implicated the liver as the target site of the chemicals detected in the Mississippi River, while this was not substantiated by the epidemiology. The lung was not implicated as a target organ, and it is not even likely to be one, yet the EDF study found a significant association between lung cancer in nonwhite males and Mississippi River water as a drinking source. These authors were of the opinion that with more refined measurement of cultural differences and level of industrialization, the significance of the water variable would be reduced even further (18). In spite of these criticisms, the Mississippi River study was useful in providing a hypothesis to be tested by further epidemiological studies.

Taking all these early studies into account, three alternative hypotheses involving water quality can be stated:

1. Cancer rates are higher in areas served by private water supplies (as opposed to public water supplies).
2. Cancer rates are higher in areas served by upland surface water.
3. Cancer rates are higher in areas served by river water.

15.4 HYPOTHESES TESTING OR REFINING STUDIES

The studies reviewed here include those in the United Kingdom that examined the quality of upland water compared with river and ground waters, or compared river water quality with ground water quality on a scale of pollution. Also reviewed are studies in the United States and elsewhere that examined different sources of drinking water with respect to a variety of measures of organic micropollutants, and those that compared different qualities of ground waters.

Ground water comprises about 30% of the drinking water supply in England and Wales. In contrast to the United States, less than 1% of all households are served by private wells. Other sources of drinking water are natural springs, of comparable quality to ground water, upland catchments, and river supplies. Two possible hypotheses were entertained in the U.K. studies concerning health risks associated with water supplies of differing qualities: (1) that cancer mortality is related to the degree of sewage effluent (pollution) in a water source; (2) that cancer mortality is related to the product of natural organics in and chlorination treatment of a water source. The first hypothesis would implicate the lowland rivers, and the second, predominantly the upland catchments. It should be noted that the chloroform to total trihalomethane ratio is highest in upland catchments. The geographical distribution of drinking water in England, Wales, and Scotland is shown in Figure 15.1. Natural spring water has been classified as ground water on the map, and in the U.K. studies reviewed here. It is notable that upland catchments predominate in Scotland, in northern England, and in Wales. The distribution of upland supplies is consequently highly confounded with weather variation and with latitude and longitude as expressed in a north-west/south-east gradient.

The aggregate population studies conducted in the United Kingdom are summarized in Table 15.2. In all the U.K. studies reviewed here, cancer mortality or morbidity indices were limited to the age group 25–74 years, and adjusted for age using 5- or 10-year age groups. This addresses the first methodological consideration cited in Table 15.1. The long lag time between response to a causative agent and occurrence of cancer calls for historical estimates of water quality. Accordingly, in each study the correlation between historical estimates and current measures of water quality was examined, to check that current measures could reasonably represent past measures in the analysis. In each study, social factors and estimates of migration were taken into account in a weighted multiple regression analysis, with weights calculated iteratively (19). The extent of geographical clustering was estimated, and taken into account where necessary using estimates of correlated residuals (20).

The first study, of cancer mortality (21), restricted the analysis to the London area, in order to reduce variation due to extraneous factors such as water hardness, noncity environment, and weather. No significant association was found between reuse and male stomach cancer mortality, or indeed any other sex-specific gastrointestinal or urinary tract cancer mortality, after adjustment had been made for age, social factors, and borough size (21). Reuse was defined as mean dry weather sewage effluent flow/mean river flow × 100, and measurements were available around 1971. For the study

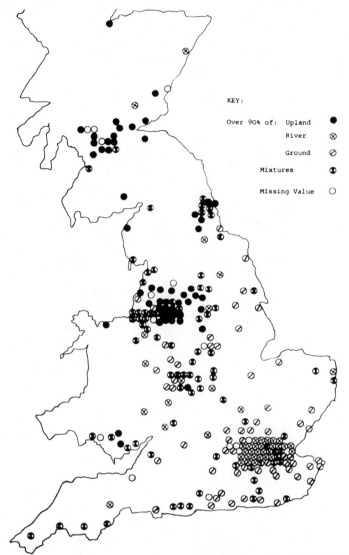

Figure 15.1. Sources of drinking water in Great Britain (1969–1973).

boroughs, this measure was found to correlate extremely highly with chemical indicators of organic pollution history that were available back to 1931.

The second study was also of cancer mortality and subdivided the non-upland sources according to their degree of pollution with sewage effluent (22). The greatest pollution problem is found in rivers that are used both as a drinking source and for transporting waste to the sea. In such rivers, water is abstracted from the river, stored, treated, and stored again before being

TABLE 15.2
U.K. Aggregate Population Studies with Weighted Analysis and Adjustment for Socioeconomic Characteristics

Authors	Location	Health Outcome {water variables}	Factors Adjusted for: (Minimum Subset From)	Cancer Sites Investigated	Sites Implicated (sex)
Beresford (21)	London area, 29 boroughs	Mortality 1968–1974 {Reuse, 1971}	*Set 1* % Manual % Semi- or unskilled Mean socioeconomic rank Population density % Overcrowding % Lack of amenities % Manufacturing industries % Unemployment % New Commonwealth % New residents % Intercensal change	All neoplasms Lung Gastrointestinal Stomach Intestinal All urinary Bladder	None
Beresford et al. (22)	Great Britain, 141 urban areas	Mortality 1969–1973 {Reuse, 1951}	*Set 2* % Overcrowding % Lack of housing amenities % Lack of inside w.c. % Manual % Semi- or unskilled	Gastrointestinal Stomach Intestinal All urinary Bladder Esophagus	Stomach (M) Intestinal (M)[a]

		Set 1	Set 2
	Mean socioeconomic rank % Higher education % Unemployment Cars per household % Large families % Manufacturing industries % Mining industries % New residents Population density % New Commonwealth		
Beresford (23)	Incidence 1968–1974 {Reuse, 1971}	Stomach Colon Rectum All urinary Bladder Esophagus	Stomach (F) All urinary (F) Bladder (F) Esophagus (M)
	South London, 14 boroughs		
Carpenter and Beresford (24)	Mortality 1969–1973 {% upland, 1971}	Stomach Intestine All urinary Bladder Esophagus	Stomach (F) Intestine (F)
	Great Britain, 238 urban areas		

Note: In all studies 1971 water variables were shown to correlate very highly with similar variables in 1951 and earlier.
[a] Negative association.

distributed to households for consumption. Sewage from humans is collected, and treated and the sludge is disposed of on land or sea, while the water from the sewage is discharged back into the river. Meanwhile, the majority of the river water has continued to flow along the river bed. This process of partial abstraction and sewage discharge can happen several times during a river's journey from source to the sea. By the time the River Thames, for example, reaches Teddington (on the outskirts of London), about 14% of the river flow is domestic sewage effluent flow. In fact, the greatest degree of reuse is to be found in the London area since, although the most polluted rivers in the United Kingdom are to be found in the north of England, they are not used as drinking sources. In contrast to the river water, ground water is relatively free from pollution. This is particularly true for the London area, where the chalk basin is protected from the pollution of London by a layer of clay that extends northwards to the Chiltern Hills and southwards to the North Downs.

Of 238 urban areas eligible for inclusion in the study, 141 were served predominantly (over 90%) by some combination of ground and river water, and had an estimate of reuse for 1951. When age, social factors, and borough size were taken into account, the association between male stomach cancer mortality and reuse in the 141 urban areas was small, but statistically significant. From the regression equations, the partial regression coefficients for water reuse and their associated 95% confidence limits (22) were used to calculate the relative risks of River Thames compared with ground water. The relative risk of male stomach cancer in urban areas served by River Thames water compared with ground water was 1.09, as is shown in Table 15.3. This is extremely close to unity. This small association is statistically significant, in contrast with the first study, possibly because of the larger number of areas in the analysis.

TABLE 15.3
Relative Risks of River Water (e.g., River Thames)
Compared to Ground Water

U.K. 141 urban areas 1968–1973	
Male stomach cancer mortality	$1.09\ (1.00,\ 1.18)^a$
London 27 boroughs 1969–1973	
Stillbirths	$1.14\ (1.02,\ 1.26)$
London 14 boroughs 1968–1974	
Female stomach cancer incidence	$1.13\ (0.98,\ 1.27)$
Female all urinary cancer incidence	$1.24\ (0.97,\ 1.51)$

Source: Reprinted from *First Atlantic Workshop Proceedings*, by permission. Copyright © 1982, The American Water Works Association.
From weighted regression analysis, after adjustment for socioeconomic factors.
[a] ( ) 95% confidence limits.

To increase the precision of cancer risks, the next study (23) examined cancer incidence rather than cancer mortality. The quality and completeness of cancer registration data varies from registry to registry. Consequently, the study was restricted to the 14 boroughs in the London area lying within the South Thames Cancer Registry, where the data quality is amongst the highest in Great Britain. The same measure of reuse as in the first London study was used here. One of the cancers found to be associated with water reuse was female stomach cancer.

The geographical distribution of the residual variation in female stomach cancer incidence, after age, social factors, and reuse had been taken into account, was examined. The maximum likelihood estimate of the correlation of residuals of neighboring boroughs (20) was zero, so no adjustment was necessary for geographical clustering. The two sex-specific cancers of the gastrointestinal and urinary tracts weakly associated with reuse after adjustments for age, social factors, and weighting were female stomach cancer and female all urinary cancer ($p = 0.08$ in each case). The corresponding relative risks of River Thames water compared with ground water were once again close to unity and are given in Table 15.3.

The study that examined the health risks associated with upland water (22,24) used information about 238 urban areas of Great Britain. These included all the county boroughs, all the London boroughs, all the Scottish "large" burghs, all municipal boroughs and urban districts in England and Wales with populations over 50,000 in 1971, and 19 additional urban areas or aggregates of urban areas included to increase the geographical spread. When mortality from specific cancers was examined, without restricting to one or the other hypothesis about water quality, the variation associated with different sources of water supply was found to be explained by percent supply from upland catchments. The raw data therefore supported the second hypothesis (concerning chlorination) over the first (concerning pollution).

As has been indicated earlier, weighted regression analyses using standardized mortality ratios and adjusting for social factors were conducted. Percent upland in 1971 was very highly correlated with percent upland in 1951 (correlation coefficient = 0.9), and was available for a larger number of urban areas, so was used in the analyses. Female stomach cancer mortality was found to be associated with percent upland independent of social factors and borough size. The relative risk of female stomach cancer in areas served by upland compared with river or ground water, calculated from the partial regression coefficient, is shown in Table 15.4. A relative risk of 1.11 means that the risk of female stomach cancer in upland areas is 1.11 times that for areas served by other sources of water supply. All these relative risks, although statistically significant, are very close indeed to unity.

Compounds in the water supply that may have long-term deleterious health effects on adults, such as cancer, are quite likely to have shorter-term health effects on human health when it is most vulnerable, namely before

TABLE 15.4
Relative Risks of Upland Water Compared
with River or Ground Water

U.K. 238 urban areas 1969–1973	
Female stomach cancer mortality[a]	1.11 (1.02, 1.21)[b]
Stillbirths	1.07 (1.02, 1.11)
Neonatal mortality	1.05 (1.00, 1.10)

Source: Reprinted from *First Atlantic Workshop Proceedings*, by permission. Copyright © 1982, The American Water Works Association.
From weighted regression analysis, after adjustment for socioeconomic factors.
[a] Adjusted also for NW/SE gradient.
[b] ( ) 95% confidence limits.

birth and in the first year of life. Accordingly, the U.K. studies also examined mortality from stillbirths, and neonatal and postneonatal mortality rates. The analyses adjusted for the social factors, and included information on percent of mothers with parity of 3 or more.

The hypothesis in relation to reuse was examined in the London area (22), where information on both reuse and stillbirths was available for 27 boroughs. The association between stillbirth rate and reuse, independent of social factors and borough size was statistically significant ($p < 0.05$). The relative risk of Thames water compared with ground water was 1.14, as shown in Table 15.3. The other infant mortality indices were not significantly associated with reuse. Table 15.4 shows the corresponding results in terms of estimated relative risk for the upland hypothesis. Postneonatal mortality rate was not associated with percent upland, and once again the relative risks for the other two indices, although significant, are extremely small (22).

The U.S. and other studies entertained hypotheses concerning quality of drinking water and cancer risk phrased in a different way. The first hypothesis was simply that surface waters were associated with an elevated cancer risk. The second hypothesis might be considered a refinement or restatement of the surface/ground water hypothesis in terms of chlorination treatment. Because surface waters tend to be more liable to pollution, the treatment process is likely to contain a higher level of chlorine. One of the by-products of chlorination is chloroform, which has been found to induce cancers in laboratory animals (25).

Several studies conducted in the United States met less than three of the methodological criteria outlined in Table 15.1, and in particular did not report they considered the use of weighted analysis. Five of these, (26–32) examining surface water versus ground water, are presented in Table 15.5. Also presented is a study from the Netherlands that did employ weighted analysis but made no adjustment for socioeconomic factors (33). Six studies, (30,34–38) not using weighted analysis, addressed comparisons of chlorinated and unchlorinated water supplies, and are presented in Table 15.6.

TABLE 15.5
Aggregate Population Studies Not Using Weighted Analyses: Surface Water versus Ground Water Hypothesis

Authors	Location	Type of Study	Factors Adjusted For	Cancer Sites (sex)
Buncher et al. (26) Kuzma et al. (27)	Ohio, 88 counties	Mortality 1950–1969	Urban/rural Income Population Manufacturing occupation Agriculture/forestry/fishery occupations	Stomach (M,F) Bladder (M) All neoplasms (M,F)
McCabe (28)	U.S. central city county areas, 135	Mortality 1949–1951	None	No associations
Kriebel and Jowett (29)	Minnesota, Wisconsin upper Michigan	Mortality 1950–1969	Urban/rural Income Foreign born (Scandinavia)	Stomach (M)
Carlo and Mettlin (30)	Erie County, New York	Incidence 1973–1976	Mobility Socioeconomic status % Nonwhite Occupation Urban/rural	Esophagus (M,F) Pancreas (M,F)
Frerichs et al. (31)[a]	Los Angeles County	Mortality 1969–1971	None	No significant associations
Kool et al. (33)[b]	Netherlands, 19 cities	Mortality 1964–1976	None	Esophagus (M) Bladder (M) Liver (M) Lung (M,F)

[a] A study of polluted versus nonpolluted ground water.
[b] This study did use weighted analysis, and included other measures of organic constituents.

TABLE 15.6
Aggregate Population Studies Not Using Weighted Analyses: Chlorination and Chlorination By-Products

Authors	Location	Type of Study	Factors Adjusted For	Water Variable	Cancer Sites (sex)
McCabe (34)	50 large cities, United States	Mortality 1969–1971	None	Chloroform[a] TTHM[a,b]	All neoplasms
U.S. EPA (35)	43 cities, United States	Mortality 1969–1971	None	Chloroform[c]	No association
Spivey and Sloss (36)	Southern California, 9 communities	Mortality 1966–1971 Incidence 1972–1974	Income Education % White collar Nondurable manufacturing Persons per room Value of owner-occupier units	Chlorination (chloroform content)	No consistent associations

Study	Location	Data/Years	Covariates	Exposure	Associations
Tuthill and Moore (37)	Massachusetts, 20 communities	Mortality 1969–1976	Income Education % Nonwhite % Foreign born % Population change % Textile % Printing % Chemical	Chlorine dose 1950 Chlorine dose 1978 TTHM 1978	No associations No associations No associations
Carlo and Mettlin (30)	Erie County, New York	Incidence 1973–1976	Socioeconomic status Mobility Urbanicity Occupation	TTHM	Pancreas (M)(white)
Kendrick (38)	24 hospital boards, New Zealand	Incidence 1961, 1971 Mortality 1971	None	Chlorination	Stomach (F) Colon[d] (M,F) Rectum (M) Pancreas (M) Uterus (F) Breast (F) Prostate (M)

[a] Data from National Organics Reconnaissance Survey (42).
[b] Total trihalomethanes.
[c] Data from Environmental Protection Agency Region V Survey (35).
[d] Negative association.

Because of the lack of some of the key methodological considerations, these 11 studies should be interpreted most cautiously.

There have been other studies in the United States that have considered at least the first four methodological issues summarized in Table 15.1. Three aggregate population studies (39–41) using different measures of organic constituents in drinking water are summarized in Table 15.7. Details of results are restricted to cancers of the gastrointestinal and urinary tracts and all neoplasms as a group. None of the aggregate population studies took into account possible geographical clustering of similar counties. Salg's study (39) looked at two variables that might be considered indirect measures of organic constituents, percent surface supply and percent prechlorination of supply. Inconsistent results were found for male esophageal cancer mortality, but consistent results were obtained for male and female rectal cancer and for male cancers of the bladder and other urinary organs. Cantor (40) examined different trihalomethanes (THMs) using data from National Organics Reconnaissance Survey (NORS) (42) and Reg. V (United States Environmental Protection Agency's Region V Survey) (35). Consistent results were again obtained for bladder cancer mortality. Hogan (41) used chloroform data from both NORS and Reg. V and investigated different types of weighted regression analysis.

Salg (39) included all 346 counties in the Ohio River Valley Basin (Illinois, Indiana, Kentucky, Ohio, Pennsylvania, Tennessee, West Virginia), and used the published cancer rates for 1950–1969 (13). She found consistent associations between use of surface water and cancer of the rectum for both males and females, weighting her analysis according to population size of the county. For white males, she also found associations with cancer of the large intestine, cancer of the gastrointestinal system (including the esophagus), cancer of the urinary tract, and cancer of the esophagus (alone). This latter was also significant in nonwhite females. The results were checked by restricting the analysis to within broad strata of population size, using lung cancer mortality as a surrogate variable for smoking habits, and restricting the analysis to counties having at least 50% of their water from a known water source. Reasonably consistent results were obtained. This was the first thorough statistical investigation of the surface water hypothesis, adjusting both for socioeconomic factors and for variations in county size.

In an additional analysis, she used percent of drinking source treated with prechlorination as a surrogate measure of possible THMs in the finished water. Again she found a consistent association with rectal cancer in white males and females. In white males, bladder cancer and prechlorination were significantly associated in the weighted analysis, but only a weak relationship existed between prechlorination and colorectal cancer as a whole (39).

Restricting his analysis to the 76 of the 923 urban United States counties where at least 50% of the population was served by a sampled supply, Cantor (40) examined the association between cancer mortality and THMs. He adjusted for several socioeconomic variables, as indicated in Table 15.7,

TABLE 15.7
U.S. Aggregate Population Studies with Weighted Analysis and Adjustment for Socioeconomic Characteristics

Authors	Location	Health Outcome {water variables}	Factors Adjusted For	Cancer Sites Investigated	Sites Implicated[a]
Salg (39)	Ohio River Valley basin, 346 counties	Mortality 1950–1969 {% surface % prechlorination}	% Urban Median income Population density Manufacturing Nondurable manufacturing Agriculture % Nonwhite % Foreign born Years education	Esophagus Stomach Colon Rectum Kidney Bladder and other Urinary organs All neoplasms 10 Other sites	M (+ % surf, – % pre) M (– % pre) M (+ % surf) M&F (+ % surf, + % pre) M (+ % surf, + % pre) M (+ % surf) 7 in various ways
Cantor et al. (40)	76 U.S. counties (or subset)	Mortality 1968–1971 {Chloroform BTHM TTHM (NORS & Reg. V)}	% Urban Median education Population size Increase of population % Manufacturing % Foreign born	Esophagus Stomach Colon Rectum Kidney Bladder All neoplasms 16 Other sites	M F (-TTHM) M (+BTHM), F (+ BTHM, TTHM) 2 in various ways
Hogan et al. (41)	All counties, 48 states of U.S. and District of Columbia	Mortality 1950–1969 {Chloroform (NORS & Reg. V)}	Population density % Urban % Nonwhite % Foreign born Median income Years of education % Manufacturing Population size Region of country	Esophagus Stomach Colon Rectum Kidney Bladder All neoplasms 7 Other sites	 M & F F F Thyroid M-

Source: Reprinted from *First Atlantic Workshop Proceedings*, by permission. Copyright © 1982, The American Water Works Association.

[a] *Key:* M, male; F, female; +, positive association; –, negative association; % surf, % surface; % pre, % prechlorination; BTHM, Bromine-containing trihalomethanes; TTHM, total trihalomethanes; NORS, National Organics Reconnaissance Survey (42); Reg. V, U.S. Environmental Protection Agency, Region V Survey (35).

and used weighted regression techniques. The most consistent finding was the association between bromine-containing THMs and bladder cancer mortality in males and females, which was statistically significant in the 25 counties where over 85% of the population was served by a sampled supply (40).

In a study of all counties in the 48 mainland states of the United States plus the District of Columbia, using the cancer mortality data for 1950–1969 (13), Hogan examined different methods of weighting according to county size (41). After adjusting for various socioeconomic factors, he found some correlation between chloroform levels in finished drinking water supplies and mortality from cancer of the large intestine, cancer of the bladder, and cancer of the rectum in males and females. The limitations of this type of study of geographical areas were discussed in detail.

Considering all these aggregate population studies together, the cancer sites that survived the hypothesis testing are stomach, colon, rectum, bladder, and all urinary, although the results are not completely consistent. The risk factors associated with organic chemicals in the water supply have variously been investigated as surface water, upland water, chlorination, reuse, and THMs.

15.5 CASE-CONTROL STUDIES

These studies are generally considered more powerful than aggregate population studies, but have some methodological limitations of their own. Concerns about the quality of retrospective information were referred to in Section 15.1. In a case-control study, not only may individuals have difficulty recalling information about previous place of residence and life style, but certain factors may have changed as a result of the disease diagnosis. The exposure history that is relevant is that prior to diagnosis, so information has to be recalled from the more distant past. These problems are further exacerbated if the case-control study uses death certificate cases, since information has to be recalled by family members or retrieved from routine data.

On the one hand it might be assumed that the scope for recall bias is severely limited with respect to residence history, the means of estimating exposure to organic micropollutants in drinking water, so that the main exposure measure is not subject to differential recall by cases and controls. On the other hand, the need to adjust for other major risk factors for cancer that may be confounded with water quality, calls for historical estimates of dietary, smoking, and occupational patterns. These are more likely to be affected by recall bias.

Once again, there are many investigators who dismiss the usefulness of case-control studies except as exploratory studies. Gehlbach (43) presents some counterarguments, and indeed there are ways to minimize bias, or estimate its extent, for example, by including more than one control group. The main advantage of the case-control study over the aggregate population study is the ability to use individual measures of exposure. Several case-control studies examining one or more of the indicators of organic micropollutants were undertaken after results from some of the aggregate population studies were available. Not all of the case-control studies collected historical information on smoking and socioeconomic status, but used current measures as approximations. None of the studies collected dietary information directly. The studies are summarized in Table 15.8.

In the first case-control study to be conducted with respect to chlorination and cancer, female cases of gastrointestinal and urinary tract cancer mortality in seven New York counties in the period 1968–1970 were compared with noncancer death controls. For urban women, Alavanja (44) found that the relative risk of gastrointestinal and urinary tract cancer for those served by chlorinated surface water compared with those served by nonchlorinated ground water was similar to the relative risk for those served by chlorinated ground water. The authors suggested that some of the associations between cancer mortality and surface water, found in aggregate studies, might be explained in terms of chlorination. This study is important in that it was able to separate the effects of surface water and chlorination. Further analysis of this study (45) found a significant excess among males of cancer of the gastrointestinal or urinary tracts in some urban areas served by chlorinated water (either surface or ground) compared with urban areas receiving nonchlorinated water. Taking all seven counties together, there was an excess of male deaths in chlorinated surface water areas from cancer of the esophagus, stomach, large intestine, rectum, liver and kidney, pancreas, and bladder, and an excess of female deaths from cancer of the stomach.

There has been one case-control study using incident cases (46), in which Wilkins took advantage of a private census conducted in 1963. Using information from the cancer registry for the period 1963–1975, incident cases of bladder, kidney, and liver cancer who resided in a defined area in Washington County were matched with two individuals of the same age, sex, and smoking history. Water samples were taken from participants' houses and analyzed for chloroform. There was little variation in chloroform levels, and no evidence that levels were higher in the homes of cancer cases than controls. This result is not surprising, considering the small number of cases available for analysis, fewer than 50 for each of liver and kidney cancer (47).

A case-control study of cancer deaths in twenty parishes in South Louisiana for the period 1969–1975 examined three water variables. Gottlieb (48,49) classified the parishes into groups according to industrial and urban versus rural characteristics. This was done in such a way that each cluster was served both by ground water and water from surface sources. A sample of cancer deaths representing 17 sites was obtained to form the cases. Con-

TABLE 15.8
U.S. Case-Control Studies

Authors	Location	Health Outcome {Water Variables}	Factors Adjusted For	Cancer Sites Investigated[a]	Sites Implicated[a] (sex)
Alavanja et al. (44,45)	New York, 7 counties	Deaths 1968–1970 {chlorination}	Urban/rural Occupation Type of source	Esophagus Stomach Colon Rectum Kidney and liver Bladder All GI and urinary tract Pancreas	M M & F M M M M M & F M
Wilkins et al. (46)	Washington Co., Maryland	Incidence 1963–1975 {chloroform (tap)}	Age Sex Smoking history	Kidney Bladder Liver	No consistent findings
Gottlieb et al. (48); Gottlieb and Carr (49)	20 South Louisiana parishes	Deaths 1960–1975 {surface/ground source life chlorine level (categorical var)}	Age Race Sex Year death Urban/rural Industrial factors	Esophagus Stomach Colon Rectum Kidney Bladder 11 Other sites	M & F (all water vars.)
Struba (50)	North Carolina	Deaths 1975–1978 {surface/ground reuse chlorinated/ unchlorinated}	Age Sex Race Region Urban/rural Socioeconomic status	Colon Rectum Bladder	M & F M & F M & F (rural areas only, all water vars.)

Reference	Population	Exposure/Water data	Covariates	Cancer sites	Results
Brenniman et al. (51)	70 Illinois counties (excluding Chicago)	Deaths 1973–1976 {chlorinated/unchlorinated ground water}	Age, Sex, Urban/rural, SMSA/non-SMSA	Stomach, Colorectal, Kidney, Bladder, Digestive tract (excl. liver), Urinary tract, Total gastrointestinal and urinary tract	F (Stomach); F (Digestive tract); F (SMSA, all urban) (Total gastrointestinal and urinary tract)
Young et al. (52); Kanarek et al. (53)	28 Wisconsin counties	Deaths 1972–1977 {daily average total chlorine dose (over 20 years, 3 categories), chlorination source depth, organic contamination, water purification}	Age, Year of death, County, Marital status, Occupation, Urbanization, Selected water vars.	Esophagus (F), Stomach (F), Colon (F), Rectum (F), Kidney (F), Bladder (F), Brain (F), 4 Other sites	F chlorine dose, chlorination organic contam., source depth; F chlorination, organic contamination
Lawrence et al. (54)	New York school-teachers	Deaths 1962–1978 {20 year individual residence and work history to obtain cumulative chloroform dose; source type, chlorine dose, chlorine residual}	Age, Year of death, Population density, Marital status, Selected water vars.	Colorectal (F), Colon (F), Rectum (F)	None[b]

[a] Key: M, male; F, female; SMSA, Standard Metropolitan Statistical Area; vars., variables.
[b] Small sample size.

trols were chosen from within the same parish cluster as the case, and matched for age at death, year of death, sex and race. Cancer deaths were excluded from the controls. There was a significant excess of surface water supplies amongst the rectal cancer cases compared with the controls (48). The association was strengthened when a crude estimate of the length of time served by a surface source was taken into account. In addition, the level of chlorine (none, low, high) used in treatment was found to be significantly associated with mortality from cancer of the rectum (49). By dividing the ground water sources into chlorinated and nonchlorinated, and the surface sources into high and low dose of chlorine, surface/ground differences at low doses of chlorine, and chlorine dose in surface sources could be examined. The association of surface/ground and rectal cancer was found to be independent of chlorine level, but chlorine dose was not significantly associated with rectal cancer independent of the surface/ground variable. The authors concluded that both chlorination and contamination of surface sources are associated with increased risk of rectal cancer death. There was no evidence of any association with colon cancer, and inconsistent results were obtained with cancers of the bladder and stomach and some other cancers (49).

In a case-control study of deaths occurring at age 45 or older in North Carolina during the period 1975–1978, Struba (50) examined three water hypotheses with respect to cancers of the colon, rectum, and bladder. Deaths whose principal cause was not one of these cancers but where the death certificate listed a cancer as a contributory or underlying cause were excluded from the control groups. Cases and controls were matched by age, race, sex, and region. There was a significant excess of both surface water supplies and chlorinated supplies for the cases of cancer of each site compared with the controls when the analysis was restricted to the rural areas. Odds ratios for urban areas were smaller and generally not significant. Expected results were obtained when the analyses were repeated within categories of migration.

Brenniman and others (51) conducted a case-control study of deaths in Illinois designed to replicate Alavanja's study in New York State (45). Deaths occurring in the period 1973–1976 were included in the study, the control group being formed, without matching, from a pool of noncancer deaths excluding perinatal deaths and deaths from mental disorder. Cook County (including Chicago) was excluded from the analysis. A total of 272 communities receiving chlorinated ground water in 1963 were matched with 270 communities receiving nonchlorinated ground water according to urban/rural and Standard Metropolitan Statistical Area (SMSA)/non-SMSA characteristics. Significantly elevated odds ratios were obtained for cancers of the colon and rectum combined in females and for the total digestive tract (excluding liver) in females. Odds ratios for cancers of the gastrointestinal and urinary tracts combined were significant, but mildly elevated, for all urban communities, SMSAs, and communities chlorinating since 1953. The odds ratios found were generally considerably smaller than those in the New York study and were of the order of 1.1 or 1.2 (51).

In a larger case-control study of cancer deaths in white Wisconsin females, which included a total of 12 sites, controls were matched by age, year of death, and county of residence (52,53). Only counties with low migration rates were included. Significant odds ratios associated with chlorination were found for female colon cancer, and these were increased to about 3.0 in communities exposed to rural runoff and water purification. These results emerged following extensive statistical analyses, which used a number of different models to express, and adjust for, water variables. There was some indication of an association of chlorination with brain cancer, but the inconsistent results led the authors to discount any evidence of brain cancer risk.

A recently reported study of New York state schoolteachers examined deaths from colorectal cancer in white women from 1962 to 1978. Lawrence and Trock (54) made considerable efforts to obtain 20-year histories of places of residence and work from both cases and noncancer controls. These were used to obtain cumulative chloroform dose, under certain assumptions such as daily water consumption. Only one water variable was positively associated with colorectal cancer risk, namely average source type (four possible categories), but this was after adjustment for cumulative chlorine dose, and was not statistically significant. The number of cases in this study was much smaller (25%) than in the other studies reported.

A number of other case-control studies using incident cases are currently underway, as reported by Cantor (5) or listed in the Directory of Ongoing Research in Cancer Epidemiology (56). Three are studying a number of cases of the same order as the study by Lawrence and Trock (54), and so may not have sufficient power to detect an association. The planned study in Wisconsin by Kanarek, Nashold, and Young is to include 800 cases of colorectal cancer (56), and the study by Cantor in 10 areas of the United States is to include 3000 bladder cancer cases (55).

15.6 PROSPECTIVE STUDIES ON INDIVIDUALS

There has been only one set of studies on individuals that were not case-control. It was conducted in Washington County, Maryland, and took advantage of a private census, which has been referred to earlier (46). A 12-year prospective study (57) of cancer incidence and mortality yielded an association between chlorination of the water supply and bladder cancer incidence in males, as is shown in Table 15.9. This analysis effectively contrasted Hagerstown supply with unchlorinated deep wells, and the relative risks of bladder cancer in males only reached statistical significance when the analysis was restricted to the male residents who had lived in their 1963 residence for at least 12 years. No significant association was found between deaths from cancer of the biliary passage or liver or from cancer of

TABLE 15.9
Prospective Studies on Individuals

Authors	Location	Health Outcome {Water Variables}	Factors Adjusted For	Cancer Sites Investigated	Sites Implicated
Kruse (57) Wilkins and Comstock (58)	Washington County, Maryland	Prospective mortality and incidence 1963–1975 {Chlorination, categorical variable}	Age Marital status Years of education Persons per room Housing adequacy Cigarette smoking Number of years in residence Frequency of church attendance	Esophagus[a] Stomach[a] Colon[a] Rectum[a] Kidney Bladder Liver Pancreas[a] Breast[a] 5 Other sites[a] 13 Other causes of death[a]	M (incidence) Mortality

[a] Mortality only, sexes combined.

the kidney and chlorination status of the water supply at census (57,58). The positive association found with breast cancer mortality in both sexes combined was not discussed by the authors, since it was not a strong prior hypothesis.

The existence of a positive finding for bladder cancer incidence analyses while none was found for mortality analyses is not surprising, considering that cancer incidence is more frequent, especially with the comparatively good 5-year-survival rates associated with bladder cancer. Although the size of the population was small (14,000), which led the authors to be cautious in the interpretation of the results in their earlier report (46), the low exposure cohort accounted for more than 58,000 person-years of observation (58). Possible limitations of the study including misclassifications of exposure and uncontrolled confounding factors were discussed in detail.

This study illustrates the considerable investment in terms of population size and length of follow-up required to conduct a prospective study of individuals. Only 80 cases of bladder cancer occurred in the 12 years of follow-up, and much of the relevant information concerning exposure and confounding variables had to be collected retrospectively. The main advantage of such a study is that it avoids differential recall between cases and noncases, since information is collected before membership in the two groups is known to anyone, and it also avoids the possible problem of change in life style following diagnosis of cancer.

15.7 HYPOTHESES THAT SURVIVE THE REVIEWS

15.7.1 Consistency of Cancer Site

Restricting the evidence to those studies without two major methodological flaws, seven studies (23,39–41,45,50,57), including two case-control studies using death certificates (45,50), and one prospective study found an association with bladder cancer, while the remaining eight (21,22,24,46,49,51,53,54) found no such association. All but one of these used mortality as the outcome. Mortality records may be less appropriate tools for studying bladder cancer, with its relatively good survival rate. Five studies (three of them case-control) (39,41,45,50,53) found an association with colon cancer, five studies (three case-control) (39,41,45,49,50) found an association with rectal cancer, one with colorectal cancer (51), one with intestinal cancer (22), one case-control, and three aggregate population studies (22–24) found an association with stomach cancer, while four studies (21,40,54,58) found no association with any gastrointestinal cancer. Taken together these 15 studies provide some consistent evidence of an association between broad measures of organic constituents in drinking water and cancers of the gastrointestinal system and urinary tract.

15.7.2   The Nature of the Water Factor

From all the studies available, the following ranking of cancer risk for different sources of water supply emerges.

Upland sources   (highest risk)
Polluted river sources
Less polluted river sources
Minimally polluted river sources
Chlorinated ground water
Unchlorinated ground water   (lowest risk)

A number of hypotheses have been entertained that are consistent with some part of this ranking. The broadest hypothesis remains surface versus ground water, but given the shortage of ground water sources, this does not provide the water industry with enough information to devise a feasible lower risk system of water supply. Hypotheses that may be considered refinements of this broad hypothesis, and that overlap but have different policy implications, are the reuse or pollution hypothesis and the chlorination hypothesis.

Refinement of the pollution hypothesis might be either in terms of domestic sewage effluent, or in terms of industrial pollutants. There have been no studies that would allow discrimination between these two hypotheses.

Refinement of the chlorination hypothesis has to date been in terms of the THMs and chloroform in particular, probably because of the animal evidence of chloroform's carcinogenicity. On the other hand, there are some reasons why the associations found between chlorinated water and cancer mortality should not be taken as evidence that THMs are the risk factor. First, THMs account for only a fraction of the total halogenated organic compounds. Second, although some volatile organics have been detected in the blood plasma of residents of New Orleans, and their drinking water exhibits the same pattern of volatile organics (59), the volatile organics only make up a small proportion of the total organics. There is great ignorance of the effects of the majority of other organic compounds in the drinking water (60). Ninety percent of the organic compounds in drinking water are nonvolatile, and only 10–20% of these have been analyzed, leaving between 80 and 90% yet to be identified. These nonvolatile organics are known as the nonpurgeable portion of total organic carbon (NPOC), and NPOC may well provide a better index of the organic contamination of the water than the THMs (61). However, the one study that has specifically examined nonvolatile organics in relation to cancer mortality, found no significant correlation between NPOC and cancer mortality (34). The analysis employed here was not very sophisticated and took no account of socioeconomic characteristics.

A tabulation of studies that either support or refute the different refined

TABLE 15.10
Studies That Support or Refute Various Hypotheses

	Support		
Hypothesis	Gastrointestinal	Urinary Tract	Refute
Chlorination per se	$(45,51,53^a)$	(45)	(24)
Chloroform or other THMs	$(24^a,41)$	(40,41)	(46)
Pollution or chlorination	$(22^a,23^a,39,49^a,50)$	$(23^a,50,58)$	

[a] Data support dose-response relationship.

hypotheses is given in Table 15.10. Many of the studies did not have the capability of distinguishing between pollution, chloroform, or other by-products of chlorination. The case-control studies conducted by Alavanja (45) and Brenniman (51), which specifically compared chlorinated and un-chlorinated ground water, provided evidence of a small increased risk associated with chlorination, independent of source type. The question must be asked, however, whether the reason for chlorinating ground water supplies was because they were believed to be subject to minor pollution or contamination. Kanarek's study (53) showed a larger increased risk from un-chlorinated to chlorinated, no organic contamination, than from chlorinated, no organic contamination to chlorinated, organic contamination. Carpenter's study (24) of drinking sources in the United Kingdom provides some evidence against the hypothesis that chlorination per se is associated with increased risk. For chlorination to be the prime risk factor, upland and river sources would be expected to be more alike with respect to cancer risk than ground and spring waters. In the analysis, however, river, ground and spring water areas differed significantly as a group from upland water areas.

The discriminating evidence available for the chloroform or other THM hypothesis is not so clear. The studies cited as supporting the hypothesis include those that used chloroform or another THM as the exposure variable. None of them examined the partial association with chloroform having adjusted either for pollution or chlorination. Wilkins' study (46) of Washington County, Maryland, failed to find an association between chloroform and cancer incidence, and indeed, within the cancer cases, there was no geographical clustering in areas with higher average chloroform. The other studies shown in Table 15.10 did not have the capability of distinguishing between the pollution or chlorination hypothesis when a positive cancer risk was detected.

In none of the studies was the relative risk of cancer in areas served with drinking water containing organic micropollutants very large. On the one hand, this increases the possibility that the elevated risk is due entirely to undetected confounding variables (62), although this explanation is less tenable, given the number of different studies conducted by different authors in different places at slightly different times. On the other hand, weak associations (those with a relative risk less than 2) may still be very important in public health terms if the exposure is rather common in the general population, say with more than 25% population exposed (62). This is undoubtedly the case in the United Kingdom where, for example, over 50% of London residents are served by reused water.

Another criterion for judging whether an association may be causal is whether or not there is evidence of a dose-response relationship. This was specifically investigated in eight studies (21–24,49,50,53,54) and those that provided evidence of a dose-response relationship have been indicated with a superscript in Table 15.10. The shape of the relationship with stomach cancer incidence found by one U.K. study is reproduced in Figure 15.2.

15.8 CONCLUSION

If the hypotheses concerning different aspects of the cancer risk associated with organic micropollutants in drinking water are considered together, many of the criteria for judging whether the association is causal (2,62) are satisfied. The association is consistent with a risk of cancer of the gastrointestinal or urinary tract, through being demonstrated repeatedly in different studies. The association is not strong, but a dose-response relationship has been found in most of the studies that looked for one. A temporal sequence is at least plausible, since several studies either obtained historical information concerning water quality or else checked that current measures were good proxies for historical water quality. The association is coherent, in that there is some biological plausibility. Chloroform has been found to be carcinogenic in animal studies (25), and more recent mutagenicity studies suggest the nonvolatile organics in drinking water may play a role in carcinogenesis (63). One criterion for which no evidence is available to date, is whether removal of the putative cause leads to reduction in the incidence of disease. If the cause were linked to chlorination of the water supply, countries that have switched to ozone treatment should show a reduction in incidence of cancer of the gastrointestinal or urinary tracts in 10 or 15 years' time.

In the light of this discussion, it might be concluded that there is sufficient evidence of a cancer risk associated with certain drinking waters for a causal relationship to be accepted. Given the high prevalence of polluted and/or chlorinated waters, in spite of the small relative risk, the public health implications are large enough for action to be indicated. This presupposes a solution to exist. Were the surface or pollution hypothesis correct, given the limited availability of ground water supplies, a solution might be to introduce dual water supplies. That is, lower quality water supplied to the bathroom and garden or yard and higher quality water to the kitchen, for

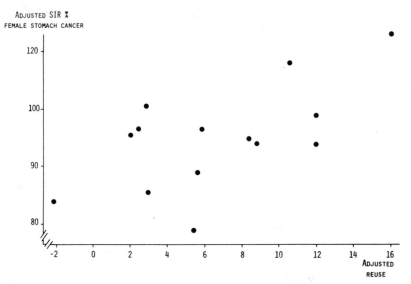

Figure 15.2. Incidence of female stomach cancer by reuse, adjusted for mean socio-economic rank.

cooking and drinking. This solution would be extremely expensive, and is not favored by the water authorities. Were the chlorination hypothesis correct, the solution might be to switch to ozone as the disinfectant treatment of choice, or to introduce carbon adsorption filters as an additional stage in the treatment process, as has already been done in several parts of the United States. A solution would be a great deal more efficient if it were targeted directly to the underlying causal factor.

The forthcoming studies using incident cases will be useful in confirming or refuting the association using a more powerful study design. However, the epidemiological studies are not going to be able to help further to differentiate between chloroform, another THM, or one of the nonvolatile organics—either a product of the chlorination process, or a pollutant from industry.

It is essential that the evidence from these epidemiological studies be put with that from toxicological studies, and chemical analyses that are discussed elsewhere in this book, before a decision is made of whether or not to control further the organic micropollutants in drinking water.

ACKNOWLEDGMENTS

I am most grateful for the guidance of Professors A.G. Shaper and W.W. Holland, and of Dr. R.F. Packham, and to Sheila Evans and Kaye Tilley, who typed the manuscript. Thanks are due to the Veterans Administration and to the Water Research Centre, under contract to the Department of the Environment, who supported this work at different stages of its development.

REFERENCES

1. WHO, *Environmental Health Criteria, 27: Guidelines on Studies in Environmental Epidemiology*, World Health Organization, Geneva, 1983, pp. 39–73.
2. Hill, A.B., *A Short Textbook of Medical Statistics*, Hodder and Stoughton, London, 1977, pp. 285–296.
3. Stocks, P., *Regional and Local Differences in Cancer Death Rates*, HMSO, London, 1947.
4. Diehl, J.C., and Tromp, S.W., First Report on the Geographical and Geological Distribution of Carcinoma in the Netherlands, *Stichtung ter Bevordering van de Psychische Physica*, vol. 1, Leiden, December, 1953.
5. Davies, R.I., and Wynne Griffith, G., Cancer and soils in the county of Anglesey, *Br. J. Cancer, 8,* 56–66, 1954.
6. Allen-Price, E.D., Uneven distribution of cancer in West Devon, *Lancet, i,* 1235–1238, 1960.
7. Cook, G.B., and Watson, F.R., The geographic locations of Missourians with multiple cancer, *Missouri Med., 63,* 997–1002, 1966.
8. Stocks, P., Mortality from cancer and cardiovascular diseases in the county boroughs of England and Wales classified according to the sources and hardness of their water supplies 1958–67, *J. Hyg. Camb., 71,* 237, 1973.
9. Clayton, D.G., personal communication, July, 1973.
10. U.S. Environmental Protection Agency, *Draft Analytical Report: New Orleans Water Supply Study*, Lower Mississippi River Facility of the Environmental Protection Agency, Slidell, LA, 1974.
11. Harris, R.H., *The Implications of Cancer-Causing Substances in Mississippi River Water*, Environmental Defense Fund, Washington, DC, Nov. 6, 1974.
12. Marx, J.L., Drinking water: Another source of carcinogens?, *Science, 186,* 809–811, 1974.
13. Mason, T.J., and McKay, F.W., *U.S. Cancer Mortality by County, 1950–1969*, DHEW Publication No. (NIH) 74-615. National Cancer Institute, Bethesda, MD, 1974.
14. Tarone, R.E., and Gart, J.J., Review of the implications of cancer-causing substances in Mississippi River water, unpublished, January 10, 1975.
15. U.S. Environmental Protection Agency, *A Report: Assessment of Health Risk from Organics in Drinking Water*, Hazardous Materials Advisory Committee, Science Advisory Board, Environmental Protection Agency, Washington, DC, May 1975.
16. DeRouen, T.A., and Diem, J.E., The New Orleans drinking water controversy—A statistical perspective, *Am. J. Publ. Health, 65* (10), 1060–1063, 1975.
17. Page, T., Harris, R.H., and Epstein, S.S., Drinking water and cancer mortality in Louisiana, *Science, 193,* 55–57, 1976.
18. DeRouen, T.A., and Diem, J.E., Relationships between cancer mortality in Louisiana drinking-water source and other possible causative agents, in Hiatt, H.H., Watson, J.D., Winston, J.A., (eds.), *Origins of Human Cancer*, Cold Spring Harbor Laboratory, Cold Spring Harbor, NY, 1977, pp. 331–345.
19. Pocock, S.J., Cook, D.G., and Beresford, S.A.A., Regression of area mortality rates on explanatory variables: What weighting is appropriate?, *Appl. Stat.* (JRSS Series C), *30* (3), 286–295, 1981.
20. Cook, D.G., and Pocock, S.J., Multiple regression in geographical mortality studies with allowance for spatially correlated errors, *Biometrics, 39* (2), 361–371, 1983.
21. Beresford, S.A.A., The relationship between water quality and health in the London area, *Int. J. Epidemiol., 10* (2), 103–115, 1981.

22. Beresford, S.A.A., Carpenter, L.M., and Powell, P., *Epidemiological Studies of Water Reuse and Types of Water Supply*, Water Research Centre Technical Report TR 216, 1984.
23. Beresford, S.A.A., Cancer incidence and reuse of drinking water, Am. J. Epidemiol., 117 (3), 258–268, 1983.
24. Carpenter, L.M., and Beresford, S.A.A., Cancer mortality and type of water source: findings from a study in the U.K., *Int. J. Epidemiol 15*, 1986. (In press.)
25. Ward, P.S., Carcinogens complicate chlorine question, *J. Water Pollut. Control Fed., 46* (12), 2638, 1976.
26. Buncher, C.R., Kuzma, R.J., and Forcade, C.M., Drinking water as an epidemiologic risk factor for cancer, in Hiatt, H.H., Watson, J.D., and Winston, J.A. (eds.), *Origins of Human Cancer*, Cold Spring Harbor Laboratory, Cold Spring Harbor, NY, 1977, pp. 347–356.
27. Kuzma, R.J., Kuzma, C.M., and Buncher, C.R., Ohio drinking water source and cancer rates, *Am. J. Publ. Health, 67* (8), 725–729, 1977.
28. McCabe, L.J., *Health Effects of Organics in Drinking Water*, Health Effects Res. Lab. U.S. Environmental Protection Agency, Cincinnati, OH, 1978.
29. Kriebel, D., and Jowett, D., Stomach cancer mortality in the north central states: High risk is not limited to the foreign-born, *Nutr. Cancer, 1* (2), 8–12, 1979.
30. Carlo, G.L., and Mettlin, C.J., Cancer incidence and trihalomethane concentrations in a public drinking water system, *Am. J. Publ. Health, 70* (5), 523–525, 1980.
31. Frerichs, R.R., Satin, K.P., and Sloss, E.M., *Water Reuse: Its Epidemiologic Impact, Los Angeles County, 1969–71*, University of California, April, 1981.
32. Nellor, M.H., Baird, R.B., and Smyth, J.R., *Health Effects Study Final Report*, County Sanitation District of Los Angeles County, Whittier, CA, March, 1984.
33. Kool, H.J., Van Kreijl, C.F., Van Kranen, H.J., and De Greef, E., Toxicity assessment of organic compounds in drinking water in the Netherlands, *Sci. Total Environ., 18*, 135–153, 1981.
34. McCabe, L.J., *Association Between Trihalomethanes in Drinking Water (NORS Data) and Mortality*, Health Effects Res. Lab., U.S. Environmental Protection Agency, Cincinnati, OH, 1975.
35. U.S. Environmental Protection Agency, *Preliminary Assessment of Suspected Carcinogens in Drinking Water* (Report to Congress), December, 1975.
36. Spivey, G.H., and Sloss, E., *Cancer and Chlorinated Drinking Water*, U.S. Environmental Protection Agency Health Effects Research Lab, May, 1977.
37. Tuthill, R.W., and Moore, G.S., Drinking water chlorination: A practice unrelated to cancer mortality, *J. Am. Water Works Assoc., 72* (10), 570–573, 1980.
38. Kendrick, B.L., A spatial, environmental and socioeconomic appraisal of cancer in New Zealand, *Soc. Sci. Med., 14D*, 205–214, 1980.
39. Salg, J., *Cancer Mortality Rates and Drinking Water Quality in the Ohio River Valley*, Ph.D. dissertation, University of North Carolina, Chapel Hill, 1977.
40. Cantor, K.P., Hoover, R., Mason, T.J., and McCabe, L.J., Associations of cancer mortality with halomethanes in drinking water, *J. Natl. Cancer Inst., 61* (4), 979–985, 1978.
41. Hogan, M.D., Chi, P.Y., Hoel, D.G., and Mitchell, T.B., Association between chloroform levels in finished drinking water supplies and various site-specific cancer mortality rates, *J. Environ. Pathol. Toxicol. 2* (3), 873–887, 1979.
42. Symons, J.M., Bellar, T.A., Carswell, J.K., DeMarco, J., Knopp, K.L., Robeck, G.G., Seeger, D.R., Slocum, C.J., Smith, B.L., and Stevens, A.A., National organics reconnaissance survey for halogenated organics, *J. Am. Water Works Assoc., 67*, 637–647, 1975.
43. Gehlbach, S.H., *Interpreting the Medical Literature*, The Collamore Press, Lexington, MA, 1982, pp. 39–53.
44. Alavanja, M., Goldstein, I., and Susser, M., *Report of Case-Control Study of Cancer Deaths in Four Selected New York Counties in Relation to Drinking Water Chlorination*, U.S. Environmental Protection Agency Health Effects Research Lab, December, 1976.
45. Alavanja, M., Goldstein, I., and Susser, M., A case-control study of gastrointestinal and

urinary tract cancer mortality and drinking water chlorination, in Jolley, R.L., Gorchev, H., and Hamilton, D.H., Jr. (eds.), *Water Chlorination, Environmental Impact & Health Effects,* Vol. 2, Ann Arbor Science Publishers, Ann Arbor, MI, 1978, pp. 395–409.

46. Wilkins, J.R., Reiches, N.A., and Kruse, C.W., Organic chemical contaminants in drinking water and cancer, *Am. J. Epidemiol. 110* (4), 420–448, 1979.

47. Williamson, S.J., Epidemiological studies on cancer and organic compounds in U.S. drinking water, *Sci. Tot. Environ., 18,* 187–203, 1981.

48. Gottlieb, M.S., Carr, J.K., and Morris, D.T., Cancer and drinking water in Louisiana: Colon and rectum, *Int. J. Epidemiol., 10* (2), 117–125, 1981.

49. Gottlieb, M.S., and Carr, J.K., Case-control cancer mortality study and chlorination of drinking water in Louisiana, *Environ. Health Perspect., 46,* 169–177, 1982.

50. Struba, R.J., Cancer and Drinking Water Quality, Ph.D. dissertation, University of North Carolina, Chapel Hill, 1979.

51. Brenniman, G.R., Lagos, J., Amsel, J., Namekata, T., and Wolff, A.W., Case-control study of cancer deaths in Illinois communities served by chlorinated or nonchlorinated water, in Jolley, R.J., Brungs, W.A., and Cummin, R.B. (eds.), *Water Chlorination, Environmental Impact and Health Effects,* Vol. 3, Ann Arbor Science, Ann Arbor, MI, 1980, pp. 1043–1057.

52. Young, T.B., Kanarek, M.S., and Tsiatis, A.A., Epidemiologic study of drinking water chlorination and Wisconsin female cancer mortality, *J. Natl. Cancer Inst., 67* (6), 1191–1198, 1981.

53. Kanarek, M.S., and Young, T.B., Drinking water treatment and risk of cancer death in Wisconsin, *Environ. Health Perspect., 46,* 179–186, 1982.

54. Lawrence, C.E., Taylor, P.R., Trock, B.J., and Reilly, A.A., Trihalomethanes in drinking water and human colorectal cancer, *J. Natl. Cancer Inst., 72* (3), 563–568, 1984.

55. Cantor, K.P., Epidemiological evidence of carcinogenicity of chlorinated organics in drinking water, *Environ. Health Perspect., 46,* 187–195, 1982.

56. Muir, C.S., and Wagner, G., *Directory of Ongoing Research in Cancer Epidemiology,* International Agency for Research in Cancer, Pub. No. 46, Lyon, 1982.

57. Kruse, C.W., *Chlorination of Public Water Supplies and Cancer–Washington County, Maryland Experience,* U.S. Environmental Protection Agency Health Effects Research Lab, January, 1977.

58. Wilkins, J.R., and Comstock, G.W., Source of drinking water at home and site-specific cancer incidence in Washington County, Maryland, *Am. J. Epidemiol., 114* (2), 178–190, 1981.

59. Dowty, B., Carlisle, D., Laseter, J.L., and Storer, J., Halogenated hydrocarbons in New Orleans drinking water and blood plasma, *Science, 187,* 4171–4175, 1975.

60. Loper, J.C., Lang, D.R., and Smith, C.C., Mutagenicity of complex mixtures from drinking water, in Jolley, R.L., Gorchev, H., and Hamilton, D.H. (eds.), *Water Chlor. Environ. Impact and Health Effects,* Vol. 2, Ann Arbor Science Publishers, Ann Arbor, MI, 1978, pp. 433–450.

61. McCabe, L.J., and Millette, J.R., Health effects and prevalence of asbestos fibers in drinking water, Proceedings American Water Works Association, June 24–29, 1979.

62. Wynder, E.L., Schlesselman, J., Wald, N., Lilienfeld, A., Stolley, P.D., Higgins, I.T.T., and Radford, E., Weak associations in epidemiology and their interpretation. Conference report, *Prev. Med., 11,* 464–476, 1982.

63. Crump, K.S., and Guess, H.A., *Drinking Water and Cancer: Review of Recent Findings and Assessment of Risks,* prepared for the Council on Environmental Quality, Washington, DC, 1980.

CHAPTER 16

Public Health Significance of Organic Substances in Drinking Water

Robert G. Tardiff
Susan H. Youngren

Environ Corporation
Washington, D.C.

16.1 INTRODUCTION

For many decades, the presence of chemicals in drinking water across the globe was often surmised and occasionally known (1). More recently, the increasing sophistication of our analytical technology and the determined application by its practitioners to drinking water have catalogued numerous substances at concentrations that varied greatly with the time and location (2–6). The origin of those substances can be traced to all forms of activity, such as agricultural practices, wastes from ecological cycles and human processes, the use and disposal of industrial compounds, and natural erosion and solubilization of physical matter (7).

The existence of such compounds in drinking water, often despite the application of advanced engineering techniques to remove contaminants from the source, has repeatedly raised concern about possible injuries to health from the ingestion of these materials. The reasons for the concern are manifold. Historically, public health officials had witnessed serious illness transmitted via previously unobserved microorganisms in the tap water. Further, biomedical scientists had come to realize that all chemicals would be toxic under appropriate conditions of exposure. And finally, drinking water was the unavoidable staple required daily to sustain life; hence, exposure was continuous from at least birth and perhaps during gestation as well.

The assessment of health risks and safe exposures to substances in drinking water has been performed numerous times over the years. Examples included the Drinking Water Standards promulgated by the U.S. Public

405

Health Service (1), similar standards issued by its regulatory successor the Environmental Protection Agency (8), and standards provided by the World Health Organization to European and developing countries (9,10). Initial regulatory initiatives focused almost exclusively on inorganic substances such as metals. Although as early as 1962 organic compounds were considered of sufficient concern in the United States for regulatory consideration, it was not until recently that authorities concluded that sufficient information existed to provide a basis for federal control (8,11).

The recent heightened interest in risk assessment is the result of numerous factors, the most obvious of which are the growing recognition of the many substances in the human environment including tap water, the need to allocate carefully limited resources to the more serious health problems, and improvements in the scientific understanding of toxic actions of chemicals on the body. As a consequence, the use of risk assessment has been encouraged by the National Academy of Sciences as a means of avoiding the oversimplification perceived by the terms "safe" and "unsafe" (2).

The purpose of this chapter is to present the scientific process by which risk to—and by contrast safe exposure conditions for—humans can be described and estimated using a variety of information. Risk assessment reveals the types of injury that might be associated with the unintended consumption of chemicals in drinking water and the probability that such injuries are likely, or unlikely, to occur under various circumstances of exposure (12). The scope of such an analysis encompasses all forms of injury, be it reversible and mild, such as gastric constriction, or irreversible and fatal, such as cancer. There exist variations in dealing with differing forms of data, and those approaches with their scientific bases will be elucidated. Having described briefly the approach and principles used to assess health risks, we will provide an overview of select compounds known to be present in some tap waters.

Risk assessment does not and can not specify the appropriate mechanisms for managing unreasonable or unacceptable risks. However, to the extent that it permits an orderly array of information from which health protection decisions can be made, its interface with some risk management options will be mentioned. One such implication noted above is the setting of standards or maximum levels of exposure in tap water. Others include the selection of more desirable sources of water, the determination of preferable treatment techniques, thresholds for public notifications, and comparison of risks and benefits (often in incommensurable units).

16.2 APPROACHES TO HEALTH RISK ASSESSMENT

Risk is the probability of injury, disease, or death under specific circumstances (12). It may be expressed in quantitative terms, taking values from zero (certainty that harm will not occur) to one (certainty that it will). In many cases, risk can only be described qualitatively, as "high," "low," or "trivial."

All human activities carry some degree of risk. Many risks are known with a relatively high degree of accuracy, because we have collected data on their historical occurrence. The risk associated with many activities, including the exposure to various chemicals in drinking water, cannot be readily assessed and quantified. Although there are considerable historical data on the risks of some types of chemical exposures (e.g., the annual risk of death from intentional overdoses or accidental exposures to drugs, pesticides, industrial chemicals), they are generally restricted to those situations in which a single, very high exposure has occurred and has resulted in an immediately observable form of injury, thus leaving little doubt about causation. Assessment of the risks of levels of chemical exposure that do not cause immediately observable forms of injury or disease (or only minor forms such as transient eye or skin irritation) is a complex task, irrespective of whether the exposure may have been either brief, extended but intermittent, or extended and continuous. It is the latter type of risk assessment activity that is reviewed in this chapter.

As recently defined by a committee of the National Academy of Sciences (12), risk assessment is:

the scientific activity of evaluating the toxic properties of a chemical and the conditions of human exposure to it in order both to ascertain the likelihood that exposed humans will be adversely affected, and to characterize the nature of the effects they may experience.

The term "safe," in its common usage, means "without risk." In technical terms, however, this common usage is misleading because science cannot ascertain the conditions under which a given chemical exposure is likely to be absolutely without a risk of any type. The latter condition—zero risk— is simply immeasurable. Science can, however, describe the conditions under which risks are so low that they are of no practical consequence. As a technical matter, the safety of chemical substances, whether in food, drinking water, air, or the workplace, has always been defined as a condition of exposure under which there is a "practical certainty" that no harm will result in exposed individuals. We note that most "safe" exposure levels established in the way we have described are probably risk-free, but science simply has no tools to prove the existence of what is essentially a negative condition.

Another concept concerns classification of chemical substances as either "safe" or "unsafe" (or, as "toxic" and "nontoxic"). This type of classification, while common, is highly problematic and often misleading. All substances, even those consumed in high amounts every day, can be made to produce a toxic response under some conditions of exposure. In this sense, all substances are toxic. The important consideration is not simply that of toxicity, but rather of risk, that is, the probability that the toxic properties of a chemical will be realized under actual or anticipated conditions of human exposure. To address the latter requires far more extensive data and evaluation of the characterization of toxicity.

There are four components to a risk assessment:

1. *Hazard Identification*. Involves gathering and evaluating data on the types of health injury or disease that may be produced by a chemical and on the conditions of exposure under which injury or disease is produced. It also involves characterization of the behavior of a chemical within the body and the interactions it undergoes with organs, cells, or even parts of cells. Data of the latter types may be of value in answering the ultimate question of whether the forms of toxicity known to be produced by a substance in one population group or in experimental settings are also likely to be produced in humans.

2. *Dose-Response Assessment*. Involves describing the quantitative relationship between the amount of exposure to a substance and the extent of toxic injury or disease. Data derive from animal studies, or less frequently, from studies in exposed human populations. There are sometimes many different dose-response relationships for a substance if it produces different toxic effects under different conditions of exposure as is often the situation. The risks of a substance cannot be ascertained with any degree of confidence unless dose-response relationships are known, even if the substance is known to be "toxic."

3. *Human Exposure Assessment*. Involves describing the nature and size of the population exposed to a substance and the magnitude and duration of their exposure. The evaluation could concern past or current exposures, or exposures anticipated in the future.

4. *Risk Characterization*. Involves integration of the data and analysis involved in parts 1, 2, and 3 to determine the likelihood that humans will experience any of the various forms of toxicity associated with a substance.

The dimensions of any risk assessment must be carefully delineated to provide guidance for the acquisition of data and in the formulation of concepts by which to interpret relevant data. A risk question may be narrowly focused (e.g., whether a substance present only in a tap water poses a carcinogenic hazard when ingested by humans) or very broad (e.g., elucidation of all health risks by multiple routes of exposure to a substance pervasive in the human environment of an entire nation). The scope of the risk assessment question may be predicated by legislative mandates, use characteristics, or risk management options. Whatever the driving force, the risk question must be carefully crafted to be relevant to the issue, answerable scientifically, and sufficiently comprehensive to avoid excluding crucial considerations but appropriately constrained to minimize cumbersome complexity.

The individual components of risk assessment are now presented in some detail.

16.2.1 Hazard Identification

The collection of data that bear on adverse biological effects of compounds in drinking water must be compatible with the objectives of the risk assessment. To ascertain human health risks, the information collected may derive from human studies (13,14) but, more commonly, from studies in nonhuman species whose responses are generally similar to those of humans (2). Some in vitro systems are also used in generating data that contribute to qualitatively assessing hazards for humans (12,15). Additional information is also obtained from the analysis of molecular structures of substances of interest (16). Ideally experimental conditions in whole animals should be comparable in route of administration and duration of exposure to those anticipated in humans in order to minimize uncertainties in predicting risks (2,17); however, other less comparable data (for example, alternative routes of exposure) may also prove useful in identifying target organs and in evaluating relative potencies. Metabolic and mechanistic information, particularly if it can be shown to be applicable across species, is of substantial value in providing fundamental understanding of the behavior of toxicants, thus permitting extension with substantial confidence to other circumstances. Examples of the latter include the toxicity, biotransformation, and mechanisms of action of methyl mercury (18) and vinyl chloride (19).

Chemicals often elicit toxic manifestations at more than one site in the body. Multiple target organs and pathological manifestations may appear at the same dose level (e.g., vinyl chloride is hepatotoxic and hepatocarcinogenic at essentially the same dose level); however, it is common to observe an increasing number of target sites as the dose rates rise [e.g., chlorinated hydrocarbon solvents, frequently observed in low concentrations in drinking water, often cause hepatotoxicity at relatively low (albeit higher than ambient concentrations) dose rates and central nervous depression at much higher doses; likewise some compounds are teratogenic at low doses but fetotoxic at somewhat higher doses and maternally toxic at still higher doses]. If a compound produces multiple toxic end points of significant concern to human health, risk and safety judgments should be provided for each separately. Such a display gives the decision maker the opportunity to discern whether the severity of the lesions might require greater or lesser attention to establish public health measures. As an example, for a substance that is expected to produce neurologic, teratogenic, fetotoxic, and hepatotoxic reactions in humans, the standard setting process might focus predominantly on one form of toxicity such as birth deformities, because of the anticipated overall societal consequences of that pathological manifestation.

Toxic responses, regardless of the organ or system in which they occur, can be of several types. For some, the severity of the injury increases as the dose increases. In other cases, the severity of an effect may not increase with dose, but the incidence of the effect will increase with increasing dose. In such cases the number of animals experiencing an adverse effect at a given dose is less than the total number, and, as the dose increases, the fraction experiencing adverse effects (i.e., the incidence of disease or injury) increases; at sufficiently high dose all experimental subjects will experience the effect. Toxic responses also vary in their degree of reversibility. In some cases, an effect will disappear almost immediately following cessation of exposure. At the other extreme, some exposures will result in a permanent injury.

Another characteristic of a toxic response that requires attention is its seriousness. Most scientists will agree that certain types of toxic damage (e.g., liver injury caused by carbon tetrachloride) are clearly adverse and are a definite threat to health. There are, however, other types of effects observed during toxicity studies that are not clearly of health significance. Determining whether biologic changes are of significance to health is one of the critical issues in assessing risk and safety that has been attempted by a committee of the National Research Council (13).

Individual investigations of the adverse effects of substances vary in design, execution, and applicability to the specified risk assessment. Their specific natures will contribute significantly to the level of confidence that can be attributed to the conclusions about health risks as well as to the characterization of hazards. To facilitate the analysis of primary data, it has been recommended that findings from individual studies be classified into four parts: unequivocal, limited, equivocal, and uninterpretable (20).

Evaluation of individual studies requires consideration of several factors associated with a study's hypothesis, design, execution, and interpretation. A valid study must address an answerable and clearly delineated hypothesis, and its basic design must be reasonably capable of testing the hypothesis. In evaluating toxicologic and epidemiologic studies, consideration must be given to many factors, including characterization of the compound(s) under study, the test species, and their relation to humans; the number of individuals in study groups; the number of study groups; the types of observations and methods of analysis; the nature of pathologic changes; alterations in metabolic responses; sex and age of test species; and routes of administration.

In general, such studies should attempt to achieve three objectives: (1) comprehensive measurement of end points, including delayed effects; (2) use of species most predictive of human responses; and (3) identification of dose-response relationships to permit quantification of adverse effects.

Following the analysis of all appropriate toxicity data, those data that will be the cornerstone for the estimation of population thresholds and levels of risk should be selected. The identification of critical data points remains a major objective of the risk assessment. Consequently, for any compound, more than one critical end point may be identified, depending upon the number and scope of the risk questions. For example, if the assessment favors exclusively an acute injury as is the case for a carbon tetrachloride spill in an aquifer, the set of data most appropriate should be quite different than if the evaluation centers on the risk of cancer.

Human data, where available, may be selected as the cornerstone to assess a specific risk. The use of human data, as is the case for vinyl chloride and benzene, has the advantage of providing the most relevant dose-response relationships, since it avoids the problems inherent in interspecies extrapolation. Even when some epidemiological data exist, their use for low-dose extrapolation is often difficult. Quantitative estimates of exposure are often difficult to make, particularly in retrospective studies where documentation of levels of exposure is often scanty. When the type of adverse effect induced by the toxicant occurs commonly, a low incidence of exposure-related toxic manifestations may not be detectable. Also, because of the long latency involved in the development of chronic toxicity, particularly cancers, epidemiological methods cannot be applied to newly introduced chemicals. Furthermore, various criteria may be needed for selecting among human data sets, if more than one is available. Criteria include factors such as the selection of the data set giving the highest estimate of relative risk, with the "best" study design, the largest study population, the highest degree of statistical significance, or the most reliable estimates of substance exposure and disease incidence.

Animal data are often selected as the governing information. Indeed, animal data are often the only kind of toxicity information available. Where there is a combination of human and animal data, then a comparison of the quality of the respective data bases is undertaken. Even with such comparisons, animal data are often identified as the critical determinant for the risk assessment. The use of animal studies for risk assessment has two major advantages: first, that exposures in such studies can be controlled more stringently and often provide more clearly defined dose-response relationships than epidemiological studies, and, second, that they do not require such long observation periods before results are obtained. However, the use of animal data introduces complex problems—and uncertainties—of interspecies extrapolation.

In using animal data as pivotal, a series of professional judgments are required. Faced with data from multiple animal models, the analyst first seeks to identify that species and strain that is most relevant to humans based on the most defensible, biological rationale. Selection of the most sensitive species, however, is generally the norm. Deciding which animal data set is most "relevant" to human risk may not be possible where a different route of exposure is involved in humans and animals, or if comparative data on metabolism and toxicokinetics are not available to permit identification of those species most "relevant" to humans. However, when appropriate biological data are available to permit selection of the animal model most likely to predict human responses, then data for this animal model are used and given more weight than data from other models.

The goal of this stage of the evaluation process is to consider all such evidence and to judge its quality and strength. It is also necessary to describe the confidence with which results obtained in one biological system (e.g., laboratory animals) are likely to pertain to another system (in our case, humans).

After evaluating the total data base, a risk assessor might reach one of the several conclusions. One scheme has proposed seven options that reflect gradients in the certainty derived from different quality of information (20).

Several factors should always be considered when selecting one of the many analytic approaches available for evaluating the total data base. The critical ones are replication of results, reproducibility of results, and concordance of results.

For a given set of data pertinent to a given toxic end point (assuming each study has been critically evaluated), it is advisable to describe the degree of replication, reproducibility, and concordance. In general, as the degree of data replication, reproducibility, and concordance increases, it becomes more certain that a substance possesses the capacity to cause the specific toxic effect under review.

With the selection of the data derived from laboratory animals to provide a statement of possible toxic effects in humans, those data must be examined further for appropriate application to humans. The use of animal studies has been strongly questioned because of the uncertainty of whether the animal model really predicts how a human will respond. There are specific biologic causes of interspecies differences in susceptibility to toxic agents, including difference in absorption, gastrointestinal flora, tissue distribution, metabolism, mechanisms and efficiencies of repair, and excretion (21).

Extrapolation between species adds considerably to the uncertainty of risk assessment, as has been discussed by many individuals in numerous forums including the National Academy of Sciences (2,22), the Interagency Regulatory Liaison Group (23), and the Scientific Committee of the Food Safety Council (24). In the preamble to its Cancer Policy (Occupational Safety and Health Administration, 45 FR 5200), OSHA discussed many of the complications of interspecies risk extrapolation and concluded:

> Extrapolation from animal data to predict risks in humans introduces many additional uncertainties. These include selection of appropriate scaling factors for size, lifespan, and metabolic rate; differences in routes of exposure, duration and schedule of exposure, absorption, metabolism, and pharmacokinetics; differences in intrinsic susceptibility and repair capabilities; intrapopulation variation and susceptibility; and exposure to other carcinogens and intrinsic and extrinsic modifying factors. At least theoretically, these factors can affect the relative response of humans and animals by many orders of magnitude.

To a great extent, certainty in such extrapolation becomes a function of the strength of evidence for concordance between species for specific toxic effects and for biological similarities and differences. Hence if a substance affects the same target organs in several species, it has a high probability of doing the same in an unstudied species. Likewise, biochemical similarity between species will enhance the probability of similar responses. Despite some uncertainties acknowledged in interspecies extrapolation, the process remains a reasonably reliable approach for anticipating many adverse effects in humans and for finding doses for which there is only minimal likelihood of injury.

The quantitative aspect of interspecies extrapolation will be addressed below.

16.2.2   Dose-Response Assessment

Three forms of dose-response relationships are often distinguished: (1) the dose-effect, or quantal, relationship in which the number of responding individuals in a population varies with dose; (2) the graded response, in which the severity of a lesion within an individual is modified with dose; and (3) the continuous response in which a biologic parameter (e.g., body weight) is altered by dose regimen (25). Each relationship is based on the receptor theory of biological response. That theory specifies that molecular receptors mediate the initiation and continuation of adverse responses and that the nature and severity of the response are a function of the time of interaction with the receptor. An important concomitant assumption is that the dose and duration at the receptor are related to the administered dose.

Depending upon the nature of the substance and its toxicity and upon the mathematical description of the data, the curve will be linear, sigmoid, rectilinear, or another shape. The slope of the curve provides valuable information to determine in part the potency of the substance and is of use in subsequent estimations of risk as well as for computing indices of comparative potency (26).

In general in dose-response studies, the intensity of the effect and the frequency of response decrease with reduction in dose; the biological reaction often reaches zero before the dose becomes equal to zero. That point has generally been referred to as an experimental threshold below which no adverse effects are anticipated, and it implies a discontinuity in the slope of a dose-response curve. Such no observed effect levels have been identified for several organic compounds in drinking water (2,22).

The experimental threshold is a function of the design of a study, particularly its statistical power, the spacing of doses, and the sensitivity of test group and the experimental methods (e.g., observing the appropriate group at risk as noted by Tomatis et al. [27]). By contrast, population thresholds, which usually refer to those thresholds that apply to a very large number of organisms (e.g., threshold dose for the insecticide parathion in the mouse,

Mus musculus). Population thresholds are dependent upon the breadth of the distribution of sensitivities among all members of the population, including the highly sensitive and the highly resistant, and also upon competing factors and substances that induce or facilitate biological damage.

There are valid biological reasons for believing that population thresholds are likely to exist, although their precise measurement for virtually all substances is beyond our technical capability. The biological bases for predicting their existence are: (1) the existence of repair processes in most tissues of all organisms and the possibility that these processes may be more efficient with mild damage; and (2) the presence of chemical detoxification processes in many tissues and, for some substances, the increased efficiency of these processes at low levels of exposure. Some examples of approaches to identify thresholds in laboratory animals for one toxic manifestation (carcinogenesis) by mechanisms of action include critical quantity of molecules to activate receptors (28), identification of proximate reactants (29), binding of toxicant to critical target molecules, and toxicodynamic and toxicokinetic functionalities that govern the formation of proximate toxicants and their detoxification (19). Similar considerations apply to other toxic effects. Population thresholds are not to be perceived as immutable, because of the likely competition by environmental factors for inducing similar adverse effects or for altering susceptibility to toxic effects.

In the estimation of risks from various levels of exposure, a discrepancy often exists between the doses in investigations—particularly studies of laboratory animals—and exposure levels for humans. Consequently, low-dose extrapolation is often necessary to obtain indications of risk or virtual safety at dosages at which effects are not observable by state-of-the-art measurement techniques. Approaches utilized for this process will be presented in the Risk Characterization section.

16.2.3 Exposure Assessment

Exposure assessment has been characterized as "the process of measuring or estimating the intensity, frequency, and duration of human exposures to an agent currently present in the environment or of estimating hypothetical exposures that might arise from the release of new chemicals into the environment. In its most complete form, it describes the magnitude, duration, schedule, and route of exposure; the size, nature, and classes of the human populations exposed; and the uncertainties in all the estimates" (12).

For substances in the environment, the extent and intensity of intake by various routes are difficult to determine precisely, particularly over a long period of time. Hence, many uncertainties exist in characterizing exposure, and each must be considered carefully. Uncertainties exist because of incompleteness of data on environmental concentrations over time, inherent errors in sample collection and analytic procedures for existing data, and the inability to identify impacted populations. Simulation models and assumptions about parameters for approximating actual exposure conditions, while useful, provide additional uncertainty because of their foundations on numerous assumptions about the behavior of substances in the environment, the patterns of environmental contamination, and the effectiveness of control technologies.

Exposure assessment of organic substances in drinking water is a particularly complex subject. Since the main concern with organic substances in drinking water is chronic injury from repeated low-dose insult, exposure assessments must consider past, present, and future exposures to provide a basis for accurate estimates of health risks. Particular difficulty exists for substances for which there is a substantial latency period between damage and manifestation of disease (e.g., cancer), thereby making incompatible, for risk assessment, simultaneous observations of environmental concentrations and disease incidence. Thus, major interests of the assessor include length of consumption of the contaminant, variations in concentrations of the substance with time, population relocations and other sociological factors that change intake characteristics, and other sources of intake or contact such as foods and ambient air or the workplace.

By contrast, the only types of information available for the analysis of exposure from compounds in tap water are the results of environmental surveys (i.e., measurements at only one brief period of time) and of environmental monitoring efforts (i.e., intermittent measurements that usually span a few years, certainly less than a lifetime). Such data describes, usually with great precision, levels of substances in an environmental medium such as drinking water; however, they are limited in being unable to describe human intake, even at the time contemporary to the time of sampling. Attempts made to reduce such deficiencies include estimates of biological accumulation for past exposures and modeling techniques for future exposures. Thus, these environmental monitoring and survey data must be extrapolated with substantial uncertainties.

Measurements of human intakes would provide the most reliable estimates of exposure characteristics. Analyses of intakes should transcend descriptions of variations in concentrations over extended periods of time (e.g., several decades and perhaps several generations) to address the following: (1) determine the quantities of those compounds consumed by evaluating (by sex and age groups) the daily oral intake of the tap water, (2) ascertain the sociologic factors that may alter intake (e.g., replacement of tap water by other beverages as the major source of liquid intake), (3) determine the variability of intake related to geographic mobility both short- and long-term, and (4) establish the extent to which other forms of contact with that same medium will contribute to systemic intake (e.g., through skin contact during bathing or inhalation of volatile substances during showering). Such an approach would eventually provide a historic data base that would greatly reduce uncertainties about the nature and magnitude of exposures to substances in tap water.

An example of direct historical data of human intake is the Food and Drug Administration's continuous survey of chemicals in the food supply performed by the Market Research Corporation of America. The results of that survey were used, for example, by the National Academy of Sciences to obtain data concerning the direct consumption of saccharin. This data gathered from more than 12,000 individuals was compared with data from a public opinion poll conducted in 1978 by Market Facts Incorporated of nearly 1500 individuals (30). These reports provided reliable information about the prevalence of use, demographic characteristics of the users, the frequency and amounts of saccharin consumed, and the increased incidences of use. This type of information allows for maximum levels of certainty but, regretably, is seldom available for use in exposure assessments.

16.2.4 Risk Characterization

Having analyzed the strengths and weaknesses of the data base, these data are to be applied either to the estimations of risks or to the derivations of the margins of safety for explicit conditions of human exposure. Either process involves many choices and assumptions, some of which may include extra-scientific considerations. The major choices focus on the selection of dose-response information appropriate to derive virtually safe doses, identification of experimental no-observed-effect-levels (NOELs), selection of relevant dosage units, low-dose extrapolation, selection of appropriate measures of human dose, and estimation of risk or computation of margins of safety.

Conventionally, two general assumptions are recognized about the form of dose-response relationships at low doses. For effects that involve alteration of genetic material (including the initiation of cancer), there are theoretical reasons to believe that effects may take place at very low dose levels; and several specific mathematical models of dose-response relationships have been proposed. For most other biological effects, it is usually assumed that "threshold" levels exist, below which adverse effects are unlikely to occur; and the estimation of population thresholds in humans is usually derived by the application of safety or uncertainty factors. Conversely, NOELs are also used to describe the margin-of-safety (MOS) under defined conditions of exposure. Details of these procedures are presented below.

16.2.4.1 *Estimated Population Thresholds for Humans*

It is widely accepted on theoretical grounds, if not definitively proved empirically, that most biological effects of chemical substances occur only after a threshold dose is achieved. In groups of laboratory animals used to identify hazards, the threshold dose is approximated by the NOEL.

It has also been widely accepted, at least in the public health standard-setting process, that the human population is likely to be much more variable in its responses to toxic agents than are the small groups of well-controlled, genetically homogeneous animals ordinarily used in experiments. Moreover, the NOEL is itself subject to some uncertainty (e.g., how can it be known that the most serious effects of a substance have been identified?). For these reasons, public health agencies seek to protect populations from substances displaying threshold effects by dividing experimental NOELs by safety or uncertainty factors to obtain an estimate of the threshold level in a population of exposed humans.

The magnitude of these factors varies according to the nature and quality of the data from which the NOEL is derived; the seriousness of the toxic effects; the type of protection being sought (i.e., are we protecting against acute, subchronic, or chronic exposures?); and the nature of the population to be protected (e.g., the general population, or populations, such as healthy adult workers, expected to exhibit a narrower range of susceptibilities). The results are referred to as estimated population threshold for humans [i.e., EPT-Hs, a term used to avoid the value-laden concept of "acceptability" contained in the acceptable daily intake (ADI)].

The choice of uncertainty or safety factors is critical in the derivation of population thresholds or EPT-Hs. Several uncertainty factors are currently used to estimate EPT-Hs for toxicants, depending on the type and quality of human or animal toxicity data. If the data are sufficiently close to the theoretical population threshold dose, then small uncertainty factors are used. Also, intimate knowledge of a chemical's mechanism of toxicity, critical effect, and/or toxicokinetics in humans and experimental animals allows for the use of smaller uncertainty factors.

An uncertainty or safety factor of 10 is used to estimate EPT-Hs with appropriate chronic human or subchronic sensitive human data and reflects intraspecies human variability to the adverse effects of a chemical (2,31). An uncertainty factor of 100 (i.e., 10×10) is used with sufficient chronic animal data, which accounts for both intra- and interspecies variability (2). Numerous substances in drinking water have been evaluated using such an approach with the resultant formulation of ADIs or suggested no-adverse-response level (SNARLs) (2,22).

Other uncertainty factors may be applied to compensate for other shortcomings in the data base. Characteristically, these additional uncertainty factors are applied to the lowest-observed-effect level (LOEL) in the absence of a reliable NOEL, when only less-than-lifetime data in animals are the basis for application to chronic exposures for humans, or when there exist other major discrepancies between experimental conditions and the circumstances of human contact. Factors ranging from 2 to 10 have been used to adjust for these differences alone and have been juxtaposed to other uncertainty factors. Recently, Dourson and Stara (25) showed that some of these factors, while somewhat simplistic, may, in fact, have some experimental support and may be protective for many chemical substances.

An uncertainty factor of 5000 has been suggested for carcinogenic substances (32); however, that approach has not been used for risk assessments because the use of low-dose extrapolation models for carcinogens is believed to incorporate more supportable scientific justification.

Estimated population threshold doses in humans are obtained by dividing the experimental NOEL (or in its absence, the LOEL) by the appropriate safety or uncertainty factors. The EPT-Hs is presented in either of two forms: A dose expressed as gravimetric measure of the substance (e.g., milligrams or micrograms) per unit of body measure generally weight (in kilograms) but occasionally in surface area (in square centimeters) per unit of time reflecting duration of exposure (e.g., lifetime); or a concentration in an environmental medium (e.g., food or water or air). The former is a general expression that assumes a direct relationship between intake and target organ dose and permits distribution of exposure among several routes of intake assuming knowledge of toxicokinetic behavior. The latter, however, implies either that the entire intake has been apportioned to one route or that some apportionment among diverse routes of intake has been predetermined (in that instance, the derived value has been called a "tolerance"). In either case, this approach is particularly applicable to the preenvironmental release of substances.

That first expression of the EPT-Hs has been referred to historically by health organizations (e.g., the World Health Organization), regulatory bodies (e.g., the U.S. Food and Drug Administration), and scientific groups (e.g., the National Research Council) as the "acceptable daily intake" or ADI. That term has been the source of some confusion, because it has referred simultaneously to either scientifically derived intake units or to exposure limits that reflect extrascientific considerations.

16.2.4.2   Margin-of-Safety

For substances for which there is actual, rather than prospective, human exposure, a modification of the EPT-Hs approach is used to determine the magnitude between the experimental threshold (usually in laboratory animals) and defined human intake. The objective of this analysis is to compare the calculated margin-of-safety (MOS) for a substance with that conventionally defined as either suitable or tolerable.

Defining the MOS requires the identification of the experimental evidence particularly appropriate to the exposure conditions, selecting the NOEL or LOEL from the most scientifically defensible study in humans or laboratory animals, comprehensively evaluating the characteristics of exposure, and dividing the NOEL or LOEL by the intake rate. Although hypothetically this method could be applied to all forms of toxicity as well as carcinogenicity and mutagenicity, it is not used for those two specialized forms of toxicity.

The resulting MOS is often compared to uncertainty or safety factors such as those used to derive EPT-Hs to establish whether the magnitude between the two values appears to have incorporated consideration of all of the known uncertainties. Such a procedure does not attempt to extrapolate between species, for example, but rather seeks to judge whether interspecies differences are incorporated in the MOS.

From such analyses, intakes are often judged as either sufficiently protective or not of the consumer's health. There are no MOS whose magnitude is prescribed; however, the 10-fold incremental uncertainty factors used to derive the EPT-Hs may serve as guides to judging the scientifically supported adequacy of the MOS. The MOS approach leads to a continuum of ratios, such that when the value is "large" there is generally little concern for adverse health effects; however, as the ratio approaches, or goes below, unity, concern for health risks heightens. It is widely recognized that extra-scientific considerations can be applied to the scientific conclusions to modify judgments about the adequacy of any MOS.

16.2.4.3 Low-Dose Extrapolation

Several mathematical models (e.g., one-hit, multistage, multihit, Weibull, probit, and time-to-tumor) exist for the extension of dose-response curves for carcinogens and have been reviewed extensively (2,24,33,34). Presently, the multistage model is believed to be the most biologically plausible, for it incorporates contemporary understanding of chemical carcinogenesis, assumes no threshold for cancer initiation (based primarily on observations in radiation carcinogenesis), and allows for the use and "best" fit of the full range of experimental data. Although constructed on sound biological information, the low-dose models nevertheless contain numerous assumptions that, when modified, lead to substantial differences in predicted risks (35) and differ markedly in their slope (36) such that risk estimates for a particular chemical may vary by many (6) orders of magnitude depending on the particular statistical model chosen (2,37,38).

Some investigators have advanced ways of modulating the models to incorporate information about a compound's metabolism and mechanisms of toxic action (39–41). Such approaches afford the best promise for advances in reliably predicting risks.

Empirical demonstration of accuracy of low-dose extrapolations is believed to be beyond technical feasibility. However, a close approximation has been reported by Ramsey et al. (42) who compared the predicted risks to ethylene dibromide exposures derived from the one-hit model applied to animal data with risks measured in epidemiologic studies. They reported that the mathematical model had overestimated the risk by one order of magnitude, a range considered by many scientists to be reasonably small in view of the variability extant in biological measurements. For those carcinogens present in tap water in the United States, the multistage model has been used to derive unit risks based on standardized individual daily intake of 2 L of water containing 1 µg of the substance per liter of water (2).

As a final step in the risk assessment, the unit risks can be converted into population risks. This process takes into account known or anticipated intake levels and the size of the impacted populations to obtain an estimate of the numbers of people that are likely to be affected.

16.3   EVALUATION OF ORGANIC SUBSTANCES IN DRINKING WATER

The health significance from the presence of organic substances in U.S. drinking water is best addressed by first providing a few select examples of the type of information available on some of these materials and of the weighting of such data in formulating risk conclusions. Second, we will provide an overview of those compounds for which sufficient data exist to estimate the relative risks or safety from their ingestion via tap water.

Four substances, two not known to be carcinogens and two carcinogens, have been selected as examples of the risk assessment methods: styrene, toluene, chloroform, and vinyl chloride. The nature of the data base supporting the analysis of the risks in drinking water for each compound is presented with special focus on those data elements that buttress the conclusions about either anticipated adverse health effects or their absence.

16.3.1   Example: Styrene

Styrene monomer, used in the manufacture of polystyrene plastics, resins, insulators, synthetic rubber, and protective coatings, has been detected in drinking water at levels less than 1 μg/L.

It does not appear to accumulate in the body, as evidenced by studies in laboratory animals. The metabolites of styrene were determined from oral administration to rabbits. Only about 2% of the administered dose was eliminated unchanged in the expired air, while 30–40% of the oral dose was hippuric acid. The metabolites of styrene were almost completely excreted 1–2 days after administration of a single dose (43).

Human subjects, exposed to styrene vapors, have experienced transient neurologic impairment and eye and nasal irritation at 375 and 600 ppm in air for short periods of time.

To measure injury from high exposure of short durations, rats and guinea pigs have been exposed to styrene vapors at various concentrations. The maximum tolerated time of exposure to styrene without serious adverse effects was 1 h at 2500 ppm while 10,000 ppm was fatal in 30–60 min (44).

That data provides a sufficient basis to derive estimated population thresholds for brief (i.e., 1 day only) exposures to humans. Applying an uncertainty factor of 10 to the human data appropriately adjusted for differences in routes of exposure from inhalation to ingestion, the concentration in tap water unlikely to produce acute toxic injury is 2000 μg/L. Studies relevant to repeated long-term exposure in humans have also been reported.

When rats were intubated 5 days/week for 28 days at various concentrations, the NOEL was determined to be 100 mg/kg bw/day (44). In a second study, rats were intubated 5 days/week for 185 days at various concentrations; and the NOAEL was determined to be 133 mg/kg bw/day (45). Increased liver and kidney weights were observed at the higher dosages. Chronic exposure by inhalation by rats, guinea pigs, rabbits, and monkeys derived NOAELs of 650–1300 ppm with some species variability (45).

An ADI (equivalent to an estimated population threshold in humans [EPT-Hs], used herein) has been calculated on the basis of the two gavage studies on rats (2). The NOAEL of 133 mg/kg bw/day was divided by an uncertainty factor of 1000 to calculate an ADI of 0.133 mg/kg bw/day. The SNARL in drinking water was calculated to be 900 μg/L (assuming: the average weight of a human adult is 70 kg, the average daily intake of water for man is 2 L; and that 20 percent of total intake is from water).

16.3.2   Example: Toluene

Toluene, formed in petroleum refining and coal tar distillation, is used in the manufacture of benzene derivatives, caprolactam, saccharin, perfumes, dyes, medicines, solvents, TNT, detergent, and as a gasoline component; it has been reported in concentrations up to 6400 μg/L in finished water (46).

Inhaled toluene is retained in body fat and hence bioaccumulates. In humans, the uptake of inhaled toluene was influenced by body fat, that is, those with the least adipose tissue had the smallest uptake. Toluene exposure may alter the metabolism, disposition, and biological effects of other agents, that is, toluene reduces benzene toxicity.

Toluene exposure has produced various effects in humans, including adverse mental changes, cardiac arrhythmias, and liver and kidney dysfunction. Chronic effects of toluene from occupational exposures have only been observed in car painters (47), but the evidence is not sufficient to establish that the effects are due to toluene alone.

The actual oral toxicity of toluene is age dependent, with young animals more susceptible than adults. The acute effects are also species dependent. A slight decrease in the liver microsomal activity of some of the enzymes associated with the mixed-function oxidase system and inhibition of liver microsomal lipid peroxidation was observed in rats after short-term, high dosage by oral administration, which was in contrast to no alterations to the liver of guinea pigs. Sleep patterns of rats were affected by inhalation of 1000–4000 ppm for 4 h (48).

Application of toluene to the skin of rats increased plasmic and lymphoid reticular cells in bone marrow and impaired leukopoiesis, at 10 g/kg, but no effect was observed at 1 g/kg (49).

Protection against acute injury from single exposures (as in the case of an accidental spill in a water supply) can be estimated using acute toxicity information. With the application of appropriate uncertainty factors to the results of laboratory animal studies, an estimated virtually safe concentration in tap water is 2500 µg/L.

Several repeated-dose studies have been reported in rats and mice (50–52). Rats exposed to toluene vapors at 1000 ppm for 8 h/day for 1 week showed metabolic acidosis. However, no significant effects were noted in male and female rats exposed to vapors at concentrations from 0 to 1000 ppm for 6 h/day, 5 days/week for 13 weeks, or in mice exposed to 4000 ppm toluene vapor, 5 times/week for 8 weeks. Eye reddening and hair loss were noted in male and female rats exposed to 0–300 ppm toluene vapor, 6 h/day, 5 days/week for 6 months.

Mice exposed to 1–1000 ppm toluene vapor for 6 h/day for 20 days showed changes in the composition of peripheral blood and behavioral deficits, and bone marrow hypoplasia was noted at 1000 ppm (53).

There are no data on the mutagenicity of toluene in bacterial systems. However, chromosome damage of bone marrow cells in rats has been observed after inhalation exposure and injection of toluene (54), but studies of workers who had been exposed to toluene for years failed to show significant increases in chromosome aberrations (55).

Toluene was applied topically twice a week for 72 weeks to the skin of 30 mice at levels of 16–20 µl. A skin carcinoma was observed in one mouse and a skin papilloma was observed in another (56). This is in contrast to a study where toluene was applied to the skin of mice three times a week for the lifetime of the mice, which failed to produce any carcinogenicity attributable to the toluene application (57).

Embryo toxicity, but no teratogenicity, was observed in pregnant rats exposed to toluene vapors at a concentration of 600 mg/m^3 for an unspecified period (58). There has been a high incidence of menstrual disorders and reported effects on embryonic and fetal developments, such as more frequent fetal asphyxia, a greater number of low birth weights, and belated sucking of the maternal breast from women occupationally exposed to toluene and other agents (59).

The National Research Council (2) determined that exposure of rats to approximately 300 ppm toluene vapor for 6 h/day, 5 days/week for 6 months caused minimal toxicity, but was the most suitable data from which to derive an EPT-H equivalent. The SNARL in drinking water was calculated to be 340 µg/L using an uncertainty factor of 1000 (assuming: the average weight of human adult is 70 kg, an average adult inhales 10 m^3/day, 30% of the inhaled toluene is absorbed from the lung into the blood, and 20% of the toluene intake each day comes from drinking water).

16.3.3   Example: Chloroform

Chloroform has been used in the United States as a refrigerant and aerosol propellant and in the synthesis of fluorinated resins. Other uses included industrial solvent, heat-transfer medium in fire extinguishers, as a pesticide, and directly in pharmaceuticals and toiletries. However, the Food and Drug Administration issued final regulations to ban chloroform as an ingredient in any human drugs or cosmetic products effective in 1976.

Chloroform is a unique contaminant of tap water, in that it is produced during chlorine disinfection of drinking water. Water supply surveys by the Environmental Protection Agency of finished chlorinated drinking water indicate that 95–100% of the finished waters surveyed contained chloroform. The highest reported concentration was 311 μg/L, with the mean concentration of 20 μg/L (3,60).

Chloroform is absorbed rapidly and distributed through body fat and tissues, and then excreted rapidly in mice, rats, and humans. Oral doses of 44.6 g and 148.3 g produced severe but nonfatal poisonings in humans, while ingestion of 296.6 g was fatal to human adults (61). The acute effects of chloroform have been studied in mice, rats and dogs using oral administration. Oral LD_{50} values range from 0.008 to 1.0 ml/kg (62,63).

Male guinea pigs and male albino rats were administered oral doses of 35 mg/kg bw and 125 mg/kg bw (2% of their respective oral LD_{50} values), respectively, for 5 months. Guinea pigs showed decreased blood albumin-globulin ratio and decreased blood catalase activity. Guinea pigs that died had fatty infiltration, necrosis, cirrhosis of the liver parenchyma, lipoid degeneration, and proliferation of interstitial cells in the myocardium. Rats had an impaired ability to develop new conditional reflexes during the fourth and fifth months (64).

Chloroform was negative in the *Salmonella typhimurium* TA-100 microsome assay. Other trihalomethanes were positive under the same conditions.

No evidence of teratogenicity was noted in rats or rabbits from oral administration of chloroform at 20, 50, or 126 mg/kg bw on days 6–18 of gestation. However, low birth weights were observed in both species (65). Daily ingestion by humans of 23–37 mg/kg/day in a 10-year clinical study resulted in some reversible liver injury.

The carcinogenic effects of chloroform have been studied in mice and rats. Gavage studies of rats and mice have noted dose-response relationships for epithelial tumors of the kidneys and renal pelvis in the rats and hepatocellular carcinomas in the mice (66,67). Also, benign thyroid tumors were seen in female mice (67). The results of these studies were determined to be the most serious toxic consequence and were selected as the basis for estimating risks to humans.

Four sets (2 mouse studies, 2 rat studies) of dose-response data (66,67) were used to estimate statistically both the life-time risk and an upper 95% confidence bound on the lifetime risk at the low-dose level. The estimates are of lifetime human risks and are corrected for species conversions on a dose-per-surface-area basis. For chloroform at a concentration of 1 μg/L, the estimated risk for humans would fall between 1.5–17.0 per million. The upper 95% confidence estimate of risk at the same concentration would fall between 3.0–22.0 per million (2).

16.3.4   Example: Vinyl Chloride

Vinyl chloride monomer is used primarily in the production of polyvinyl chloride resins for the building and construction industries. The sale of propellants and all aerosols containing vinyl chloride were banned in 1974, because it was determined to be carcinogenic to humans and animals. The occupational standard for atmospheric vinyl chloride is 1 ppm (ceiling) for 8 h/day, 5 days/week. Vinyl chloride is slightly soluble in water and has been observed in tap water at concentrations as high as 50 μg/L (68).

Studies on the metabolism of vinyl chloride indicate a dose-dependent fate after inhalation or oral administration in rats with elimination more complete and more rapid at lower than at higher doses. The primary mechanism of detoxification of vinyl chloride involves conjugation with hepatic glutathione. This is consistent with a noted decrease in hepatic nonprotein sulfhydryl groups in rats exposed to vinyl chloride (69).

Acute effects of vinyl chloride in humans have shown central nervous system dysfunction, sympathetic-sensory polyneuritis, and organic disorders of the brain. Lesions of the skin, bones, liver, spleen, and lungs have been reported as a result of chronic exposure in humans. Also, 48 cases of hepatic angiosarcoma have been diagnosed in industrial vinyl chloride workers around the world, with high exposure concentrations (70,71).

Vinyl chloride has been shown to produce lung congestion and some hemorrhaging, blood-clotting difficulties, and congestion of liver and kidneys in experimental animals after inhalation exposure (72).

Chronic studies of vinyl chloride administered in diet and in soy oil to laboratory animals have indicated decreased blood-clotting time, enlargement of liver and spleen, and development of angiosarcomas, hepatocellular carcinomas and adenocarcinomas. Vinyl chloride also produced tumors in mice, hamsters, and rats by inhalation. Exposures ranged from 1 to 30,000 ppm in air, with the lowest dose to have a carcinogenic effect being 50 ppm in air for an unspecified exposure time in rats (73).

Vinyl chloride was mutagenic in several microbial mutagenicity test systems, in one in vitro mammalian mutagenesis assay, and in a mouse micronucleus test. One animal species showed chromosome aberrations and sister chromatid exchange. Occupationally exposed workers showed increases in the incidence of chromosome aberrations of lymphocytes (74).

Maternal toxicity, but no teratogenicity, was noted in mice, rats, and rabbits exposed to vinyl chloride (75). The data on humans is not sufficient to reach conclusions about this cause-effect relationship.

The International Agency for Research on Cancer (76), after reviewing data on humans, concluded that exposure to vinyl chloride results in an increased carcinogenic risk to humans. Liver, brain, lung, and hemato- and lymphopoietic systems are most likely target organs to be affected.

The National Research Council (2) used a rat gavage study and a rat inhalation study (77,79a) to estimate carcinogenic risk. The dose-response data from the rat gavage study were used to estimate statistically both the lifetime risk and an upper 95% confidence bound on the lifetime risk at the low-dose level. The estimates are of lifetime human risks and are corrected for species conversion on a dose-per-surface-area basis. For vinyl chloride, at a concentration of 1 μg/L, the estimated risk for man is 3.0/million. The upper 95% confidence estimate at the same concentration is 4.7/million.

16.4 SUMMARY

Having reviewed the scientific derivation of health risks and safety, we will now consider compounds—of the few hundred organic substances reported in U.S. drinking water (2)—for which there is sufficient data to draw inferences about the health consequences of exposure. The compilation is presented in Table 16.1. For some substances, only qualitative occurrence information is available; however, for those substances whose concentrations have been measured, the highest reported levels are listed. If a substance is a carcinogen either in humans or in animals taken to be predictive of human responses, a standardized unit risk (i.e., 95% confidence estimate) is presented. For toxicants not known to be carcinogens, the EPT-Hs for chronic exposure is listed. (That latter value may have been described by another term, such as ADI or SNARL, by various organizations; however, an analysis of the definitions indicates that their meaning is virtually identical.) These estimated values are presented, when available, for comparison with maximum concentration levels in tap water and with existing national or international standards or guidelines.

The maximum concentrations reported have at times been suggested as an indication of population exposure. However, to use such data as reflective of general exposure is likely to be highly misleading, because of the large variances in both occurrence and concentration of these compounds in tap water. Consequently, these data do not identify how many people are exposed to which combination of substances, the average lifetime daily dose obtained from tap water, or the frequency of exposure. Nevertheless, these data are useful in comparing risks for priority setting. For example, those compounds whose estimated risk exceeds acceptable levels based on maximum concentrations could be given higher consideration for remedial action than those whose risk is tolerable.

The results presented in Table 16.1 indicate that standardized (i.e., per microgram of substance per liter of water, assuming daily consumption of 2 L of water by a 70-kg individual) unit risks vary considerably among compounds. Caution must be exercised in interpreting these data, because they are predicated on lifetime exposure at that concentration. Any changes in dose rate may have a profound influence on the estimated risk. For example, if exposure occurs over a small percent of the lifespan (particularly late in life), the estimated increase risk may be well within tolerable levels when compared to lifetime exposure to that same dose rate. The estimates of risk are presented for individual compounds and do not take into account

TABLE 16.1
Risk Assessment of Organic Substances Identified in Water

Organic Compound	Highest Observed Concentration in Water (µg/L)	Upper 95% Confidence Estimate of Lifetime Cancer Risk per µg/L [a]	Chronic EPT-H in Drinking Water (µg/L)	Standard/Guideline	Authority [g]
2,4, D	0.04[b]		90[b]	100 µg/L	EPA (8)
2,4,5-T			700[b]		EPA (8)
2,4,5-TP	0.21[b]		5.25[b]	10 µg/L	EPA (8)
Acetaldehyde	0.1[b]				
Acetone	3000[f]				
Acetonitrile[d]					
Acrolein[b]				320 µg/L	EPA (78)
Acrylonitrile		1.3×10^{-6c}		0.058 µg/L	EPA (78)
Alachlor	0.29[b]		700[b]		
Aldicarb			70[b]		
Aldrin	0.006[b]	Insufficient data[b]		17 µg/L	PHSAC (82)
Amiben			1750[b]		
Anthracene/phenathrene	21.0[f]				
Atrazine	5.1[b]		150[b]		
Azinphosmethyl			88[b]		
BHC-	6[f]	$0.7–3.5 \times 10^{-6b}$			
BHC-β	3.8[f]	$2.5–5.8 \times 10^{-6b}$			
Beneton[b]					
Benzene	330[f]	4.4×10^{-6f}		0.66 µg/L	EPA (78)
Benzo[a]pyrene	Detected[b]				
Bis(2-Chloropropyl)ether[c]					
Bromacil			86[b]		

Compound					
Bromobenzene	Detected[b]				
Bromodichloromethane	116[c]			100 μg/L as THM	EPA (8)
Bromoform	280[f]			100 μg/L as THM	EPA (8)
Butachlor	1.21[b]	70[b]			
tert-Butyl alcohol	Detected[b]				
Butyl benzyl phthalate	38.0[f]				
e-Caprolactam	Detected[b]				
Captan		35[b]			
Carbaryl		574[b]			
Carbofuran[e]					
Carbon tetrachloride	400[f]		1.9×10^{-6f}	0.42 μg/L	EPA (78)
Carbon disulfide	Detected[b]				
Catechol[c]	5[b]				
Chloral					
Chloramines[c]					
Chloramino acids[c]					
Chlorate[c]	0.1[b]				
Chlordane			$0.96\text{--}18 \times 10^{-5b}$	0.00046 μg/L	EPA (78)
Chlorine dioxide		380[c]			
Chlorite[c]					
Chlorobenzen	5.6[b]		2.13×10^{-7b}		
bis(2-Chloroethyl)ether	0.5[b]		1.2×10^{-6b}		
Chloroform	700[f]		$3.0\text{--}22.0 \times 10^{-7b}$	100 μg/L as THM	EPA (8)
Cyanazine	Detected[b]				
Cyanogen chloride	Detected[b]				
Cyclohexane	540[f]				
DDT	0.144[b]	4200[b]	$0.65\text{--}20.0 \times 10^{-6b}$	0.000024 μg/L	EPA (78)
Di(2-ethylhexyl)phthalate	170[f]	770[b]			
Di-n-Butylphthalate	470[f]	14[b]			
Diazinon					
Dibromochloromethane	317[f]			100 μg/L as THM	EPA (8)

TABLE 16.1
(Continued)

Organic Compound	Highest Observed Concentration in Water (μg/L)	Upper 95% Confidence Estimate of Lifetime Cancer Risk per μg/L[a]	Chronic EPT-H in Drinking Water (μg/L)	Standard/Guideline	Authority[g]
1,2-Dibromo-3-chloropropane	137[f]	2.0×10^{-4f}			
Dicamba			9[b]		
o-Dichlorobenzene	3[b]		94[b]	400 μg/L	EPA (78)
p-Dichlorobenzene			94[b]		
Dichlorodifluoromethane			5600[c]		
1,1-Dichloroethane	7[f]	1.5×10^{-4f}			
1,2-Dichloroethane	250[f]	1.0×10^{-6f}			
1,1-Dichloroethylene	280[f]		100[e]	0.033 μg/L	EPA (78)
1,2-Dichloroethylene-*cis*	69[g]			0.033 μg/L	EPA (78)
1,2-Dichloroethylene-*trans*	3.3[g]			0.033 μg/L	EPA (78)
2,4-Dichlorophenol	36[c]		700[b]	3,090 μg/L	EPA (78)
Dieldrin	0.122[b]	$1.9–2.4 \times 10^{-4b}$		17 μg/L	PHSAC (82)
Diethylpthalate	4.6[f]				
Dimethylformamide[d]					
2,4-Dimethylphenol	Detected[b]		70[d]		
Dinitrophenols			39[e]		
Dinoseb					
Dioxane	2100[f]	3.9×10^{-7f}			
Diphenylhydrazine	1[b]	Insufficient data[b]		0.00042 μg/L	EPA (78)
Disulfoton			0.7[b]		
EDB	300[f]	4.8×10^{-4f}			
ETU		2.2×10^{-6b}			
Endrin	0.214[b]	Insufficient data[b]		0.2 μg/L	EPA (8)

Epichlorohydrin	2000[f]				
Ethylbenzene			1100[b]		
Folpet[c]					
Glyoxal[c]					
Glyoxylic acid[c]					
Heptachlor	0.0031[b]	$3.5–4.8 \times 10^{-5b}$		18 µg/L	PHSAC (82)
Heptachlor epoxide	0.008[b]	Insufficient data[b]		18 µg/L	PHSAC (82)
Hexachlorobenzene	0.010[b]	1.85×10^{-6d}			
Hexachloroethane	0.5[b]		7[b]		
Hexachlorophene	0.01[b]		8[b]		
	(drinking water)				
Iodide			1190[c]		
Isopropyl benzene	290[f]				
Kepone		$1.4–8.0 \times 10^{-5b}$			
Lindane	22[f]	$5.6–13 \times 10^{-6b}$		4 µg/L	EPA (8)
MCPA			9[b]		
Malathion			160[b]		
Maneb			350[b]		
Methomyl			175[b]		
Methoxylchlor			700[b]	100 µg/L	EPA (8)
o-Methoxyphenol	Detected[b]				
Methyl glyoxal[c]					
Methyl chloride	44[f]				
Methyl methacrylate	1[b]				
Methyl parathion			700[b]		
Methylene chloride	3000[f]				
Nicotine	3[b]				
Nitralin[b]					
Nitrobenzene[d]				19,800 µg/L	EPA (78)
Nonanal[c]					
Octanal[c]					

429

TABLE 16.1
(Continued)

Organic Compound	Highest Observed Concentration in Water (μg/L)	Upper 95% Confidence Estimate of Lifetime Cancer Risk per μg/L[a]	Chronic EPT-H in Drinking Water (μg/L)	Standard/Guideline	Authority[g]
PCBs	3.0[b]	3.1 × 10⁻⁶[b]			
Paraquat			60[b]		
Parathion	4.6[f]	2.9 × 10⁻⁵[f]	30[b]		
Pentachloronitrobenzene		1.4 × 10⁻⁷[b]			
Pentachlorophenol	1.4[b]		20[b]	1010 μg/L	EPA (78)
Phenylacetic acid	4[b]				
Phorate			0.7[b]		
Phthalic anhydride	Detected[b]				
Picloram			1050[e]		
Popachlor			700[b]		
Propanil			140[b]		
Propazine	Detected[b]		320[b]		
Propylbenzene	0.05[b]				
Resorcinol[c]					
Rotenone	Detected[b]		14[e]		
Simazine	1[b]		1505[b]		
Styrene			900[b]		
TCDD			0.007[b]		

1,1,1-Tetrachlorethane	1^b				
Tetrachloroethylene	1500^f	9.3×10^{-7f}		0.8 µg/L	EPA (78)
Thiram			35^b		
Toluene	6400^f		340^b	14,300 µg/L	EPA (78)
Toxaphene			8.6^b	0.00071 µg/L	EPA (78)
Trichlorobenzene	Detected^b				
1,1,2-Trichloroethane	20^f	1.6×10^{-6f}			
1,1,1-Trichloroethene	5440^f	2.98×10^{-8c}	3800^c		
Trichloroethylene	$27,300^f$	3.0×10^{-7f}		2.7 µg/L	EPA (78)
Trichlorofluoromethane	13^f				
2,4,6-Trichlorophenol					
Trifluorotrichloroethane	135^f				
Trifluralin	Detected^b	4.7×10^{-7b}	700^b		
Vinyl chloride	50^f			500 µg/L	EPA (78)
Xylene	300^f		$11,200^b$		
Zineb			350^b		
Ziram			88^b		

[a] Assumes consumption of 2 L of water per day
[b] Ref. 2.
[c] Ref. 22.
[d] Ref. 79.
[e] Ref. 80.
[f] Ref. 68.
[g] Ref. 81.
EPA United States Environmental Protection Agency.
PHSAC Public Health Services Advisory Committee.

any interactive effects that might influence the overall risk estimate. It has been postulated that substances that have biologically similar mechanisms of action (e.g., genetoxic carcinogens) are likely to possess additive toxic properties. The empirical evidence for that hypothesis remains incomplete and controversial; consequently, the decision to use assumptions about additivity, synergism, or antagonism remain for the most part the province of risk management.

The expression of finite risk estimates for carcinogens at all concentrations necessitates concepts by which to judge the acceptability, or lack thereof, of exposures. For several years, federal regulatory agencies have addressed this issue with varying degrees of success. The EPA has adopted as an operational principle for judging acceptability that exposure should not increase the background cancer rate by more than one cancer in 100,000 people exposed for a lifetime. By this definition, the data in Table 16.1 indicate that trihalomethanes, dichloroethane, and ethylene dibromide would exceed that tolerable limit, if the maximum concentrations represented average exposures, a situation unlikely except for trihalomethanes. The trihalomethanes represent a unique situation in which these products of chlorine disinfection are unwanted but obligate contaminants whose risks are considered to be much smaller than the contrasting public health consequences without that disinfectant.

Comparing the estimated population thresholds for humans for noncarcinogens to the maximum exposure concentrations indicate that they vary considerably among compounds, depending upon their potency. These units also are predicated on lifetime exposure; hence under different conditions of exposure, the EPT-Hs will likely vary. As for the carcinogens, these virtually safe levels do not take into account chemical interactions. However, there is no evidence for toxic interactions for these compounds at these concentrations. The results further indicate that 1,2-dichloroethylene, toluene, and 1,1,1-trichloroethene may have exceeded the EPT-Hs, if the maximum exposures are representative of continuous dosing.

The risk estimates and EPT-Hs serve as the basis of various public policy options including the setting of drinking water standards. Such activities entail many risk management considerations, such as technological feasibility, public acceptability, and implementation costs. When such factors are successfully balanced, such decisions generate public health policies that maximize societal benefits.

REFERENCES

1. U.S. Public Health Service, Public Health Service Drinking Water Standards Revised 1962, U.S. Department of Health, Education, and Welfare, Public Health Service, Washington, DC, 1962.
2. National Research Council, *Drinking Water and Health*, Vol. 1, A report of the Committee on Safe Drinking Water, National Academy Press, Washington, DC, 1977.

3. Environmental Protection Agency, Preliminary Assessment of Suspected Carcinogens in Drinking Water, Report to Congress. U.S. Environmental Protection Agency: Office of Toxic Substances, Washington, DC, 1975a.
4. Environmental Protection Agency, Region V Joint Federal/State Survey of Organics and Inorganics in Selected Drinking Water Supplies, U.S. Environmental Protection Agency, Washington, DC, 1975b.
5. Kraybill, H.F., Global distribution of carcinogenic pollutants in water, *Ann. N.Y. Acad. Sci., 298,* 80–88, 1977.
6. Junk, G.A., and Stanley, S.E., *Organics in Drinking Water Part I. Listing of Identified Chemicals,* National Technical Information Service, Publication No. IS-3671, Springfield, VA. 1975.
7. Tardiff, R.G., and Deinzer, M., Toxicity of organic compounds in drinking water, paper presented at the Fifteenth Water Quality Conference, University of Illinois, Champagne-Urbana, February 7–8, 1973.
8. Environmental Protection Agency, National Interim Primary Drinking Water Regulations, 40 Code of Federal Regulations 141, June 24, 1977.
9. World Health Organization, European Standards for Drinking Water, World Health Organization, Geneva, 1970.
10. World Health Organization, International Standards for Drinking Water, World Health Organization, Geneva, 1971.
11. Environmental Protection Agency, National Primary Drinking Water Regulations; Volatile Synthetic Organic Chemicals, 49 *Federal Register* 24330–24355. June 12, 1984.
12. National Research Council, *Risk Assessment in the Federal Government,* A report of the Committee on the Institutional Means for Assessment of Risks to Public Health, National Academy Press, Washington, DC, 1983.
13. National Research Council, *Principles for Evaluating Chemicals in the Environment,* A report of the Committee for the Working Conference on Principles of Protocols for Evaluating Chemicals in the Environment, National Academy Press, Washington, DC, 1975.
14. Interagency Regulatory Liaison Group, Epidemiology Work Group, Guidelines for documentation of epidemiologic studies, *Am. J. Epidemiol., 114,* 609–613, 1981.
15. Tardiff, R.G., In vitro methods of toxicity evaluation, *Ann. Rev. Pharmacol. Toxicol., 18,* 357–370, 1978.
16. Einslein, K., Computer-assisted predictions of toxicity, in Rodricks, J.V., and Tardiff, R.G. (eds.), *Principles for the Evaluation of Toxic Hazards to Human Health,* Plenum Publishing Co., New York, 1985. (In press.)
17. World Health Organization, Principles and Methods for Evaluating the Toxicity of Chemicals, Part I., Environmental Health Criteria 6, World Health Organization, Geneva, 1978.
18. Nelson, N., Byerly, T., Kolbye, A., Kurland, L., Shapiro, R., Shibko, S., Stickel, W., Thompson, J., Van Den Berg, L., and Weissler, L., Hazards of mercury, *Environ. Res., 4,* 1–69, 1971.
19. Gehring, P.J., Watanabe, P.G., and Park, C.N., Resolution of dose-response toxicity data for chemicals requiring metabolic activation: Example—vinyl chloride, *Toxicol. Appl. Pharmacol., 44,* 581–591, 1978.
20. Tardiff, R.G., and Rodricks, J.V., Comprehensive risk assessment, in Rodricks, J.V., and Tardiff, R.G. (eds.), *Principles for the Evaluation of Toxic Hazards to Human Health,* Plenum, New York, 1985. (In press.)
21. Calabrese, E., *Principles of Animal Extrapolation,* John Wiley & Sons, New York, 1983.
22. National Research Council, *Drinking Water and Health,* Vol. 3, A report of the Committee on Safe Drinking Water, National Academy Press, Washington, DC, 1980.
23. Interagency Regulatory Liaison Group, Work Group on Risk Assessment, Scientific bases for identification of potential carcinogens and estimation of risk, *J. Natl. Cancer Inst., 63,* 168–241, 1979.
24. Food Safety Council, Proposed System for Food Safety Assessment, A report of the Scientific Committee, Food Safety Council, Washington, DC, June, 1980.

25. Dourson, M.J., and Stara, J.F., Regulatory History and Experimental Support of Uncertainty (Safety) Factors, *Regulatory Toxicol. Pharmacol., 3,* 224–238, 1983.
26. Klaassen, C.D., and Doull, J., Evaluation of safety: Toxicologic evaluation, in Doull, J., Klaassen, C., and Andur, M. (eds.), *Cassarett and Doull's Toxicology The Basic Science of Poisons,* 2nd ed., Macmillan Publishing Co., New York, 1980.
27. Tomatis, L., Hilfrich, J., and Turusov, V., The occurrence of tumors in F_1, F_2, and F_3 descendents of BD rats exposed to *N*-nitrosomethyl urea during pregnancy, *Int. J. Cancer, 15,* 385–390, 1975.
28. Claus, G., Krisko, I., and Bolander, K., Chemical carcinogens in the environment and in the human diet.: Can a threshold be established? *Fd. Cosmet. Toxicol., 12,* 737–746, 1974.
29. Newmann, H., Ultimate electrophilic carcinogens and cellular nucleophilic reactants, *Arch. Toxicol., 32,* 27–38, 1974.
30. National Research Council, *Saccharin: Technical Assessment of Risks and Benefits,* A report of the Committee on Technical Assessment of Risks and Benefits of Saccharin, National Academy Press, Washington, DC, 1978.
31. Lehman, A.J., and Fitzhugh, O.G., 100-Fold margin of safety, *Q. Bull. Assoc. Food Drug Officials, 18,* 33–35, 1954.
32. Weil, C., Statistics vs. safety factors and scientific judgment in the evaluation of safety for man, *Toxicol. Appl. Pharmacol., 21,* 454–463, 1972.
33. Brown, C., Approaches to intra-species extrapolation, in Rodricks, J.V., and Tardiff, R.G. (eds.), *Principles for the Evaluation of Toxic Hazards to Human Health,* Plenum Publishing Co., New York, 1985. (In press.)
34. Van Ryzin, J., and Rai, K., The use of quantitative response data to make predictions, in Witschi, H. (ed.), *The Scientific Basis of Toxicity Assessment,* Elsevier Publishing Co., New York, pp. 273–290, 1980.
35. Whittemore, A., Mathematical models of cancer and their use in risk assessment, *J. Environ. Pathol. Toxicol., 3,* 353–362, 1980.
36. Krewski, D., Clayson, D., Collins, B., and Munro, I.C., Toxicological procedures for assessing the carcinogenic potential of agricultural chemicals, in Fleck, R.A., and Hollander, A. (eds.), *Genetic Toxicology,* Plenum Publishing Corp., New York, 1982, pp. 461–497.
37. Munro, F., and Krewski, D., Risk assessment and regulatory decision making, *Fd. Cosmetic Toxicol., 19,* 549–560, 1981.
38. Gaylor, D., The use of safety factors for controlling risk, *J. Toxicol. Environ. Health, 11,* 329–336, 1983.
39. Gehring, P., Watanabe, P., Young, J., and Lebeau, J., Metabolic thresholds in assessing carcinogenic hazards, in *Chemicals, Human Health and the Environment: A collection of Dow Scientific Papers,* Vol. 2, 1977, pp. 56–70.
40. Burlingame, A., Straub, K., and Baillie, T., Mass spectrometric studies on the molecular basis of xenobiotic-induced toxicities, *Mass Spectrometry Rev., 2,* 331–387, 1983.
41. Hoel, D., Kaplan, N., and Anderson, M., Implications of nonlinear kinetics and risk estimation in carcinogenesis, *Science, 219,* 1032–1037, 1983.
42. Ramsey, J., Park, C., Ott, M., and Gehring, P., Carcinogenic risk assessment: Ethylene dibromide, *Toxicol. Appl. Pharmacol., 47,* 411–414, 1978.
43. El Masri, A.M., Smith, J.N., and Williams, R.T., *Biochem. J., 68,* 199–204, 1958.
44. Spencer, H.C., Irish, D.D., Adams, E.M., and Rowe, V.K., The response of laboratory animals to monomeric styrene, *J. Ind. Hyg. Toxicol., 24,* 295–301, 1942.
45. Wolf, M.A., Rowe, V.K., McCollister, D.D., Hollingsworth, R.L., and Oyen, F., Toxicological studies of certain alkylated benzenes and benzene: experiments on laboratory animals, *Arch. Ind. Health, 14,* 387–398, 1956.
46. Council on Environmental Quality, Contamination of Ground Water by Toxic Organic Chemicals, Washington, DC, 1981.
47. Raipta, Husman, C.K., and Tossavainen, A., Lens changes in car painters exposed to a mixture of organic solvents, *Albrecht von Graefes Arch. Klin. Ophthalmol., 200,* 149–156, 1976.

48. Takeuchi, Y., and Hisanaga, N., The neurotoxicity of toluene: EEG changes in rats exposed to various concentrations, *Br. J. Ind. Med., 34*, 314–324, 1977.

49. Yushkevich, L.B., and Malysheva, M.V., Study of the bone marrow as an index of experimentally-induced poisoning with chemical substances (such as benzene and its homologs), *Sanit.-Toksikol. Metody Issled. Gig., 36*–41, 1975.

50. Tahti, H., Ruusha, J., and Vapaatalo, H., Toluene toxicity studies on rats after one week inhalation exposure, *Acta Pharmacol. Toxicol. Suppl., 41*, 78, 1977.

51. Rhudy, R.L., Lindberg, D.C., Goode, J.W., Sullivan, D.J., and Gralla, E.J., Ninety-day subacute inhalation study with toluene in albino rats, Abstr. 150, *Toxicol. Appl. Pharmacol., 45*, 284–285, 1978.

52. Bruckner, J.V., and Peterson, R.G., Evaluation of toluene toxicity, utilizing the mouse as an animal model of human solvent abuse, Abstr. 713, *Pharmacology, 18*, 244, 1976.

53. Inoue, K., Studies on occupational toluene poisoning. (2) An animal experiment using inhalation of toluene vapor in mice, *Osaka Shiritsu Daigaku Igaku Zasshi, 24*, (10-12), 791–803, 1975.

54. Lyapkalo, A.A., Genetic activity of benzene and toluene, *Gig. Tr. Prof. Zabol., 17*, 24–28, 1973.

55. Forni, A., Pacifico, E., and Limonta, A., Chromosome studies in workers exposed to benzene or toluene or both, *Arch. Environ. Health, 22*, 373–378, 1971.

56. Lijinsky, W., and Garcia, H., Skin carcinogenesis tests of hydrogenated derivatives of anthanthrene and other polynuclear hydrocarbons, *Z. Krebforsch. Klin. Onkol., 77*, 226–230, 1972.

57. Poll, W.E., Skin as a test site for the bioassay of carcinogens and carcinogen precursor, The First International Conference on the Biology of Cutaneous Cancer, *Natl. Cancer Inst. Monogr., 10*, 611–631, 1963.

58. Hudak, A., Rodics, K., Stuber, I., Ungvary, G., Krasznai, G., Szomolanyi, I., and Csonka, A., The effects of toluene inhalation on pregnant CFY rats and their offspring, *Munkavedelem 23* (1-3, Suppl.), 25–30, 1977.

59. Syrovadko, O.N., Working conditions and health status of women handling organosilicon varnishes containing toluene, *Gig. Tr. Prof. Zabol., 12*, 15–19, 1977.

60. Environmental Protection Agency, A report: Assessment of Health Risks from Organics in Drinking Water, Ad Hoc Study Group to the Hazardous Materials Advisory Committee, Science Advisory Board, U.S. Environmental Protection Agency, Washington, DC, 1975c.

61. Van Oettingen, W.F., The halogenated hydrocarbons of industrial and toxicological importance, Elsevier Publishing Co., New York, 300 pp. 1964.

62. Kimura, E.T., Ebert, D.M., and Dodge, P.W., Acute toxicity and limits of solvents residue for sixteen organic solvents, *Toxicol. Appl. Pharmacol., 19*, 699–704, 1971.

63. Hill, R.N., Clemens, T.L., Liu, D.K., Vesell, E.S., and Johnson, W.D., Genetic control of chloroform toxicity in mice, *Science, 190*, 159–161, 1975.

64. Miklashevskii, V.E., Tugarinova, V.M., Rakhmanina, N.I., and Yakovleva, G.P., Toxicity of chloroform administered perorally, *Hyg. Sanit.* (USSR), *31*, 320–322, 1966. (Transl. from Russian.)

65. Thompson, D.J., Warner, S.D., and Robinson, V.B., Teratology studies on orally administered chloroform in the rat and rabbit, *Toxicol. Appl. Pharmacol., 29*, 348–357, 1974.

66. Roe, F.J.C., Unpublished report, Preliminary report of long-term tests of chloroform in rats, mice, and dogs, 1976.

67. National Cancer Institute, Report on carcinogenesis bioassay of chloroform, 1976.

68. Council on Environmental Quality, Contamination of Ground Water by Toxic Organic Chemicals, Washington, DC, 1981.

69. Hefner, R.E., Jr., Watanabe, P.G., and Gehring, P.J., Preliminary studies of the fate of inhaled vinyl chloride monomer in rats, *Ann. N.Y. Acad. Sci., 246*, 135–148, 1975.

70. Anonymous, How hazardous to health is vinyl chloride? *J. Am. Med. Assoc., 228*, 1355, 1974.

71. Makk, L., Delmore, F., Creech Jr., J.L., Ogden, II, L.L., Fadell, E.H., Songster, C.L.,

Clanton, J., Johnson, M.N., and Christopherson, W.M., Clinical and morphologic feature of hepatic angiosarcoma in vinyl chloride workers, *Cancer 37* (1), 149–163, 1976.

72. Mastromatteo, E., Fischer, A.M., Christie, H., and Danziger, H., Acute inhalation toxicity of vinyl chloride to laboratory animals, *Am. Ind. Hyg. Assoc. J., 21,* 394–398, 1960.

73. Maltoni, C., and Lefemine, G., Carcinogenicity bioassays of vinyl chloride: Current results, *Ann. N.Y. Acad. Sci., 246,* 195–218, 1975.

74. Anderson, D., Richardson, C.R., Purchase, I.F.H., Evans, H.J., and O'Riordan, M.L., Chromosomal analysis in vinyl chloride exposed workers: Comparison of the standard technique with the sister-chromatid exchange technique, *Mutat. Res. 83,* 137–144, 1981.

75. John, J.A., Schwetz, B.A., Leong, B.K.J., Smith, F.A., Nitschke, K.D., Haberstroth, H.D., Murray, F.J., Balmer, M.F., and Gehring, P.J., The effects of maternally inhaled vinyl chloride on embryonal and fetal development in mice, rats and rabbits, Report, Dow Chemical Company, 1975.

76. IARC, *Monographs on the Evaluation of the Carcinogenic Risk of Chemicals to Man,* Vol. 19: Some monomers, plastics and synthetic elastomers, and acrolein, Lyon, 1979, pp. 377–401.

77. Viola, P.L., Bigotti, A., and Caputo, A., Oncogenic response of rat skin, lungs, and bones to vinyl chloride, *Cancer Res., 31,* 516–522, 1971.

78. Environmental Protection Agency, Water Quality Criteria Documents: Availability, 45 *Federal Register* 79317–79379, November 28, 1980.

79. National Research Council, *Drinking Water and Health,* Vol. 4, A report of the Committee on Safe Drinking Water, National Academy Press, Washington, DC, 1982.

79a. Maltoni, C., Cilberti, A., Gianni, L., and Chieco, P., The oncogenic effects of vinyl chloride administered by oral route in the rat, *Gli Ospedali della Vita, 2* (6), 65–66, 1975.

80. National Research Council, *Drinking Water and Health,* Vol. 5, A report of the Committee on Safe Drinking Water, National Academy Press, Washington, DC, 1984.

81. Environmental Protection Agency, The Occurrence of Volatile Synthetic Organic Chemicals in Drinking Water. U.S. Environmental Protection Agency: Office of Drinking Water, Washington, DC, December, 1981.

82. U.S. Public Health Service Advisory Committee, Report of the Secretary's Commission on Pesticides and Their Relationship to Environmental Health, E.M. Mrak, Chairman, U.S. Department of Health, Education, and Welfare, Washington, DC, 1969.

CHAPTER 17

Unresolved Issues in Risk Assessment

Edward J. Calabrese
Charles E. Gilbert

School of Health Sciences
Division of Public Health
University of Massachusetts
Amherst, Massachusetts

17.1 INTRODUCTION

Attempting to estimate human risk from exposure to organic carcinogens in drinking water is a task with many uncertainties. The fields of toxicology and epidemiology have been asked to provide society with answers about the extent of human cancer risks that result from environmental exposure to carcinogens. Often the demands made by society far exceed the capacity of modern day science to provide accurate estimates of cancer risk. Several decades ago, the issue concerning carcinogens was considered much more qualitative in nature. In other words, is the chemical agent carcinogenic or not? Today, society not only demands to know if the chemical is carcinogenic but also what is the likelihood of developing cancer at each level of exposure. While it is quite possible to provide society with answers to such questions, the issue is: How good are the answers? This chapter will identify and evaluate some of the limitations of modern toxicology and epidemiology facing the field of risk assessment in the earnest attempt to estimate quantitatively human cancer risks from drinking water.

17.2 Quantification of Exposure

Two of the greatest limitations of epidemiological investigations are the precise estimation of human exposure to the chemical carcinogens in question and the influence of potentially confounding variables. This is most evident in studies of an ecologic nature where data is estimated on the group but not on individuals. Most of the initial epidemiological studies concerning the practice of chlorination and cancer risk were exclusively of an ecological nature. These studies compared cancer rates in communities with chlorinated surface water versus communities whose drinking water was derived from ground water. For the most part, these studies did not include the historical levels of trihalomethanes (THMs) or other carcinogens. They also did not have any evidence of water consumption patterns, occupational exposures to carcinogens, nor early childhood exposures to harmful agents or medications, dietary consumption of natural carcinogens, or the level of dietary promoters such as fats. Ecological studies usually assume that these factors are similar amongst the comparison communities, but the assumption usually is never verified, let alone reconstructed, in historical profiles reaching back over 3–4 decades. One very limited, though creative, attempt to improve estimates of exposure was published by Moore et al. (1), who developed a mathematical model for estimating past chloroform levels in drinking water. However, even such an advance is only an initial step toward quantifying exposures. Given the long latency between initiation of tumorigenesis and the clinical manifestation of the cancer, epidemiological studies are always going to find it especially difficult to characterize exposure sufficiently for use in quantitative risk assessment. In addition, the dynamic nature of the U.S. citizenry, with large percentages of the population moving from one location to another—both intra- and interstate migrations—contributes to the difficulty of associating exposure to drinking water with chronic health effects.

Other problems involving the quantification of exposure include a general lack of biological measure of past exposure. Most organic agents have a relatively short half-life and thus can be eliminated quickly. This leads to indirect estimates of exposure.

A new dimension has been added to the role of drinking water and its contribution to volatile organic contaminant (VOC) exposure. It is becoming evident that a significant percentage of VOCs in shower water become stripped from the water and are available for inhalation (2–4). Additionally, organic carcinogens from potable water possess some degree of lipophilicity. This supports the hypothesis that bathing in such water may result in dermal absorption. A theoretical study by Brown et al. (5) argued that dermal absorption of organic contaminants in drinking water may result in a severalfold greater exposure over the absorption from water consumption. The precise magnitude of dermal absorption is contingent on the specific organics under consideration (6).

As a result of the need to determine personal chemical exposure before any risk can be estimated and the general deficiencies of most earlier studies in the area of organic carcinogens in drinking water, there has been a strong orientation for ecological studies (i.e., hypothesis generating studies) to be followed by investigations in which considerable data are collected on the level of the individual. This orientation should lead to the development of greater opportunities to deal more effectively with confounding variables.

A major development in the area of epidemiology is to utilize the biomedical/toxicological sciences so that quantification of exposure can be more accurately determined. Since many of these indicators of exposure are of a biochemical or molecular nature, this new development is referred to as biochemical or molecular epidemiology. Although these approaches represent improvements in estimations of exposure, they must address issues of sensitivity and specificity as well as the likelihood that such exposure quantification is relatively expensive, may be restricted to one evaluation per individual, and may not provide an adequate integration of exposure over a lifetime or during a sensitive period in carcinogen susceptibility (6a).

17.3   HIGH-RISK GROUPS

The human population is highly heterogeneous with respect to numerous traits, including susceptibility to environmental disease such as cancer. A variety of host factors such as one's age, sex, diet, genetic makeup, and predisposing disease conditions have been reported in the literature as affecting susceptibility to chemically induced cancer (7–10).

It is well-known that only about 10% of persons considered heavy smokers actually develop lung cancer. Why then do some people seem to develop a smoking-related cancer while others are by-passed? There is clearly something going on that involves more than just the potential of a chemical to cause cancer and that this risk is merely a function of exposure. Smoking activity seems to be able to cause this effect, lung cancer, significantly more quickly in some but not other people. Considerable research has now pointed to the fact that one's sex (10), diet (7), and genetic characteristics (9) are factors affecting susceptibility to cigarette smoking-induced lung cancer. The extent to which these host factors affect susceptibility are not trivial. For example, males display a 2.5- to 4.5-fold greater risk than females to cigarette smoking-related lung cancer after confounding variables are controlled. People who have a high capacity for lymphocyte aryl hydrocarbon hydroxylase (AHH) inducibility have a 32-fold greater relative risk of developing cigarette-induced bronchogenic carcinoma (11). Carcinogens in drinking water, like those in cigarette smoke, also affect some susceptible individuals more than others in a nonrandom but unknown way.

That people may differ in their response to environmental agents is not a new idea and was, in fact, strongly endorsed as far back as 1938 by the famous geneticist J.B.S. Haldane (12). There have been several national conferences and governmental studies concerning better identification and quantification of those considered at enhanced risk (13–15). Table 17.1 provides a listing of genetic high-risk groups including their estimated occurrence in the U.S. population and the agents to which persons with those

TABLE 17.1
Responses of Oncogenic Synergism and Antagonism[a]

Modifying Agent	Carcinogen, Acted Upon	Effect[b]	Remarks
Polycyclic aromatic hydrocarbons, e.g.,			
Benzanthracenes, benzfluorenes	9,10-Dimethyl-1, 2-benzanthracene	—	Inhibition of skin carcinogenesis
Benzanthracene	7,12-Dimethylbenzanthracene	—	In hamster embryonic fibroblasts (in vitro)
1,2,5,6-Dibenzofluorene	Methylcholanthrene	—	Delayed cutaneous response
Various noncarcinogenic-hydrocarbons:			
Chrysene, 1,2-benzanthracene	3-Methylcholanthrene, dibenzanthracene	—	Anti-initiating effect
Polyhydrodibenzanthracenes	3-Methylcholanthrene, dibenzanthracene	—	Partial anticarcinogenic effect
Phenanthrene	9,10-Dimethyl-1,2-benzanthracene	—	Inhibited tumor initiation and sarcoma production, antagonizes
Phenanthrene	3,4-Benzopyrene, dibenzanthracene	—	tumorigenesis by competing with the carcinogen for binding sites on tissue macromols
Related hydrocarbons from polluted air fractions	3,4-Benzopyrene	—	Diminished responses to varying degrees
Methylcholanthrenes and relatives	Methylated aminoazo dyes	—	Protection; liver protein-azo dye binding reduced; increased binding of azo dye to DNA
Various agents: methylcholanthrene, benzpyrene, chrysene, Sudan III, IV	7,12-Dimethyl-1,2-benzanthracene	—	Inhibition of adrenal necrosis and mammary gland tumorigenesis, traceable to enzyme inductive action
3-Methylcholanthrene	2-Acetylaminofluorene	—	Inhibition of hepatoma and mammary gland tumor formation

3-Methylcholanthrene	Dimethylnitrosamine (DMNA)	+	Mutual syncarcinogenic effects between both carcinogens
Noncarcinogenic azo dyes	Carcinogenic analogs	−	Delayed hepatocarcinogenesis
Nitrogen mustards	Aminoazo dyes	−	Inhibition of hepatoma initiation
3'-Methyl-4-dimethylaminoazobenzene	2-Acetylaminofluorene	+	Supra-additive hepatocarcinogenic response ~ inhibition of esterification of hepatic N-hydroxy-2-acetyl-amino-fluorene
4-Nitroquinoline-N-oxide	3,4-Benzopyrene	−	Skin carcinogenesis inhibition
4-Nitroquinoline-N-oxide	20-Methylcholanthrene	+	Synergistic cutaneous reactions
Methylcholanthrene, naphthylisocyanate	2,4-Diaminotoluene	−	Hepatotumorigenesis reduced
Butylated hydroxytoluene (antioxidant)	2-Acetylaminofluorene	−	Prevents hepatoma- and adeno-carcinoma induction
Phenobarbital	2-Acetylaminofluorene	−	Simultaneous treatment, reduced carcinogenic response
p,p'-DDT	4-Dimethylaminoazobenzene Dimethylbenzanthracene	−	Suppression of mammary tumor induction
o,p'-DDD (TDE)	9,10-Dimethyl-1,2-benzanthracene	−	Antitumorigenic action (breast cancer)
Tetrachloromethane (CCl_4)	2-Acetylaminofluorene	+	Oncogenic synergism, reflected in enhanced tumor initiation and higher hepatoma yield
CCl_4 CCl_4	4-Dimethylaminoazobenzene Dimethylnitrosamine (DMNA)	+ +	Enhanced tumor incidence in liver and kidney after CCl_4 pretreatment, reduced DMNA hepatotoxicity
CCl_4; ethanol Disulfiram	3,4-Benzopyrene Dimethylnitrosamine	± −	Syncarcinogenic effect with CCl_4 Reduces hepatoxicity and RNA alkylating ability

TABLE 17.1
(Continued)

Modifying Agent	Carcinogen, Acted Upon	Effect[b]	Remarks
Polychlorinated biphenyls	Aminoazobenzenes, 2-acetylaminofluorine, diethylnitrosamine	–	Inhibited development of nodular hyperplasia and liver tumorigenesis
CCl$_4$ and miscellaneous hepatotoxins, e.g., allylformate, bromobenzene	3,4-Benzopyrene and allied carcinogens	+	Increased carcinogenic response, related to altered metabolism of the carcinogens
Paraoxon	3,4-Benzopyrene (BP)	±	Delays BP clearance from blood and tissues; inhibition of BP metabolism
cis-aconitic acid	3,4-Benzopyrene (BP)	–	Drastic inhibition of tumorigenesis
8-Hydroxyquinoline	2-Acetylaminofluorene	–	Inhibition of carcinogenicity and hepatotoxicity
Acetanilide	2-Acetylaminofluorene	–	Suppressed toxicity and carcinogenicity
Chloramphenicol	2-Diacetylaminofluorene	–	Prefeeding prevents hepatocarcinogenicity, due to competition for binding sites
Chloramphenicol	3'-Methyl-4-dimethylaminoazobenzene		
α-Benzene hexachloride (α-BHC)	3'-methyl-4-dimethylaminoazobenzene	–	Counteraction of hepatic nodulation
Indole	2-Acetylaminofluorene	+	Synergism of urinary bladder tumorigenesis
Antithyroideal agents	7,12-Dimethylbenzanthracene	–	Incidence of mammary tumorigenesis, reduced in hypothyroidism
3-Amino-1,2,4-triazole	2-Acetylaminofluorene and N-Hydroxy derivative	–	Inhibition of carcinogenic metabolite formation

[a] Source: Ref. 52.
[b] –, Antagonistic response; +, synergistic response.

conditions may be or are known to be at increased risk. This table provides genetic subgroups that are susceptible to either noncarcinogenic or carcinogenic agents. Despite the impressiveness of this listing, which includes some 50 or so human genetic conditions affecting enhanced susceptibility, there is only a handful of conditions where causal associations exist and none for which precise quantitative relationships exist, including the extent to which such people are at increased risk. This is especially true of some of the genetic conditions predisposing for environmental cancers such as high AHH inducibility, differential capacity to acetylate aromatic amines, and DNA repair diseases.

This evidence indicates that the identification of genetic factors and other host factors that may predispose the occurrence of job-related disease is a science truly in its infancy. Nevertheless, the foundations of this concept are on sound biological grounds. Although it appears that host factors may contribute in important ways in explaining the variability of responses to environmental agents, insufficient evidence exists to estimate accurately the extent of the total variability in cancer incidence that may be explained by such host factors.

17.4 ANIMAL EXTRAPOLATION ISSUES

17.4.1 Mouse Hepatoma and Extrapolative Relevance

In the standard cancer bioassay performed by the National Toxicology Program (NTP), two animal models are used, the B6C3F1 mouse strain and usually the Fischer 344 rat strain. Over the course of literally several hundred cancer bioassays, it has become evident that the B6C3F1 responds much more sensitively than the rat to the development of pollutant-induced hepatomas (16). For example, in a review of 85 chronic exposure assays, the mouse model developed hepatomas in 45 studies while the rat developed such tumors in 15 studies. This discrepancy between the responses of these two models has created considerable regulatory concern. The issue emerges as to which model best predicts the risk of human liver cancer.

As a result of such interspecies differences in response as well as the possibility that the mouse is uniquely sensitive to developing hepatomas, the Nutrition Foundation (16) organized a highly prestigious ad hoc review panel to assess the predictive relevance of the mouse hepatoma. One of the important conclusions derived for their appraisal is that the mouse, in contrast to the rat, develops tumors in the NTP bioassay from agents that have been found to be nongenotoxic in mutagenicity assays. While this area remains under intense investigation, the panel offered the suggestion that this mouse model may possess preinitiated cells in that "nongenotoxic" carcinogens may actually be performing as cancer promoters and not as initiators and that this may explain the negative findings in rats.

17.4.2   Differences in Enterohepatic Circulation

Other assessments have suggested that mice and rats may be at enhanced risk to developing cancer caused by carcinogenic agents such as benzo[a]pyrene and polychlorinated biphenyls (PCBs), which are conjugated by glucuronidation and excreted via the bile (17). There is predicted increased risk because the mouse and rat display higher β-glucuronidase activity in the proximal small intestine than humans. Mouse β-glucuronidase activity is 60,000 times greater than humans and rat β-glucuronidase activity is 15,000 times greater than humans. Presumably the mouse should be much more capable of hydrolyzing the conjugated pollutant and thereby release the parent compound, allowing it to be reabsorbed by the gastrointestinal tract into the bloodstream. If this were the case, the mouse, and to a lesser extent the rat, would receive a much higher biological dose than the human even though all may receive the same exposure on a mg/kg basis.

17.5   QUANTITATIVE RISK ASSESSMENT

This subsection of the chapter will address several of the most pressing scientific issues associated with quantitative risk assessment and its application by regulatory agencies such as the U.S. Environmental Protection Agency (EPA) in the estimation of risk to organic carcinogens in drinking water.

17.5.1   Specific Problems of the Multistage Model

17.5.1.1   Selection of Stages

Doll (18) has asserted that if the multistage model of carcinogenesis were correct, then the value of k, or the number of stages of cellular transformation, would generally be from 4 to 6. However, the algorithm employed in the calculation of cancer risk by the EPA using the multistage model requires k to assume a value not greater than the number of dose levels used in the study. Since three dose levels are usually employed; the maximum tolerable dose (MTD), one-half MTD, and controls, the value of k is restrained to be at maximum equal to three. Furthermore, the MTD may be overestimated which may result in this dose killing so many animals that it would be of little use for risk assessment calculations. When excessive mortality occurs, this dose would not be selected for use in the model, thereby reducing the k from 3 to 2. Allowing the value to be determined by the number of dose levels in a study, rather than on an understanding of the process of carcinogenesis, may result in predictions of the multistage model that have diminished biological relevance.

17.5.1.2 Biological Relevance Dose

It is now becoming recognized that the quantitative estimates of cancer risk at low exposure levels can be greatly benefited by information on the delivered dose to the critical tissue (19–22). Going one step further, Hoel et al. (23) stated that it is likely to be "biologically more meaningful to relate tumor response to concentrations of specific DNA adducts in the target tissue than it is to relate tumor response to administered dose of a chemical." Despite the hundreds of cancer bioassays that have been conducted, such quantitative relationships between administered and delivered doses exist for less than a handful of agents (i.e., vinyl chloride [24], dimethylnitrosamine, benzo[a]pyrene [23], and formaldehyde [25]).

Given the fact that the data needed to construct biologically relevant doses for cancer risk assessments are lacking, it is usually assumed that the administered dose is proportional to this measure of exposure. The treatment dose is assumed to be a "valid" linear surrogate for the unknown delivered dose, varying from it by not more than a constant scaling factor.

This assumption was severely criticized by Hoel et al. (23), who showed that this approach was extremely conservative in that it overstates the cancer risk at low doses when the delivered dose to the administrative dose relationship is sublinear.

17.5.1.3 Differential Susceptibility and Quantitative Risk Assessment

While the need to incorporate information on the kinetics of the process of carcinogenesis has been emphasized above, it is also important to recognize that there is considerable human interindividual variation with respect to susceptibility to environmentally induced cancer. This range of susceptibility in the highly heterogeneous human species is thought to be enormously greater than that occurring in the highly inbred rodent strains used in cancer testing. Tolerance distribution models such as the logit and probit recognized the occurrence of differential susceptibility in a very general way. Knowledge of the distribution of susceptibilities to carcinogenic agents in the population, like the genetic data above, need to be acquired and then incorporated mathematically into the quantitative risk assessment procedures. The most biologically plausible approaches are going to be those models that can incorporate the salient features of the process of carcinogenesis as well as the occurrence of differential susceptibility. Thus, cancer must be seen not only as a disease in an individual process but also as a disease that operates within a population. Isolation of just one, as is common today, is not adequate. Thus, when the EPA justified its use of the multistage model because of its biological plausibility, one should recognize that this is true only for the process of carcinogenesis in an individual and not in the sense that the distribution of population sensitivities are accounted for. The caveat is, even if this hoped-for new generation of more biologically plausible models could be developed, can we validate their prediction?

17.5.1.4   Validation of Quantitative Risk Assessment Predictions

Although the previously discussed problems of quantitative risk assessment are formidable, there appear to be reasonable courses of action to resolve these issues in some large measure. However, the problem of trying to verify low-dose cancer risk predictions of, say, 10^{-5} or 10^{-6} is not economically possible given the present resources. In a number of other instances, when a national regulation is established it may be at least reasonably testable, such as the impact on selected public health outcomes of lowering the speed limit to 55 or reducing the drinking age limit from 21 to 18. However, validating that a cancer risk of 10^{-6} is off by two orders of magnitude in either direction is probably not verifiable with current epidemiologic methodologies nor with so-called megamouse studies (26). Nevertheless, this issue has such public health and economic implications that it is always going to be pushed to the forefront. Although the present chapter does not pretend to offer any unambiguous resolution to this knotty issue, validation of quantitative risk predictions has not been given the research priority it requires.

17.5.1.5   Less Than Lifetime Exposure

One of the most common situations in which the EPA and state health departments find themselves is when a relatively small number of people may be exposed to drinking water with elevated levels of several carcinogens for up to several years. Thus, one is faced with the question of what would the excess lifetime risk (ELR) be for that population if it is exposed for but a small fraction of their normal lifespan. The implications of such a situation are profound since one's estimations of risk may markedly affect which, if any, remedial actions may be followed (i.e., drilling new wells, maintaining present exposure levels).

While cancer risk estimates are usually performed with respect to determining excess lifetime risks for a lifetime exposure, the Carcinogen Assessment Group (CAG) of the EPA has adapted the multistage-Weibull model to address this issue. The EPA's approach to date customarily assumes that the earlier in life exposure starts, the greater the ultimate risk. It also assumes that even if the exposure is stopped the population will still continue to accrue incremental risk. Such assumptions have been challenged by observations that postexposure accrual of risk may be markedly diminished (i.e., smoking-related lung cancer, nickel-related nasal cancer, DDT-induced liver cancer in mice and benzo[a]pyrene-induced skin cancer in mice). Freni (27) stated that "a model that does not recognize the potential of reduced additional incremental risk accrued after cessation of exposure is a model with an inherent safety factor of unknown magnitude.'' Given the widespread practical importance of this problem, more research is needed on evaluating the toxicological validity of this approach and in determining what possible refinement may be necessary in the EPA's CAG model.

17.6 SAFETY FACTORS FOR EPIGENETIC CARCINOGENS

The use of safety factors for determining acceptable levels of carcinogens in drinking water is not employed by the EPA nor has it been recommended by the National Academy of Sciences in their ongoing reports to the agency. Since it has been believed by many that there is no safe level of exposure to carcinogens and, therefore, no threshold for response exists, the use of safety factors [even large ones of up to 5000-fold (28)] was rejected in favor of the use of mathematical models for quantitative risk assessment.

Despite this regulatory posture by the EPA, a major issue is whether some carcinogens may act by a threshold response. Those agents that clearly cause cancer and yet do not apparently cause any quantifiable DNA alterations have been eloquently agreed as having a demonstrable threshold for response (29). The implications of these findings are of considerable importance since the threshold approach is likely to permit a considerably greater carcinogen level in water.

Those opposed to this perspective argue that negative studies on the current battery of genotoxicity by these so-called epigenetic carcinogens do not unequivocally prove that the agent is not acting on the DNA or that it is acting by a threshold response. They argue that the lack of sensitivity of current tests to discern the capacity of all carcinogens to initiate DNA should not be an excuse to create a new belief system which clearly does not err on the side of safety. The proponents of the epigenetic position counter by saying that is is clearly impossible to prove a negative and that because of philosophical orientation, such scientists would still not be satisfied if a dozen or two of the new genotoxicity tests were negative.

This issue may go unresolved, but it remains important with major economic and public health implications. If the epigenetic perspective were sustained, then the issue of what would be an acceptable safety factor for carcinogens would emerge.

17.7 THE HORMESIS CONCEPT

Hormesis, according to Luckey (30), is the positive stimulation by subharmful quantities of any agent to any biological system. This term was employed in 1943 by Southern and Ehrlich (31) to describe the stimulation of fungal growth by low concentrations of oak bark extracts that inhibited fungal growth at physiologic doses. Hormesis is a restatement of the Arndt-Schulz (32) law that "small doses of poisons are stimulating." The agent may be physical, chemical, or biological.

Hormetic responses in biological systems have been described by three types of concentration-response curves (Fig. 17.1). The α-curve is the familiar pattern found for the adverse effects of "toxic" agents. In this case, no effect is seen up to a certain level of exposure and a progressive inhibition occurs above the threshold. In contrast are the various postulated types of hormetic dose-response curves. The most frequent hormetic curve (i.e., β-curve) is described by a single stimulatory peak at doses immediately below those found to be inhibitory. The γ-δ-curves are thought to be much less frequent and therefore less studied.

Hormetic responses have been reported for a variety of outcomes, including growth and development, fecundity, cancer, disease resistance,

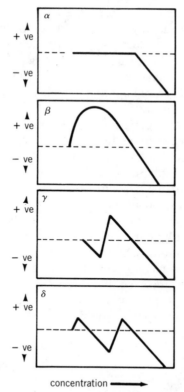

Figure 17.1. Various types of concentration-response curves identified by Townsend and Luckey. The α curve is the one generally found in growth experiments with toxicants, but other types of curves showing hormesis are also known. However, the most common form of hormetic curve is the β-curve. (Source: Ref. 33.)

wound healing, and lifespan (30). These hormetic associations have been studied extensively with respect to both radiation and chemical agents including a wide range of inorganics (e.g., Pb, Cd, Cu) and organics [polycyclic aromatic hydrocarbons (PAH), pesticides]. Examples of the effects of various agents on the growth of organisms and populations have been summarized by Stebbing (33) and are presented in Figures 17.2–17.4. The range of agents illustrated in these figures is quite broad, including radiation, pesticides, heavy metals, and antibiotics. The range of different types of organisms affected is also quite broad, involving worms, oysters, duckweed, crickets, chickens, crabs, salmon, mice, rats, and guinea pigs amongst others.

It is not questioned that radiation exposure at elevated doses is an animal model and human carcinogen, however, evidence exists that hormetic dose-response relationships for radiation may also exist. These low-dose radiation exposures may reduce tumor induction and its subsequent growth and development. Several examples of animal studies may help illustrate this phenomenon. In an experiment by Murphy and Marton (34), the following protocol was used: (1) spontaneous tumors were removed from three groups of mice; (2) the animal, the tumor, or neither were lightly irradiated; and (3) the excised tumor was regrafted to the groin of each mouse. The researchers found that whole-body irradiation-treated animals were more resistant to both implanted and new tumors. Resorption of implanted tumors occurred in 50% of the irradiated mice and only 3.4% of the controls. Forty-eight new palpable tumors were observed in 29 control mice whereas only 21 such tumors appeared in 52 irradiated mice. Irradiation of the tumors only gave results similar to those reported for the controls. These findings were replicated by Lisco et al. (35) and Ong (36), with methyl-cholanthrene-induced tumors, and extended by Lorenz (37), Sacher (38), and Mewissen and Rust (39). In addition, Grahn et al. (40) reported that low doses of ionizing radiation given to mice resulted in a five-fold decrease in the incidence of leukemia below that observed in controls.

In a similar approach, Roe et al. (41) submitted the results of lifetime cancer studies with chloroform exposure to three species (mice, rats, dogs) and found that at very high doses chloroform was indeed a carcinogen, but at low levels tumors were absent and/or life was significantly extended. This led Roe to speculate that low levels of chloroform in drinking water may have a beneficial effect!

Human epidemiological cancer studies have also found the hormetic dose-response for radiation in evidence. After carefully controlling for numerous possible confounding variables, researchers have found a consistent negative correlation between background radiation and deaths attributed to cancer (Fig. 17.5). This has been suggested by Frigerio et al. (42), Sanders (43), Cohen (44), and Hickey et al. (45, 46) with U.S. populations and by the High Background Radiation Research Group (47) with the Chinese. Furthermore, epidemiologic studies on leukemia incidence display a

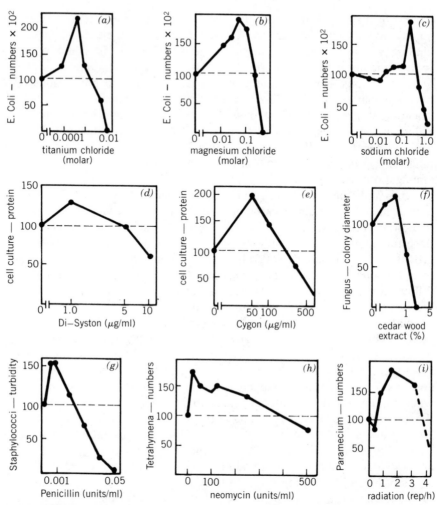

Figure 17.2. The effect of various agents on the growth of organisms and populations drawn from data or figures in the literature (*a, b,* and *c*). The effect of titanium chloride, magnesium chloride, and sodium chloride on the growth of cultures of *Escherichia coli.* (*d* and *e*) The effect of Di-Syston and Cygon on the growth of cultured mouse liver cells (*f*) The effect of an extract of western red cedar (*Thuja plicata*) heartwood on the growth of a fungus (*Fomes officinalis*). (*g*) The effect of penicillin on the growth of cultures of *Staphylococcus* (N.C.T.C. No. 6571). (*h*) The effect of neomycin on the growth of cultures of a protozoan (*Tetrahymena gelii*). (*i*) The effect of low levels of beta radiation on the growth of cultures of a protozoan (*Paramecium caudatum*). (Source: Ref. 33.)

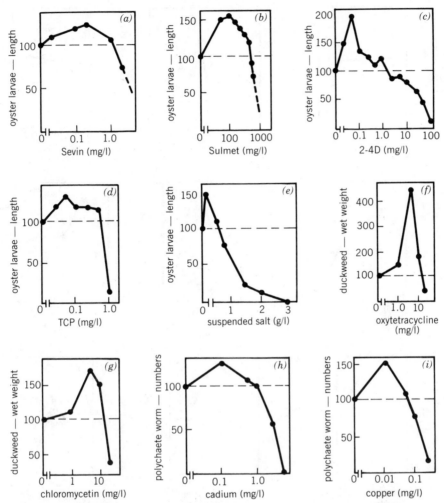

Figure 17.3. The effect of various agents on the growth of organisms and populations drawn from data or figures in the literature. (*a, b, c, d,* and *e*) The effect of Sevin, Sulmet, 2-4D, TCP, and suspended silt on the growth of oyster larvae (*Crassostrea virginica*). (*f* and *g*) The effect of oxytetracycline and chloromycetin on the growth of populations of an angiosperm (*Lemna minor*). (*h* and *i*) The effect of cadmium and copper on the growth of populations of a polychaete (*Ophryotrocha diadema*). (Source: Ref. 33.)

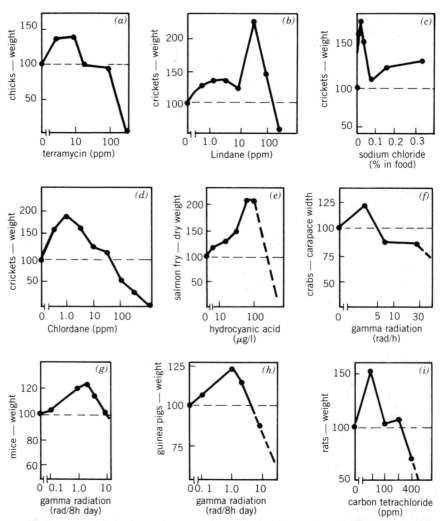

Figure 17.4. The effect of various agents on the growth of organisms and populations drawn from data or figures in the literature. (a) The effect of terramycin (oxytetracycline) on the growth of chicks. (b and d) The effect of Lindane and Chlordane on the growth of crickets (*Acheta domesticus*). (c) The effect of sodium chloride on the growth of crickets (*Acheta domesticus*). (e) The effect of hydrocyanic acid on the growth of salmon fry (*Salmo salar*). (f) The effect of gamma radiation on the growth of crabs (*Callinectes sapidus*). (g and h) The effect of gamma radiation on the growth of mice and guinea pigs. (i) The effect of carbon tetrachloride on the growth of rats. (Source: Ref. 33.)

452

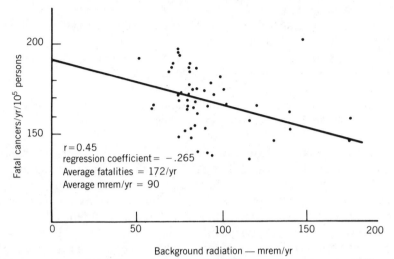

Figure 17.5. Epidemiologic inverse correlation between cancer morbidity, normalized cancer rate, and background radiation of the 48 contiguous United States. The regression equation is: FC $= 191.03 - 0.265$ BR ($P = 0.0013$). (Source: Ref. 44.)

comparable pattern (42, 48). Yet, at higher doses, leukemia is induced by radiation. In addition, Jacobson et al. (49) and Paniker (50) found no statistically harmful effects in the population of Karala, India, where the average background radiation is 1.3 rem/yr or about 10-fold greater than the average for most regions. Finally, epidemiologic evidence revealed that people living in areas of moderately high radiation, such as Denver, have a lower incidence of leukemia than people residing in regions with lower levels (42).

17.8 CHEMICAL INTERACTIONS

Estimating human cancer risk from animal studies has numerous complicating features. One such recurring factor is that animal studies involve the use of clean animals in environmentally controlled situations where few or no other carcinogens are present, and where the genetic background and diet are well characterized. In contrast, many humans smoke, breathe various air pollutants, such as ozone and sulfur dioxide, consume alcohol, take medication, and eat many kinds of preserved foods. Thus, our human experience is usually quite different from that of the animal model. With respect to drinking water, multiple contaminants are common, yet such agents are usually studied separately in animals.

Murphy (51) defined a toxicological interaction "as a condition in which exposure to two or more chemicals results in a qualitatively or quantitatively altered biological responses relative to that predicted from the action of a single chemical. Such multiple chemical exposures may be simultaneous or sequential in time and the altered response may be greater or smaller in magnitude." According to Murphy (51) there are several ways to try to get a toxicological handle on this problem:

1. Search existing literature for evidence of interactions.
2. Initiate laboratory/field studies for interactive effects.
3. Use knowledge of toxicokinetics and toxicodynamic characteristics of individual chemicals to develop predictive models.

Of these approaches, Murphy believes that the third approach is likely to be of greatest potential, since it requires the application of basic principles that underlie the mechanisms of chemicals. The fundamental principles for understanding the kinetics of interactions include: (1) relative affinities of the individual chemicals for sites of action, (2) relative affinities for sites of activation (i.e., detoxifying enzymes, storage sites, pathways of execution) and de(in)activation (i.e., sites of loss), and (3) the intrinsic activity of the compounds at their sites of action. Since there are limited numbers of sites of action or sites of loss within any individual, Murphy (51) claims that there will be a limiting dosage to observe synergism and antagonism.

The gist of Murphy's (51) thesis and that of others such as deBruin (52) is that the number of possible combinations of exposures is multiplicative and far exceeds the capacity of the toxicological community to assess even a very small percentage of the permutations for but a few toxic end points. The most plausible way out of this societal dilemma is to gain a better understanding of cellular mechanisms and chemical disposition of individual agents so that biologically and chemically realistic models for predictions can be developed and tested for their validation. Murphy (51) concluded that "creative application of data obtained in the molecular toxicology laboratory may enable us to examine the" laundry list of chemicals from varied dump sites and make reasonable predictions as to whether there is increased risk because of likelihood of toxicological interactions.

Although Murphy (51) has laid out a general research agenda, the OSHA (53) carcinogen hearings tried to establish the potential importance of carcinogen interactions. Many leading scientists testified at these hearings on the widespread occurrence of synergism amongst carcinogens. For example, Dr. Richard Griesemer of the National Cancer Institute (53) testified that synergism amongst carcinogens was not only widespread but was also a frequent phenomenon. However, it must be understood that carcinogen interaction may also result in antagonistic effects and, as in the case of synergisms, antagonisms are widespread and also frequent. For example, the suppressive effects of polycyclic aromatic hydrocarbons (PAHs) on carcinogenicity are very well known (52). Table 17.1 provides a partial summary of evidence identifying synergistic and antagonistic effects of carcinogens.

Given the fact that carcinogen potency can be markedly affected by the presence of other carcinogens and that for humans multiple exposures are commonplace, the need to predict accurately the occurrence and direction of the interaction is of considerable health concern.

Predicting the potential interaction of multiple chemical agents has been viewed as a major regulatory concern by international and U.S. advisory and regulatory agencies such as the OSHA (53, 54), American Conference of Governmental Industrial Hygienists (ACGIH) (14), World Health Organization (WHO) (55), and, most recently, the EPA (56). In their perspective on the subject the EPA (56) identified several areas of concern.

1. While most information known about toxicant interactions is based on acute toxicity studies in animal models with mixtures of two compounds, the EPA has questioned the toxicological framework whereby chronic responses can be inferred from the acute interaction studies. The EPA claimed that the major problem in this regard was the need to prove that the mechanism(s) of the interaction from the acute study would apply to the low-dose chronic exposure.
2. The EPA believes that the use of information from two component mixtures to predict the interactions of greater than two compounds has problems as far as developing an inherent mechanistic perspective. They stated that if two agents interact because of the effects of one agent on the kinetics of the other, the addition of a third compound that modified the kinetics of one of the two agents may markedly change the degree of toxicologic interaction.
3. The Agency expressed concern that interspecies differences are difficult enough to understand when only a single toxicant is used. It is possible that the magnitude of any interaction in animals will be significantly different in humans.
4. The EPA asserted that none of the models employed for describing chemical interactions are able to predict the magnitude of toxicant interactions without considerable data.

17.9 DISCUSSION

This chapter has not tried to address all the areas of possible uncertainty in the risk assessment process for carcinogens. However, enough examples with documentation were shown to allow the reader to develop the perspective that the science of regulatory toxicology, while having made great strides, has a long way to go before highly reliable estimates of human risk from carcinogen exposure, particularly from drinking water, can be made.

The degree of scientific uncertainty in the process of predictive toxicology is often unknown but thought to be potentially large. This lack of knowledge that confronts the regulatory agencies such as the EPA has enormous implications from public health and economic perspectives. The Agency needs to develop an aggressive basic research program designed to influence the majority of toxicologists to make better qualitative and quantitative cross-species extrapolations. These activities are important, especially when they concern thoughtful judgments on the extrapolative relevance for human disease processes and risk of those diseases. In addition, the Agency has never addressed in a systematic way the occurrence of human interindividual variation in response to toxic and/or carcinogenic agents. Most information on these areas must be gleaned from allied areas of scientific research, particularly with respect to the area of drug metabolism. The targeted research that the Agency often carried out for specific areas of "current" concern will be of less help ultimately to regulatory decision makers in developing a sound extrapolation methodology than the broader, more basic perspective described above. Both types of research are needed, but what usually happens is that the concern for the most recently recognized carcinogen hazards tends to divert limited resources away from providing the type of research needed to establish a firmer foundation for animal extrapolation.

This chapter has spent considerable effort on the limitations of inter- and intraspecies extrapolation, but one must not lose sight of the fact that current testing methodologies present their own scientific concerns. For example, the NTP frequently employs the use of gavage studies in their cancer bioassays. While such oral exposure simulates the same route of exposure to drinking water agents, it has several serious problems. These include the application of extremely large quantities of suspect carcinogens to the same spot on repeat treatments. This has lead animal cancer researchers often to question the relevance of front stomach tumors in rodents during gavage studies. Other methodological concerns involve the typical use of vehicles or carriers of the carcinogen during gavage studies. For example, most organic carcinogens need to be dissolved in oil for delivery to the front stomach in order to maximize the dose delivered. The oil carrier is necessary because of the organic carcinogens' lipophilic nature. However, this exposure process differs markedly from the human experience where we are exposed to these agents via drinking water. Recent EPA-sponsored studies (56) have shown that the toxicity and/or carcinogenicity can be markedly affected by the presence of the oil vehicle. These concerns about the testing methodology itself must be superimposed on the other types of uncertainties already of concern to the Agency.

The use of biostatistical models in low-dose cancer predictions will remain a major issue of contention. Since the estimated risk based on the model drives, in large measure, the course of action by the EPA, the Agency needs to aggressively pursue attempts to improve the biological plausibility of the various biostatistical models for low-dose cancer predictions. This will be markedly enhanced by developments in other areas, such as in understanding the relevance of animal models for predicting human responses and the occurrence of human heterogenicity in response to carcinogenic agents.

Approaches used by the EPA currently assume that the animal model responses are similar to the human and that the distribution of human sensitivity to the carcinogenic responses is like that displayed by the highly inbred animal models. Both of these considerations are biologically untenable and distort accurate estimates of human risk based on animal model predictions. Most emphasis on improving the biological plausibility of the models have focused on the process of carcinogenesis in the individual, usually animal, model. As important as these considerations are, they represent only one area where progress is needed to improve the biological plausibility of risk assessment models.

The Agency must also allow itself to be intellectually honest enough to consider in their research programs the study of phenomena that may undercut some of their basic assumptions for risk assessment. More specifically, the hypothesis that some (maybe all) carcinogens may act in a hormetic manner needs to be addressed in a formal way. If true, the findings would have enormous implications since it would challenge the fundamental tenet that no level of carcinogen exposure is safe.

Attempts to validate low-dose cancer predictions should also be approached in a systematic sense via the use of epidemiological studies, especially of an occupational nature. Such an approach is subject to numerous limitations, however it allows decision makers to gain some understanding of the capabilities of animal studies to predict human cancer risk. For this approach to be realized effectively, it is necessary that greater cooperation between the EPA and National Institute of Occupational Safety and Health (NIOSH) be established.

Because the stakes are so high in the area of environmental cancer risk assessment, the incentive to improve our methodologies for providing better risk assessment is obvious. It is hoped that this documenting of where the areas of most uncertainty lie in our predictive schemes will help provide a perspective of not only how much confidence one should or should not have in our estimations of cancer risk but also where we need to direct future research efforts to acquire appropriate data to assist the Agency in making better predictions of human cancer risk from consumption of water with organic carcinogens.

REFERENCES

1. Moore, G.S., Tuthill, R.W., and Polakoff, D.W., A statistical model for predicting chloroform levels in chlorinated surface water supplies from routinely measured water characteristics, *J. Am. Water Works Assoc.*, 37–39, January, 1979.
2. Calabrese E.J., DiNardi, S.R., and Decker, D., The volatilization of chloroform during the showering process, presented at the EPA-sponsored Conference on Drinking Water and Non-ingestion Routes of Exposure, University of Pittsburgh, School of Public Health, Pittsburgh, PA, April 25, 1985.
3. Andelman, J.B., Non-ingestion exposures to chemicals in potable water, Center for Envi-

ronmental Epidemiology, Graduate School of Public Health, University of Pittsburgh, Pittsburgh, PA, 1984.

4. Zagraniski, R., Volatilization field studies: Interpretation and policy development, with emphasis on benzene, presented at Workshop of Non-ingestion Exposures to Contaminants of Potable Water, University of Pittsburgh, School of Public Health, Pittsburgh, PA, April 25, 1985.

5. Brown, H.S., Bishop, D.R., and Rowan, C.A., The role of skin absorption as a route of exposure for volatile organic compounds (VOCs) in drinking water, *Am. J. Public Health, 74* (5), 479–484, 1984.

6. Flynn, G., Cutaneous absorption of waterborne pollutants, presented at the Workshop on Non-ingestion Exposures to Contaminants of Potable Water, University of Pittsburgh, School of Public Health, April 25, 1985.

6a. Calabrese, E.J., *Pollutants and High Risk Groups,* John Wiley & Sons, New York, 1978, 260 pp; Calabrese, E.J., *Methodological Approaches to the Development of Environmental and Occupational Health Standards,* John Wiley & Sons, New York, 1978, 402 pp.

7. Calabrese, E.J., *Nutrition and Environmental Health: The Influence of Nutritional Status on Pollutant Toxicity,* Vol. I, *The Vitamins,* John Wiley & Sons, New York, 1980, 585 pp.

8. Calabrese, E.J., *Nutrition and Environmental Health: The Influence of Nutritional Status on Pollutant Toxicity,* Vol. II, *The Mineral and Macronutrients,* John Wiley & Sons, New York, 1981, 460 pp.

9. Calabrese, E.J., *Ecogenetics,* John Wiley & Sons, New York, 1984, 340 pp.

10. Calabrese, E.J., *Toxic Susceptibilities: Male and Female Differences,* John Wiley & Sons, New York, 1985, 350 pp.

11. Kellermann, G., Shaw, C.R., and Luyten-Kellermann, M., Aryl hydrocarbon hydroxylase inducibility and bronchogenic carcinogens, *N. Engl. J. Med., 280,* 934–937, 1973.

12. Haldane, J.B.S., *Heredity and Politics,* George Allen and Unwin, London, 1938, pp. 179–180.

13. Calabrese, E.J., Conference Proceedings on the Effects of Pollutants on Human High Risk Groups, *Environ. Health Perspect., 29,* 1–77, 1979.

14. American Conference of Governmental Industrial Hygienists (ACGIH), TLV's: Threshold limit values for chemical substances and physical agents in the work environment with intended changes for 1983–1984, Cincinnati, Ohio, 1983, p. 58.

15. Office of Technology Assessment (OTA), *The Role of Genetic Testing in the Prevention of Occupational Disease,* Washington, DC, 1983.

16. Nutrition Foundation, The relevance of mouse liver hepatoma to human carcinogenic risk, A Report of the International Expert Advisory Committee, Washington, DC, 1983, p. 34.

17. Calabrese, E.J., *Principles of Animal Extrapolation: Predicting Human Responses from Animal Studies,* John Wiley & Sons, New York, 1983, 600 pp.

18. Doll, R., The age distribution of cancer: Implication for models of carcinogenesis, *J. R. Statist. Soc., A13,* 133–166, 1971.

19. Brown, C.C., Mathematical aspects of dose-response studies in carcinogenesis—the concept of thresholds, *Oncology, 33,* 62–65, 1976.

20. Cornfield, J., Carcinogenic risk assessment, *Science, 198,* 693–698, 1977.

21. Crump, K.S., Dose response problems in carcinogenesis, *Biometrics, 35,* 157–167, 1979.

22. Anderson, M.W., Hoel, D.G., and Kaplan, N.L., A general scheme for the incorporation of pharmacokinetics in low-dose risk estimation for chemical carcinogenesis: Example—vinyl chloride, *Toxicol. Appl. Pharmacol., 55,* 154–161, 1980.

23. Hoel, D.G., Kaplan, H.L., and Anderson, M.W., Implication of nonlinear kinetics on risk estimation in carcinogenesis, *Science, 219,* 1032–1037, 1983.

24. Gehring, P.J., Watanabe, P.G., and Park, C.N., Resolution of dose-response toxicity data for chemicals requiring metabolic activation: Example—vinyl chloride, *Toxicol. Appl. Pharmacol., 44,* 581–591, 1978.

25. Starr, T.B., and Buck, R.D., The importance of delivered dose in estimating low-dose cancer risk from inhalation exposure to formaldehyde, *Fund. Appl. Toxicol., 4,* 740–753, 1984.

26. Hughes, D.H., Bruce, R.D., Hart, R.W., Fishbein, L., Gaylor, D.W., Smith, J.M., and Carlton, W.W., A report on the workshop in biological and statistical implications of ED_{o1} study and related data bases, *Fund. Appl. Toxicol., 3,* 129–136, 1983.

27. Freni, S.C., Issues in the use of cancer risk estimates: An epidemiologic approach, *Risk Analysis,* 1985. (Submitted for publication.)

28. Weil, C.S., Statistics vs. safety factors and scientific judgment in the evaluation of safety for man, *Toxicol. Appl. Pharmacol., 21,* 454–463, 1972.

29. Williams, G.M., Liver carcinogenesis: The role for some chemicals of an epigenetic mechanism of liver tumor promotion involving modification of the cell membrane, *Food Cosmet. Toxicol., 19,* 577–583, 1981.

30. Luckey, T.D., *Hormesis with Ionizing Radiation,* CRC Press, Boca Raton, FL 1980.

31. Southern, C.M., and Ehrlich, J., Effects of extract of western red-cedar heartwood on certain wood decaying fungi in culture, *Phytopathology, 33,* 517, 1943.

32. Schulz, H., *Pflüger Arch. Ges. Physiol., 42,* 517, 1888.

33. Stebbing, A.R.D., Hormesis—the stimulation of growth by low levels of inhibitors, *The Science of Total Environment, 22,* 213–234, 1982.

34. Murphy, J.B., and Marton, J.J., The effect of roentgen rays on the rate of growth of spontaneous tumors in mice, *J. Exp. Med., 22,* 800, 1915.

35. Lisco, H., Ducoff, H.S., and Baserga, R., The influence of total body x-irradiation on the response of mice to methylcholanthrene, *Bull. Johns Hopkins Hosp., 103,* 101–113, 1958.

36. Ong, S.G., Cosmic radiation and cancer. I. Influence of cosmic radiation at sea level on induced cancer, *Scientia Sin., 12,* 1760–1761, 1963.

37. Lorenz, E., Some biologic effects of long-continued irradiation, *Am. J. Roentgenol., 63,* 176–185, 1950.

38. Sacher, G.A., On the statistical nature of mortality, with special reference to chronic radiation mortality, *Radiology, 67,* 250–258, 1956; Sacher, G.A., Survival of mice under duration of life exposure to x-rays at various rates, Office Tech. Pub. U.S. Department of Commerce, Washington, DC, 1956, pp. 435–463.

39. Mewissen, D.J., and Rust, J.H., Reticuloendothelial neoplasms in C57 black mice after fast neutron irradiation at low doses, in *Hanford Biology Symposium on Radiation and the Lymphatic System,* U.S. Atomic Energy Commission, Oak Ridge, TN, AI7:255518, 1976, pp. 230–238.

40. Grahn, D., Sacher, G.A., Rust, J.H., and Fry, R.J.M., Duration of life with daily ^{60}Co gamma irradiation: Report on survival and incidence of tumors, Argonne National Laboratory, Argonne, France *Annual Report* ANL-7635, 1969, pp. 1–8.

41. Roe, F.J.C., Palmer, A.K., Worden, A.N., and VanAbbe, N.J., Safety evaluation of toothpaste containing chloroform. I, Long-term studies in mice, *J Environ. Pathol. Toxicol., 2,* 799–819, 1979.

42. Frigerio, N.A., Eckerman, K.F., and Stowe, R.S., Carcinogenic hazard from low-level, low rate radiation, The Argonne Radiological Impact Program, Part I Rep. ANL/ES-26 (Argonne National Laboratory, Argonne, France), 1973.

43. Sanders, B.S., Low-level radiation and cancer deaths, *Health Phys., 34,* 521–583, 1978.

44. Cohen, J.J., Natural background as an indicator of radiation induced cancer, *Proc. 5th Int. Congr. of the Int. Radiat. Protect. Assoc.,* Jerusalem, Israel, March 1980.

45. Hickey, R.J., Clelland, A.B., Clelland, R.C., Zemel, B.S., and Bowers, E.J., Ionizing radiation and human mortality: Multivariate analysis of U.S. state data, *Am. J. Phys. Anthrop., 54,* 233, 1981.

46. Hickey, R.J., Bowers, E.J., Spence, D.E., Zemel, B.S., Clelland, A.B., and Clelland, R.C., Low level ionizing radiation and human mortality: Multi-regional epidemiological studies, *Health Phys., 40,* 625–641, 1981.

47. High Background Radiation Research Group, Health survey in high background areas in China, *Science, 209,* 877–880, 1980.

48. Eckhoff, N.D., Shultis, J.K., Clack, R.W., and Ramer, E.R., Correlation of leukemia mortality rates with altitude in the United States, *Health Phys., 27,* 377–380, 1974.

49. Jacobson, A.P., Plato, F.A., and Frigerio, N.A., The role of natural radiations in human leukemogenesis, *Am. J. Public Health, 66,* 31–37, 1976.
50. Paniker, P.G.K., Resources not the constraint on health improvement, a case study, *Econ. Pol. Week., 14,* 1803–1809, 1979.
51. Murphy, S.D., General principles in the assessment of toxicity of chemical mixtures, *Environ. Health Perspect., 48,* 141–144, 1983.
52. deBruin, A., Synergism and antagonism between organicals, in *Biochemical Toxicology of Environmental Agents,* Elsevier, NY, 1976, pp. 383–419.
53. Occupational Safety and Health Administration (OSHA), *Carcinogen Hearing,* published by Commerce Clearing House, Inc., Chicago, IL, 1980.
54. Occupational Safety and Health Administration (OSHA), General Industry Standards, Subpart 2, Toxic and Hazardous Substances, Code of Federal Regulations, *40,* 1910.1000(d)(2)(i), Chapter XVII—Occupational Safety and Health Administration, 1983, p. 667.
55. World Health Organization (WHO), Health effects of combined exposures in the work environment, WHO Tech. Rept. Series No. 662, 1981.
56. U.S. Environmental Protection Agency (EPA), Proposed guidelines for the health risk assessment of chemical mixtures and request for comments, *Federal Register, 50*(6), 1170–1178, January 9, 1985.

PART FIVE

Legislative and Regulatory Aspects of Organic Contaminant Control

CHAPTER 18

The Relative Risks
Associated with
Ingesting Organic Contaminants

Barbara B. Taylor

Cambridge Analytical Associates, Inc.
Boston, Massachusetts

18.1 INTRODUCTION

In the past few decades it has appeared that a major impact of our increasingly complex technological society has been the imposition of new and copious hazards to the environment in general and to public health in particular (1–8). Questions about the nature and severity of these hazards are being asked with ever more frequency and urgency by both the public at large and governmental regulatory agencies. One such hazard that has stimulated enormous concern is the exposure of humans to small amounts of environmental contaminants, such as the ingestion of minute concentrations of organic substances (9–17). How one examines the risks associated with this hazard, how these risks compare with the others we are subjected to in daily life, and why this topic has caught the attention of the public, the scientific community, and the government make up the focus of this chapter.

18.2 BACKGROUND

The reasons for the increased public and governmental concern about environmental hazards can be traced to a complex interweaving of related and unrelated events. Much of this is connected, in general, to changes in public perceptions about science and technology which began primarily in the 1960s (18,19). During the 1960s and early 1970s, an era of major cultural change during which the environmental movement was conceived and nurtured, there was a growing awareness that the results of technology were not always as beneficial as was once thought (4,18–20). There was, simultaneously, an atmosphere of technological euphoria, in which the public's attitude was most aptly described as the "War-on-X" syndrome—if enough money could be spent on a particular problem, it would be solved (18).

During this time, colleges and universities increased course offerings in the environmental field, consumer and environmental advocacy groups blossomed, and increased pressure for environmental and consumer protection legislation was exerted and heeded, leading to new laws and regulations as well as to the establishment of such governmental agencies as the Office of Technology Assessment (OTA) and the Environmental Protection Agency (EPA). All of these actions served to heighten the awareness of an already awakened public to possible environmental hazards.

Furthermore, the frequent reporting of environmental hazards in the newspaper and on the radio or television became commonplace. For example, by the mid-1970s media attention focused on between 40 and 50 different hazards each year (21). There were, and continue to be, a seemingly endless array of potential chemical hazards reported: DDT, thalidomide, cigarette smoke, alcohol, cyclamates, saccharin, vinyl chloride, arsenic, mercury, lead, mirex, kepone, chlorinated drinking water, dioxin, monosodium glutamate (MSG), urea formaldehyde, chlorofluorocarbons, sulfiting agents, and ethylene dibromide (EDB) (3,13,22–39).

Technical developments in computer facilities and analytical equipment combined with the increased media attention to give rise to the reporting of very small amounts of potentially hazardous substances in water, soil, air, and consumer products. Concurrently, increased attention in the scientific community was focused on the effects of such low levels of exposure, as evidenced by an increase in experiments studying the subtle impacts of environmental hazards, such as cancer—a topic that itself was gaining more and more public attention. The resulting emotional response of the public to cancer risks associated with chemicals in the environment undoubtedly was related to such factors as: the greater media attention given to possible environmental carcinogens, the apprehension over the fact that years of research had produced neither any generalized cures for cancers nor detailed information on its specific causes, and an attitude held by many people, albeit erroneously, that cancer was almost always fatal (7,8,24,40–44).

With this public attitude and all the media attention, it was not surprising that regulatory bodies within the government were pressed to "do something" about environmental carcinogens. Indeed, as far back as 1958, with the insertion of the Delaney Amendment to the Food and Drug Act, governmental concern was being expressed over the ingestion of possible carcinogens in food. As stated in the amendment:

> no additive shall be deemed to be safe if it is found . . . after tests which are appropriate for the evaluation of the safety of food additives to induce cancer in man or animal. (45)

The Delaney Amendment was structured with the best of intentions. With the benefit of hindsight, however, it is easy to see the difficulty inherent in a regulation so ripe with well-meaning naiveté. Among the problems in this amendment were the difficulty in determining whether or not a particular substance would cause cancer, the validity of animal studies for estimating human cancer potential, the omission of any reference concerning what doses should be used in the studies to "induce" cancer, and the very sticky questions as to what was "deemed to be safe" and by whom this was to be done. Thus, while attempting to clarify an issue in the regulatory decision-making process, the Delaney Amendment "in some ways confounded it" (2).

This amendment provoked a controversy over the process of regulating environmental contaminants that continues through to the present and will likely persist for some time in the future. Determining what substances should be regulated and to what extent they should be reduced or eliminated are at the center of this dispute. Although some people remain faithful to the idealistic notion that no risk should be tolerated (as voiced in the Delaney Amendment), others maintain the more pragmatic philosophy that a regulation should be set at some level of reasonable risk. Hence, the present demand for more accurate assessments of environmental hazards on which to base legislative, judicial and personal decisions has evolved.

18.3 RELATIVE RISK ASSESSMENT

One way by which we can begin to study and try to understand the myriad hazards we all face every day is to measure the individual risks against one another to see how they compare. This is the basis for risk assessment analyses and relative risk comparisons.

As observed in the previous section, three terms are used frequently in such discussions. These are hazards, risks, and safety. All of us are familiar with these terms, if not in a directly technical or professional capacity, at least in a casual one. Nevertheless, there is considerable confusion with the use of these terms when discussing hazards in the environment. Much of this confusion stems from the common usage of the term "safe" as meaning completely free from any harm. Since it is clear that there is a possibility (however minute) of harm (however minor) related to even the most seemingly innocuous or mundane of activities, it is obvious that to be absolutely safe is an impossibility (2). Thus, instead of asking, "Is it 'safe'?" and expecting a definitive "yes" or "no" answer, it appears to be more appropriate to ask, "How 'safe' is it?" or "Is it 'safe enough'?".

To avoid this confusion, especially concerning what is or is not safe, a set of definitions for hazards, risks, and safety have evolved in the professional literature (1,2,21,46–52). While the exact definitions may vary somewhat from source to source, the following composite is typical and adequate for the purposes of this discussion. (Consult Chapter 16 for a slightly different view of this topic.)

Hazard: An exposure, event or activity that poses a threat to people or to the environment.

Risk: A quantitative measure that combines the nature of the harm resulting from a hazard and the probability of that hazard occurring.

Safety: The quality or state of having an acceptable level of risk.

These crisply defined terms help to clarify the discussion of environmental hazards considerably, particularly the notion that while nothing is absolutely safe, an exposure, event, or activity may have a risk that is deemed acceptable and therefore can be considered safe enough. Thus, the questions about environmental hazards become:

What are the risks of those hazards?
Which risks are deemed acceptable?
How can we limit or eliminate the hazards that have unacceptable risks?

To start with, the determination of an acceptable risk, whether for personal decision-making or regulatory purposes, implies the ranking of the relative risks of activities, events, or exposures to various hazards, at the very least into the categories of risks that are acceptable and those that are not. Thus, the idea that the probabilistic determinations of environmental hazards should be carried out with the exact numerical results used for a variety of decision-making purposes is a very natural one. This has provided the technical community with the need to come up with quantitative methods for the evaluation and ranking of hazards, leading directly to risk assessment.

Risk assessment for the determination of relative risks can be viewed in terms of three main components (21,38,46–50,53,54):

1. Identification of the hazards.
2. Estimation of the respective risks.
3. Comparative risk evaluation.

A closer examination of these three parts of risk assessment indicates that enormous difficulties may arise when one attempts to assess the risks of ingesting low concentrations of organic contaminants and to compare them to the other hazards faced in life.

18.4 IDENTIFICATION OF THE HAZARDS

The identification of the hazards is not as easy as it appears at first glance. For example, when trying to identify the type of harm that might result from a particular hazard it is a simple matter if the hazard results in a major or gross injury, such as blindness, the loss of a limb, or death. It is another matter entirely if the potential hazard may have only the most subtle of impacts, such as carcinogenicity, mutagenicity, or teratogenicity, slight behavioral changes, minor skin or functional irritations, small decreases in mental ability, mild nervous system disorders, changes in reproductive systems, and so forth.

Thus, while the harm that results when a pedestrian is hit by a fast-moving truck is obvious, the harm from ingesting extremely low levels of organic substances, unfortunately, most likely is not. Here is the case wherein the impacts of a suspected hazard, if there are any at all, will almost certainly be of the most subtle nature, perhaps so subtle as to appear not to exist.

Additionally, there is the problem of the time lags involved between the exposure to a hazard and the observation of the result of that exposure. Again, in the case of the truck accident, the results of the crash are immediate or known in a relatively short period of time. On the other hand, cancer resulting from the ingestion of small amounts of contaminants is unlikely to be observed for 10–40 years (24,40,43,55). Thus, it often is extremely hard to determine, at or near the time of exposure, what the ultimate health impacts, if any, will be.

Related to this problem is the difficulty of proving a particular cause resulted in a specific effect. In the accident example, it is clear that the effect (death) was a result of a particular cause (a direct hit). However, establishing that a particular substance or group of substances resulted in cancer is extremely difficult. Certainly, this has been done in a number of cases, most notably the causal relationship between cigarette smoking and lung cancer, as well as in some cases of occupational exposure to certain substances, including asbestos, vinyl chloride, arsenic, and benzene, among others (24,25,38,43,56). Nevertheless, these examples are based on only a very small number of the thousands of substances to which we are subjected and are cases where exposure levels tend to be much higher than in the environment and causal relationships more evident.

Since there is a risk, even a very small one, involved in every activity we undertake, it is difficult to believe that sufficient evidence will be forthcoming to prove, beyond a doubt, what harms are associated with the ingestion of small amounts of contaminants and which harms are the results of which contaminants. Thus, simply identifying the potential hazards we might encounter from such minute exposures is an arduous task.

18.5   ESTIMATION OF RISKS

The second task, that of estimating the risks related to the hazards, is even more difficult. By definition, the risks related to the ingestion of organic contaminants in minute quantities are given in terms of probabilities of harm. Once the nature of the harm is established, the next question to ask is, How can these probabilities best be determined? Data relating the frequency and nature of the harm to the type and level of exposure clearly would seem a likely way to begin.

Statistics concerning accidental deaths indicate that this strategy can work very well. As in the example of the truck accident referred to previously, we know there are data available to give us such information as how many pedestrian/truck accidents there were in the past year, how many deaths occurred, where the accidents took place, and so on. With this information, it is a relatively simple matter to calculate a person's probability of death in a given year as a result of being hit by a truck. Unfortunately, assessing the risks of ingesting low concentrations of contaminants is vastly more complicated, especially since the harms that may result are typically very subtle and may not be evident for an extended period of time, as in the case of cancer. How then can these risks be estimated at all?

The first step is to try to ascertain what the effects will be (or most likely be) from the exposure to very small amounts of various substances. While no single technique or experiment currently available can prove unambiguously the effects of such minute exposures, there are several techniques that can be used to provide a good indication of what type of harm can be expected and how likely that harm is (57–60). Each of the methods has distinct advantages and disadvantages when compared with the others, although they all tend to suffer, to some extent, from particular uncertainties and ambiguities inherent in their natures.

For the purposes of this discussion, the only harm examined will be that of cancer since it is of such widespread concern, is likely to be at least one of the harms associated with the ingestion of minute amounts of organic contaminants, and has generated sufficient data so as to be meaningful. The techniques used to gather information about human cancer risks can be grouped into the three general categories shown below (2,14,36,59–70):

1. Animal experiments.
2. Epidemiological studies.
3. Short-term microbiological assays.

The first technique, experiments on animals, has been the classic way to examine the gross effects (like death) from acute exposure to toxic substances. However, when conducting animal studies employing significantly sublethal doses of substances to ascertain the risk of cancer in humans exposed to minute concentrations of organic contaminants, there are several major uncertainties concerning how the experiments are conducted and how conclusions are drawn from the data collected.

For example, one of the problems with animal studies is that while the amounts of test chemical to which the animals are exposed are below lethal doses, they are well above the amounts that humans would ingest. Thus, in order to draw any conclusions from the results of the animal experiments, it is necessary to extrapolate the data from the relatively high doses used in the studies to the very low doses to which humans actually are exposed.

The necessity for using such high doses in animal studies is clear (36,58,71). If a particular substance caused 2300 cancers per year in the United States, its cancer incidence rate would be about 0.001% (or each year, a person in the United States would have one chance in 100,000 of getting cancer from that substance). To detect a cancer with such a low incidence, an experiment using the exposure level of humans would have to be run on an enormous number of test animals, the higher the test population the higher the statistical confidence in the findings.

However, if an animal experiment was conducted with significantly higher doses, increasing the chances for cancer to develop, then a relatively small number of animals would have to be used. Thus, for statistical validity, animal experiments must be conducted on either small numbers of animals subjected to very high doses or immense numbers of animals subjected to the very low actual human doses (36). Given the obvious time, money, and space problems associated with testing large numbers of animals, if animals are to be used at all for testing carcinogens, there really seems to be no choice but to use the much higher than actual exposure levels.

Nevertheless, the use of these high doses causes some problems (36,55,57,58). For one thing, the high doses, along with the media attention given to the increasing numbers of substances that have been found to be animal carcinogens, leads some people to question whether or not almost any substance could result in some form of cancer if given in a high enough dose. This idea has been refuted by others as being "completely fallacious" for a number of reasons (55,71). First, animal experiments have been conducted on a number of industrial chemicals and pesticides at high dose levels with the majority showing no carcinogenic response. In addition, many of the substances selected for animal studies are already suspected of being cancer-causing agents, thereby increasing the likelihood that they would produce cancer (55).

Another problem of the high doses is that they might be high enough to overwhelm or significantly alter possible detoxification or repair mechanisms, which would protect against the formation of cancer at lower doses, and to change metabolic pathways (55,59,72,73). Related to this is the controversy over whether or not there are threshold values for chemical carcinogens (36,55,57,59,72,74). If there are no thresholds, any amount of contaminant, no matter how tiny, may be hazardous. On the other hand, if there are thresholds, there will be no hazard below a specified amount of contaminant. To provide the greatest protection for public health, given the uncertainties, some people believe that regulations should be based on the assumption that no thresholds for carcinogenic substances exist. However, this rather conservative approach to the regulation of possible carcinogens may well be in error (57,73,74). For example, nickel and chromium have been shown to be human and animal carcinogens in high doses, yet they are also known to be necessary dietary factors when consumed in trace amounts (58).

Furthermore, a major problem with the conclusions drawn from animal studies lies in the inherent assumption that data concerning animals such as rats and mice can be extrapolated to humans (40). The problems with this assumption include the differences between humans and test animals with respect to size, rates of adsorption of materials, distribution and storage systems, excretion and readsorption of materials, metabolism, and receptor sites (36).

Additionally, animal experiments typically test only one substance at a time, whereas humans are exposed to hundreds of different substances each day. Thus, possible synergistic or antagonistic interactions will be missed completely in conventional animal experiments. Also, the test animals are a genetically homogenous group that lives in a rigidly controlled environment, whereas people are genetically heterogeneous and lead lives that may be totally dissimilar with respect to diets, life styles, occupations, and so forth (36). Finally, concern has been voiced over the fact that animal experiments often use routes of exposure, such as direct injection or force feeding using a tube directly to the stomach, other than simple ingestion and, therefore, provide an inappropriate comparison (71,75).

Given these problems and uncertainties, there is no assurity that a chemical deemed a carcinogen as a result of animal studies will cause cancer in humans nor is there assurity that cancer risk estimates for humans from such studies will be accurate (40,70). Indeed, the different mathematical models on which risk estimates for human cancers from animal data are based often give widely varying results and it is impossible to determine which is the most accurate (38,72,76). This problem was addressed by Gori (40), "In our present culture, arcane and even trivial mathematical formulations often gain persuasiveness beyond their real intentions." Thus, while animal experiments do give considerable information about human cancer risk, particularly with respect to comparative or relative risks, they cannot be viewed as delineating actual risks.

The second method used for carcinogenic risk assessments, epidemiological surveys, avoids some of the problems of animal experiments in that these studies directly survey people and their respective exposures so that there is no need for interspecies extrapolation of effects or the extrapolation from very high doses to the low levels of environmental exposure (4,36,63). Essentially, epidemiological studies attempt to establish that a particular human health effect, in this case cancer, can be attributed to a specific environmental factor.

As with animal studies, there are some serious problems and uncertainties with the use of epidemiological studies. For example, we know that there is typically a latency period of between 10 and 40 years between the ingestion of substances at such low levels and the onset of cancer (24,40,43,55). Thus, one of the problems is that by the time a substance has been shown to be a carcinogen by epidemiological studies, it has been in the environment for a significant amount of time and countless people may have been exposed to it (40). Additionally, the types and quantities of substances ingested by people change significantly over time, as do other confounding factors such as life styles, occupations, diets, personal habits, and where people live (43,67,77). Thus, establishing causality—proving that a cancer was caused by a particular agent—and correctly estimating the risk are extremely difficult (63,76).

Because of these substantial uncertainties in epidemiological studies, it is best to look at a relative risk factor—the ratio of the number of people who have cancer and were exposed to a particular environmental agent to the number of people who have the same cancer but were not exposed to that agent. The higher this factor is above 1.0, the greater certainty there is that the environmental agent under scrutiny caused the cancer. Typically, to establish any causality at all, this factor should be at least 1.3–1.5 (69). In some cases this factor can be much higher, thereby lending greater credibility. For example, the relative risk factor for cigarette smokers who get lung cancer compared with nonsmokers who get lung cancer is 10–25, depending on how much one smokes, (or smokers are 10–25 times more likely to develop lung cancer than are nonsmokers) (25,69). Such high relative risk factors have not been observed when examining the cancer risks from the ingestion of minute concentrations of contaminants (70). (For additional information on epidemiological studies and the risks of micropollutants in drinking water, see Chapter 15.)

The third method used to gather information about human cancer potential is by short-term microbiological assays. There are several possible techniques that can be employed including the use of bacteria, yeasts, and mammalian cells or body fluids (57,70,78). However, the most common of these is some variation of the *Salmonella* bioassay, developed by Dr. Bruce Ames (40,60). This in vitro test in bacteria gives an indication as to whether or not a particular substance is, or groups of substances are, mutagenic. Since there is evidence that a substance will be a carcinogen if it is a mutagen, this test has been used as a screening technique to examine environmental contaminants for carcinogenic potential (41,60,79,80). However, even if a substance found to be a mutagen was not a carcinogen, the fact that it is a mutagen is sufficient to warrant public health interest.

The *Salmonella* bioassay, and other similar tests, have distinct advantages over the first two methods in that the experiments can be done within a few days and inexpensively. Both animal studies and epidemiological studies can take several months or years to complete and are very expensive (40,55,60,81).

However, these short-term assays are not without their own problems. For example, there are some substances that have shown false positives (the test indicated that a substance was a mutagen when it really was not) and false negatives (the test indicated that a substance was not a mutagen when it really was) when compared with animal data (60,83). Additionally, it is difficult to ascertain a dose-related response in humans from the results of such tests and, like animal studies, there are questions about the effects of relatively high doses and the extrapolation of those effects to humans.

Furthermore, there are a number of substances for which microbiological assays do not work very well. For example, this type of test may be inappropriate for important contaminants such as chloroform (74,83). Also, there is the problem of correlating the potency of a substance as a carcinogen and its activity as a mutagen (70). Finally, the validity of the results of microbiological assays is often based on animal studies, which themselves have been questioned.

Thus, it is clear that regardless of the technique used to ascertain the human cancer risk from ingesting small amounts of organic contaminants, there are considerable uncertainties about the conclusions (40,84). In addition, concerns have been expressed in recent years regarding improper testing procedures, poor laboratory practices, and fraudulent data generation in some toxicology laboratories (27,60,81,85,86). Despite all the problems cited, some scientists are optimistic about the future of risk determinations. For example, a recent draft report by the National Institute of Environmental Health Sciences' Task Force III, composed of 85 scientists, suggests that we are "about to enter an era of major developments that will better delineate the hows and whats of chemical risk" (87). (See Chapter 16 for a further discussion of risk estimation techniques.)

18.6   COMPARATIVE RISK EVALUATION

Even if the methodologies for human risk assessment worked well enough to give an accurate picture of the risks involved, there still would be problems when comparing those risks for both governmental and personal decision-making. However, since there are considerable uncertainties in the assessments of human cancer risks, decisions about them are even more difficult. One issue has to do with the selection of what is and what is not an acceptable risk. The other has to do with the perception, understanding, and acceptance of risks.

Examining the first, it is clear that the selection of an acceptable risk, for regulatory or personal purposes, always involves a value judgment. It is hoped, of course, that with respect to governmental decisions, judgments as to which risks should be accepted or should not be accepted will be based on considerable expertise and will be completely unbiased. However, as has been illustrated, the estimation of risks is fraught with uncertainties and is based on so many assumptions that judgments about these risks are not at all clear-cut. Furthermore, to expect that judgments will be made without any personal input, however inadvertent or unconscious, is to expect a great deal. This is observed quite dramatically during the testimony of technical experts concerning, for example, some potential environmental hazard.

Given the same data, one expert takes one side (it is not safe enough) while another expert takes the opposite view (it is safe enough) (18,20,29,34,52, 88,89). How can this happen?

For one thing, studies have indicated that once an opinion is reached, evidence to the contrary is often times given little or no credence (89,90). Thus, once a person has developed an opinion about a potential environmental hazard, it will be difficult to change that opinion. Since we all have opinions and since the uncertainties and assumptions in risk assessment leave large gaps, it is no surprise that there are major disparities in expert judgments. This divergence of opinions, not to mention vested interests, has led to a decreased credibility given to experts by the American public leading to an erosion of the myth of scientific objectivity (18,88,91). As stated by Brooks (88), "Scientists today are listened to much more but believed much less than they were."

One of the problems, of course, is that many people, including politicians, still want simple "yes" or "no" answers concerning safety (88,91). While they may be given information about relative risk estimates and may be told by experts what those estimates might mean, what they really want to be told is whether or not it is safe (6,72,91). Even if everyone was comfortable with the notion that safe means an acceptable level of risk, the selection of what is safe would still be problematic.

Simply stated, people view hazards, and their respective risks, differently and there are several apparent anomalies in those views (1,2,25). What appears to be a very low risk to one person may be viewed as an overwhelming risk to another, a hazard that has a relatively high risk of injury may be viewed with less alarm than a hazard that has a much lower risk of harm, and some people are willing to accept (or even seek) high levels of risk, while others will not.

Some of the disparities in these views can be explained by the differences in how people perceive voluntary hazards as opposed to nonvoluntary hazards (2,92). Generally, if we are allowed to make a choice, i.e., the risk is voluntary, we are willing to accept higher levels of risk [perhaps up to 1000 times greater (92)] than if a hazard is imposed on us. Examples of this include the differences in the perceptions of the risks involved between smoking cigarettes versus breathing the air in a major urban area and consuming alcohol versus drinking tap water that has been treated with chlorine. People will often voluntarily smoke cigarettes and drink alcohol while at the same time expect that the air they breathe and the water they drink be essentially risk-free.

Similarly, there are other factors that affect how hazards are perceived including the differences between: risks from discrete events versus risks from continuous events, a known harm versus an unknown or uncertain harm, dreaded effects versus nondreaded effects, whether or not alternatives exist that can reduce or eliminate the hazards, and what benefits can be attributed to the risks (2,92).

In addition, there is a tendency for people to estimate the risks of highly publicized, unfamiliar, or catastrophic hazards too high and to estimate the risks of everyday or familiar hazards too low—airplane accidents versus auto accidents and the hazards of EDB versus the hazards of poor dietary habits are examples of this. Many people also overestimate risks when the hazards are essentially unknown (living near a well-managed hazardous waste facility) and underestimate risks when the hazards are well-known (living in the flood plain beneath a major dam). Therefore, estimates of risks often grossly conflict with known mortality statistics or other data (2,21,25,89). Also, recent public opinion polls regarding food safety (17) indicate that consumers are more and more likely to trust their own judgments rather than information given them by either governmental or industrial sources (in 1978 39% relied on their own judgments and by 1984 this rose to 48%).

All of these problems are involved when trying to compare the relative risks of cancer from ingesting low concentrations of organic substances with other risks we face daily. Thus, the final step in the assessment of such risks really turns out to be more difficult than one might have expected. It simply is not the case that one merely can look at the probabilistic risk values and assign a ranking among them or can state which risks should be accepted by the public at large and which risks should not.

The preceding discussion about risk assessment points out a number of very serious problems with its use as an accurate tool for decision-making. While this is often admitted (40,48,93–96), the fact that it is really the only method available that begins to examine environmental hazards with any degree of structure or consistency, leaves some people with the idea that it must be employed. This is fine as long as risk assessment is viewed objectively and with a knowledge of its limitations and uncertainties.

However, there is a tendency to forget the inherent uncertainties and ambiguities and to accept calculated risk estimates as fact. This can cause serious problems as noted by Sobel (96), "Beware of giving the 'ball park' estimate . . . as . . . credible estimates are propagated as facts." This reliance on the numbers generated results because of risk assessment's seeming objectivity and strictly analytical basis. The ranking of numbers, or ranges of numbers, which were calculated on the basis of some type of controlled scientific study, appears so neat and clean that it lends credibility to the decision-making process. There is, however, a major misperception in accepting such rankings as true representations of reality. All we really do know is that there is some potential for harm and based on the best information available we have derived an estimate of what the likelihood of that harm will be.

We can look at this in terms of an analogy expanded from that described by Smith (97) concerning a professor at London University contemplating the use of aerial searchlights during the World War II Battle of Britain. The searchlight, sweeping an arc across the sky in order to make London safe from air attack by facilitating the destruction of enemy planes, will work only if:

1. The enemy plane exists.
2. The enemy plane will be within the illuminated arc of the searchlight.
3. Someone knows how to identify an enemy plane.
4. Someone will be able to destroy the enemy plane without adverse impacts on the population of the city.

In the case of the risks of ingesting low levels of organic contaminants, it is clear that some type of harm exists (quite possibly cancer), but it is not clear that our methods of searching for the harm are sufficient nor is it clear that we will be able to tell which factors cause which harms at what levels, and it is unclear which decisions concerning the risks will provide the greatest level of safety for the public.

Finally, there is one other problem with risk assessment related to personal decision-making. When considering a potentially hazardous event, exposure or activity, many people are not very concerned with what the relative risk is estimated to be. Whether the estimate indicates that four people out of 1000 or four people out of 10 million will get cancer is not the concern, rather the concern is—will they be among the four people who will get cancer. Thus, risk assessment analyses do not indicate exactly who will be affected or when, they only estimate a general cancer incidence.

18.7 THE RISKS OF INGESTING ORGANIC CONTAMINANTS

As discussed, the first step in estimating the risks associated with the ingestion of organic substances in minute concentrations is to identify the hazards. These potential hazards come as a result of the consumption of foods, beverages, medicines or drugs, and drinking water. Essentially, the sources of these hazards could be from naturally occurring substances, from human-derived contaminants, and from substances formed in the consumed products prior to or during ingestion. A major problem here, however, is that there are so many different substances in the environment that could pose some type of hazard to humans.

For example, estimates of the number of commonly used or marketed chemicals range from about 63,000 to 100,000, with an estimated 200–4000 new chemicals added to the marketplace each year (21,98–102). Of these chemicals, approximately 900 have been shown to exhibit teratogenic or other reproductive effects, 3200 have been listed as mutagenic and 3100 as tumorigenic (102). These numbers seem startlingly large, however the total number of known chemicals is even more impressive and intimidating. As of the end of 1977, over four million different substances were listed in the American Chemical Society's (ACS) registry of chemicals, with an average weekly increase of new entries running about 6000 (99). Obviously, to examine all known chemicals, or even those in common use, for potential hazards would be a gargantuan task. Fortunately, the overwhelming majority of the chemicals listed or added to the ACS registry occur in nature or are used in research, rather than in commerce or consumer products (99). Nevertheless, human-synthesized chemicals are not the only ones of interest when comparing environmental risks from ingestion.

In looking at the potential hazards in life, it is clear that naturally occurring substances are not necessarily safe (71,103–108). It is well known, for example, that certain chemicals are released from some plants to act as natural herbicides against other plants (104). Such allelopathic agents, as well as natural insecticides like the rotenoids and pyrethroids, clearly pose a natural hazard to other plants and insects (104,105).

Natural substances not only can be harmful to plants and insects, but also can be hazardous to animals, including humans. For example, consuming an excessive amount of apples or apple cider can be toxic because of the malonic acid in apples, overconsumption of almonds or apricot kernels has proved fatal for some children, and eating too much rhubarb can be toxic (106). Fortunately, the amounts of toxic materials found in these and other plants that contain natural poisons are typically very small.

However, there are a number of naturally occurring substances which are known or likely to be carcinogenic, mutagenic, teratogenic or to produce sublethal health effects in humans (103,107–113). To illustrate the ubiquity of these materials, Table 18.1 lists some naturally occurring substances and their known or suspected hazards.

In addition, there are a wide variety of hazardous fungal metabolites, including the aflatoxins (known to be toxic and carcinogenic) associated with peanuts, maize, and cottonseeds and the ochratoxins (known to be toxic and possibly carcinogenic and teratogenic) found in legumes and cereals (108,112,113). Animal and fish products also contain natural substances that are hazardous. For example, animals that eat plants containing pyrrolizidine alkaloids or other natural toxins or are given cottonseed meal as a protein supplement may well pass on to human consumers potentially hazardous material in their meat and milk products (108,113,114). Many species of puffer fish, considered a delicacy in Japan, contain tetrodotoxin, a very strong neurotoxic agent which causes "convulsions and death in humans in 1.5–8 hours" (108).

Just as some potentially hazardous substances are found naturally in foods, so too do some occur naturally in bodies of water. This should be expected, given the vast numbers of chemicals found in small quantities in raw waters (115), many of which are a result of natural processes. Natural waters in Nevada, for example, often have fluoride levels well above the 1–2 mg/L level recommended daily dose for humans (116). Similarly, nitrosamines, carcinogenic substances formed when nitrites combine with secondary or tertiary amines, may be formed in clean waters or soils during natural decomposition processes (108,117). Also, the ubiquitous polycyclic aromatic hydrocarbons, a group of over 100 substances of which more than 20 are known carcinogens (113), can be synthesized by algae, and higher plants and

TABLE 18.1
Natural Substances in Foods, Sources, and Potential Hazards

Substances	Likely Sources	Potential Hazards[a]
Alkaloids	Tomato/potato vines and foliage	Digestive problems, nervous disorders
Allyl isothiocyanate	Mustard, horseradish	C, M
Anagyrine	Legumes, milk from lupine eating animals	C, M, T
Canavine	Alfalfa sprouts	Toxic, M
Caratoxin	Carrots, celery	Neurotoxin
Convicine, vicine	Broad/fava beans	Hemolytic anemia
Cyanogenetic glycosides	Fruit kernels, lima beans, cassava root	Toxic, chronic cyanide poisoning
Flavanoids	Most fruits and vegetables	M
Furocoumarins	Celery, parsnips, figs, parsley	C, M
Glycoalkaloids	Herbs, herb teas	Toxic, C, M, T
Goitrogens	Cabbage, brussel sprouts, kale, onions, mustard, turnips	Hypothyroidism, goiters
Gossypol	Cottonseed	C, M
Hemagglutinins	Sweet peas and beans	Agglutination of red blood cells
Hydrazines	Mushrooms	C, M
Lathyrogens	Chick/sweet peas	Neurological disease
Malvic and sterculic acids	Plant oils, okra, products from animals fed cottonseed	C
Oxalates	Variety of garden vegetables, teas, cocoa, rhubarb	Toxic, kidney damage
Phorbol esters	Herb teas	Cancer promoters
Phytoalexins	Damaged or fungal associated peas, beans, and potatoes	Lung disease
Quinones and precursor phenols	Rhubarb, coffee, many common plants	C, M
Safrole and related substances	Sassafras, black pepper, oils of nutmeg and cinnamon	C, M
Sequiterpene lactones	Lettuce	M
Solanine	Potatoes	Neurotoxin
Theobromine	Tea, cocoa powder	Promotes DNA damage
	Bracken fern	Gastric cancer

Source: Refs. 82, 103, 107–112.
[a] C, carcinogenicity; M, mutagenicity; T, teratogenicity.

organisms (113) or may enter waters as a result of fallout from forest or prairie fires (108).

Thus, it is not necessarily the case that sources of potentially hazardous substances must come directly or indirectly from human activities. Nevertheless, anthropogenic chemicals appear to be the ones that generate the most interest with regard to possible environmental risks. Whether or not this emphasis on human-added substances is warranted provokes a heated debate (103,118–122).

For example, when it was discovered that the nitrites used to cure meats and fish products might form nitrosamines before or during consumption, these food preservatives were subjected to considerable regulatory scrutiny and public concern (24,118). Yet, nitrates, which convert to nitrites in the digestive system, are found in significant enough quantities in drinking waters and vegetables that they may account for 5–70 times more nitrites being ingested than from an average consumption of cured foods (118).

As suggested by Stich et al. (82), "the intake of mutagens and by implication carcinogens through regular diet may exceed by far the amount derived from man-made sources such as industrial pollution, food additives, and pesticides." This idea has been at the heart of the debate over which environmental factors are responsible for the highest human cancer risks. Estimates have been made that from 60 to 90% of all human cancers can be attributed to various environmental factors (8,24,40,98,119,120). For some people, the term environmental factors is "synonymous with chemicals, specifically man-made chemicals . . . (thereby) falsely implicating man-made chemicals as the principle cause of human cancers . . . (or mistakenly implying) that workplace or industry activities are responsible for 60%+ cancers in America" (119).

However, when concerns over the role of environmental factors in causing cancer first became prominent, back in the 1960s, the intent was not to indict "man-made chemicals" (43,120), rather it was to elucidate those cancers that, theoretically, could be avoided or prevented. This is indicated by the following quote from the 1964 World Health Organization report, "Prevention of Cancer" (121) (emphasis added):

> The potential scope of cancer prevention is limited by the proportion of human cancers in which *extrinsic factors* are responsible. These include all *environmental carcinogens*. . . . The categories of cancer that are thus influenced, directly of indirectly, by *extrinsic factors* . . . account for more than *three-quarters of human cancers*. It would seem, therefore, that the majority of human cancer is potentially *preventable*.

Unfortunately, the terms "extrinsic factors" (more commonly expressed as "environmental factors") and "environmental carcinogens" often have been misrepresented to or misinterpreted by the public. Those "potentially preventable" cancers to which the report referred are primarily attributed to differences in diets, life styles, and personal habits (particularly smoking) (24,43,119). For example, it is well documented that smoking, a voluntary action, is a major cause of lung cancers, as well as being associated with other cancers. Lung cancer is the most prevalent cause of cancer deaths in the United States [in 1976 age-adjusted mortality rates for lung cancer were 66.7/100,000 for white males and 17.8/100,000 for white females (122) and in 1982 an estimated 90,000 people died as a result of cigarette-related lung cancer (25)]. This one voluntary action is the single most important environmental factor related to human cancer risks (24,25,40,43).

Because of the preponderence of smoking-related cancer mortalities, as well as those associated with the consumption of alcohol and related to particular dietary habits, it has been estimated that only about 5% of all cancer deaths in the United States should be attributed to inadvertant and occupational exposure to human-produced chemicals (40,43). This estimate also has been the source of controversy because it can, according to Epstein and Swartz (122), "denigrate the role of occupational and environmental carcinogens and the need for effective regulation."

Furthermore, it has been suggested that the 5% estimate neglects "the role of multiple causal agents, such as asbestos and smoking," does not account for the fact that "current cancer rates reflect exposures 20–30 years ago when production levels of occupational carcinogens were a small fraction of the present," and does not take into consideration the "very limited nature of the data base on exposure to occupational carcinogens" (122).

It is clear from the preceding discussion that the role of environmental factors with respect to cancer incidence is a complex and provocative issue. Nevertheless, there is, and will likely continue to be, grave concern about the cancer hazards of human-derived chemicals. Much of this concern is focused on the substances added, inadvertantly or on purpose, to foods and beverages or formed during food and beverage processing, including drinking water treatment.

The types of cancer hazards encountered in foods from human-derived substances range from the obvious (food additives and pesticides) to the arcane [nitrosamines released into baby formula from contaminated rubber nipples (123) and polychlorinated biphenyls (PCBs) migrating into breakfast cereal from packaging made of recycled paper (13)]. As indicated by Table 18.2, recent surveys (17) show that the public seems more concerned about food hazards from human-produced pesticides and herbicides than from the cholesterol, salt, or sugar content of the food they eat.

Given that there are thousands of pesticide formulations based on about 600 active ingredients currently in use, that halogenated organic pesticides tend to be extremely persistent in the environment, and that a number of pesticides have been shown to be animal carcinogens, it is no wonder that there is such public interest (17,32,113). One pesticide that has been the focus of considerable media and regulatory attention in recent years is EDB, a soil and food fumigant frequently found in foods and waters (17,37,124,125). Widely varying estimates for EDB exposure have been

TABLE 18.2
Food Concerns of Consumers[a]

Item in Food	Serious Hazard[a]	Moderate Hazard[a]	No Hazard[a]
Pesticide and herbicide residues	77	18	2
Cholesterol	45	48	5
Salt	37	53	9
Sugar	31	53	15

Source: Ref. 17.

[a] The original questions asked of consumers were: "How concerned are you about the following item being in food? Would you say that (each item was read) is/are a serious health hazard, somewhat of a hazard, or not a hazard at all?"

[b] Percent of consumers responding to the questions.

made by the Occupational Safety and Health Administration (OSHA), the EPA, and the Grocery Manufacturers of America (GMA) (37,124,125) and are shown in Table 18.3.

It has been suggested that these estimates be interpreted "cautiously" because of the different assumptions made and predictive models used (124). Furthermore, when discussing the risks of EDB in the environment, it should be noted that of the estimated 170,000–180,000 tons produced in the United States in 1981, only about 1000 tons were used for pesticide pur-

TABLE 18.3
Estimated Cancer Risks from EDB Exposure

Exposure	Cancer Risk Estimates (deaths/100,000 people)
OSHA	
Occupational (20 ppm)	7000–11,000
(0.1 ppm)	20–60
EPA	
Drinking water (1 ppb)	200
Wheat products	10
Citrus fruits	
(in states requiring fumigation)	20
(in states not requiring fumigation)	2
Lifetime exposure to residues	30–300
Crops grown in EDB fumigated soils	1
GMA	
Grain products (children)	0.025
Grain products (adults)	0.0083

Source: Refs. 37, 124, 125.

poses; most of the rest (about 115,000 tons) was used as an additive in leaded gasolines (124).

Another interesting human-derived cancer hazard comes from the ingestion of polycyclic aromatic hydrocarbons (PAHs). PAHs were first traced to human cancer back in 1775 when scrotal cancer was found to be prevalent among chimney sweeps in constant contact with soot. Years later, in the 1800s, high skin cancer rates were identified in tar workers. Following this discovery, confirmation of the carcinogenicity of tar via animal experiments was obtained. Finally, in the 1930s one carcinogenic component of coal tar was identified as benzo[a]pyrene (B[a]P) (117).

Of the many PAHs, B[a]P is probably one of the most potent carcinogens and the most widespread in the environment, having been observed in clean and polluted waters, air, food, as well as being a major carcinogenic component of tobacco smoke (108,113,117). Because of their ubiquity, PAHs are found in a wide variety of raw and unprocessed foods. And, since PAHs are formed during incomplete combustion and pyrolysis, they are common inadvertant additions to foods as a result of processing and preparation. Smoking, grilling, broiling, or caramelizing foods all produce significant amounts of PAHs and, for all practical purposes, represent unavoidable contamination (108,109,117,126).

Other potentially hazardous contaminants in foods and beverages include nitrosamines, chemical additives (sweeteners, flavoring or enhancing agents, bleaching agents, food colors, antioxidants, and antimicrobial agents), vinyl chloride, and diethylstilbestrol (DES) (17,24,27,30,31,33,35, 41,79,109,113,117,126).

It is clear from the preceding discussion that there is an abundance of known and potentially hazardous substances in foods and beverages as a result of human activity. Even narrowing the focus down to just one beverage, drinking water, does little to dispel the burdens of an overwhelming number of possible hazards and a wealth of scientific uncertainties.

The human-derived hazards in drinking water result from industrial, domestic, and agricultural wastes, atmospheric fallout of pollutants, soil leaching, road and land runoff, motorboat engine exhausts, and spilled or leaked chemicals. But perhaps the source of drinking water cancer risks that has been receiving the most attention in the past 10 years is water treatment, particularly the disinfection of water by chlorination (10–12,14,16,36,57,63, 127–138).

A large amount of toxicological data has been generated concerning chlorinated drinking waters using animal experiments, epidemiological surveys, and short-term microbiological assays (16,36,57,61–63,128–149). (See Part 4 for more detailed information on this topic.) While toxicological information has been collected on treated drinking waters and individual substances in drinking waters, particular attention has been given to the trihalomethanes (TMHs), especially chloroform. This is evidenced by the decision of the EPA to mandate a maximum contaminant level (MCL) of 100 ppb for total THM levels in finished drinking waters (14).

The reasons for the regulatory emphasis on THMs were severalfold (14,59,115):

1. More toxicological data existed for THMs (especially chloroform) than for other organic materials in drinking waters.
2. THMs had been found in all treated drinking waters, frequently in concentrations higher than other organic substances.
3. Analytical techniques for the rapid and accurate determination of minute amounts of THMs recently had been developed and were becoming readily available.
4. THMs could be used as "indicator substances" since they were likely symptomatic of the presence of other organic compounds formed during the chlorination process.
5. It was believed that THMs (particularly chloroform) imposed a significant cancer risk on the public.
6. There were several available techniques to reduce THM levels prior to distributing the water to consumers.

Although the evidence from the various studies on chloroform (and other THMs) indicates that it does pose a cancer risk, there is considerable controversy as to the magnitude of that risk. There also are discrepencies concerning which organs are the most likely targets for chloroform (or THM)-induced cancer from the ingestion of drinking water. Initial risk estimates using various mathematical models were made by Tardiff (59) showing that chloroform could impose from "no risk or negligible risk of toxicity" to a risk of 0.084 cancers/100,000 people exposed each year, depending on the species of animal tested, location of the cancer observed and mathematical model used. These estimates are shown in Table 18.4.

Other cancer risk estimates associated with chloroform (and total THMs) have been made (36,136,147–149). They, like the estimates in Table 18.4, have shown considerable divergence. For example, some experts have suggested that the EPA's cancer risk estimate of 1×10^{-4} to 1×10^{-5}/ lifetime for chloroform is too low, while others maintain that it is too high. The upper and lower bounds of these different opinions result in cancer risk estimates ranging from about 1×10^{-2} to 1×10^{-7}/lifetime. Such a great variation, while striking in its magnitude, should not be unexpected since, as has been discussed, there are many assumptions and uncertainties inherent in risk estimations.

The maximum dose used for the risk estimates shown in Table 18.4 (with the exception of the two margin of safety models) is equivalent to a 155-lb person consuming each day about 7.5 quarts of drinking water containing 100 ppb of chloroform. This amounts to a total daily dose of 0.70 mg of chloroform. It is interesting to note here that prior to the 1976 Food and

TABLE 18.4
Cancer Risk Estimates from the Ingestion of Chloroform

Model	Basis (cancer)	Annual Cancer Risk per 100,000 People[a]	Total Annual Excess Cancers[b]
Margin of Safety			
(10×)	Human: liver injury	0-negligible	
(5000×)	Rat (kidney)	0-negligible	
Probit-log	Rat (kidney)	Less than one per	Less than 2.3
(slope = 1)	Mouse (liver)	100 million	Less than 2.3
Probit-log	Rat (kidney)	0.0016–0.0040	3.68–9.2
(actual slope)	Mouse (liver)	0.0016–0.0683	3.68–157.1
Two-step	Rat (kidney)	0.0267	61.4
	Mouse (liver)	0.0283	65.1
Linear	Rat (kidney)	0.042	96.6
(one-hit)	Mouse (liver)	0.084	193.3

Source: Ref. 59.
[a] Maximum estimated cancer risks are based on a maximum dose of 0.01 mg of chloroform per kg of body weight per day, except for the two margin of safety estimates which are 0.02 mg/kg/day for 10× and 0.03 mg/kg/day for 5000×.
[b] Estimated excess cancers per year resulting from the ingestion of chloroform at the doses indicated are based on a U.S. population of 230,000,000.

Drug Administration (FDA) regulations restricting its use, chloroform was routinely used in a wide variety of common consumer products including drugs, cough drops and cough syrups, toothpaste, mouthwash, and cosmetics. In fact, during that time a daily dose of a typical cough medicine contained about 160 mg of chloroform (146). This is approximately 230 times more chloroform than the maximum dose for cancer risk estimation as specified in Table 18.4.

Clearly, for the majority of adults in the United States, the total amount of ingested chloroform has been reduced markedly since the mid-1970s. It follows then that the current risk from chloroform is much lower than it was 10 years ago. Nevertheless, chloroform is only one of many substances in drinking waters and foods that might pose a hazard when ingested. A comparison of the estimated cancer risks associated with the ingestion of chloroform and other organic substances is shown in Table 18.5 (36,63).

While the substances in Table 18.5 are listed according to their estimated cancer risks, based on 1 μg of substance per liter of water, this does not necessarily mean that the human cancer risk from ingesting those substances can be ranked accordingly. For one thing, the concentrations to which people are exposed through drinking water ingestion must be taken into account, as is shown in the column labeled Consumption B. Also, it must be kept in mind that for several of the materials listed the primary exposure route is *not* via the consumption of drinking water. EDB is a good example

TABLE 18.5
Cancer Risk Estimates for the Ingestion of Organic Contaminants in Water

Substance	Lifetime Cancer Risk[a]	Cancers/100,000 People[b]	
		Consumption A	Consumption B
Dieldrin	2.6×10^{-4}	52.0	416.0 (8)
Kepone	4.4×10^{-5}	8.8	
Heptachlor	4.2×10^{-5}	8.4	
Hexachlorobenzene	2.9×10^{-5}	5.8	
Chlorodane	1.8×10^{-5}	3.6	0.36 (0.1)
DDT	1.2×10^{-5}	2.4	
Lindane	9.3×10^{-6}	1.86	0.0186 (0.01)
EDB	9.1×10^{-6}	1.82	
β-BHC	4.2×10^{-6}	0.84	
PCB	3.1×10^{-6}	0.62	1.86 (3.0)
ETU	2.2×10^{-6}	0.44	
Chloroform	1.7×10^{-6}	0.34	124.0 (366)
γ-BHC	1.5×10^{-6}	0.30	
Acrylonitrile	1.3×10^{-6}	0.26	
Bis(2-chloroethyl)ether	1.2×10^{-6}	0.24	0.1 (0.42)
1,2-Dichloroethane	7.0×10^{-7}	0.14	
Vinyl chloride	4.7×10^{-7}	0.094	0.94 (0.1)
PCNB	1.4×10^{-7}	0.028	
Tetrachloroethylene	1.4×10^{-7}	0.028	
Carbontetrachloride	1.1×10^{-7}	0.022	0.11 (5)
Trichloroethylene	1.1×10^{-7}	0.022	0.011 (0.5)

Source: Refs. 36 and 63.
[a] Upper 95% confidence estimate of lifetime cancer risks per ppb (microgram per liter) of substance ingested.
[b] *Consumption A:* Excess cancers per 100,000 people based on a daily consumption of 2 L of water containing one ppb of substance (total dose of substance is 2 μg per day). *Consumption B:* Excess cancers per 100,000 people based on a daily consumption of 2 L of water containing the highest concentration of substance (shown in micrograms per liter in parentheses) observed in treated drinking waters, where data are available.

of a case where this is true. Table 18.5 shows only the estimated risk resulting from consumption of drinking water containing EDB whereas the average residues found in many foods may be high enough to result in a significantly greater overall exposure than for drinking water (124).

The estimates concerning EDB provide another example, similar to that of chloroform, of how calculated cancer risks vary greatly from source to source. Based on 1 ppb of EDB in drinking water, the EPA estimated (124) a lifetime cancer risk of 2×10^{-3}, whereas the National Research Council (63) estimated that risk at 1.82×10^{-5} (based on an average consumption of 2 L of water per day). Thus, as has been illustrated with both chloroform and

EDB, comparing cancer risk estimates oft times becomes rather tricky since there may be several different estimates for a single substance, at a given dose level.

It becomes even more tricky if cancer risk estimates are compared with those of other hazardous exposures, events or activities. Nevertheless, this type of comparison has become popular in recent years and has been carried out in a number of ways (6,46,47,72,136). For example, items have been compared according to expected deaths per population, probability of occurrence, events that increase the chance of death by a specified value, and anticipated life expectancy losses. Regardless of how such comparisons are set up, the object is to be able to examine a wide variety of hazards within a common framework. Table 18.6 is a composite of several such indices listing various activities or exposures and their respective estimated risks of death.

As expected, occupational hazards, as a group, tend to have the highest risks. While there is considerable variation in risks among the different occupations listed, with the exception of professional boxing and coal mining, all risks are within a factor of 10. The second group of hazards, those associated with leisure-time activities, generally have lower levels of risk than occupational hazards. Again, the variation in risks of death, on a per hour basis, is about a factor of 10. Natural hazards show much lower risks of death than those associated with either occupations or leisure-time activities and exhibit even less variation.

But, perhaps the most interesting hazards are the ones listed under home and habits. Here there is a much greater variation among the levels of risk than observed in any of the three previous groups, with the highest risks associated with the voluntary actions of smoking cigarettes and drinking alcohol. Accidents related to necessary (involuntary) day-to-day activities, such as driving or riding in a motor vehicle, climbing stairs and inadvertant falling, also show relatively high risks.

Finally, for the other factors listed under home and habits, the risks tend to be quite low in comparison to all but the risks related to natural events. For example, the risk of death from chloroform listed here is about 100 times less than moderate wine drinking, almost 2000 times less than smoking 10 cigarettes each day, and only twice the risk of death from lightning. Clearly, from these data drinking treated tap water does appear to be a relatively low-risk activity.

However, if higher risk estimates for chloroform were used in this index (say, 10–100 times the level indicated), the risk of chloroform ingestion would still be relatively low in comparison to occupational hazards, leisure-time activities, smoking, and car accidents, but would be significantly higher than most of the other hazards listed. Herein lies the problem of using such indices for decision-making purposes: The numerical risk estimates, while easy to compare and rank according to some criteria, may well be inaccurate. And, unless the risk estimates are accurate, the use of comparative or relative risks indices must be viewed with caution.

TABLE 18.6
Risk Estimates for Various Activities or Exposures

Activity or Exposure	Risk of Death/ 100,000 People/yr
Occupations	
Professional boxing	728
Coal mining	130
Coal mining (from accidents only)	60
Firefighting	80
Construction	60
Farming	60
Police duty	20
Truck driving	10
Manufacturing	8
Leisure-time activities (40 h/year)	
Canoeing	40
Motorcycle riding	25
Swimming	12
Mountain climbing	10.8
Hunting	3.8
Skiing	3.0
Natural events	
Floods	0.06
Tornadoes	0.06
Lightning	0.05
Animal bite or sting	0.02
Home and habits	
Drinking:	
1 L of wine/week (death from cirrhosis)	10.4
1 can of diet soda/day (saccharin)	0.92
1 pint of milk/day (aflatoxin)	0.20
2 L of treated water/day (chloroform)	0.10
Eating:	
1 tablespoon of peanut butter/day (aflatoxin)	0.92
100 charcoal-broiled steaks (B[a]P)	0.10
Living:	
Smoking 10 cigarettes/day	182
Motor vehicle accidents	20
Falls	6
Fires	3
Climbing stairs (40h/year)	2.2
Living with a person who smokes	0.6
Living in a stone or brick building (radiation)	0.6

Source: Refs. 6, 46, 72, 136.

18.8 SUMMARY

For a variety of reasons, there has been considerable recent attention focused on the nature of risks from environmental hazards, particularly those from ingesting minute amounts of contaminants. While this effort has resulted in much better techniques for the detection of both the contaminants and their potential effects, there still are serious problems when trying to estimate the risks of exposure from these low concentrations of materials. Frequently, data are sparse, controversial, or conflicting, leading to major disagreements about their interpretation. In addition to the problems with the data, there are enough uncertainties in the assumptions and disparities in the techniques employed by different people to assure that attempts to calculate and compare risks will be tenuous.

Even so, there tends to be an overwhelming desire to at least have some scientific basis for decision-making concerning environmental hazards. The use of relative risk comparisons, or risk compendia, is one way to accomplish this desire. Nevertheless, while these comparisons of risk estimates can be informative, thought provoking, and even entertaining, they fall far short of indexing the true risks of life.

At best, risk assessment for decision-making can be viewed as an imprecise tool to examine and better understand the myriad of hazards we face each day, especially when we approach the problem by examining comparative or relative risks. However, because risk assessment is a composite of scientific techniques and mathematical models that rely on professional judgments and individual or public policy decisions, it will not tell us the actual risks we are subjected to nor will it result in information which lends itself well to clear-cut decisions.

REFERENCES

1. Salem, S.L., Solomon, K.A., and Yesley, M., Issues and Problems in Inferring a Level of Acceptable Risk, U.S. Department of Energy, R-2561-DOE, August 1980.
2. Lowrence, W.W., *Of Acceptable Risk,* William Kaufmann, Inc., Los Altos, California, 1976.
3. Neely, W.B., *Chemicals in the Environment,* Marcel Dekker, Inc., New York, 1977.
4. Teich, A.H. (ed.), *Technology and Man's Future,* 2nd ed., St. Martin's Press, New York, 1977.
5. Coates, J.F., *The Futurist, 5,* 225, 1971.
6. Wilson, R., *Technol. Rev., 82* (2), 41, 1979.
7. Regenstein, L., *America the Poisoned,* Acropolis Books, Ltd., Washington, DC, 1983.
8. Agran, L., *The Cancer Connection,* Houghton Mifflin Company, Boston, Massachusetts, 1977.
9. Kessler, D.A., *Science, 223,* 1034, 1984.
10. Harris, R.H., and Brecher, E.M., *Consumer Reports, 39,* 436, 1974.
11. Harris, R.H., and Brecher, E.M., *Consumer Reports, 39,* 538, 1974.
12. *Science, 186,* 809, 1974.
13. Fenner, L., *FDA Consumer, 18* (8), 1984.

14. Baum, B., Drinking water chlorination and the regulation of organics, in *The Harvard Environmental Law Review,* Vol. 3., 1979, pp. 399–413.
15. Epstein, S.S., *The Politics of Cancer,* Anchor/Doubleday, New York, 1979.
16. Glatz, B.A., Chriswell, C.D., Arguello, M.D., Svec, H.J., Fritz, J.S., Grimm, S.M., and Thompson, M.A., *Am. Water Works. Assoc., 70,* 465, 1978.
17. Lecos, C., *FDA Consumer, 18* (6), 12, 1984.
18. Brooks, H., *Optics News,* Spring, 1976.
19. Morison, R.S., Science and social attitudes, in Teich, A.H. (ed.), *Technology and Man's Future,* 2nd ed., St. Martin's Press, New York, 1977, pp. 35–51.
20. Marshall, E., *Science, 205,* 281, 1979.
21. Hohenemser, C., Kates, R.W., and Slovic, P., *Science, 220,* 378, 1983.
22. Carson, R., *Silent Spring,* Houghton Mifflin Company, New York, 1962.
23. Cartwright, F.F., and Biddles, M.D., *Disease and History,* Thomas Y. Crowell Company, New York, 1965.
24. *NRDC Newsletter, 5* (2), 1, 1976.
25. *Consumer's Research, 67* (4), 11, 1984.
26. Lecos, C., *Consumers' Research, 67* (6), 35, 1984.
27. *Chemical Week, 135* (4), 10, 1984.
28. *Chemical Week, 135* (5), 13, 1984.
29. Ashford, N.A., Ryan, C.W., and Caldart, C.C., *Science, 222,* 894, 1983.
30. Sanders, H.J., *Chem. Eng. News, 49,* 16, 1971.
31. Smith, R.J., *Science, 201,* 887, 1978.
32. Thompson, R.C., *FDA Consumer, 18* (6), 6, 1984.
33. Ando, M., and Sayato, Y., *Water Res., 18* (3), 315, 1984.
34. Marshall, E., *Science, 222,* 906, 1983.
35. Lecos, C., *FDA Consumer, 18* (3), 23, 1984.
36. National Academy of Sciences, *Drinking Water and Health,* Vol. 1, Washington, DC, 1977.
37. Sun, M., *Science, 223,* 464, 1984.
38. Miller, S., *Environ. Sci. Technol., 17* (5), 199, 1983.
39. Long, J.R., and Hanson, D.J., *Chem. Eng. News, 61* (23), 23, 1983.
40. Gori, G.B., *Chem. Eng. News, 60* (36), 25, 1982.
41. Bowman, M.C., *Carcinogens and Related Substances,* Marcel Dekker, Inc., New York, 1979.
42. Cairns, J., *Cancer: Science and Society,* W. H. Freeman and Company, San Francisco, California, 1978.
43. Doll, R., and Peto, R., *The Causes of Cancer,* Oxford University Press, New York, 1981.
44. Lilenfield, A.M., Levin, M.L., and Kessler, I.I., *Cancer in the United States,* Harvard University Press, Cambridge, Massachusetts, 1972.
45. Delaney Amendment, Public Law 85–929, 72 Stat. 1784, 1958.
46. Whyte, A.V., and Burton, I. (eds.), *Environmental Risk Assessment,* John Wiley & Sons, New York, 1980.
47. Fischoff, B., Lichentstein, S., Slovic, P., Derby, S.L., and Keeney, R.L., *Acceptable Risk,* Cambridge University Press, New York, 1981.
48. Nicholson, W.J. (ed.), *Management of Assessed Risk for Carcinogens, Ann. N. Y. Acad. Sci., 363,* New York, 1981.
49. Thompson, P.B., *J. Environ. Syst., 13* (2), 137, 1983–84.
50. Strauch, R., Risk Assessment as a Subjective Process, The Rand Paper Series # P-6460, March 1980.
51. Maki, A.W., Design and conduct of hazard evaluation programs for the aquatic environment, in Jolley, R.L., Brungs, W.A., and Cumming, R.B., (eds.), *Water Chlorination: Environmental Impact and Health Effects,* Vol. 3, Ann Arbor Science, Ann Arbor, Michigan, 1980, pp. 949–960.

52. Cairns, J., Jr., Regulating hazardous chemicals in aquatic environments, *Boston College Environmental Affairs Law Review*, Vol. 11, Number 1, October 1983, 1–10.
53. The Regulatory Council, *Regulation of Chemical Carcinogens*, September 28, 1979.
54. Rall, D.P., Issues in the determination of acceptable risk, in Nicholson, W.J. (ed.), *Management of Assessed Risk for Carcinogens, Ann. N.Y. Acad. Sci., 363*, New York, 1981.
55. Maugh, T.H., II, *Science, 201*, 1200, 1978.
56. *Chem. Eng. News, 56* (31), 20, 1978.
57. Stokinger, H.E., *J. Am. Water Works Assoc., 69* (7), 399, 1977.
58. Maugh, T.H., II, *Science, 202*, 37, 1978.
59. Tardiff, R.G., *J. Am. Water Works Assoc., 69* (12), 658, 1977.
60. Smith, A.M., *New Engineer*, April 25, 1977.
61. Bull, R.J., Robinson, M., Meier, J.R., and Stober, J., Use of biological assay systems to assess the relative carcinogenic hazards of disinfection by-products, *Environ. Health Perspect., 46*, 215–227, 1982.
62. Cantor, K.P., Epidemiological evidence of carcinogenicity of chlorinated organics in drinking water, *Environ. Health Perspect., 46*, 187–195, 1982.
63. National Academy of Sciences, *Drinking Water and Health*, Vol. 3, Washington, DC, 1980.
64. Marx, J.L., *Science, 201*, 515, 1978.
65. Loper, J.C., Overview of the use of short-term biological tests in the assessment of the health effects of water chlorination, in Jolley, R.L., Brungs, W.A., and Cumming, R.B. (eds.), *Water Chlorination: Environmental Impacts and Health Effects*, Vol. 3, Ann Arbor Science, Ann Arbor, Michigan, 1980, pp. 937–948.
66. Kraybill, H.F., Animal models and systems for risk evaluation of low-level carcinogenic contaminants in water, in Jolley, R.L., Brungs, W.A., and Cumming, R.B. (eds.), *Water Chlorination: Environmental Impacts and Health Effects*, Vol. 3, Ann Arbor Science, Ann Arbor, Michigan, 1980, pp. 973–982.
67. Pike, M.C., Epidemiological methods for determining human cancer risks from exposure to chlorinated by-products, in Jolley, R.L., Brungs, W.A., and Cumming, R.B. (eds.), *Water Chlorination: Environmental Impacts and Health Effects*, Vol. 3, Ann Arbor Science, Ann Arbor, Michigan, 1980, pp. 1019–1028.
68. Cumming, R.C., Mutagenicity and water chlorination: Prospect and perspective, in Jolley, R.L., Gorchev, H., and Hamilton, D.H., Jr. (eds.), *Water Chlorination: Environmental Impact and Health Effects*, Vol. 2, Ann Arbor Science, Ann Arbor, Michigan, 1978, pp. 411–416.
69. Cantor, K.P., and McCabe, L.J., The epidemiological approach to the evaluation of organics in drinking water, in Jolley, R.L., Gorchev, H., and Hamilton, D.H., Jr. (eds.), *Water Chlorination: Environmental Impact and Health Effects*, Vol. 2, Ann Arbor Science, Ann Arbor, Michigan, 1978, pp. 379–394.
70. Bartsch, H., Tomatis, L., and Malaveille, C., Qualitative and quantitative comparisons between mutagenic and carcinogenic activities of chemicals, in Heddle, J.H. (ed.), *Mutagenicity: New Horizons in Genetic Toxicology*, Academic Press, New York, 1982, pp. 36–72.
71. Epstein, S.S., *Nature, 228*, 816, 1970.
72. Doull, J., Food safety and toxicology, in Roberts, H.R. (ed.), *Food Safety*, John Wiley & Sons, New York, 1981, pp. 295–316.
73. Reitz, R.H., Quast, J.F., Stott, W.T., Wantanabe, P.G., and Gehring, P.J., Pharmacokinetics and macromolecular effects of chloroform in rats and mice: Implications for carcinogenic risk estimation, in Jolley, R.L., Brungs, W.A., and Cumming, R.B. (eds.), *Water Chlorination: Environmental Impacts and Health Effects*, Vol. 3, Ann Arbor Science, Ann Arbor, Michigan, 1980, pp. 983–994.
74. Reitz, R.H., Fox, T.R., and Quast, J.F., Mechanistic considerations for carcinogenic risk estimation: Chloroform, *Environ. Health Perspect., 46*, 163–168, 1982.

75. Sontag, J.M., Introduction, in Sontag, J.M. (ed.), *Carcinogens in Industry and the Environment*, Marcel Dekker, Inc., New York, 1981, pp. 1–6.

76. Schneiderman, M.A., Regulation of carcinogens in an imprecise world, in Nicholson, W.J. (ed.), *Management of Assessed Risk for Carcinogens*, The New York Academy of Sciences, New York, 1981, pp. 217–232.

77. Alavanja, M., Goldstein, I., and Susser, M., A case control study of gastrointestinal and urinary tract cancer mortality and drinking water chlorination, in Jolley, R.L., Gorchev, H., and Hamilton, D.H., Jr. (eds.), *Water Chlorination: Environmental Impact and Health Effects*, Vol. 2, Ann Arbor Science, Ann Arbor, Michigan, 1978, pp. 395–410.

78. Albertini, R.J., Sylwester, D.L., and Allen, E., The 6-thioguanine-resistant peripheral blood lymphocyte assay for direct mutagenicity testing in humans, in Heddle J.H. (ed.), *Mutagenicity: New Horizons in Genetic Toxicology*, Academic Press, New York, 1982, pp. 305–335.

79. Lotlikar, P.D., Nitrogeneous carcinogenic industrial chemicals, in Sontag, J.M. (ed.), *Carcinogens in Industry and the Environment*, Marcel Dekker, Inc., New York, 1981, pp. 345–438.

80. Ames, B., McCann, J., and Yamasaki, E., Methods for detecting carcinogens and mutagens with the *Salmonella*/mammalian-microsome mutagenicity test, in Kilbey, B.J., Legator, M., Nichols, W., and Ramel, C. (eds.), *Handbook of Mutagenicity Test Procedures*, Elsevier, Amsterdam, 1979, pp. 1–17.

81. Brown, A.S., *Chemical Business*, 5 (6), 1983.

82. Stich, H.F., Rosen, M.P., Wu, C.H., and Powrie, W.D., The use of mutagenicity testing to evaluate food products, in Heddle, J.H. (ed.), *Mutagenicity: New Horizons in Genetic Toxicology*, Academic Press, New York, 1982, pp. 117–142.

83. Zoeteman, B.C.J., Hrubec, J., de Greef, E., and Kool, H.J., Mutagenic activity associated with by-products of drinking water disinfection by chlorine, chlorine dioxide, ozone and UV-irradiation, *Environ. Health Perspect., 46*, 197–205, 1982.

84. Schneiderman, M., Risk assessment of the health effects of water chlorination, in Jolley, R.L., Gorchev, H., and Hamilton, D.H., Jr. (eds.), *Water Chlorination: Environmental Impact and Health Effects*, Vol. 2, Ann Arbor Science, Ann Arbor, Michigan, 1978, pp. 509–518.

85. Fox, J.L., *Science, 222*, 1217, 1983.

86. Tokay, B.A., *Chemical Business, 6* (2), 12, 1984.

87. *Chemical Week, 135* (3), 24, 1984.

88. Brooks, H., Expertise and politics—Problems and tensions, *Proc. Am. Philos. Soc., 119*, 257–261, 1975.

89. *Chem. Eng. News, 60* (43), 15, 1982.

90. Steinbruner, J.D., *The Cybernetic Theory of Decision*, Princeton University Press, Princeton, New Jersey, 1974.

91. Ruckelshaus, W.D., *Science, 221*, 1026, 1983.

92. Starr, C., *Science, 165*, 1232, 1969.

93. Budiansky, S., *Environ. Sci. Technol., 14* (11), 1281, 1980.

94. *Chem. Eng. News, 61* (37), 23, 1983.

95. Clive, D., Comparative chemical mutagenicity: Can we make risk estimates?, in De Serres, F.J., and Shelby, M.D. (eds.), *Comparative Chemical Mutagenesis*, Plenum Press, New York, 1981, pp. 1039–1066.

96. Sobels, F.H., Establishment of requirements for estimation of risk for the human population, in De Serres, F.J., and Shelby, M.D. (eds.), *Comparative Chemical Mutagenesis*, Plenum Press, New York, 1981, pp. 1067–1101.

97. Smith, H., *Forgotten Truth: The Primordial Tradition*, Harper Colophon Books, Harper & Row, New York, 1977.

98. Weinstein, M.C., *Science, 221*, 17, 1983.

99. Maugh, T.H., II, *Science, 199*, 162, 1978.

100. Walsh, J., *Science, 202,* 598, 1978.
101. Shaikh, R.A., and Nichols, J.K., *Ambio, 13* (2), 88, 1984.
102. Milligan, J.E., Sarvaideo, R.J., and Thalken, C.E., *J. Environ. Health, 46* (1), 19, 1983.
103. Ames, B.N., *Science, 221,* 1256, 1983.
104. Putnam, A.R., *Chem. Eng. News, 61* (14), 34, 1983.
105. Jacobson, M., and Crosby D.G. (eds.), *Naturally Occurring Insecticides,* Marcel Dekker, Inc., New York, 1971.
106. Huxtable, R., The actions of poisons, *Defenders of Wildlife News,* Fall 1971, pp. 324–330, through Giddings, J.C., and Monroe, M.B. (eds.), *Our Chemical Environment,* Harper & Row, New York, 1972, pp. 32–41.
107. Roberts, H.R., Food safety in perspective, in Roberts, H.R. (ed.), *Food Safety,* John Wiley and Sons, New York, 1981, pp. 1–14.
108. Rodricks, J.V., and Pohland, A.E., Food hazards of natural origin, in Roberts, H.R. (ed.), *Food Safety,* John Wiley & Sons, New York, 1981, pp. 181–238.
109. Sugimura, T., and Nagao, M., The use of mutagenicity to evaluate carcinogenic hazards in daily life, in Heddle, J.A. (ed.), *Mutagenicity: New Horizons in Genetic Toxicology,* Academic Press, New York, 1982, pp. 73–87.
110. *Chem. Eng. News, 61* (15), 37, 1983.
111. Rogers, A.E., Naturally occurring carcinogens in higher plants, in Sontag, J.M. (ed.), *Carcinogens in Industry and the Environment,* Marcel Dekker, Inc., New York, pp. 519–534.
112. Tartakow, I.J., and Vorperian, J.H., *Foodborne and Waterborne Diseases,* AVI Publishing, Westport, Connecticut, 1981, pp. 196–217.
113. Grice, H.C., Clegg, D.J., Coffin, D.E., Lo, M., Middleton, E.J., Sandi, E., Scott, P.M., Sen, N.P., Smith, B.L., and Withey, J.R., Carcinogens in food, in Sontag, J.M. (ed.), *Carcinogens in Industry and the Environment,* Marcel Dekker, Inc., New York, 1981, pp. 439–518.
114. Berardi, L.C., and Goldblatt, L.A., Gossypol, in Liener, I.E. (ed.), *Toxic Constituents of Plant Foodstuffs,* Academic Press, New York, 1969, pp. 212–266.
115. 43 *Federal Register* 5, 756, 5, 759, 1978.
116. *Chemical Business, 5* (7), 23, 1983.
117. Woo, Y., and Arcos, J.C., Environmental chemicals, in Sontag, J.M. (ed.), *Carcinogens in Industry and the Environment,* Marcel Dekker, Inc., New York, 1981, pp. 167–282.
118. Roberts, H.R., Food additives, in Roberts, H.R. (ed.), *Food Safety,* John Wiley & Sons, New York, 1981, pp. 239–294.
119. Hurst, E.H., Needed: A practical cancer policy?, in Nicholson, W.J. (ed.), *Management of Assessed Risk for Carcinogens, Ann. N. Y. Acad. Sci., 363,* New York, 1981, pp. 79–88.
120. Commoner, B., Chemical carcinogens in the environment, in Keith, L.H. (ed.), *Identification & Analysis of Organic Pollutants in Water,* Ann Arbor Science, Ann Arbor, Michigan, 1971, pp. 49–72.
121. World Health Organization, Prevention of Cancer, Technical Report Series 276, Geneva, Switzerland, 1964, through Doll, R., and Peto, R., *The Causes of Cancer,* Oxford University Press, New York, 1981, p. 1197.
122. Epstein, S.S., and Swartz, J.B., *Nature, 289,* 127, 1981.
123. Miller, R.W., *FDA Consumer, 18* (2), 17, 1984.
124. Brown, A.F., Jr., *J. Environ. Health, 46* (5), 220, 1984.
125. Coorsh, R., *Consumers' Research, 67* (7), 4, 1984.
126. Freidman, L., and Shibko, S.I., in Liener, I.E. (ed.), *Toxic Constituents of Plant Foodstuffs,* Academic Press, New York, 1969, pp. 410–448.
127. Rook, J.J., *Water Treatment and Examination, 23,* 234, 1974.
128. Page, T., Harris, R.H., and Epstein, S.S., *Science, 193,* 55, 1976.
129. National Academy of Sciences, *Drinking Water and Health,* Vol. 2, Washington, DC, 1980.

130. National Academy of Sciences, *Drinking Water and Health,* Vol. 4, Washington, DC, 1982.

131. Jolley, R.L. (ed.), *Water Chlorination: Environmental Impact and Health Effects,* Vol. 1, Ann Arbor Science, Ann Arbor, Michigan, 1978.

132. Jolley, R.L., Gorchev, H., and Hamilton, D.H., Jr. (eds.), *Water Chlorination: Environmental Impact and Health Effects,* Vol. 2, Ann Arbor Science, Ann Arbor, Michigan, 1978.

133. Jolley, R.L., Brungs, W.A., and Cumming, R.B. (eds.), *Water Chlorination: Environmental Impacts and Health Effects,* Vol. 3, Ann Arbor Science, Ann Arbor, Michigan, 1980.

134. Wade, N., *Science, 196,* 1421, 1977.

135. Miller, S., *Environ. Sci. Technol., 14* (11), 1287, 1980.

136. Crouch, E.C.A., Wilson, R., and Ziese, L., *Water Resources Res., 19* (6), 1359, 1983.

137. Tiernan, J.E., *J. Environ. Health, 46* (3), 118, 1983.

138. Munson, A.E., Sain, L.E., Sanders, V.M., Kauffmann, B.M., White, K.L., Jr., Page, D.G., Barnes, D.W., and Borzelleca, J.F., Toxicology of organic drinking water contaminants: Trichloromethane, bromodichloromethane, dibromochloromethane and tribromomethane, *Environ. Health Perspect., 46,* 117–126, 1982.

139. Cotruvo, J.A., Introduction: Evaluating the benefits and potential risks of disinfectants in drinking water treatment, *Environ. Health Perspect. 46,* 1–6, 1982.

140. Balster, R.L., and Borzelleca, J., Behavioral toxicity of trihalomethane contaminants of drinking water in mice, *Environ. Health Perspect., 46,* 127–136, 1982.

141. Jorgenson, T.A., Rushbrook, C.J., and Jones, D.C.L., Dose-response study of chloroform carcinogenesis in the mouse and rat: Status report, *Environ. Health Perspect., 46,* 141–149, 1982.

142. Pereira, M.A., Lin, L.C., Lippitt, J.M., and Herren, S.L., Trihalomethanes as initiators and promoters of carcinogenesis, *Environ. Health Perspect., 46,* 151–156, 1982.

143. Savage, R.E., Jr., Westrich, C., Guion, C., and Pereira, M.A., Chloroform induction of ornithine decarboxylase activity in rats, *Environ. Health Perspect., 46,* 157–162, 1982.

144. Gottlieb, M.S., and Carr, J.K., Case-control cancer mortality study and chlorination of drinking water in Louisiana, *Environ. Health Perspect., 46,* 169–177, 1982.

145. Kanarek, M.S., and Young, T.B., Drinking water treatment and risk of cancer death in Wisconsin, *Environ. Health Perspect., 46,* 179–186, 1982.

146. *Chem. Eng. News, 54* (10), 6, 1976.

147. Reitz, R.H., Gehring, P.J., and Park, C.N., *Fd. Cosmet. Toxicol., 16,* 511, 1978.

148. Page, T., Harris, R., and Bruser, J., Waterborne carcinogens: An economist's view, in Crandall, R.W., and Lave, L.B. (eds.), *The Scientific Basis of Health and Safety Regulation,* The Brookings Institution, Washington, DC, 1981, pp. 197–228.

149. Hoel, D.G., and Crump, K.S., Waterborne carcinogens: A scientist's view, in Crandall, R.W., and Lave, L.B. (eds.), *The Scientific Basis of Health and Safety Regulation,* The Brookings Institution, Washington, DC, 1981, pp. 173–196.

CHAPTER 19

Technical and Enforcement Aspects of Organic Contamination in Drinking Water

Jerome J. Healey

Water Supply Branch
Water Management Division
Environmental Protection Agency, Region I
Boston, Massachusetts

19.1 INTRODUCTION

The need to take precautions with drinking water to protect public health was recognized as many as 4000 years ago. M. N. Baker (1) provides historical accounts of water supply practices with one of the earliest references of water treatment described by Francis Evelyn Place, who while in India in 1905, wrote, "It is good to keep water in copper vessels, to expose it to sunlight, and filter through charcoal." She credited this quotation to "Ousruta Sanghita," a collection of medical lore in Sanskrit, approximately 2000 B.C., Chapter XIV, verse 15. Baker then details the development of water supply practices to recent times. This recommended action is very appropriate when considering the problem of organic contamination of drinking water.

The origins of the current Federal National Interim Primary Drinking Water Regulations (NIPDWRs) began with the National Quarantine Act of 1878, "An Act To Prevent The Introduction of Infectious or Contagious Diseases In The United States." Additions to this Act continued until the first formal drinking water standards were developed in 1914, known as the Public Health Service Drinking Water Standards. Because they were devel-

Note: The views expressed in this paper are those of the author. They do not necessarily reflect the policy of the Environmental Protection Agency.

oped as part of the Interstate Quarantine Regulations, they pertained only to water systems providing drinking water to vessels or conveyances operated in interstate commerce. The first standards were concerned with the microbiological quality of the water. Subsequent revisions were undertaken in 1925, 1943, 1946, and 1962, in which parameters for physical and chemical quality were added. In 1914, there were some 9000 water systems covered by these standards. By 1974, when the Safe Drinking Water Act was passed, this total had dropped to 700 systems. However, all of the states, either formally or informally, adopted these Standards as the basis for evaluating the quality of their water supplies. The 1963 Inventory of Municipal Water Facilities (2) reported that there were an estimated 19,236 water systems across the country. This total has increased to nearly 60,000 systems in 1985.

Enforcement of the 1962 standards varied from state to state. The results of a nationwide Community Water Supply Study (CWSS) during 1969 in eight geographically distributed standard metropolitan statistical areas (SMSAs) and the state of Vermont provided the first representative measure of the enforcement by the states of the Standards (3). Of the 969 systems evaluated, only 59% met the 1962 Public Health Service Drinking Water Standards.

Thirty six percent of the 2595 water samples taken from the taps of consumers of these systems contained one or more constituents with a concentration that exceeded the Standards. The results of this study and others conducted in the early 1970s led to the passage of the Safe Drinking Water Act (SDWA) in 1974.

19.2 ORGANIC CHEMICALS IN DRINKING WATER

The evolution of concern for organics in drinking water began with the development of analytical methods to measure trace quantities of organic compounds in water. Bellar et al. (4) of the U.S. Environmental Protection Agency (EPA) were the first to identify volatile organics in drinking water in the United States. This report found that chloroform and other trihalomethanes (THMs) occur consistently in chlorinated drinking water. These compounds are formed by the reaction of chlorine used for disinfection and naturally occurring humic acids.

As a result of these findings EPA then began a series of national studies to investigate the problem of THMs in water supplies. The National Organics Reconnaissance Survey (NORS) (5) of 80 water systems was undertaken after EPA found THMs in the water supply of New Orleans, Louisiana. The results of NORS indicated that THMS were present in many chlorinated drinking water supplies as a result of chlorination. This occurred when surface water was the source and the raw water was chlorinated sufficiently to leave a free chlorine residual of more than 0.4 mg/L.

A second study, The National Organics Monitoring Survey (NOMS) (6), was conducted in 1976 and 1977 with 113 community water systems partici- pating. This study considered compounds included in NORS, as well as 15 others including such compounds as trichloroethylene, benzene, and vinyl chloride. It was a study of finished drinking water with sampling carried out three times in a 12-month period. Over 60 additional organic compounds were detected in this survey.

The results were similar to those of the NORS study with regard to THMs. However, it was found that compounds other than TTHMs were generally present in the finished water due to raw water contamination. Also, THM values increased when there was a delay in time before analysis (3–6 weeks rather than 1–2 weeks). This allowed for more complete THM formation and indicated that additional THM production took place in the water distribution system. EPA conducted additional surveys, the National Screening Program and the Community Water Supply Survey, both of which indicated that organic contaminants were present in drinking water at low levels.

A much more comprehensive study, the Ground Water Supply Survey (GWSS) (7) was conducted by EPA in 1980 and 1981. A total of 466 water systems were chosen at random and an additional 479 systems were selected by state agencies. The latter group was considered nonrandom because most of these systems were believed to have some contamination from volatile organics. However, no data existed to confirm this belief. The results for each group were separated according to sample size.

Of the 466 systems selected randomly, 280 served fewer than 10,000 people and 16.8% contained one or more volatile organic contaminants (VOCs). Twenty-eight percent (28%) of the systems serving more than 10,000 people contained one or more volatile organic contaminants. As was expected, the 479 nonrandomly selected systems exhibited higher levels of contamination. Some 22.4% of the 321 systems serving fewer than 10,000 people exhibited one or more contaminants while 37.7% of the systems serving more than 10,000 people showed one or more of the volatile organic contaminants.

Since that time more information has been developed. Chin (8) has collected information on the organic contamination of water systems in the six New England states as of January 1, 1984 (Table 19.1).

Pesticides and herbicides represent another group of organic compounds of potential concern in drinking water. NIPDWRs contain MCLs for four insecticides—endrin, lindane, methoxychlor, and toxaphene—and two her- bicides—2,4-D and 2,4,5-TP. Since the promulgation of the regulations in 1975, there have been virtually no cases of noncompliance with these six standards. However, a number of other synthetic organic chemicals have been detected in drinking water; pesticides such as ethylene dibromide (EDB) have been found in water supplies in Florida, Hawaii, Washington, Connecticut, and Massachusetts while aldicarb has been found in water systems in New York and Rhode Island.

TABLE 19.1
Occurrence of Organic Chemicals in New England Drinking Water

State	Number of Contamination Incidents Involving Private Supplies	Number of Contamination Incidents Involving Industrial Supplies	Number of Contamination Incidents Involving Public Supplies
Connecticut	15	11	27
Maine	14	0	1
Massachusetts	10	4	20
New Hampshire	21	0	4
Rhode Island	13	0	3
Vermont	15	1	2
Total	88	16	57

EPA is currently considering revisions to the NIPDWRs (9) and has requested information on a number of synthetic organic chemicals that have been found in drinking water, that are registered for use in or around drinking water, or that have the potential for entering drinking water.

19.3 FEDERAL LEGISLATION ADDRESSING ORGANIC CONTAMINATION

There are various federal programs that include provisions for the protection of surface or underground water bodies from organic or hazardous waste contamination. Most of the environmental laws enacted over the past 2 decades have addressed specific types of contamination and proposed regulations or procedures to mitigate the concerns. Organic contamination comes from a number of different sources. The Congressional Office of Technology Assessment in a recent report (10) compiled data on various types of ground water contamination sources. Among the more significant are septic systems (organic cleaning solvents for tanks), injection wells (hazardous or toxic waste, oil-field brine, and municipal sewage wastes), landfills and surface impoundments of wastes, abandoned hazardous waste sites, industrial waste sites, and underground storage tanks of petroleum and hazardous chemicals. Perhaps the most pervasive sources of organic contamination are those associated with agricultural practices such as fertilizers and herbicides.

The following is a review of the major environmental laws designed to address various problems. The laws in most cases place EPA in a reactive rather than a proactive role.

19.3.1 The Safe Drinking Water Act

With the passage of the Safe Drinking Water Act of 1974 (SDWA) (11), provisions were finally made to establish drinking water standards that applied to all public water systems. The Act also clearly defined the responsibilities of the federal and state governments as well as the water purveyor in providing adequate quantities of potable water.

The federal government, through EPA, has the responsibility for establishing national drinking water standards. In addition EPA is authorized to provide technical and financial assistance to states and technical assistance to public water systems. As part of this assistance EPA is to conduct research and studies regarding health, economic and technological problems of drinking water supplies. Two sets of water quality standards are specified in the SDWA. The NIPDWRs are developed to protect public health and are mandatory requirements for public water systems. The National Secondary Drinking Water Regulations relate to aesthetic concerns and are optional at the discretion of the states.

The states have a major responsibility to enforce their own laws and regulations covering public drinking water systems. Each of the 57 states and territories must adopt laws and regulations at least as stringent as those developed by EPA in order to assume this enforcement responsibility. To date all but three states (Indiana, Oregon, and Wyoming) and the District of Columbia have assumed this role. A state program, in addition to having sufficiently stringent laws and regulations, must implement adequate surveillance and enforcement procedures.

The public water systems must comply with the NIPDWRs. Providing adequate quantities of potable water has become an increasingly difficult task. Water system operators are constantly learning of new contaminants that may be found in water delivered to their customers. Those professionals responsible for designing and/or managing water treatment systems must continue to review the effectiveness of treatment techniques. They must anticipate the ability of these treatment methods to remove other unknown contaminants.

Although the SDWA is 10 years old, the revisions of the NIPDWRs are still continuing. The development of regulations for volatile and synthetic organics is currently underway. This development process will continue even when standards are established for the contaminants currently under consideration. As health studies are completed, new standards will be proposed or existing ones will be revised. Many of the organic substances found in drinking water supplies are not presently regulated; however, EPA has developed health advisories for a number of organics. These advisories have provided guidance to states and to water systems as to the advisability of using a contaminated source of supply.

Another major provision of the SDWA calls for the protection of underground sources of drinking water by the development of a regulatory program to control deep well injection of waste that might endanger underground sources of drinking water. The regulations require that an injection well must not allow waste water to move into a potable aquifer. Also, wells located in close proximity to the injection well must be monitored to insure that the injected waste fluid does not migrate through wells which penetrate the injection zone.

Finally, the SDWA contains one additional provision, the Sole Source Aquifer Program, which should provide protection of underground sources of water supply that serve as the principal drinking water supply for a region. Any aquifer designated as a sole source receives protection from projects in the recharge zone that are funded by federal monies. EPA has the authority to deny federal funding for any project that may create a significant hazard to public health. As of April, 1985, 21 designations have been made. Among the areas included in the group are Long Island, New York, and Cape Cod, Massachusetts. The veto power available to EPA in federal project reviews is significant and therefore insures that most applicants will make an extra effort to prevent contamination problems. While this law does not pertain to nonfederally funded projects, it supports the state and local governments faced with reviewing those projects which may be a problem. Massachusetts has adopted its own Environmental Policy Act (MEPA) (12) which includes a provision for review of similar projects not funded with federal monies.

19.3.2   The Federal Insecticide, Fungicide, and Rodenticide Act

While much attention is focused upon point sources of groundwater contamination, perhaps the most widespread form of organic chemical contamination is associated with the use of pesticides. The EPA has the authority to regulate these chemicals because of the Federal Insecticide, Fungicide, and Rodenticide Act (FIFRA) (13).

The FIFRA was initially passed in 1947 to protect pesticide users from products that were ineffective. By 1972, Congress had enough information about the concerns of potential health and environmental risks of the chemicals to amend the Act. These amendments required all applicants to provide comprehensive data for new pesticide registrations. Furthermore, all registrants were required to do the same for old products. With this data, EPA could then make a decision on use of the product to ensure that the pesticide would not cause unreasonable risk to health or the environment. Additionally the pesticide amendment of the Federal Food, Drug, and Cosmetic Act (FFDCA) provided for the establishment of limits on the amount of pesticide residues that may remain on a food or feed crop after harvesting.

The FIFRA registration process is a complicated and time-consuming process for a new pesticide and may take as long as 10 years. The early years are spent determining patent requirements and need for and effectiveness of the pesticide and toxicological testings. Next the company field tests the product after obtaining an experimental use permit (EUP) from EPA in order to determine effects on crops, residues remaining on the crop, and best application method. In addition all health and environmental studies required by EPA are conducted in accordance with EPA guidelines. Finally the information is submitted to EPA for intensive review leading to a decision on registration.

Registration or reexamination of old pesticides approved before the 1972 amendments to the FIFRA is a major task because of the numbers involved. Between 1947, when the FIFRA was passed by Congress, and 1970, when EPA inherited responsibility for the FIFRA and part of the FFDCA from the U.S. Department of Agriculture, there were an estimated 600 active ingredients and 50,000 different pesticide products registered and marketed in the United States.

To cope with the backlog, EPA has identified chemicals thought to be a problem, updated whatever data are available on the chemicals, and made new judgments on the approved uses. The Agency has developed a list of effects that are deemed to be unreasonable health or environmental risks. A registrant can rebut the decision and a special review process, including risk analysis and a review of benefits as well as public comments is undertaken.

The problem of chemicals approved prior to this Act is underscored by the closure of wells contaminated by such pesticides as aldicarb (Long Island and Rhode Island) and EDB (Massachusetts, Connecticut, Florida, Washington, and Hawaii) percolating directly to the groundwater table. The review process employed in the applicability of such pesticides must consider the potential for groundwater contamination. Banning or limiting use of the pesticide over sensitive aquifers to protect the groundwater from organic contamination is an appropriate decision in some cases.

19.3.3   The Toxic Substances Control Act

Under the Toxic Substances Control Act (TSCA) (14), EPA is directed to regulate chemical substances and mixtures that present an unreasonable risk of injury to health and the environment. Most other laws give EPA authority to control toxics only after damage occurs. The Agency can operate only in a reactive rather than a proactive role. TSCA gives the Agency the opportunity to review the health and environmental effects of all new chemicals before they are produced and made available to consumers.

EPA has approached the law in four major ways. First the Agency has gathered information about the chemicals in the environment. When the law was passed there was no definitive information about these chemicals. Such items, as the number of chemicals, the quantities produced, location of production, and by-products needed to be defined. A 1979 inventory published by EPA from information provided by chemical manufacturers, importers, and processors indicated that nearly 50,000 commercial chemical substances are manufactured or imported into the United States. This total does not include the millions of research and development chemicals not available commercially. Development of this inventory is important because any chemical not on the list will then be reviewed by the EPA for environmental and health effects before being allowed for commercial use.

Second, the Agency, recognizing that sufficient data is necessary to evaluate the impact of chemicals on public health and the environment, requires under TSCA that those who manufacture or process these commercial chemicals develop this information in accordance with EPA rules and procedures. TSCA provides for the creation of the Interagency Testing Committee (ITC) which is made up of representatives of the Council on Environmental Quality, the Department of Commerce, the National Science Foundation, the National Institute of Environmental Health Sciences, and the National Institute for Occupational Safety and Health. The group identifies and recommends those chemicals whose effects should be considered for health and environmental impact.

EPA is authorized to seek existing health or safety information from the industries or others who might be of assistance. From this, EPA can identify information gaps. When insufficient information is available or the data show that the chemical may pose an unreasonable risk to health or the environment, testing is undertaken.

As a third step, if the data indicate that the risks of a chemical outweigh the benefits to public health and the environment, among the variety of actions EPA may undertake are those ranging from labeling requirements to an outright ban on the manufacturing or processing of the chemical. Whatever action EPA chooses to take, the Agency must publish a notice in the Federal Register. This notice explains the basis for the decision including health and environmental concerns and limits, economic concerns, and any available alternative products. Under this third step, EPA has banned the manufacture, processing, distribution, and use of polychlorinated biphenyls (PCBs).

The fourth and final step pertains to new chemicals. TSCA requires that chemical manufacturers or processors notify EPA before making or importing any new chemical not on the original inventory. Information supplied to EPA includes health and environmental effects of the chemical, proposed uses, exposure data, and disposal processes. EPA must then make a judgment within 90 days or require additional information.

19.3.4   The Resource Conservation and Recovery Act

During the late 1960s and early 1970s it was apparent that state and local governments needed assistance in disposing of municipal and industrial solid wastes. Studies carried out by EPA during the early 1970s led to the conclusion by Congress that action was needed in the matter of dealing with the hazardous wastes. As a result, the Resource Conservation and Recovery Act (RCRA) (15) was passed in 1976. Some of the major provisions included the prohibition of future uncontrolled open dumping on land, and the requirement for conversion of existing open dumps to controlled and managed nonhazardous facilities. In addition the Act provided for the regulation of treatment, storage, transportation, and disposal of hazardous wastes that have adverse effects on public health and the environment. One additional important provision included the requirement for guidelines for solid waste collection, transportation, separation, recovery, and disposal practices.

With RCRA, the statutory framework for comprehensive federal and state regulation of hazardous waste is established. The Act requires the identification and listing of the type of hazardous waste, taking into account such factors as toxicity, persistence and degradability in nature, the potential for accumulation in tissue, and other characteristics. It directs promulgation of such standards for generators of hazardous waste as may be necessary to protect human health and the environment. Standards are also developed for the transporters of hazardous waste. The owners and operators of hazardous waste treatment, storage, and disposal facilities must also follow performance standards. When appropriate, these standards distinguish between requirements for new facilities and for already existing facilities. Among the requirements are those for record keeping, reporting, monitoring, and inspection, as well as requirements for the design and construction of facilities and provisions for compliance with permits for treatment, storage, and disposal of the hazardous waste.

As mentioned earlier, the Act recognized the need for a cooperative effort between the state and federal programs. The Act provides that EPA will promulgate guidelines to assist the states in the development of their own hazardous waste programs so that the states will be able to operate in place of the federal program. The Act provides a wide range of inspection and enforcement tools for EPA and the states. These provisions apply to hazardous waste handlers including the generators, transporters, and facilities where hazardous wastes are or have been stored or treated.

The second major part of RCRA provides for the development of methods for the disposal of solid wastes which are environmentally safe and which conserve valuable resources. These objectives are to be accomplished through federal technical and financial assistance to states and regional authorities for comprehensive planning. Each state or authority is to develop its own solid waste management plan meeting certain minimum requirements which emphasize the closing or upgrading of all open land dumps and a prohibition on the formation of new ones. EPA has adopted regulations to assist the states in developing solid waste management plans. One of the prime requirements is that the solid waste facility will not contaminate current or potential underground drinking water sources beyond the solid waste disposal site boundary. While each state is asked to develop this plan and to assume the responsibility for carrying it out, EPA has no legal authority to require the state to follow the guidelines.

19.3.5 The Comprehensive Environmental Response, Compensation, and Liability Act (Superfund)

While the RCRA provided the regulatory framework to deal with known hazardous waste generation, treatment, and disposal operations, it became quickly apparent that some mechanism was needed to deal with the uncontrolled or abandoned hazardous waste sites. The Comprehensive Environmental Response, Compensation, and Liability Act (CERCLA) (16), more commonly known as Superfund, was passed by Congress in 1980, after thousands of these sites were discovered across the country. In many cases, responsible parties could not be found; moreover, financially capable parties were rarely available.

The Clean Water Act enabled the federal government to take action when oil or certain hazardous substances are discharged into navigable waterways. However, it did not permit the government to act when substances were released elsewhere in the environment, such as to groundwater. Although most environmental laws provide the federal government with the authority to take legal action against those responsible for these discharges, there is not enough time for legal proceedings to take place before some mitigation efforts must be undertaken. Superfund has become the vehicle to approach this problem by giving EPA authority to respond to the cleanup of these sites and emergency spill situations. Congress provided a $1.6 billion fund as part of a 5-year program to lead a federal and state effort to respond to the release of these hazardous substances and to eliminate the most serious threats to public health and the environment.

CERCLA provides for a Hazardous Substances Response Fund which was financed primarily by taxes on petroleum products and chemical industries. The Fund provides monies for abatement and cleanup of these hazardous substances. In addition, liability provisions are included allowing EPA to assess those liable for the costs of cleanup. Finally, the law requires all parties owning facilities, now or in the past, that handle hazardous substances to notify EPA of the location of the facility and amount and type of substances found there. All spills of hazardous substances in excess of specified quantities must be reported to EPA. When there is a release of a hazardous substance, a National Contingency Plan outlines procedures and methods to be followed for cleanup.

Many states have established their own programs for spill response or the cleanup of uncontrolled waste disposal sites, however the problem is not easily abated. CERCLA encourages private parties to handle the problems of site cleanup, and EPA will not use Superfund money if the owner or responsible party is undertaking appropriate cleanup actions. Immediate removal of hazardous substances from a site is undertaken by EPA only to bring the problems under control. Remedial or long-term cleanup actions are much more involved. States must agree to pay for a portion of the cost (10–50%) and to maintain the site once response work is completed.

Finally, the law requires EPA to consider the public health and environmental benefits of remedial actions at one site versus another to maximize the use of the funds for the country as a whole. There is a finite amount of funds available for the cleanup of an as yet to be defined amount of work necessary to abate the total problem.

19.3.6   The Clean Water Act

The major focus of Clean Water Act (17) has been to protect surface water quality. States are required to adopt surface water quality standards and to develop water quality management programs. Perhaps the most well known portion of this Act is the construction grants program which provides funds to municipal governments to finance the construction of waste water treatment facilities to comply with the requirements of the Act. The Act also provides for the funding of state program activities related to carrying out the goals of the legislation.

There are some provisions of the law that apply to organic contamination, and the water quality standards adopted by states include organic parameters. Direct dischargers of pollutants (point discharge sources) must comply with the National Pollutant Discharge Elimination System (NPDES) which incorporates and applies effluent limitations in permits granted to municipal and industrial discharges. Organic parameters are included in these permits and limits are set based on stream classification and downstream water use. Finally, nonpoint sources of pollution, such as from agricultural activities, must be addressed by the states in their water quality management plans.

19.4   STATE ACTIVITIES ADDRESSING ORGANICS CONTAMINATION

The federal programs discussed above were developed to address various public health and environmental concerns without focusing specifically on the problem of organic contamination. EPA recognized the need to coordinate the efforts of these programs and has taken steps to do so in the area of groundwater. As recent history shows, most organic contamination problems occur in groundwater. This is especially true with the volatile and synthetic organics. Therefore, the development of the "Ground-Water Protection Strategy" (GWPS) (18) by EPA is an attempt to coordinate groundwater efforts among the programs at the federal level as well as at the state and local levels. While most programs at the federal level concerned with organic contaminants are administered by EPA, this is not the situation in many states where various agencies administer these programs.

To understand better the complexities of administering public health and environmental protection programs at the state level, a review of these programs in one area of the country is helpful. The six New England states, Connecticut, Maine, Massachusetts, New Hampshire, Rhode Island, and Vermont, have as many as four agencies responsible for these programs. Four of the six states have placed the public water supply supervision programs in an agency other than those responsible for administering the other environmental programs. Although some coordination mechanisms do exist through interagency agreements—formal and informal—a comprehensive evaluation of the total impacts of organic contamination is difficult. States are faced with the problem of accommodating competing economic, health, and environmental interests and, therefore, the coordination of efforts is extremely important. A recent EPA report on "State Ground-Water Programs" (19) shows that nationally, as many as six types of agencies are responsible for administering the RCRA program, five different groups manage the Underground Injection Control (UIC) program, and four different agencies administer the Public Water System Supervision (PWS) activities.

The state water supply agencies generally have primary responsibility when organic contamination is found in public water supplies. Most states do not have drinking water standards for organics other than those for the six herbicides and pesticides, and TTHMs included in the NIPDWRs. States generally rely upon EPA to provide health advisories when organics have appeared in water samples taken from water supplies. Many states have developed toxicological capability to evaluate the impacts of contamination. This resource coupled with the assistance provided by EPA has enabled the states to determine whether a contaminated water supply source should be used.

An example of this type of activity is the groundwater contamination of the well serving the Cannongate Condominium complex in Tyngsboro, Massachusetts (20). In the summer of 1981, analytical results from water samples taken by the Massachusetts Department of Environmental Quality Engineering (MA DEQE) indicated that the two wells serving the complex were contaminated by a number of organic chemicals. Well #1 was contaminated by elevated levels of methyl ethyl ketone (i.e., levels exceeding the draft EPA short-term health advisory) and trace amounts of other organics, some of which are potential carcinogens (i.e., trichloroethylene and benzene). Well #2 also indicated the presence of organic chemicals with concentrations not as high as those in well #1 and none exceeding any EPA health advisory level.

Relying on the EPA draft health advisory for methyl ethyl ketone, the Office of Criteria and Standards of the MA DEQE determined that well #1 was unsuitable for human consumption. Well #2, although it did not represent a public health emergency, would be unacceptable on a long-term basis on account of the uncertainties associated with the ingestion of the chemicals. At the request of the Tyngsborough Selectmen, the EPA evaluated the situation and came to a conclusion similar to the State's. A series of meetings were held between the EPA, the DEQE, and town officials to evaluate and develop short-term and long-term alternatives.

When additional sampling of well #2 during the spring of 1982 indicated an increase in the concentrations, as well as an increase in the number of contaminants, the decision was made to close well #2 for drinking and domestic purposes by the end of July. In the interim, a tank truck was supplying water to the residents.

The Charles George Landfill, adjacent to the complex, was designated a Superfund site, placed on the National Priority List, and implicated as a source of contamination. With the help of Superfund money, DEQE decided to bring in a temporary above ground supply line from the North Chelmsford Water District to serve the residents of the Cannongate Condominium complex. A long-term solution for providing water to the area is currently under consideration.

19.5 TREATMENT TECHNIQUES FOR ORGANICS REMOVAL

Organics are most commonly removed from water by adsorption and aeration. EPA has compiled information on such treatment as it applies to individual compounds (21). Adsorption processes have been used to reduce turbidity, taste, and odor, recently to minimize the organic precursors associated with THM production in chlorinated surface waters. These techniques are now being utilized to remove organic substances from ground water. Aeration of water has traditionally been used to oxidize iron and manganese and in some cases to remove carbon dioxide. More recently the practice of aeration has been applied to remove volatile organic compounds from groundwater. Information recently compiled by EPA's Office of Drinking Water provides data on treatment efficiency and prevalence of use for the various types of aeration and adsorption techniques as shown on Table 19.2.

Aeration or air stripping to reduce the concentration of volatile organics in drinking water is dependent upon several factors, which include the molecular properties of the contaminant, the air-to-water ratio, and the capacity to apply large amounts of air to a thin film of water. A packed tower aerator usually consists of a column 3–10 ft in diameter and 15–30 ft high. The packing, which can be made of glass, ceramic, or plastic and is available in many shapes, facilitates the transfer of the contaminant from the aqueous to the gas phase. The water enters the top of the column and flows through the packing. Air is forced counter current to the flow of the water either by pressure or suction. The design of the interior walls of the column forces the water through the packing and prevents water from running along the walls of the column. Tray-type aerators force the water to trickle over trays to form water drops thereby enhancing air contact with the water. A mixture of contaminants will be removed by aeration and the removal efficiency is usually very constant with set operating conditions.

TABLE 19.2
Organics Treatment Information

Type of Treatment	Estimated Removal Efficiency (%)	Number of Installations Currently in Operation in the U.S. for VOC Removal
Aeration		
a. Packed tower aeration	90–99.9	27
b. Slat tray aeration	50–90	6
c. Diffused aeration	50–90	1
d. Spray aeration	75–90	1
e. Air lift pumping	Unknown	2
f. Cascade	Unknown	1
Adsorption		
a. Granular activated carbon	>99	4
b. Powdered activated carbon	50–90	1
c. Synthetic resins	>99	0

While aeration is successful in removing organics from water, there is concern about the potential air quality problems created by the gases formed during this process. These concerns are often not significant, but they must be considered at any installations using aeration for organics removal. Table 19.3 developed from information collected by Love et al. (22) presents a summary of treatment information for packed tower aerators used to remove trichloroethylene (TCE), tetrachloroethylene (PCE), and 1,2-dichloroethane (1,2-DCE).

Theoretically, removal of better than 99% of the volatile organic contamination is possible under ideal conditions using packed tower aeration. However, the EPA studies indicate that 95–99% removal is achieved under normal practices for groundwater sources of supply. Minimal data for removal of organic contaminants from surface water sources is available. The percent removal may be some what less because of water temperature variations and the higher total organic carbon of surface water. The ultimate concern is the quality of the finished water. If the concentration of VOCs in the raw water varies then the concentrations in the finished water will vary also because of the percentage removal characteristic of packed tower aeration. Design criteria must include provisions for these concerns.

Adsorption of volatile organic contaminants in water supplies has been an effective form of treatment using granular or powdered activated carbon (GAC or PAC) or different types of synthetic resins as the adsorbent. Adsorption has been used primarily for the removal of the precursors found in surface water supplies that react with chlorine to form THMs. The effectiveness of adsorption depends upon the type of adsorbent, the type and

TABLE 19.3
Packed Tower Aeration Performance

Location	Contaminant Influent (µg/L)	Air/Water Ratio	Removal (%)	Effluent Concentration (µg/L)
Rockaway, NY	TCE (50–220)	114:1	99	<1
Rock Hill, NJ	TCE (45–95)	83:1	99	<1
Tacoma, WA	TCE (54–130)	62:1	95	7
Hartland, WI	TCE (175)	50:1	99	<2
Warrington, PA	TCE (130)	40:1	97	4
Wurtsmith, AFB, MI	TCE (50–8000)	25:1	99	4–8
Glen Cove, NY	TCE (117–277)	15:1	85	18–42
Upper Merion, PA	TCE (3–20)	11:1	94	<1
Tacoma, WA	PCE (1.6–5.4)	62:1	95	<1
Glen Cove, NY	PCE (33–207)	15:1	90	3–21
Tacoma, WA	1-2 DCE (30–100)	62:1	95	2–5
Glen Cove, NY	1-2 DCE (33–79)	15:1	75	10–20

concentration of the contaminant, and contact time of the contaminated water with the adsorbent, and the background concentration of total organic carbon (TOC). These factors combine to make adsorption a more complicated process than aeration.

Laboratory and pilot studies must be undertaken so that appropriate treatment is utilized. Such studies are important because the concentration of contaminants in surface waters may vary greatly compared with that found in groundwaters. TOC content of surface waters is generally higher than that of groundwater. High TOC water can rapidly exhaust GAC filter beds thereby allowing the VOCs to pass through to the finished water. This result is known as breakthrough. Another important factor in the removal of organic compounds is the molecular weight. Generally, higher molecular weight compounds are more readily adsorbed on GAC. Adsorption isotherms are developed to measure the amount of compound adsorbed per amount of carbon adsorbent.

Removal of VOCs from water by adsorption is accomplished by passing the water through filter beds (pressure or gravity flow). The empty bed contact time (EBCT), that is, volume of carbon bed divided by the hydraulic flow rate, is an important design criteria. Actual contact time between contaminated water and GAC is generally one-half the EBCT and depends upon the GAC porosity. EPA studies (21) have found a range of 5–44 min in recent practices. Another important factor is the carbon usage rate, that is, the mass of carbon used per unit volume of water treated. This rate depends upon the VOCs, the type of carbon, and other organics present.

Eventually, breakthrough of the contaminants to the finished water will occur. The time to breakthrough varies, but as mentioned earlier, it depends primarily on the TOCs. Groundwater sources have lower TOCs than surface waters. Table 19.4 presents results for five studies conducted by EPA. When breakthrough occurs, the GAC must be replaced or regenerated. This decision is based on economics with larger systems more likely to regenerate the carbon.

The use of synthetic resins, while successful in experimental use, is no longer possible because the product is not presently available commercially because of the expense. PAC has lower adsorption capability and therefore is not used at this time for VOC removal.

The secondary effects of GAC adsorption centers on two areas. Reactivation of the spent carbon through furnaces (gas or oil fired) can result in air pollution problems with the release of contaminants from the adsorbed carbon. Generally, with pollution control equipment, this problem can be minimized. Waste GAC can be disposed of in a manner similar to that used in wastewater treatment. Backwash water can be handled in a similar manner. Finally, the problem of microbiological growth on the filters can be minimized by frequent backwashing and postfiltration chlorination before the finished water enters the distribution system.

TABLE 19.4
Breakthrough Performance of Granular Activated Carbon

State	Influent	EBCT	Effluent	Approximate Time to Detection
New Jersey	194 μg/L PCE	10.5 min	<1 μg/L	22 mos.
Pennsylvania	20–30 μg/L TCE	7.5 min	<1 μg/L	20 mos.
New Hampshire	120–276 μg/L TCE	9 min	<1 μg/L	18 mos.
New Jersey	23 μg/L TCA	18 min	<1 μg/L	13 mos.
Connecticut	1–214 μg/L TCA	8.5 min	<1 μg/L	12 mos.

While the problem of organic contamination in drinking water has evolved into a major concern for the water purveyors and consumers of this nation, the solutions to the problem are now being developed. The health studies necessary to make decisions and establish standards are well underway. Added information will be provided through the rest of this decade. Additional treatment processes are being considered and existing ones continue to be refined. State and federal enforcement agencies must utilize this information to move problem water systems into compliance with the drinking water standards established for organics. Finally, the various state and federal programs charged with protecting the environment must work together to ensure the coordination necessary to prevent further contamination to our nation's water supply.

REFERENCES

1. Baker, M.N., *The Quest for Pure Water*, American Water Works Association, 1948.
2. Statistical Summary of Municipal Water Facilities in the U.S., U.S. Department of Health, Education and Welfare, Public Health Service, Washington, DC, January 6, 1963.
3. Community Water Supply Study, Analysis of National Survey Findings, U.S. Department of Health, Education and Welfare, Public Health Service, Washington, DC, January, 1970.
4. Bellar, T.A., Lichtenberg, J.J., and Kroner, R.C., *The Occurrence of Organohalides in Chlorinated Drinking Water*, U.S. Environmental Protection Agency, Cincinnati, Ohio, November, 1974.
5. Symons, J.M., Bellar, T.A., Carswell, J.K., DeMarco, J., Kropp, K.L., Robeck, G.G., Seeger, D.R., Slocum, C.J., Smith, B.L., and Stevens, A.A., National Organics Reconnaissance Survey for Halogenated Organics, *J. Am. Water Works Assoc., 67*, 634, November, 1975.

6. National Organic Monitoring Survey, U.S. Environmental Protection Agency, Cincinnati, Ohio, 1977.

7. Westrick, J.J., Mello, J.M., and Thomas, R.F., *Ground Water Supply Survey, Summary of Volatile Organic Contaminant Occurrence Data,* U.S. Environmental Protection Agency, Cincinnati, Ohio, 1982.

8. Chin, D., Environmental Engineer, U.S. Environmental Protection Agency, Region I, personal communication, January, 1984.

9. 48 Federal Register 45502 (October 5, 1983) and 49 Federal Register 24330 (June 12, 1984).

10. Protecting the Nation's Groundwater From Contamination, Congress of the United States, Office of Technology Assessment, Washington, DC OTA-0-233, October, 1984.

11. Safe Drinking Water Act, 42. U.S.C. Sections 300f *et seq.* as amended 1980.

12. Massachusetts Environmental Policy Act, General Laws, Chapter 30, Sections 62 through 62H.

13. Federal Insecticide, Fungicide and Rodenticide Act, 7 U.S.C. Sections 135 *et seq.* Also Federal Environmental Pesticides Control of 1972, as amended by the Federal Pesticide Act of 1978, 7 U.S.C. Sections 136.

14. Toxic Substances Control Act, 15 U.S.C. Sections 2601 *et seq.*

15. Resource Conservation and Recovery Act of 1976, enacted as an amendment to the Solid Waste Disposal Act, 42 U.S.C. Sections 6901–6987 as amended 1980, 1984.

16. Comprehensive Environmental Response, Compensation and Liability Act, 42 U.S.C. Sections 9601 *et seq.*

17. Clean Water Act, 33 U.S.C. Sections 1251 *et seq.*

18. Ground-Water Protection Strategy, U.S. Environmental Protection Agency, Washington, DC, August, 1984.

19. Overview of State Ground-Water Program Summaries, Vols. 1 and 2, U.S. Environmental Protection Agency, Washington, DC, March, 1985.

20. Chow, C., U.S. Environmental Protection Agency, Region I, personal communication, January, 1983.

21. Love, O.T., Jr., Miltner, R.J., Eilers, R.G., and Fronk-Leist, C.A., Treatment of Volatile Organics Compounds in Drinking Water, U.S. Environmental Protection Agency, Cincinnati, Ohio, May, 1983.

22. Love, O.T., Jr., Feige, W.A., Carswell, J.K., Miltner, R.J., Clark, R.M., and Fronk, C.A., Draft Interim Report on Aeration to Remove Volatile Organic Compounds from Ground Water, U.S. Environmental Protection Agency, Cincinnati, Ohio, March, 1984.

CHAPTER 20

Regulatory Significance
of Organic Contamination
in the Decade of the 1980s

Joseph A. Cotruvo
Susan Goldhaber
Craig Vogt

Criteria and Standards Division
Office of Drinking Water
Environmental Protection Agency
Washington, D.C.

20.1 INTRODUCTION

The goal of the Safe Drinking Water Act is to control those contaminants that may have an adverse effect on health. Organic contaminants in drinking water may contribute small health risks, and EPA is currently working to ensure that these organic contaminants are controlled so that drinking water will continue to be as safe and free from unnecessary contamination as can be feasibly achieved.

As EPA works to revise the interim primary drinking water regulations, according to statutory requirements under the Safe Drinking Water Act, numerous important issues have arisen. These issues involve policy questions that cannot be totally separated from scientific issues. These include toxicological procedures for assessing chronic toxicants, quantitative risk assessment, societal judgments on what constitutes "acceptable" risk, policy considerations on strength of evidence of available data, and the use of occurrence information to set regulations. Each of these considerations plays a role in determining the direction EPA will take in setting national policy regarding drinking water.

Note: The views expressed in this paper are those of the authors. They do not necessarily reflect the policy of the Environmental Protection Agency.

511

20.2   OCCURRENCE OF ORGANIC CHEMICALS
IN DRINKING WATER

Hundreds of chemicals have been detected in drinking water supplies in the United States. The public health significance of these substances in drinking water at trace levels has not been resolved. Over the years, the concerns and major emphases have shifted, as analytical methodology advanced. Initial concerns in the 1970s focused on synthetic organic chemicals detected in water systems using surface water sources. These chemicals were detected primarily as the result of industrial waste water discharges and the reaction between chlorine and natural organic matter which forms trihalomethanes (THMs) and other products.

THMs were found to be ubiquitous contaminants in drinking water. These chemicals, which include chloroform and bromoform, were first reported in drinking water in 1974. The National Organics Reconnaissance Survey (NORS), a study of 80 water utilities, confirmed that trihalomethanes were formed during chlorination in the drinking water treatment process. The National Organics Monitoring Survey (NOMS) was initiated in 1975 and studied 113 cities. The results showed that THMs could form in the water after it had entered the distribution system and that concentrations of THMs were greater than other synthetic organic chemicals.

Data from NORS, NOMS, the National Screening Program (NSP), which examined 166 water supplies between 1977 and 1981, and the Community Water Supply Survey (CWSS), which was conducted in 1978, have demonstrated the presence of organic contaminants in drinking water, generally at levels much less than 10 µg/L. State monitoring data have generally shown higher levels of contamination than noted in EPA surveys. This is because state sampling is generally in response to a specific problem such as a spill or investigations around hazardous waste sites. Table 20.1 indicates the concentrations of six volatile organic chemicals detected in NOMS, NSP, CWSS, and state data.

In the late 1970s reports of contaminated ground water increased and concern over ground water as an invaluable resource intensified. Volatile synthetic organic chemicals, which are commonly used as solvents, were detected in some drinking water supplies using ground water as a source. These chemicals have been detected around hazardous waste sites and they have also been associated with leaking underground storage tanks.

In 1982, EPA conducted the Ground Water Supply Survey, which consisted of a survey of 1000 drinking water supplies that used ground water as a source; 500 supplies were selected at random and 500 were selected by states as having high potential for contamination by volatile organic chemicals. The results showed that, in the random portion of the survey, approximately 21% of the systems had one or more volatile organic chemicals at detectable levels (primarily in the sub µg/L range). In the nonrandom portion of the survey, higher frequencies of occurrence were found at all levels.

TABLE 20.1
Occurrence of Volatile Organic Chemicals in Drinking Water

Survey	Number Sampled	Number Positive	Range of Positives
Trichloroethylene			
State data	2894	810	Trace, 35,000 μg/L
NOMS	113	23	0.2–49.0 μg/L
NSP	142	36	Trace, 53 μg/L
CWSS	452	15	• 0.5–210 μg/L
Tetrachloroethylene			
State data	1652	231	Trace, 3000 μg/L
NOMS	113	48	0.2–3.1 μg/L
NSP	142	24	Trace, 3.2 μg/L
CWSS	452	22	0.5–30 μg/L
Carbon tetrachloride			
State data	1659	370	Trace, 170 μg/L
NOMS	113	19	0.2–29 μg/L
NSP	142	32	Trace, 30 μg/L
CWSS	452	19	0.5–2.8 μg/L
1,1,1-Trichloroethane			
State data	1611	370	Trace, 401,300 μg/L
NOMS	113	19	0.2–1.3 μg/L
NSP	142	32	Trace, 21 μg/L
CWSS	452	19	0.5–650 μg/L
1,2-Dichloroethane			
State data	1212	85	Trace, 400 μg/L
NOMS	113	2	0.1–1.8 μg/L
NSP	142	2	Trace, 4.8 μg/L
CWSS	451	4	0.5–1.8 μg/L
Vinyl chloride			
State data	1033	73	Trace, 380 μg/L
NOMS	113	2	0.1–0.18 μg/L
NSP	142	7	Trace, 76 μg/L
CWSS		Did not look for this compound	

The six compounds that occurred most frequently in the samples analyzed during the survey are: trichloroethylene, tetrachloroethylene, 1,1,1-trichloroethane, *cis*- and *trans*-1,2-dichloroethylene, 1,1-dichloroethylene, and carbon tetrachloride. Table 20.2 presents results of the random portion of the survey for these seven compounds.

In the 1980s, the occurrence of pesticides in drinking water has become an issue of increasing concern. Various pesticides, including ethylene dibromide (EDB), aldicarb, and chlordane, have been detected in drinking water using both surface waters and ground waters. To gain a better under-

TABLE 20.2
Summary of Ground Water Supply Survey Occurrence Data

Chemical	Quantification Limit (μg/L)	Positives		Median (μg/L)	Maximum (μg/L)
		Number	Percent		
Tetrachloroethylene	0.2	34	7.3	0.5	23
Trichloroethylene	0.2	30	6.4	1	78
1,1,1-Trichloroethane	0.2	27	5.8	0.8	18
1,2-Dichloroethane	0.2	18	3.9	0.5	3.2
1,2-Dichloroethylenes (cis and/or trans)	0.2	16	3.4	1.1	2
Carbon tetrachloride	0.2	15	3.2	0.4	16

Random sample, $n = 466$.

standing of the extent of the problem, EPA is currently planning a national pesticides survey that will study approximately 1500 public water systems and private wells using ground water for a large number of pesticides. This survey is scheduled to be completed in 1988.

20.3   STATUTORY REQUIREMENTS

The Safe Drinking Water Act (1) (SDWA or the Act) requires the EPA to publish primary drinking water regulations which:

1. Apply to public water systems.
2. "Specify(s) contaminants which in the judgment of the Administrator, may have any adverse effect on the health of persons" [Section 1401(1), 42 U.S.C. 300g-1].
3. Specify for each contaminant either (a) maximum contaminant levels (MCLs) or (b) treatment techniques.

A treatment technique requirement would only be set if "it is *not* economically or technologically feasible" to ascertain the level of a contaminant in drinking water.

In the revised primary drinking water regulations, recommended maximum contaminant levels (RMCLs) must also be specified. RMCLs are nonenforceable health goals. RMCLs are to be set at a level which, in the Administrator's judgment, "no known or anticipated adverse effects on the health of persons occur and which allows an adequate margin of safety."

The primary drinking water regulations must also set MCLs; MCLs are the enforceable standards. MCLs must be set as close to RMCLs as is feasible. Feasible means "with the use of the best technology, treatment techniques and other means, which the Administrator finds are generally available (taking costs into consideration)" [Section 1412(b)(3)].

In addition, the SDWA specifies that primary drinking water regulations contain criteria and procedures to assure a supply of water that complies with the MCLs (i.e., monitoring and reporting requirements) [Section 1401(1)(D) 2 U.S.C. 300f(1)(D)].

The SDWA also requires that the revised primary drinking water regulations be reviewed every 3 years and amended whenever changes in technology, treatment techniques, or other factors permit greater health protection.

The SDWA provides that if a system will not be able to comply with an MCL after installation and/or use of the "best technology, treatment techniques, or other means which the Administrator finds to be generally available," taking costs into consideration, the system may apply for a variance [Section 1415(a)(1)(A), 42 U.S.C. 300g-4(a)(1)(A)].

On June 19, 1986 the 1986 Amendments to the Safe Drinking Water Act became law. Among other provisions, the new law requires EPA to regulate more than 80 contaminants in drinking water within three years, and, after that at least 25 more by 1991. In addition, the law requires certain water systems using surface water to use filtration treatment under appropriate circumstances.

20.4 REVISED PRIMARY DRINKING WATER REGULATIONS

The National Interim Primary Drinking Water Regulations were promulgated in 1975 and contain MCLs for 10 inorganic chemicals, 6 organic chemicals (endrin, lindane, methoxychlor, toxaphene, 2,4-D and 2,4,5-TP), microbiological contaminants (coliform bacteria), and radionuclides. Since that time, EPA has promulgated several additions and modifications to the interim regulations, including a regulation for TTHMs, defined as the sum of the concentrations of chloroform, bromoform, dibromochloromethane, and dichlorobromomethane.

EPA is currently in the process of revising the interim regulations. This process consists of a reexamination of all the interim regulations and either proposing revised values for these chemicals or concluding that due to inadequate health effects or occurrence, the chemical will no longer be regulated. In addition, new chemicals will be considered for regulation, based upon the criteria outlined in Section 20.4.1. The development of the revised regulations will be accomplished in four phases:

- Phase I—Volatile synthetic organic chemicals (VOCs).
- Phase II—Synthetic organic chemicals (SOCs), inorganic chemicals (IOCs), and microbiological contaminants.
- Phase III—Radionuclides.
- Phase IV—Disinfectant by-products including THMs.
- Other SOCs/pesticides and IOCs not considered previously will be added in subsequent updates.

In general, the approach for each phase will be similar and will follow the following steps:

- Initially, an advance notice of proposed rulemaking (ANPRM) will be published followed by a comment period and a public meeting.
- RMCLs will then be proposed followed by a public comment period and a public hearing(s).
- RMCLs will then be promulgated and proposals published for MCLs, monitoring, and reporting, and other requirements followed by a public comment period and a public hearing(s). Technologies will be identified that were used as the basis of determining the MCLs; in addition, generally available treatment technologies will be identified for use in compliance with the MCLs and the issuance of variances.
- The MCLs, monitoring, and reporting, and other requirements including generally available treatment technologies will be promulgated.

An ANPRM for Phase I (volatile synthetic organic chemicals) was issued on March 4, 1982 (2). RMCLs were proposed for nine VOCs in the *Federal Register* on June 12, 1984 (3). The final RMCLs and proposed MCLs were published in the *Federal Register* on November 13, 1985 (4).

An ANPRM for Phases II and III was published on October 5, 1983 (5), thereby initiating the regulatory reassessment of the interim regulations. On November 13, 1985 RMCLs were proposed for the Phase II chemicals (6). In addition, within Phase II, regulations for fluoride have been promulgated separately in response to a petition filed by the State of South Carolina (7).

Phase IV of the revised primary drinking water regulations will address THMs and other disinfection-related contaminant issues, since regulations for these substances have not been in effect for a sufficient time for a reevaluation and revision to be feasible at this time. It is expected that by 1987, after additional data on implementation and other experience are gathered, including new data on the nature and toxicology of alternate disinfectants and their by-products, EPA will review those regulations and determine appropriate revisions.

The basic questions being considered in each of the phases of the revised regulations are as follows:

For which contaminants should regulations be set?
What levels for the RMCLs and MCLs would be appropriate?
What monitoring and reporting requirements would be appropriate?

Each of these issues will be discussed in turn, focusing on the Phase I RMCL proposal but also dealing with all phases of the revised regulations.

20.4.1   Criteria for Selection of Contaminants for Regulation

EPA is currently revising the interim regulations, a process that consists of assessing the chemicals which are currently regulated and also considering additional chemicals for regulation. Problems arise concerning which additional chemicals should be considered for regulation because it is impossible to consider every contaminant that may be detected or that has the potential for adverse health effects under unlikely circumstances. The SDWA provides little guidance on how to select which contaminants should be considered for regulation, other than the directive that EPA is to establish RMCLs for "each contaminant which, in the Administrator's judgment. . . may have any adverse effect on the health of persons" [Section 1412(b)(1)(B)]. A primary drinking water regulation is to be established for each contaminant for which an RMCL is established [Section 1412(b)(2)].

To determine which contaminants should be considered for possible regulation under the SDWA, EPA has developed a set of selection criteria. These criteria are used to prioritize chemicals such that a reasonable number of contaminants of sufficient concern can be addressed in the regulations. The most relevant criteria for selection of contaminants are: (1) the analytical ability to detect a contaminant in drinking water, (2) the potential health risk, and (3) occurrence or potential for occurrence in drinking water.

A number of subissues are considered as part of each of the three key criteria. These issues are as follows:

- Analytical methods. Analytical methods must be available such that the presence of the chemicals in water can be validly determined. This factor is an important part in determining whether the substance can be regulated and whether an MCL or a treatment technique regulation should be promulgated. The SDWA states that MCLs are appropriate if "it is economically and technologically feasible to ascertain the level of such contaminant in water in public water systems" [Section 1401]; if not, a treatment technique is to be specified.

  A number of factors are taken into consideration in evaluating if analytical methods are available, including such factors as:
  - Method validity (reliability)
  - Laboratory experience with method
  - Sampling techniques and preparation including volume of sample, preservation, and time of transport
  - Laboratory availability/capabilities
  - Precision and accuracy
  - Detection limits
  - Costs of analysis
- Health effects. Consideration of the potential health effects of a chemical encompasses the (1) suitability of the available data for assessing the toxicology of the chemical and (2) the possibility of human health concern from exposure in drinking water. The human health concerns relate to acute and chronic toxicities, carcinogenic effects including effects in animals or humans, and other toxicological concerns such as if a contaminant is a mutagen or teratogen. Assessment of the potential health effects also considers the assessments made by scientific

bodies such as the International Agency for Research on Cancer (IARC) and the National Academy of Sciences (NAS).

- Occurrence in drinking water:
  a. Actual occurrence. Consideration of occurrence data encompasses both the frequency of occurrence, the level of occurrence, and the extent of the population exposed. An examination of the available data in regard to its representativeness is made along with an evaluation of the quality of the data.

  EPA has conducted a number of national sampling surveys to assess potential occurrence of certain contaminants in drinking water across the country. In addition, a number of states have conducted surveys of public water systems for certain contaminants. These two sources constitute the best sources of available data for occurrence of contaminants in drinking water. However, there are limitations of this information for certain contaminants, such as:

    — Extent of sampling
    — Samples analyzed for a limited number of contaminants
    — Representativeness of sampling sites
    — Reliability of resultant data (quality assurance, limits of detection)

  The extent and quality of the available data varies for each of the contaminants under consideration. Thus, EPA must sometimes base its decision on appropriate regulatory action (or no action) for certain contaminants on an imperfect data set.

  b. Potential for occurrence. For contaminants that have been detected in drinking water but for which data are limited, an analysis of the potential for drinking water contamination is conducted. Factors considered in this analysis in decreasing order of priority are the following:

    1. Occurrence in drinking water other than community water supplies. Certain contaminants have been detected in private wells but not in public water systems, usually because of limited sampling programs. For the most part, this factor deals with pesticides that have been detected during certain studies of pesticide usage and drinking water contamination.

    2. Direct or indirect additives. Numerous contaminants are in drinking water as a result of direct addition as a water treatment chemical or indirectly through such actions as leaching from pipe coatings or corrosive actions on piping materials. Pesticides registered for use in or around drinking water fall into this category.

    3. Ambient surface water or ground water. Contaminants detected in surface waters or in ground waters through vari-

ous water quality surveys or in sampling around hazardous waste sites have the potential for contaminating drinking water.

4. Present in liquid or solid waste. Contaminants known to be in industrial or municipal waste water effluents or in waste ponds or known to be in solid waste being disposed in landfills have potential to migrate to drinking water intakes.

5. Mobile to surface water (run-off) or ground water (leaching). The physical/chemical characteristics of contaminants are examined to determine their potential for movement to a drinking water supply. This is essentially an analysis of the fate and transport of contaminants looking toward the potential for contamination of drinking water sources.

6. Widespread dispersive use patterns. This evaluation assesses the characteristics of the use of a contaminant and the locations of that use that would contribute to potential widespread contamination problems in drinking water.

7. Production rates. An assessment of the amount of contaminant being produced annually to assess if the potential exists for significant contamination.

Figure 20.1 summarizes the decision logic discussed above. Each chemical is evaluated based upon these criteria, with a final decision resulting in the proposal of a regulation, nonregulatory guidance (health advisory), or no action.

20.4.2   Levels for RMCLs

20.4.2.1   Phase I: VOCs

RMCLs were proposed on June 12, 1984, for the following VOCs: trichloroethylene, tetrachloroethylene, 1,2-dichloroethane, carbon tetrachloride, vinyl chloride, benzene, 1,1-dichloroethylene, 1,1,1-trichloroethane, and *para*dichlorobenzene. For substances not considered to be carcinogens (1,1,1-trichloroethane and *para*dichlorobenzene), RMCLs were proposed at levels based upon the acceptable daily intake (ADI) for each substance. The ADI was determined using standard toxicological procedures for assessing noncarcinogens. The methodology is as follows:

- Identify the highest no-observed-adverse-effect-level (NOAEL) based upon an assessment of human or animal data (usually animal).
- Determine the ADI. The NOAEL is divided by an appropriate "uncertainty" or "safety" factor to accommodate for extrapolation of animal data to the human, for existence of weak or insufficient data, and for individual difference in human sensitivity to toxic agents. ADIs are reported in mg/kg/body weight/day.

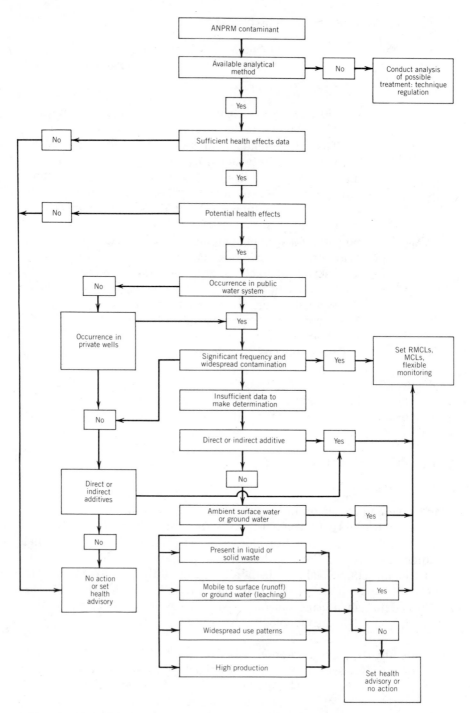

Figure 20.1. Decision logic to determine for which contaminants RMCLs should be set.

- Adjust the ADI to mg/L (AADI) by factoring in the weight of the consumer (i.e., 70 kg) and the amount of water consumed per day (i.e., 2 L).
- Determine the RMCL. Multiply the AADI by the estimated contribution of drinking water to total daily intake.

RMCLs for substances considered to have carcinogenic effects (trichloroethylene, tetrachloroethylene, 1,2-dichloroethane, carbon tetrachloride, vinyl chloride, benzene, and 1,1-dichloroethylene) were proposed at the zero level. This was derived directly from the legislative history of the SDWA (8) which stated that,

> . . . the recommended maximum [contaminant] level must be set to prevent the occurrence of any known or anticipated adverse effect. It must include an adequate margin of safety, unless there is no safe threshold for a contaminant. In such a case, the recommended maximum contaminant level should be set at zero level.

The following is a summary of the proposed RMCLs and the AADIs that were determined for each substance. AADIs were calculated for carcinogens as well as noncarcinogens in order to add some perspective on the chemical's total toxicity including noncarcinogenic end points. However, AADIs were not used to determine the RMCLs for carcinogens.

VOC	Proposed RMCL (mg/L)	AADI (mg/L)
Tetrachloroethylene	zero	0.085
Trichloroethylene	zero	0.26
1,2-Dichloroethane	zero	0.26
Carbon tetrachloride	zero	0.025
Vinyl chloride	zero	0.06
Benzene	zero	0.025
1,1-Dichloroethylene	zero	0.35
1,1,1-Trichloroethane	0.2	1.0
para-Dichlorobenzene	0.75	3.75

Public comment was requested on all areas of the proposal and specifically on several major issues that have a bearing on future regulatory activities, as well as the VOC proposal. These issues are:

Should RMCLs for carcinogens be set at zero? If RMCLs are set at zero, what guidance, if any, should be provided on the actual attainable target levels in drinking water?

RMCLs were proposed at the zero level for carcinogens because the SDWA specifies that RMCLs are to be set at the "no-effect" level incorporating an adequate margin of safety. Since science cannot presently demonstrate the existence of a threshold level for carcinogens, it may conservatively be assumed that no threshold or "no-effect" level exists. RMCLs set at zero are thus in accord with the nonthreshold philosophy on carcinogens.

RMCLs set at zero would present a general philosophy that as a goal, carcinogens should not be present in drinking water.

The argument against zero as the RMCL level is that zero is unachievable and unmeasurable. Due to limitations in analytical techniques, it is impossible to say with uncertainty that the carcinogen is not in the drinking water. In addition, the argument was advanced that RMCLs at zero do not provide practical guidance to public water systems on achievable goals for contaminant removal and may be perceived as a moving target as analytical capabilities increase.

Should RMCLs be set at the analytical detection limit? Since RMCLs at zero are theoretically unattainable, basing the levels upon defined state-of-the-art analytical detection limits would provide measurable goals for carcinogens in drinking water. The analytical detection limit is, for all practical purposes, the functional equivalent of zero and thus would present the same philosophy as a zero RMCL. However, analytical detection limits are also moving targets as the state-of-the-art in chemistry progresses.

Should RMCLs be set at a nonzero level based upon excess cancer risk calculations? Mathematical models have been used to project the risk of excess cancer cases occurring in a population as a result of a chemical of specified concentration being present in the drinking water. These models are based upon the assumption that health effects in humans at low dose levels can be extrapolated from results observed in animals at high dose levels. Many areas of uncertainty exist in respect to high- to low-dose extrapolation, including (1) the heterogeneity of sensitivity of exposed populations, (2) the pharmacokinetic behavior of the toxic agent, and (3) mechanisms of action. These areas of uncertainty lead to large variations in risk estimates based upon the specific assumptions of the model, and can provide at best only a rough estimate of the carcinogenic hazard.

Many different models have been used to estimate carcinogenic risk. The EPA has generally employed the linear multistage model; a model that is linear at low doses and does not exhibit a threshold. This model estimates the upper bound of the excess cancer risk at a specific exposure level for a 70-kg adult, consuming 2 L of water per day over a 70-year lifespan. None of the models is experimentally verifiable, as there is no scientifically valid method for determining the actual risks at low environmental exposure levels.

Theoretically, RMCLs could be set based upon an excess cancer risk level low enough to be considered a "virtually safe" level. Such a level is not really very different than zero and could be argued to fit the requirement that the RMCL be set at the no-effect level with an adequate margin of safety. RMCLs set at finite levels would present an actual target for water systems to strive for and would present guidance on the relative risks of the contaminants.

The selection of a target risk for setting the RMCLs is problematic. Such a decision must be based upon a societal judgment on what constitutes a negligible risk and is not strictly based upon science. Federal regulations for environmental contaminants have generally fallen in the 10^{-4} to 10^{-6} lifetime risk range, taking costs and feasibility into consideration. Thus, EPA considered two risk levels as possibly representing an upper limit of risk; 1 in 100,000 (10^{-5}) probability per 70 years of exposure and 1 in 1,000,000 (10^{-6}) probability.

How strong should the scientific evidence be to justify regulating a substance for carcinogenicity? RMCLs were proposed at zero for trichloroethylene, tetrachloroethylene, 1,2-dichloroethane, carbon tetrachloride, vinyl chloride, benzene, and 1,1-dichloroethylene. Two of these substances, vinyl chloride and benzene, are known human carcinogens and 1,2-dichloroethane and carbon tetrachloride have strong evidence of animal carcinogenicity. Trichloroethylene, tetrachloroethylene, and 1,1-dichloroethylene have only "limited" evidence of animal carcinogenicity. This classification as "limited" evidence is based upon several factors, including lack of replication in multiple experiments and questions concerning the indicators of carcinogenicity.

The evidence in question was the use of mouse liver tumors as indicators of carcinogenicity for trichloroethylene and tetrachloroethylene. The concern exists due to the fact that several strains of laboratory mice appear to develop a high proportion of liver tumors with or without exposure to the chemicals. This may indicate that the mouse liver contains a significant proportion of initiating tumor cells which do not appear to be present in the human liver or it may indicate bias in the test protocol currently used by the National Toxicology Program (corn oil gavage). Thus, the relevance of mouse liver tumors to human carcinogenic risk and classifying these compounds as carcinogens based upon this evidence have been questioned.

The extent of the evidence needed to justify regulating a substance as a carcinogen is an issue upon which there is no scientific consensus. In addition to the mouse liver tumor issue, questions remain concerning the pathological interpretation of tumor data, the statistical procedures used to analyze tumor incidence data, and the weight that should be given to positive studies versus negative studies.

However, several scientific groups have classified the strength of evidence of carcinogenicity for specific compounds. The IARC (9) has developed a classification system for carcinogens based upon strength of evidence. The classification system is based upon a qualitative review of the relevant information bearing upon whether or not an agent may pose a carcinogenic hazard. Information included in the assessment are short-term tests, long-term animal studies, human studies, pharmacokinetic studies, comparative metabolism studies, structure-activity relationships, and other relevant toxicological studies.

The IARC assesses studies as showing (1) sufficient evidence of carcinogenicity, (2) limited evidence of carcinogenicity, or (3) inadequate evidence of carcinogenicity. The chemicals are then placed in three groups:

Group 1—Chemical is carcinogenic to humans (sufficient evidence from epidemiological studies).

Group 2—Chemical is probably carcinogenic to humans.

Group 2A—At least limited evidence of carcinogenicity to humans.

Group 2B—Usually a combination of sufficient evidence in animals and inadequate data in humans.

Group 3—Chemical cannot be classified as to its carcinogenicity to humans.

EPA has proposed a categorization scheme for strength of evidence of carcinogenicity based upon the IARC criteria (10). The primary difference between the two approaches is that IARC does not distinguish between those chemicals with inadequate animal evidence of carcinogenicity and those chemicals with no evidence of carcinogenicity, while the EPA scheme makes that distinction. EPA's proposed categorization for carcinogens is as follows:

Group A—Human carcinogen (sufficient evidence from epidemiological studies).

Group B—Probable human carcinogen.

Group B1—At least limited evidence of carcinogenicity to humans.

Group B2—Usually a combination of sufficient evidence in animals and inadequate data in humans.

Group C—Possible human carcinogen (limited evidence of carcinogenicity in animals in the absence of human data).

Group D—Not classified (inadequate animal evidence of carcinogenicity).

Group E—No evidence of carcinogenicity for humans (no evidence for carcinogenicity in at least two adequate animal tests in different species or in both epidemiological and animal studies).

How should the degree of evidence of carcinogenicity be factored in RMCL determinations? As previously discussed, the strength of evidence of carcinogenicity varies greatly among those chemicals classified as carcinogens for determining the RMCLs. The issue was raised that dividing all compounds into two groups does not consider strength of evidence and the likelihood that a particular substance may actually be a carcinogenic threat to humans. Classifying all compounds together and setting RMCLs at zero suggests that the evidence for each compound is equally strong and that the risk from exposure to each is the same. It has been argued that this is not scientifically defensible, as chemicals exhibit varying degrees of evidence of carcinogenicity and varying risks associated with their exposure. In particular, having the RMCLs for benzene and vinyl chloride, both known human carcinogens, being equal to the RMCLs for trichloroethylene and tetrachloroethylene, with limited animal evidence of carcinogenicity, was questioned.

Should an RMCL and an MCL be set for total VOCs? In addition to setting RMCLs/MCLs for individual contaminants, an RMCL/MCL could be set for total VOCs. Such a level would protect against situations in which many contaminants were found in drinking water, all just below the MCL. In such an instance, without an MCL for total VOCs, no action need be taken.

The difficulty in setting an RMCL for total VOCs revolves around the basis upon which the level would be set. Additivity cannot be assumed among chemicals, as antagonistic or synergistic effects may also be seen. Scientifically, it is very difficult to justify a level for a mixture of compounds.

Final RMCLs were promulgated on November 13, 1985 for all of the VOCs except tetrachloroethylene. The public comment period on tetrachloroethylene was reopened because of new data, with the final RMCL to be promulgated after the closing of the comment period.

The final RMCLs for the eight VOCs were set using a three-category approach based upon the EPA and IARC criteria as follows:

Category I—Known or probable human carcinogens: Strong evidence of carcinogenicity. RMCLs set at zero.
  • EPA Group A or Group B
  • IARC Group 1, 2A, or 2B
Category II—Equivocal evidence of carcinogenicity. RMCLs set based upon the AADI with an additional uncertainty factor or upon a lifetime risk calculation.
  • EPA Group C
  • IARC Group 3
Category III—Noncarcinogens: Inadequate or no evidence of carcinogenicity in animals. RMCLs set based upon the AADI.
  • EPA Group D or E
  • IARC Group 3

The following is a summary of the final RMCLs for the VOCs based upon their categorization in the three-category approach:

Compound	RMCL	Category
Benzene	Zero	I
Vinyl chloride	Zero	I
Carbon tetrachloride	Zero	I
1,2-Dichloroethane	Zero	I
Trichloroethylene	Zero	I
1,1-Dichloroethylene	7 μg/L	II
1,1,1-Trichloroethane	200 μg/L	III
p-Dichlorobenzene	750 μg/L	III

20.4.3 Levels for MCLs

The approach to setting MCLs involves evaluating the availability and performance of technologies for control of contaminants in drinking water, and assessing the costs of the application of technologies to achieve various levels that would be close to the RMCL. The process followed is:

- Identify the available analytical methods.
- Determine what are reasonable expectations of technical performance at levels considered for the MCLs.
- Determine the costs of analysis.
- Identify technologies for control of a contaminant.
- Determine the performance of the technologies (i.e., what levels are achievable by varying intensities of treatment) and assess other feasibility factors (e.g., limits of analytical detection).
- Determine the costs of the technology achieving various levels.
- Assess the economic impact of applying the technology to achieve various levels.
  - Public water system costs impacts: capital and operating costs, increases in monthly family water bills.
  - National cost impacts: capital and operating (i.e., annual) costs.
- Judgment on the MCL level: based on performance, feasibility factors and costs.

The judgment on the MCL can be influenced by any of the key decision criteria pertinent to that particular compound. These criteria include:

- Limits of technology performance.
- Limits of analytical detection.
- Other cost and feasibility factors.

MCLs were proposed on November 13, 1985 for the eight VOCs. EPA determined that three analytical methods were "economically and technologically feasible" for compliance with one or more of the proposed MCLs. EPA examined the performance of laboratories using the proposed analytical methods to determine the practical quantitation level (PQL). The PQL is the lowest level that can be reliably achieved within specified limits of precision and accuracy. The PQL thus represents the lowest level achievable by good laboratories within specified limits under routine laboratory conditions. The PQL is determined through interlaboratory studies, such as performance evaluation studies.

Treatment technologies for the removal of VOCs were examined. EPA determined that all eight VOCs could be removed from drinking water using packed tower aeration or activated carbon adsorption. The costs for packed tower aeration were determined to be 5¢–15¢/1000 gallons. For granular activated carbon, costs of 10¢–90¢/1000 gallons for medium to large systems and 40¢–$1.50/1000 gallons for small systems were determined.

The MCLs were proposed as follows:

Compound	MCL mg/L
Trichloroethylene	0.005
Carbon tetrachloride	0.005
Vinyl chloride	0.001
1,2-Dichloroethane	0.005
1,1,1-Trichloroethane	0.20
p-Dichlorobenzene	0.75
Benzene	0.005
1,1-Dichloroethylene	0.007

20.4.4   Monitoring and Reporting Requirements

20.4.4.1   Revised Regulations

The Interim Regulations require monitoring to assess compliance with the MCLs at set frequencies for certain contaminants; for example, monitoring for inorganic compounds must be conducted at least once per year or once per 3 years for supplies using surface or ground water sources, respectively. While monitoring once a year or every 3 years does not seem to be overly demanding, this can be a burden upon small systems, and upon those states that conduct monitoring for certain systems (e.g., small systems) within their boundaries. States have reported that certain of these inorganic compounds have not been detected at significant levels in the drinking water in many systems and the probability of future contamination is very slight. Monitoring has shown that little change in concentrations occurs over time for certain contaminants, primarily inorganics in ground water. In addition, some contaminants such as the six pesticides in the Interim Regulations have been found only rarely since compliance monitoring requirements went into effect.

To provide for more efficient use of state and local resources, flexibility in monitoring requirements will be a general principle in development of the revised regulations. In addition, to assure detection and control of intermittent contaminants or those that are not homogeneously distributed, more specific monitoring requirements will be designed.

A three-tiered approach has been developed for determining whether and in what manner to regulate specific contaminants. Drinking water contaminants would be divided into three tiers for regulatory purposes:

Tier I—Those that occur with sufficient frequency and are of sufficient concern to warrant national regulation (MCLs) and consistent monitoring and reporting.

Tier II—Those that are of sufficient concern to warrant national regulation (MCLs) but occur at limited or predictable frequency, justifying flexible national minimum monitoring requirements to be applied by state authorities and expanded as needed.

Tier III—Those that would not warrant development of a regulation but for which nonregulatory health guidance could be provided to States or water systems.

Monitoring and reporting requirements were proposed for the eight VOCs as part of the MCL proposal. The VOCs were included in the second tier of the three-tiered approach. EPA proposed a monitoring scheme that called for monitoring requirements to be phased in depending upon the size of the systems, a differentiation between ground water and surface water systems, and state discretion on repeat monitoring.

20.4.4.2 Unregulated Contaminants

The extent of contamination is currently unknown for many contaminants because there has not been a systematic approach to determine if contaminants are present in drinking water supplies across the country. The revised regulations may contain monitoring requirements for many contaminants that are not presently being monitored for, however the implementation of the regulations are still several years away.

EPA has proposed a monitoring regulation for unregulated contaminants in drinking water as part of the Phase I MCL proposal. This regulation will require public water systems to sample their drinking water and analyze for potential contamination, supplying much needed information on the quality of the nation's drinking water supplies.

20.4.5   Phase II and Phase IV Regulations

20.4.5.1   Pesticides and Synthetic Organic Chemicals

The greatest portion of organic contaminants in drinking water will be encountered in the second phase of the ongoing process for regulation of drinking water quality.

Industrial synthetic organic chemicals (SOCs) and numerous pesticides have been proposed for regulation or health advisory development under the auspices of Phase II of the Revised Primary Drinking Water Regulations (6).

Health advisories are nonregulatory guidance for inorganic and organic contaminants for noncarcinogenic endpoints of toxicity. These evaluations are considered to be exposure levels that would not result in adverse health effects over a specified short-term time period. Draft health advisories were prepared for 51 compounds in 1985.

Table 20.3 summarizes the pesticides and SOCs proposed for regulation and health advisory development as part of the Phase II proposal. Other pesticides and SOCs will be considered for inclusion in later iterations of the revised regulations.

Pesticides have been detected in drinking water supplies across the country. These compounds have been detected in both surface and ground

TABLE 20.3
Phase II: Pesticides and Synthetic Organic Chemicals Proposed for
Regulation and Health Advisory Development

Pesticides (regulations)

Alachlor	Heptachlor epoxide
Aldicarb	Hexachlorobenzene
Carbofuran	Lindane
Chlordane	Methoxychlor
2,4-D	Pentachlorophenol
Ethylene dibromide (EDB)	2,4,5-TP
Heptachlor	Toxaphene

Synthetic Organic Chemicals (regulations)

Dibromochloropropane (DBCP)	Monochlorobenzene
1,2-Dichloropropane	PCBs
ortho-Dichlorobenzene	Styrene
Cis- and *trans*-1,2-dichloroethylene	Toluene
Epichlorohydrin	Xylene
Ethylbenzene	

*Pesticides and Synthetic Organic Chemicals
with Only Health Advisories*

Atrazine	Endrin
meta-Dichlorobenzene	Hexachlorobenzene
2,3,7,8 TCDD (Dioxin)	

waters, and the presence of the chemicals appears to be dependent upon the pesticide application mode and geological factors.

Synthetic organic chemicals occur in drinking water as the result of contamination from hazardous waste sites and industrial discharges. Ground water contamination from hazardous wastes is of primary concern, for it is difficult to detect and once contamination occurs it is difficult to remedy the situation.

20.4.5.2 Disinfectants

The disinfection process is an area of increasing interest. Questions have been raised on the possible relationship between drinking water disinfectants and cancer, hypertension and cardiovascular diseases. Chlorine and other by-products, which include chloramines, dihaloacetonitriles, chlorinated phenols, trichloroethanol, trichloroacetaldehyde and chlorine dioxide, chlorite, and chlorate will be examined under Phase IV of the revised regulations. In addition, alternate disinfectants such as ozone, bromine, and iodine will be examined.

20.5   CONCLUSIONS

As EPA moves forward through the 1980s, we will be faced with new scientific and practical challenges. These challenges will touch upon the forefront of scientific knowledge, as our analytical capabilities increase and toxicology and other biological sciences move forward. Practical challenges such as the bearability of costs and accessibility of appropriate treatment technologies for all water supplies, but in particular for small community systems, will require new strategies. Developing and implementing practical monitoring programs, such as the three-tiered approach for the revised regulations, will remain a high-priority area for the years ahead. Controlling organic contamination at the source and through the regulatory process will remain a prime issue for the 1980s.

REFERENCES

1. The Safe Drinking Water Act, 42 U.S.C. s300f *et seq.* as amended through December 1980.
2. 47 *Federal Register* 9350, March 4, 1982.
3. 49 *Federal Register* 24330, June 12, 1984.
4. 50 *Federal Register* 46880, November 13, 1985.
5. 48 *Federal Register* 45502, October 5, 1983.
6. 50 *Federal Register* 46936, November 13, 1985.
7. 51 *Federal Register* 11396, April 2, 1986.
8. H.R. Rep. No. 93-1185, 93d Cong., 2d Sess. 20 (1974).
9. The International Agency for Research on Cancer, IARC Monographs on the Evaluation of the Carcinogenic Risk of Chemicals to Humans, Supplement 4, 1982, pp. 11–14.
10. EPA Proposed Guidelines for Carcinogen Risk Assessment, 49 FR 46294.

INDEX

Acceptable daily intake (ADI), 417, 519, 521
ACL (average contaminant level), 96
Activated carbon:
  as adsorbent, 237
  and adsorption, 113
  granular, 12, 246–248
  powdered, 246–248
Activity coefficient, 104
ADI (acceptable daily intake), 417, 519, 521
Adsorption:
  and breakthrough curve, 250, 258
  and carcinogen detection, 112–113, 122
  competitive, 239, 241
  concepts of, important:
    competitive adsorption, 239, 241
    feed composition, variability in, 241,
      243–244
    microbial activity, 244–245
    PAC vs. GAC, 246–248
    regeneration, 245–246
    service time, 238–239, 245, 248, 250
    THM precursors vs. SOCs, 239
  data acquisition for process design, 256–260
  design considerations for:
    backwashing, 252
    EBCT and application rate, 251–252
    examples of, 252–253, 258
    and ground water, 255–256
    prechlorination, 251
    process, 248–249
    THMs vs. THM precursors removal,
      249–250
  objectives of, 237
  process costs, 260–262
  and removal of organic compounds, 505–
    508
  of SOCs, 239
  surface diffusitivities in, 257
  of THM precursors, 239
  and TOC breakthrough, 241, 262
Aeration, 506–507
Aggregate population studies, 373–374
Air stripping, 506–507
Alternative disinfection processes, see
    Disinfection, alternatives to
    chlorination

Alum coagulation:
  coagulation mechanisms in, 207–208
  and direct filtration, 230–231
  effectiveness of, 233–234
  function of, 199
  and NPTOC removal, 208
  and pH, 203–204
  and stoichiometry, 204, 206–207
  and THM precursor removal, 208
Ames test:
  description of, 303–304
  and DHANs, 6
  as genotoxicity test, 300
  and structure-activity relationships, 298
Amino acids, 51, 76–79
Analytical schemes, comprehensive, 17. See
    also MAS (Master Analytical Scheme)
Analytical studies
  bias in, 328–329
  and case-control studies, 326–327
    incidents of, 329–334
  confounding in, 327–328
  effect modification in, 329
  measure in, 325
Animal extrapolation, 443–444
Animal husbandry, 307
Aquatic humic substances. See also Humic
    substances
  definition of, 63, 67–68
  description of, 66–68
  in ground water, 68–69
  and isolation procedure, 67
  in lake water, 72–73
  in river water, 70–72
  in seawater, 69–70
  in wetland water, 73–74
Aqueous chlorine:
  chemistry of:
    and chloramines, 38–41
    and inorganic substances, 37–38
    and nitrogenous organic compounds, 41
    and organic compounds, 40–41
    reactivity of, 36–37
    in water, 34–36
  and chlorination reactions, specific:
    amino acids, 51–52

531

Aqueous chlorine, and chlorination reactions, specific (*Continued*)
  aromatic ring, 47–49
  carbon-carbon double bonds, 47
  haloform, 49–50
  combined, 39
  distribution of forms in natural waters, 35
  electrophilicity reactivity of, 36
  free, 38–39
  nature of, 34
  odor control by, 44–45
  as oxidant, 45
  redox potential of, 36
  taste control by, 44–45
  and treatment of water supplies:
    and chlorination, 43–44, 45–47
    disinfection by chlorine, 42–43
  use of, 33–34
  see also Chlorine
Arochlor number, 6
Aromatic compounds, 47–49
Artifacts:
  and adsorption methods, 112–113
  contamination, 108
  and ion-exchange methods, 116
  matrix interference, 120
  problems of, 101
  system blanks, 108
Assessment, *see* Epidemiologic assessment;
    QRA; Risk assessment; Toxicologic
    assessment
Average Contaminant Level (ACL), 96

Background response, 347–348
Backwashing, 252
Batch equilibrium tests, 257–258
Bias, 328–329
Biochemical oxygen demand (BOD), 10–11
Biocidal activity, and disinfectants:
  chloramines, 266–270
  chlorine dioxide, 266–270
  extrapolation, from laboratory to field,
    271–272
  free chlorine, 266–270
  ozone, 266–270
  purpose of, 266
Biodegradation, 244–245
Biologic plausibility, 336
BOD, *see* Biochemical oxygen demand, 10–
  11
Breakpoint reaction, 39, 45
Broad spectrum analysis, 97–99, 112, 121
Bromine, 277, 287
Bromine chloride, 277, 287

Cancer:
  latency period of, 467
  site, consistency of, 397
  studies of, 374, 376
    case-control, 390–391, 394–395
    historical perspective, 375–376
    methodology, 373–374

Mississippi River study, 376–377
  New Orleans water supply, 376
  prospective, on individuals, 395, 397
  results of, 397–400
  in United Kingdom, 377–379, 382–384,
    388
  in United States, 388, 390
  and trace organic contaminants, 21
  and water supplies, 398–400
  see also Carcinogens; Risk assessment
Canton plant, *see* Grasse River
Carbohydrates, 79–81
Carboxylic acids, 74–76
Carcinogenesis:
  chloroform, 356–359, 361, 363
  epigenetic, safety factors for, 447
Carcinogenicity testing:
  description of, 294–295
  and EPA, 312
  problems with *in vivo:*
    agents modifying, 307–310
    interpretation of testing results, 310–313
    tumor promoters, 310
  programs, 294–295
  see also Carcinogens
Carcinogens:
  analysis of:
    error sources in, 101
    objectives of, 96
    recovery of, 108, 109, 111, 115–116,
      119
    sensitivity of, 105, 108
  in beverages, 479–481
  from chlorination, 481–485
  exposure to, 21
  in foods, 479–481
  in ground water, 318
  from involuntary actions, 485
  from occupations, 485
  potency of, 21, 26
  testing for, 294–295
  from voluntary actions, 485
  and vulnerability of individual, 26–27
  see also Cancer; Organic compounds; Risk
    assessment; Trace organic
    contaminants; *specific identification
    methods*
Case-control studies:
  and epidemiologic assessment, 374, 390–
    391, 394–395
  incident:
    choice of study population, 331–332
    exposure assessment, 332–333
    hospital-based *vs.* population based
      studies, 330–331
    questionnaire design, 332
    size of study population, 330
    summary of, 333–334
    types of, 329
  on mortality records, 326–327
Cationic polyelectrolytes, *see* Cationic
    polymers

Cationic polymers:
  and coagulation of humic substances, 209–210
  and direct filtration, 225–226, 228
  dosage selection of, 225–226
  at Glenmore Reservoir, 228
  at Grasse River, 226, 228
CERCLA, *see* Comprehensive Environmental Response, Compensation, and Liability Act, 502–503
Charge Neutralization/Precipitation model, 209–210
Chemical interactions, 453–455
Chloramines:
  application of, as disinfectants, 285
  and aqueous chlorine, 41
  and biocidal activity, 269–270
  by-products of, 279
  and disinfection, 43–44, 46
  formation of, 38–40, 51
  and organic halogen, 283
  and THM formation, 272–275
Chlorinated organic compounds, 40
Chlorinated phenols, 365. *See also* Phenols
Chlorination:
  biological applications of, other, 43–44
  by-products of:
    chlorinated phenols, 365
    chlorine, 367–368
    haloacetonitriles, 363–365
    halogenated phenols, 366
    identifying, 6
    ketones, 366
    THMs, 200, 353, 356–359, 361, 363
  and carcinogens, 318, 481–485
  chlorine used in, 7–8
  concerns of, other, 354–356
  and disinfection, 43
  and haloform, 46
  and odor control, 44–45
  and organic compounds, 3
  for oxidation, 45
  practice, 45–47
  reactions, specific:
    amino acids, 51–52
    aromatic ring, 47–49
    carbon-carbon double bonds, 47
    haloform, 49–50, 52
  safety of, 353–354
  and taste control, 44–45
  and THMs, 200, 353, 356–359, 361, 363
  use of, first, 5
  see also Disinfection
Chlorine:
  application of, as disinfectant, 284–285
  by-products of, 278–279
  demand, 40
  disinfection by, 42
  electrophilicity reactivity of, 36
  hydrolysis of, 34–36
  ingestion of, 367–368
  and organic halogen, 281–282

  redox potential of, 36–37
  see also Aqueous chlorine; Free chlorine
Chlorine dioxide:
  application of, as disinfectant, 286
  by-products of, 279
  and disinfection, 43
  and organic halogen, 283
  and THM formation, 275–276
Chloroform:
  carcinogenesis, 356–359, 361, 363
  and carcinogens, 481–485
  and risk assessment, 422–423
Chloroorganic compounds, 6–7
Chromatogram, 134
Chromatographic methods:
  gas, 154–157
  liquid, 157–158, 160–163, 165
  see also GC/MS (gas chromatography/mass spectroscopy)
Clean Water Act, 503–504
Closed loop stripping analysis, *see* CLSA
CLSA (closed loop stripping analysis), 107–109, 121–122
Coagulant aids, 199
Coagulants, 199
Coagulation:
  background of, 201–202
  and cationic polyelectrolytes, 209–210
  and jar test procedures, 202–203
  mechanisms, 207–208
  see also Alum coagulation; Humic substances, coagulation of
Community Water Supply Study (CWSS), 183, 494–495
Compliance monitoring, 177
Composite sampling, 95–96
Compounds, *see specific compounds*
Comprehensive Environmental Response, Compensation, and Liability Act (CERCLA), 502–503
Concentration/isolation methods:
  for analysis of organics in water, 102–104
  analytical approaches to:
    adsorption methods, 112–113
    CLSA, 107–109
    distillation methods, 109
    HPLC, 110–112
    ion-exchange methods, 114–116
    membrane methods, 116–119
  for organics in water, 102–104
  purge and trap technique, 106–107
  QA/QC (Quality Assurance/Quality Control), 99–102
  residue analysis, 119–120
  static headspace analysis, 104–106
  types of, 97–99, 112
  and (Limit of Detection), 99–101
  purpose of, 99
  and sampling environment, 95–96
Confounding, 327–328, 373–374
Contaminants, *see* Organic compounds; Trace organic contaminants

Contamination, 108
Conventional water treatment:
   and coagulation of humic substances, 201–
      204, 206–210
   functions of, 199–201
   and plant performance:
      discussion of, 219–220
      plant monitorings, 214, 217, 219
      UV as surrogate parameter, 220, 222–
        224
      water plants, 212–214
      water supplies, 211–212
Cresols, 5

Data quality objectives (DQOs), 177–178
Desorption, 243
Detection limits, *see* LOD
DHANs (dihaloacetonitriles), 3, 5–6, 13,
   52
Dialysis membranes, 116
Dichloracetic acid, 13, 51
Dichloramine, 39
Dihaloacetonitriles (DHANs), 3, 5–6, 13,
   52
Direct filtration:
   and alum, 230–231
   and cationic polymer dosage selection, 225–
      226
   description of, 199–200, 224
   in-line, 228–230
   and NPTOC removal, 231–233
   and performance with cationic polymers,
      226, 228–230
   studies of, pilot plant, 225
   and THM precursor removal, 231–233
Direct injection techniques, 113
Disease rates, 334
Disinfectants:
   application of, 284–288
   and biocidal activity:
      alternatives to free chlorine, 270–271
      chloramines, 266–270
      chlorine dioxide, 266–270
      extrapolation of, from laboratory to
        field, 271–272
      free chlorine, 266–270
      ozone, 266–270
      purpose of, 266
   bromine, 277, 287
   chemistry of, 268–270
   chloramines, 272–275, 285
   chloride, 277, 287
   chlorine, 284–285
   chlorine dioxide, 275–276, 286
   ferrate ion, 277, 287
   free chlorine, 265–266
   iodine, 277, 286–287
   microorganism effects of, 268
   as oxidants, 354
   ozone, 276–277, 285–286
   permanganate, 277, 287
   pH, 287
   purpose of, 266

   regulations for, revised, 529
   silver, 277, 287
   UV radiation, 277, 288
   see also Disinfection
Disinfection:
   and alternatives to chlorination
      bromine, 277
      chloramines, 272–275, 279
      chloride, 277
      chlorine dioxide, 275–276, 279
      ferrate ion, 277
      hydrogen peroxide, 277
      iodine, 277
      ozone, 276–277, 280
      permanganate, 277
      silver, 277
      UV radiation, 277
   and application of disinfectants, 284–
      288
   and biocidal activity:
      alternatives to free chlorine, 270–271
      chlorine dioxide, 266–270
      extrapolation of, from laboratory to
        field, 271–272
      free chlorine, 266–270
      ozone, 266–270
      purpose of, 266
   by-products of, other than THM:
      inorganic, 280
      organic, 278–280
      organic halogen, 280–284
   by chlorine, 42, 278–279
   of drinking water supplies, 3, 5–8
   types of, 43
   see also Chlorination; Disinfectants
Dissolved organic compounds, *see* DOCs
   (dissolved organic compounds)
Distillation methods, 109, 122
Distribution coefficient, 103, 114
DNA excision repair, 304
DOCs (dissolved organic compounds):
   and amino acids, 76–79
   and aquatic humic substances:
      definition of, 63
      description of, 66–68
      ground water, 68–69
      and isolation procedure, 67
      lake water, 72–73
      river water, 70–72
      seawater, 69–70
      wetland water, 73–74
   and carbohydrates, 79–81
   and carboxylic acids, 74–76
   continuum for, 58–60
   histogram of, 62–66
   and hydrocarbons, 81–86
   knowledge about, 55
   in natural waters, 60–62
   terms for, 56–58
   understanding, 86–87
Dose-response relationships, 335–336, 408,
      413–414
DQOs (data quality objectives), 177–178

Drinking Water Standards, 405–406
Drinking water supplies:
  and cancer risks, 398–400
  chlorination of, 5
  legislation for standards of, 18–21
  microorganics in, 8, 20–21
  organic compounds in:
    assessment of risk of, 405–406
    chloroform, 422–423
    measurement of, 494–495, 497
    occurrence of, 512–514
    styrene, 420–421
    toluene, 421–422
    vinyl chloride, 424–425, 432
  trace organic contaminants in, 3–4, 21
  treatment of:
    aqueous chlorine, 44–45
    chlorination, 43–44, 45–47
    disinfection by chlorine, 42–43
    from utilities, studies of, 14
  see also specific water treatment processes
Dynamic column tests, 258–260
Dynamic headspace technique, 106, 107

EBCT (empty bed contact time), 238–239,
  251–252, 258
Ecologic studies, 373
EDF (Environmental Defense Fund):
  and Mississippi River study, 376–377
  and New Orleans water supply, 376
Effect modification, 329
Electrophiles, 36
Empty bed contact time (EBCT), 238–239,
  251–252, 258
Enforcement aspects, *see* Legislation;
  Regulatory significance
Environmental Defense Fund, *see* EDF
  (Environmental Defense Fund)
Environmental hazards:
  background of, 463–464
  from human activities, 478–481
  naturally occurring, 476, 478
Environmental Protection Agency, *see* EPA
EPA (Environmental Protection Agency):
  and carcinogenesis, 356–357
  and carcinogenicity, 312
  challenges facing, 530
  and drinking water standards, 497, 511
  and ground water, 14, 504
  and MCLs, 19
  and New Orleans water supply, 376
  and purge and trap techniques, 106
  and QA project plans, 178
  and regulation of contaminants in drinking
    water, 9
  and revised primary drinking water
    regulations, 515
  and THMs, 200, 481–482
  and trace organic contaminants, 4
  and VOCs, 19
Epidemiologic assessment:
  and analytic studies
    bias, 328–329

    case-control studies, 326–327
    confounding, 327–328
    effect modification, 329
    historical cohurt studies, 325–326
    incident cases and controls, 329–334
    measures in, 325
  evaluation of results of, guidelines for:
    associations, strength of, 335
    biologic plausibility, 336
    disease rates, temporal considerations of,
      334
    dose-response relationships, 335–336
    findings, consistency of, 335
  factors influencing, 318–320
  and indirect studies:
    definition of, 321
    exposure estimates, 324
    mathematical model, 322–323
    morbidity records, 321–322
    mortality records, 321–322
    regression considerations, 323–324
    strengths of, 324–325
    weaknesses of, 324–325
  of micropollutants, organic:
    case-control studies, 390–391, 394–395
    historical perspective of, 375–376
    hypotheses testing, 377–379, 382–384,
      388, 390
    methodological context of, 373–375
    Mississippi River study, 376–377
    prospective studies on individuals, 395,
      397
    results of testing, 397–400
  origin of, 317–318
  risk in, measures of, 320
Evaporation and freezing technique, 119–120,
  122
Exploratory activities, 176–177
Exposure:
  assessment, 408, 414–416
  limitations of, 437–439
Extractors, continuous, 111

Federal Insecticide, Fungicide, and
  rodenticide Act (FIFRA), 499–500
Federal National Interim Primary Drinking
  Water Regulations (NIPDWRs), 493,
  515–516, 22–25, 18–21
Ferrate ion, 277, 287
FIFRA (Federal Insecticide, Fungicide, and
  Rodenticide Act), 499–500
Flocculation, 228–230
Fluorescence, 11
Fractionation, 17
Free chlorine:
  alternatives to, 270–271
  and biocidal activity, 266–270
  and THM formation, 265–266
  see also Chlorine
FTIR spectroscopy:
  with GC, 155–157
  and HPLC, 162–163, 165
Fulvic acid, 62, 64, 70

GAC (granular activated carbon), 12, 246–248
Gas chromatographic methods, *see* GC (gas chromatographic) methods
Gas chromatography/mass spectroscopy, *see* GC/MS (gas chromatography/mass spectroscopy)
GC (gas chromatographic) methods:
  and capillary column, 135–137
  with conventional detectors, 154
  with FTIR detectors, 155–157
  with spectrometric detectors, 154–155
GC/MS (gas chromatography/mass spectroscopy):
  alternatives to, 153–154
  application of, 131
  and DOCs, 75–76
  and future developments, 169–171
  and organic compounds, 98
  problems of, 131–133
  qualitative analysis of:
    in capillary columns, 135–137
    and CI mass data, 142–146
    with computer, 137–139
    and low-resolution accurate mass data, 142–146
    and MS information, supplementary, 141–142
    pitfalls of, 139
    procedure for, 139–141
    purpose of, 135
  quantitative analysis of:
    and internal standard selection, 147–149
    and multicomponent analysis, 149–150
    principles of, 147
    procedures for, 146–147
  terminology of:
    identification, 133
    internal standard, 134, 147–149
    reconstructed ion-current profile, 134
    response factor, 135
    response ratio, 135
    selected ion monitoring, 135
  and trace organic contaminants, 4, 9
Genotoxicity tests, 300–303
Geographical clustering, 373
Germicidal potency, 42–43
Glenmore Reservoir:
  and alum, 230–231
  cationic polymers at, 226, 228
  description of, 212–214
  monitorings of, 217, 219
  studies at, 225, 233
  THM precursor removal at, 219–220, 232
  and UV as surrogate parameter, 232
  water supplies in, 211–212
Grab sampling, 95–96
Granular activated carbon (GAC), 12, 246–248
Grasse River:
  and alum, 230
  cationic polymers at, 226, 228
  cationic polymer dosage selection at, 225–226

  description of, 212
  and flocculation, 229
  and in-line direct filtration, 229
  monitorings of, 214, 217
  NPTOC removal at, 208
  stoichiometry at, 204, 206, 208
  studies at, 225, 233
  THM precursor removal at, 208, 219–220, 232
  and UV as surrogate parameter, 232
  water supplies in, 211–212
Gross parameters, 167–169
Ground water:
  adsorption in treatment of, 255–256
  aquatic humic substances in, 68–69
  contaminants of, 14, 318
  DOCs in, 60–61
Ground-Water Protection Strategy (GWPS), 504
Ground Water Supply Survey (GWSS), 183, 495, 512
GWPS (Ground-Water Protection Strategy), 504
GWSS (Ground Water Supply Survey), 183, 495, 512

Haloacetonitriles, 363–365
Haloform, 13, 46, 49–50, 52
Halogenated aldehydes, 366
Hazard, 466. *See also* Hazard identification; *specific types of hazards*
Hazard identification:
  definition of, 408–413
  in risk assessment, 466–467
Health risks, *see* Epidemiologic assessment; Risk assessment
Henry's law, 104, 106
High-molecular-weight organics, 160–162
High-performance liquid chromatography, *see* HPLC (high-performance liquid chromatography)
High-risk groups, 439, 443
Historical cohurt studies, 325–326, 374
Hormesis, 447–449, 453
HPLC (high-performance liquid chromatography):
  as alternative to GC, 157
  and analysis:
    of high-molecular-weight compounds, 160—162
    of low-molecular-weight compounds, 158–160
  application of, 157
  with FTIR, 162–163, 165
  microbore, 165
  with MS, 162–163, 165
  with NMR detectors, 162–163, 165
  and trace organic contaminants, 9
Humic acid, 62, 64
Humic substances:
  and adsorption, 241, 255
  coagulation of:
    and alum coagulation, 203–204, 206–208
    background of, 201–202

and cationic polyelectrolytes, 209–210
and jar test procedures, 202–203
definition of, 63–64
divisions of, 64
and XAD resins, 65
see also Aquatic humic substances; Soil
  humic substances
Hydrocarbons, 81–86
Hydrogen peroxide, 277, 287
Hydrophilic acids, 65–66, 70
Hypochlorite, 35–37, 43
Hypochlorous acid:
  and aromatic compounds, 47–48
  and chlorine hydrolysis, 34
  electrophilicity reactivity of, 36
  germicidal potency of, 43
  and inorganic compounds, 38–39
  redox potential of, 36–37

IARC (International Agency for Research
  Cancer), 312, 517–518, 523–524
Identification, 133. *See also specific methods
  of identification*
Identification methods, *see* Concentration/
  isolation methods; GC methods; GC/
  MS (gas chromatography/mass
  spectroscopy); HPLC (high-
  performance liquid chromatography);
  Nonchromatographic methods; QA
  (quality assessment) programs
Indirect studies:
  definition of, 321
  exposure estimates, 324
  mathematical model, 322–323
  morbidity records, 321–322
  mortality records, 321–322
  regression considerations, 323–324
  strengths of, 324–325
  weaknesses of, 324–325
Inorganic compounds:
  and aqueous chlorine, 37–38
  national survey for, 192
Insecticides, 18
Instantaneous TTHMs, 223–224
Internal standards (IS), 134, 147–149
International Agency for Research Cancer
  (IARC), 312, 517–518, 523–524
In vivo bioassay, 304–307
Involuntary hazards, 485
Iodine:
  application of, as disinfectant, 286–287
  and THM formation, 277
Ion-exchange methods, 114–116, 122, 158
IS (internal standard), 134, 147–149
Isolation methods, *see* Concentration/
  isolation methods

Jar tests, 202–203, 225–226

Ketones, 366
Kuderna-Danish evaporation, 98

Laboratory certification program, 174–
  175

Lake water:
  aquatic humic substances in, 72–73
  DOCs in, 62
Legislation:
  for drinking water standards, 18–21
  federal:
    CERCLA, 502–503
    Clean Water Act, 503–504
    FIFRA, 499–500
    RCRA, 501–502
    SDWA, 497–499
    TSCA, 500–501
  and measure of organic chemicals, 494–
    495
  need for, 493–494
  state, 504–505
  and treatment techniques for removal of
    organics, 505–509
  see also Regulatory significance
Limited-scope activities, 176
Limit of detection, *see* LOD (limit of
  detection)
Liquid chromatographic techniques, *see*
  HPLC (high-performance liquid
  chromatography)
Liquid-liquid extraction, 110–112, 122
LOD (limit of detection):
  and adsorption, 113
  and CLSA, 108
  and exchange-ion methods, 115
  and QA/QC, 99, 101
LOEL (lowest-observed-effect level), 417–418
Low dose extrapolation, 344–347
Lowest-observed-effect level (LOEL), 417–
  418

Margin-of-safety (MOS), 418–419
MAS (Master Analytical Scheme):
  and adsorption, 113
  and CLSA, 108–109
  description of, 17
  and ion-exchange methods, 115
  and liquid-liquid extraction, 111
  and QA/QC, 99
  and static headspace analysis, 105
Massachusetts Environmental Policy Act
  (MEPA), 479
Mass spectrometry, *see* MS (mass
  spectrometry)
Master Analytical Scheme, *see* MAS (Master
  Analytical Scheme)
Maximum contaminant levels, *see* MCLs
  (maximum contaminant levels)
Maximum tolerated dose (MTD), 305–307
MCLs (maximum contaminant levels):
  and EPA, 19
  and grab samplings, 96
  for insecticides, 18
  for pentachlorophenol, 5
  for pesticides, 18, 237
  for phenol, 5
  revised regulations of, 525–527
  for THMs, 237
Membrane methods, 116–119, 122

MEPA (Massachusetts Environmental Policy Act), 479
Metabolic overload, 306–307
Microbial activity, 244–245
Microextraction, 111
Micropollutants:
  and adsorption, 239
  epidemiologic assessment of:
    case-control studies, 390–391, 394–395
    historical perspective of, 375–376
    hypotheses testing, 377–379, 382–384, 388, 390
    methodological context, 373–375
    Mississippi River study, 376–377
    prospective studies on individuals, 395, 397
    results of testing, 397–400
  see also SOCs (synthetic organic chemicals)
Mississippi River study, 376–377
MOS (margin of safety), 418–419
MS (mass spectrometry):
  and aquatic organics, 161
  chemical ionization (CI), 142–146
  and HPLC, 162–163, 165
  low-resolution accurate, 142–146
  supplementary information of, 141
  see also GC/MS (gas chromatography/mass spectroscopy)
MTD (maximum tolerated dose), 305–307
Multicomponent analysis, 149–150
Multiple ion detection, 135
Multistage model:
  biological relevance dose, 445
  cellular transformation, 444
  differential susceptibility, 445–446
  exposure, less than lifetime, 446–447
  selection of states, 444
  validation of QRA, 446
Mutagens, 14

National Cancer Institute (NCI), 295, 356–358
National Inorganic and Radionuclides Survey (NIRS), 192
National Interim Primary Drinking Water Regulations (NIPDWRs), 493, 515–516
National Organics Monitoring Survey (NOMS), 12, 494–495, 512
National Organics Reconnaissance Survey (NORS), 12, 494
National Pollutant Discharge Elimination System (NPDES), 504
National Quarantine Act of 1878, 493
National Screening Program, 17, 495, 512
National surveys:
  for inorganic contaminants, 192
  for QA programs, 175–176
  for radionuclide contaminants, 192
  for trace organic contaminants, 9
  for VOCs, 183–186, 189–190, 192
National Toxicology Program (NTP), 294–297, 523

Natural waters, see DOCs (dissolved organic compounds); specific types of natural waters
NCI (National Cancer Institute), 295, 356–358
New Orleans water supply, 376
NIPDWRs (National Interim Primary Drinking Water Regulations), 493, 515–516
NIRS (National Inorganic and Radionuclide Survey), 192
Nitrogenous organic compounds, 41
Nitrogen trichloride, 39
NMR (nuclear magnetic resonance) spectroscopy, 165
NOAEL (non-adverse-effect level), 519
NOEL (no-observed-effect level), 416–418
NOMS (National Organics Monitoring Survey), 12, 494–495, 512
Non-adverse-effect level (NOAEL), 519
Nonchromatographic methods:
  disadvantages of, 165, 167
  and tests that measure gross organic levels, 167–169
Non-purgeable organic halide (NPOX), see NPOX (non-purgeable organic halide)
Non-purgeable total organic carbon, see NPTOC (non-purgeable total organic carbon)
Nonvolatile organics, 15–16, 114, 116, 119
No-observed-effect level (NOEL), 416–418
NORS (National Organics Reconnaissance Survey), 12, 494
NPDES (National Pollutant Discharge Elimination System), 504
NPOX (non-purgeable organic halide), 11, 15–16
NPTOC (non-purgeable total organic carbon):
  removal of:
    and alum coagulation, 208
    general, 231–232
    at Glenmore Reservoir, 219–220
    at Grasse River, 219–220
    and temperature, 220
  as surrogate parameters, 12
NTP (National Toxicology Program), 294–297, 523
Nuclear magnetic resonance (NMR) spectroscopy, 165
Nuisance organisms, 43

Occupational hazards, 485
OECD (Organization for Economic Cooperation and Development), 300–301
Oneida plant, see Glenmore Reservoir
Organic compounds:
  analytical approach to detecting:
    adsorption methods, 112–113, 122
    AQ/QC, 99, 101
    broad spectrum analysis, 97–99, 112, 121
    CLSA, 107–109, 121–122

concentration/isolation methods, 102–104
distillation methods, 104, 122
HPLC, 110–112, 122
ion-exchange methods, 114–116, 122
membrane methods, 116–119, 122
purge and trap, 106–107, 121
residue analysis, 119–120, 122
specific compound analysis, 97–99, 112, 121
static headspace analysis, 104–106, 121
and aqueous chlorine, 41–42
chlorinated, 40
in drinking water supplies, 512–514
evaluation of:
    chloroform, 422–423
    styrene, 420–421
    toluene, 421–422
    vinyl chloride, 424–425, 432
human, 4–8
measurement of, in drinking water, 494–495, 497
natural, 4
nitrogenous, 41
and public health:
    approaches to, 406–420
    assessment of, 405–406
    evaluation of, 420–425, 432
treatment techniques for removal of:
    adsorption, 505–508
    aeration, 506–507
    air stripping, 506–507
see also DOCs (dissolved organic compounds; Organic contaminant ingestion; Trace organic contaminants)
Organic contaminant ingestion:
background of, 463–465
focus on, 463
hazard identification of, 466–467
risk assessment of, 465–466
risks of:
    estimation of, 467–472
    evaluation of, comparative, 472–475
    list of, 475–476, 478–485
Organic halogen, 280–284
Organic micropollutants, see Micropollutants
Organization for Economic Cooperation and Development (OECD), 300–301
Oxidants, 354
Oxidation, 45
Ozonation, 161, 255, 276
Ozone:
application of, as disinfectant, 285–286
and biocidal activity, 270
by-products of, 280
and THM formation, 276–277

PAC (powdered activated carbon), 246–248
PAHs (polynuclear aromatic hydrocarbons), 6, 13, 481
Particulate organic carbon (POC), 57–58
Partition coefficient, 103–104, 110–111, 114
PCBs (polychlorinated biphenyls), 6, 13, 479
Permanganate, 277, 287

Pesticides, 18, 528
pH:
in alum coagulation, 203–204
application of, as disinfectant, 287
in coagulation experiments, 202–203
in jar tests, 202–203
and THM formation, 277
Phase rule, 103
Phenols, 5, 48, 158. See also Chlorinated phenols
POC (particulate organic carbon), 57–58
Polar organics, 109, 114, 116, 119, 158
Polychlorinated biphenyls (PCBs), 6, 13, 479
Polynuclear aromatic hydrocarbons (PAHs), 6, 13, 481
Polyurethane foams, 112–113
Potassium permanganate, see Permanganate
POX (purgeable organic halide), 11
PQL (practical quantitative level), 526
Practical quantitative level (PQL), 526
Prechlorination, 251
Preozonation, 255
Protective coatings, 5
Public health, see QRA; Risk assessment
Public Water System (PWS) Supervision activities, 504
Purgeable organic halide (POX), 11
Purgeable organics, 106
Purge and trap techniques, 106–107, 121
PWS (Public Water System) Supervision activities, 504

QA programs:
and analytical methods:
    importance of, 181
    levels of method validation, 181–182
    quality control provisions, 182–183
assessing:
    data quality objectives, 177–178
    project plans, 178–179
    standard operating procedures, 179–181
and case histories, specific:
    national surveys for inorganic and radionuclide contaminants, 192
    national surveys for VOC, 183–186, 189–190, 192
    smaller-scale projects, 192–194
importance of, 173
levels of required:
    compliance monitoring, 177
    exploratory activities, 176–177
    limited-scope activities, 176
    national surveys, 175–176
purpose of, 173–175
and studies of trace organics, 123
QA/QC, 99–102. See also QA programs
QRA (quantitative risk assessment):
background of, 340–341
and background response, 347–348
issues of, 339–340
research in, current, 348
statistical models used in:
    stochastic, 343–344

QRA (quantitative risk assessment) statistical models used in (*Continued*)
tolerance distribution, 342–343
unknown parameters in, estimate of, 348, 350–351
unresolved issues in, 444–447
  see also Risk assessment
Quality assurance programs, *see* QA programs
Quality assurance project plan, 178–179
Quality assurance/quality control, *see* QA/QC
Quality control, *see* QA/QC; QA programs
Quantitative risk assessment, *see* QRA

Radionuclide contaminants, 192
Rate tests, 257–258
RCRA (Resource Conservation and Recovery Act), 501–502
Recommended maximum contaminant levels, *see* RMCLs
Reconstructed ion-current profile, 134
Redox process, 36–37
Regeneration:
  costs of, 260–261
  and design considerations, 248
  and GAC, 245–246, 253
  process of, 245–246
Regulation, *see* Legislation; Regulatory significance
Regulatory significance:
  of EPA, 511
  and occurrence of organic compounds, 512–514
  and revised regulations:
    interim, 515–516
    for MCLs, 525–527
    for monitoring, 527–528
    Phase II, 528–529
    Phase IV, 528–529
    for reporting, 527–528
    for RMCLs, 519, 521–525
    for selection of contaminants, 516–519
  of SDWA, 511, 514–515
  and statutory requirements, 514–515
Residue analysis, 119–120, 122
Resins, 114–115. *See also* XAD resins
Resource Conservation and Recovery Act (RCRA), 501–502
Response factor, 135
Response ratio, 135
Reverse osmosis (RO), 116–117
Risk:
  characterization, 408, 416–420
  definition of, 320, 466
  estimation of, 467–472
  evaluation of, comparative, 472–475
  measures of, epidemiologic, 320
  of organic contaminant ingestion:
    from chlorination, 481
    and chloroform, 482–485
    hazard identification, 475–476
    from human activities, 478–481
    from involuntary actions, 485

natural substances, 476, 478
from occupations, 485
preventing, 478–479
and THMs, 481–482
from voluntary actions, 485
see also QRA; Risk assessment
Risk assessment:
  approaches to:
    dose-response assessment, 408, 413–414
    exposure assessment, 408, 414–416
    hazard identification, 408–413
    risk characterization, 408, 416–420
    terms of, 406–408
  components of, 466
  unresolved issues in:
    animal extrapolation issues, 443–444
    chemical interactions, 453–455
    discussion of, 455–457
    exposure, 437–439
    high-risk groups, 439, 443
    Hormesis, 447–449, 453
    overview of, 437
    QRA, 444–447
    safety factors for epigenetic carcinogens, 447
  see also QRA; Risk
River water:
  aquatic humic substances in, 70–72
  DOCs in, 61
RMCLs:
  for organic chemicals, 18
  regulations for, revised, 519, 521–525
RO (reverse osmosis), 116–117
Rotary evaporation, 115, 119

Safe, 406–407
Safe Drinking Water Act of 1974, *see* SDWA (Safe Drinking Water Act)
Safe Drinking Water Act of 1984, 19
Safety, 466
Salmonella mutation assay, *see* Ames test
Salting out, 104, 108
Sample integrity:
  and microbiological degradation, 119
  preservation of, 107
  problems of, 101
Sampling:
  composite, 95–96
  considerations, 107
  grab, 96–97
  for nonvolatile organic compounds, 120
SAR (structure-activity relationship), 298–300
Scale-up principles, 258
SCE (sister-chromatid exchange), 361
SDWA (Safe Drinking Water Act) of 1974:
  and federal responsibility, 497
  goal of, 511
  and ground water, 498
  and health, human, 18
  and quality control, 174
  regulations of, 9, 497–499
  and regulations for SOCs, 498
  regulations for VOCs, 498
  requirements of, 514–515

and Sole Source Aquifer Program, 498
and state responsibility, 498
Seawater:
    aquatic humic substances in, 69–70
    DOCs in, 60
Secondary carcinogenesis, 306–307
Selected ion monitoring, 135
Semivolatile organics, 110, 112
Service life, see Service time
Service time:
    concept of, 238–239
    and design criteria, 253
    and EBCT, 251–252
    and regeneration, 245–248
    and THM precursor removal, 250
Short-term testing, 300
Silver, 277, 287
Silver filtration, 57
Sister-chromatid exchange (SCE), 361
SMSAs (standard metropolitan statistical
    areas), 494
SNARLs (suggested no-adverse-response
    levels), 417
SOC (suspended organic carbon), 57–58
SOCs (synthetic organic chemicals):
    adsorption of, 237
    desorption of, 243
    regulations for, revised, 528–529
    removal of, 262
    and SDWA, 498
    types of, 237
    and water treatment, 5
Soil humic substances, 63, 68. See also Humic
    substances
Sole Source Aquifer Program, 498
SOPs (standard operating procedures), 179–
    181
Specific compound analysis, 97–99, 112, 121
Spectrometric detectors, 154–155
Standard metropolitan statistical areas
    (SMSAs), 494
Standard operating procedures (SOPs), 179–
    181
State Ground-Water Programs, 504
Static headspace analysis, 104–106, 121
Stochastic models, 343–344
Stoichiometry, 204, 206–207
Structure-activity relationship (SAR), 298–
    300
Styrene, 420–421
Suggested no-adverse-response levels
    (SNARLs), 417
Superfund, 502–503
Surface diffusivities, 257
Surrogate parameters:
    removal of, 237
    and trace organic contaminants, 9–12
    UV as:
        in direct filtration, 232
        plant monitoring of TOC and THM
            precursors, 220, 222–223
        predicting inst TTHMs, 223–224
Suspended organic carbon (SOC), 57–58
Synthetic chloroorgranic compounds, 7

Synthetic ion-exchange resins, 114
Synthetic organic chemicals, see SOCs
    (synthetic organic chemicals)
System blanks, 108

TDMs (tolerance distribution models), 340,
    342–343
Technical aspects, see Legislation; Regulatory
    significance
THM (trihalomethane) precursors:
    formation of, 239
    removal of:
        by adsorption, 249–253
        by coagulation of humic substances, 201–
            204, 208
        general, 231–232
        at Glenmore Reservoir, 219–220
        at Grasse River, 208, 219–220
        and temperature, 220
    see also THMs (trihalomethanes)
THMs (trihalomethanes):
    in adsorption, 249–253
    and aromatic compounds, 48–49
    and chlorination, 200, 353, 356–359, 361,
        363
    and chloroform carcinogenesis, 356–359,
        361, 363
    and coagulation of humic substances, 201
    data on, 4
    description of, 6
    formation of, other than by chlorination:
        by bromine, 277, 287
        by chloramines, 272–275, 285
        by chlorine dioxide, 275–276, 286
        in drinking water, 265–266
        by ferrate ion, 277, 287
        by iodine, 277, 286–287
        by ozone, 276–277, 285–286
        by permanganate, 277, 287
        by silver, 277, 287
        by UV radiation, 277, 287
    forms of, 356
    and fulvates, 50
    and haloform, 49–50, 52
    and humates, 50
    and nonvolatile compounds, 15
    removal of, 262
    and risk assessment, 438
    source of, 5
    see also THM (trihalomethane) precursors;
        specific water treatment processes
TOC (total organic carbon):
    and adsorption, 255
    and batch equilibrium tests, 257
    and coagulation of humic substances,
        201
    removal of, 244
    term of, 57–58
    and trace organic contaminants, 10–12
TOCl (total organic chlorine), 11–12
Tolerance distribution models (TDMs), 340,
    342–343
Toluene, 421–422
Total ion-current (TIC), 134

Total organic carbon, *see* TOC (total organic carbon)
Total organic chlorine (TOCl), 11–12
Total organic halogen, *see* TOX (total organic halogen)
Total trihalomethane formation potential, *see* TTHMFP (total trihalomethane formation potential)
TOX (total organic halide):
  as gross parameter, 167–169
  measuring, 280–284
  and nonvolatile compounds, 15
  as surrogate parameter, 11
Toxicologic assessment:
  carcinogenecity testing programs, 294–295
  chemicals studied for, 295–298
  genotoxicity testing:
    Ames test, 303–304
    guidelines for, 300–303
    in vivo bioassay, 304–307
  and problems with *in vivo* carcinogenicity testing
    agents modifying, 307–310
    interpretation of testing results, 310–313
    tumor promoters, 310
  short-term testing, 300
  structure-activity relation correlates in, 298–300
  studies for, 293–294
  see also Carcinogens
Toxic Substances Control Act (TSCA), 500–501
Trace organic contaminants:
  and cancer, 21, 26
  human, 4–8
  legislation for, 18–21
  level and composition of:
    comprehensive analytical schemes, 17
    identification of, 8
    individual components identification, 12–15
    national surveys for, 9
    nonvolatile compounds, 15–16
    and surrogate parameters, 9–12
  measurement of, 3–4
  natural, 4
  perspectives on, 21, 26–27
  risks of, 4
  sources of, 4–8
  types of, 3–4
  see also specific identification methods; specific water treatment processes
Trichloracetic acid, 13, 51
Trihalomethanes, *see* THMs (trihalomethanes)
TSCA (Toxic Substances Control Act), 500–501
TTHMFP (total trihalomethane formation potential):
  and chlorination, 46
  and trace organic contaminants, 11

Tumor promoters, 310
Turbidity, 199–200, 224

UIC (Underground Injection Control), 504
Ultrafiltration (UF) membranes, 116, 118
Underground Injection Control (UIC), 504
Unsafe, 406–407
UV:
  application of, as disinfectant, 288
  as surrogate parameters:
    in direct filtration, 232
    in jar tests, 203
    plant monitoring of TOC and THM precursors, 220, 222–223
    predicting Inst TTHMs, 223–224
  and THM formation, 277

van't Hoff equation, 117
Vapor/vapor extraction, 109
VHO (volatile halogenated organics), 255
Vinyl chloride, 424–425, 432
Virtual safe dose, *see* VSD (virtual safe dose)
VOCs (volatile organic compounds):
  and static headspace analysis, 104
  and EPA, 19
  national surveys for, 183–186, 189–190, 192
  and purge and trap techniques, 106
  RMCLs for, revised, 519, 521–525
  and SDWA, 498
  and trace organic contaminants, 11–12, 14–15
Volatile halogenated organics (VHO), 255
Volatile organic compounds, *see* VOCs (volatile organic compounds)
Voluntary hazards, 485
VSD (virtual safe dose):
  and background response, 347–348
  and low-dose extrapolation, 344–347
  and QRA, 341

Water supplies, *see* Drinking water supplies; *specific types of natural waters*
Water treatment processes, *see* Adsorption; Aqueous chlorine; Conventional water treatment; Disinfection
Wetland water:
  aquatic humic substances in, 73–74
  DOCs in, 62
WHO (World Health Organization), 20–21
World Health Organization (WHO), 20–21

XAD resins:
  and adsorption, 113
  and aquatic humic substances, 67
  description of, 63–64
  and DOCs, 65–66
  and humic substances, 65
  and ion-exchange methods, 115
  and organic compounds, 98

Zero risk, 406–407, 521–522

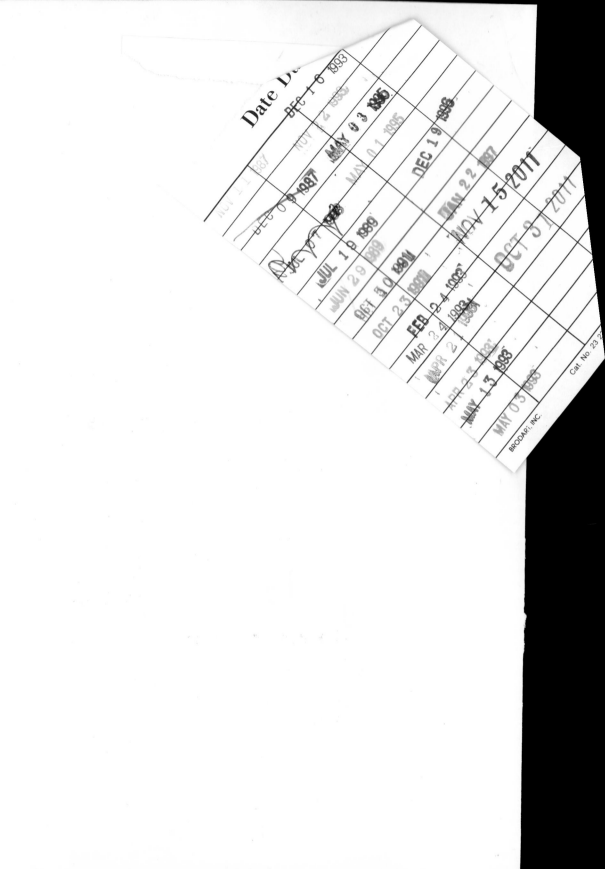